Texts in Applied Mathematics 14

W0232306

Texts in Applied Mathematics

1. *Sirovich:* Introduction to Applied Mathematics.
2. *Wiggins:* Introduction to Applied Nonlinear Dynamical Systems and Chaos.
3. *Hale/Koçak:* Dynamics and Bifurcations.
4. *Chorin/Marsden:* A Mathematical Introduction to Fluid Mechanics, 3rd edition
5. *Hubbard/West:* Differential Equations: A Dynamical Systems Approch, Part I: Ordinary Differential Equations.
6. *Sontag:* Mathematical Control Theory: Deterministic Finite Dimensional Systems.
7. *Perko:* Differential Equations and Dynamical Systems.
8. *Seaborn:* Hypergeometric Functions and Their Applications.
9. *Pipkin:* A Course on Integral Equations.
10. *Hoppensteadt/Peskin:* Mathematics in Medicine and the Life Sciences.
11. *Braun:* Differential Equations and Their Applications: An Introduction to Applied Mathematics, 4th edition
12. *Stoer/Bulirsch*: Introduction to Numerical Analysis, 2nd edition
13. *Renardy/Rogers:* A First Gradute Course in Partial Differential Equations.
14. *Banks:* Growth and Diffusion Phenomena: Mathematical Frameworks and Applications.

Robert B. Banks

Growth and Diffusion Phenomena

Mathematical Frameworks and Applications

With 216 Figures

Springer-Verlag Berlin Heidelberg GmbH

Robert B. Banks
Former President and Professor
of Environmental Engineering
Asian Institute of Technology
Bangkok 10501, Thailand

Editors

F. John
Courant Institute
of Mathematical Sciences
New York University
New York, NY 10012
USA

J.E. Marsden
Department
of Mathematics
University of California
Berkeley, CA 94720
USA

L. Sirovich
Division
of Applied Mathematics
Brown University
Providence, RI 02912
USA

M. Golubitsky
Department
of Mathematics
University of Houston
Houston, TX 77004
USA

W. Jäger
Department
of Applied Mathematics
Universität Heidelberg
Im Neuenheimer Feld 294
69120 Heidelberg, FRG

Cover figure. Among the many growth and diffusion models examined by the author is one relating to the temporal-spatial growth of cities. London is utilized as an illustration. The growth of London comensing 1840 is examined in detail in Section 7.5.6

Mathematical Subject Classification (1991):
60j60, 92-01, 90-01, 90A16, 92D40, 92C30, 92H10, 92H20, 94A05

ISBN 978-3-642-08140-8

Library of Congress Cataloging-in-Publication Data
Banks, Robert B.: Growth and diffusion phenomena: mathematical frameworks
and applications/Robert B. Banks.   p.   cm. – (Texts in applied mathematics; 14)
Includes bibliographical references and indexes.
ISBN 978-3-642-08140-8   ISBN 978-3-662-03052-3 (eBook)
DOI 10.1007/978-3-662-03052-3
1. Diffusion processes.   2. Growth–Mathematical models.
I. Title.   II. Series.   QA247.75.836   1994   003'.7–dc 20

© Springer-Verlag Berlin Heidelberg 1994
Originally published by Springer-Verlag Berlin Heidelberg New York in 1994
Softcover reprint of the hardcover 1st edition 1994

Typesetting: Springer TEX in-house system

41/3140 – 5 4 3 2 1 0 – Printed on acid-free paper

To Gunta, Steven and Erik

Series Preface

Mathematics is playing an ever more important role in the physical and biological sciences, provoking a blurring of boundaries between scientific disciplines and a resurgence of interest in the modern as well as the classical techniques of applied mathematics. This renewal of interest, both in research and teaching, has led to the establishment of the series: *Texts in Applied Mathematics (TAM)*.

The development of new courses is a natural consequence of a high level of excitement on the research frontier as newer techniques, such as numerical and symbolic computer systems, dynamical systems, and chaos, mix with and reinforce the traditional methods of applied mathematics. Thus, the purpose of this textbook series is to meet the current and future needs of these advances and encourage the teaching of new courses.

TAM will publish textbooks suitable for use in advanced undergraduate and beginning graduate courses, and will complement the *Applied Mathematical Sciences (AMS)* series, which will focus on advanced textbooks and research level monographs.

Preface

To coincide with the beginning of a book perhaps it is appropriate to mention how the book had its beginning.

Quite a long time ago, when I was a first-year graduate student in environmental engineering, my adviser indicated that I needed more knowledge of the life sciences. So, for starters, I enrolled in an undergraduate course in bacteriology; I was the only engineering student in the class. Furthermore, as the only graduate student in the course, the professor informed me that I would have the additional task of preparing a term paper. So I wrote one entitled "Applications of mathematics in bacteriology". Essentially it consisted of solutions to all the problems in my book on differential equations dealing with "prey-predator" phenomena. The professor evidently liked the paper; I think I got an A.

During the ensuing years my research interests were directed to topics of air and water pollution; aspects of diffusion and dispersion were salient features of these problems. For quite a long period I was involved with graduate students in studies along these lines.

Later on, at universities in Mexico and Thailand, I became interested in innovation diffusion and technology transfer. This activity accelerated during my assignment at ISNAR (International Service for National Agricultural Research) in The Hague. On numerous occasions I received reports dealing with technology transfer, prepared by staff at CIMMYT (International Maize and Wheat Improvement Center) in Mexico, at IRRI (International Rice Research Institute) in the Philippines and at a number of other agricultural R & D institutions.

Over a period of time I observed, as others have, that in many instances virtually identical problems involving growth, transfer and diffusion were being examined by people working in quite different fields of endeavor. It seemed to me that it might be useful to try to help in the enlargement of "information diffusion" among the various fields.

So, with that lengthy preamble, I simply indicate that a major purpose of this book is an attempt to amalgamate some of the many advances made by people working in numerous diverse disciplines on topics relating to growth and diffusion phenomena. Abundantly clear to me is that there are many people much more qualified than I to do this.

The book is arranged as follows. An introductory Chapter 1 is presented which deals with some of the historical aspects of growth and diffusion. Chapters 2 and 3 are devoted to an examination of the numerous frameworks for analysis of growth, transfer and diffusion processes. This is followed, in Chapters 4 and 5, with a number of topics dealing with time dependent growth coefficients and carrying capacities.

The subject of discrete and distributed time delays is presented in Chapter 6 and phenomena involving spatial diffusion are considered in Chapter 7. Some of the many other aspects and problems of spatial–temporal processes are introduced in Chapter 8. A comprehensive list of references is provided at the end of the book.

Throughout the book there are many numerical examples and illustrations. For the latter, an attempt has been made to select specific topics drawn from the numerous fields of endeavor. The following list indicates the various disciplines and respective numbers of illustrations.

Agriculture	6
Economics	7
Biology and physiology	13
Physical sciences and engineering	8
Demography and geography	12
Technology transfer	8
Ecology	6
Other areas	4

It is hoped that the book will be useful to people engaged in planning, analysis and evaluation activities in the various disciplines. It is hoped also that the book will be useful to advanced undergraduate and graduate students in the physical, biological and social sciences as well as to those in some areas of engineering and medicine. Readers should have a course in calculus; introductory differential equations would be helpful.

The entire treatment of the subject features deterministic models with continuous variables as contrasted with stochastic models with discrete variables. For the most part, attention is restricted to "single-species" considerations.

I want to express my gratitude to the numerous people who kindly reviewed portions of the manuscript and offered their corrections, suggestions for improvement and words of encouragement. I thank my former students for their many contributions. I acknowledge, with sincerest appreciation, two people who helped me launch this project: Dr. William K. Gamble, founding Director-General of the International Service for National Agricultural Research in The Hague, and Dr. M. Nawaz Sharif, former Director of the UN/ESCAP Regional Centre for Technology Transfer in Bangalore.

The editorial staff of Springer-Verlag in Heidelberg is warmly thanked. I am grateful also to Dr. Angela Lahee who typeset the book, and Mr. Frank

Ganz, the production editor. I also want to thank Mrs. Pamela Saftler, Head of Cartographics at Texas A & M University, for preparing all the figures. I am much indebted to Dr. M.J.H. Mogridge of University College London for his invaluable help with my illustration concerning the growth of London.

Finally, I want to indicate that my wife, Gunta, struggled with great patience and good humor throughout. During several typings of the manuscript, she occasionally and tactfully brought my attention to the fact that, in her view, the market would be quite limited without a somewhat better plot and a good deal more passion and excitement in the subject matter. In these aspects I know I failed; but in selecting a wonderful lady, I succeeded admirably.

Robert B. Banks   La Jolla, California
                                                                                                   Spring 1993

Table of Contents

1 **Introduction** 1

2 **Some Basic Frameworks** 5
  2.1 Exponential Function 7
    2.1.1 The Exponential Function and Its Properties 7
    2.1.2 Doubling Times 10
    2.1.3 An Illustration: Population of the United States 11
    2.1.4 Exponential Function with Migration 13
    2.1.5 Power Law Exponential Function 15
    2.1.6 An Illustration: Population of the World 16
    2.1.7 Combinations of Exponential Functions 19
    2.1.8 Solutions and Properties of the Equations 21
    2.1.9 An Illustration: Oxygen Distribution in a River 24
  2.2 Logistic Distribution 27
    2.2.1 The Differential Equation and Its Solution 27
    2.2.2 Properties of the Logistic Distribution 29
    2.2.3 An Illustration: Technology Substitution 34
    2.2.4 An Illustration: Diffusion of Improved Pasture
          Technology in Uruguay 38
    2.2.5 An Illustration: Growth of Prime Mover Horsepower
          in the U.S. 44
  2.3 Confined Exponential Distribution 52
    2.3.1 Scope of Applications of the Distribution 52
    2.3.2 The Differential Equation with Constant
          Coefficients 53
    2.3.3 The Differential Equation with Variable Coefficients 53
    2.3.4 Variable Transfer Coefficient 55
    2.3.5 Variable Equilibrium Value 60
    2.3.6 An Illustration: Oxygen Transfer
          Across a Water Surface 65
    2.3.7 An Illustration: Biochemical Oxygen Demand 67
    2.3.8 An Illustration: Growth of Humans 69
    2.3.9 An Illustration: Public Interest in a News Event 73
  2.4 Combination of the Logistic Distribution and the Confined
    Exponential Distribution 76

2.4.1 Comparison of the Two Distributions 76
2.4.2 Phenomena in Industrial Technology Transfer 77
2.4.3 Phenomena in Social Innovation Diffusion 78
2.4.4 Phenomena in Chemical Reaction Kinetics 79
2.4.5 Phenomena in the Psychology of Learning 80
2.4.6 The Differential Equation, Its Solution
       and Properties 82
2.4.7 An Illustration: Adoption of a
       Tornado Warning Device 86
2.4.8 Combination of the Exponential Function
       and the Confined Exponential Distribution 88
2.4.9 An Illustration: Population of California 89
2.5 Normal Probability Distribution 92
2.5.1 The Normal Probability Function and Its Features .. 92
2.5.2 Relationship Between the Normal Probability
       Function and the Error Function 94
2.5.3 Approximate Expressions for the Normal Function
       and Its Inverse 95
2.5.4 Comparison of the Logistic and
       Normal Probability Distributions 96
2.5.5 An Illustration: Adoption of Herbicides
       by Mexican Barley Farmers 103
2.6 Power Law Logistic Distribution 105
2.6.1 The Differential Equation and Its Features 105
2.6.2 The Growth Curve and Its Properties 106
2.6.3 The Richards Function 110
2.6.4 An Illustration: Sale of Development Property 110
2.6.5 An Illustration: Growth of Pine Trees
       in New Zealand 112
2.6.6 An Illustration: Adoption of Hybrid Corn
       in the United States 114
2.7 Logistic Growth with Migration 117
2.7.1 Immigration and Emigration 117
2.7.2 Logistic Growth with Constant Stocking 117
2.7.3 Logistic Growth with Constant Harvesting 119
2.7.4 Logistic Growth with Variable Harvesting 121
2.7.5 An Illustration: Fish Harvesting 122
2.7.6 An Illustration: The Sandhill Crane 123
2.8 Epidemics and Technology Transfer 125
2.8.1 Simple and General Epidemics 125
2.8.2 The Differential Equations and Phase Plane Display 126
2.8.3 Solutions to the Differential Equations 129
2.8.4 Logarithmic Form of the Solutions 133
2.8.5 A Technology Transfer Analogy 134

2.8.6   An Illustration: Bombay Plague of 1905–1906 135
2.9   Some Modifications of the Logistic Distribution 136
   2.9.1   Use of Taylor Series 136
   2.9.2   First Order Differential Equations 137
   2.9.3   An Illustration: Growth of Water Fleas and Trees ... 138
   2.9.4   An Illustration: Sale of Development Property
       Revisited 140
   2.9.5   Second Order Differential Equations 141
   2.9.6   An Illustration: Multiplier-Accelerator Model of a
       National Economy 144

3   Some Additional Frameworks 149
3.1   Gompertz Distribution 149
   3.1.1   The Gompertz Distribution and Its Features 149
   3.1.2   An Illustration: Growth of Plant Leaves 151
   3.1.3   An Illustration: Dynamics of Tumor Growth 155
3.2   Weibull Distribution 156
   3.2.1   The Weibull Distribution and Its Features 156
   3.2.2   An Illustration: Substitution of Diesel and
       Electric Locomotives for Steam Locomotives
       in the United States 159
3.3   A Generalized Distribution 162
   3.3.1   Cumulative and Density Distribution Functions 162
   3.3.2   A Generalized Symmetrical Function 163
   3.3.3   Extreme Maximum Value Distribution 167
   3.3.4   Extreme Minimum Value Distribution 169
   3.3.5   An Illustration: Dose Response Analysis
       of Beetle Mortality Data 172
3.4   Hyperlogistic Distribution 174
   3.4.1   The Differential Equation and Some Examples 174
   3.4.2   Solution to the Differential Equation 176
   3.4.3   Some Properties of the Hyperlogistic Equation 178
   3.4.4   Numerical Examples of the Hyperlogistic Equation .. 180
   3.4.5   An Illustration: Adoption of a Tornado Warning
       Device Revisited 182
   3.4.6   Coalition and Modified Coalition Growth Models ... 185
   3.4.7   An Illustration: Population of the World 188
   3.4.8   An Illustration: Growth of the Public Debt
       of the United States 191
3.5   Various Other Distributions 194
   3.5.1   Comparison of Distribution Functions 194
   3.5.2   Arctangent - Exponential Distribution 195
   3.5.3   Pearson Type VII Distribution 196
   3.5.4   Arctangent Distribution 197

	3.5.5	Gamma Distribution	199
	3.5.6	Generalized Gamma Distribution	201
	3.5.7	An Illustration: Generation Times of Cells	201
	3.5.8	An Illustration: Population of Great Britain	205

4 Phenomena with Variable Growth Coefficients **209**

	4.1	Linearly Variable Growth Coefficient	210
		4.1.1 The Growth Curve and Its Properties	210
		4.1.2 An Illustration: Growth and Decline of U.S. Sailing Vessels	212
	4.2	Hyperbolically Variable Growth Coefficient	214
		4.2.1 The Growth Curve and Its Properties	214
		4.2.2 Relationship to Power Law Exponential Growth	215
		4.2.3 An Illustration: Population of the Great Plains States	216
	4.3	Exponentially Variable Growth Coefficient	218
		4.3.1 Extreme Maximum Value Distribution	218
		4.3.2 Extreme Minimum Value Distribution	220
		4.3.3 An Illustration: Survival of Rats	222
	4.4	Sinusoidally Variable Growth Coefficient	224
		4.4.1 Some Examples of Oscillatory Phenomena	224
		4.4.2 Simple Harmonic Growth Coefficient	224
		4.4.3 Exponentially Decreasing Growth Coefficient: Type I	225
		4.4.4 An Illustration: Growth of a Species of Land Snails .	227
		4.4.5 Exponentially Decreasing Growth Coefficient: Type II	230
		4.4.6 An Illustration: Growth of Cell Populations	231
		4.4.7 Sinusoidally Variable Growth Coefficient in a Power Law Exponential Equation	237
		4.4.8 An Illustration: Number of Patents Issued for Inventions	237

5 Phenomena with Variable Carrying Capacities **241**

	5.1	Exponentially Variable Carrying Capacity	242
		5.1.1 The Growth Curve and Its Properties	242
		5.1.2 An Illustration: Farm Population of the United States	245
	5.2	Logistically Variable Carrying Capacity	249
		5.2.1 Some Previous Studies	249
		5.2.2 The Growth Curve and Its Properties	249
		5.2.3 Relative Values of Growth Parameters	252
		5.2.4 An Illustration: Enrollments in Universities in the United States	254

5.3 Linearly Variable Carrying Capacity 256
    5.3.1 The Growth Curve and Its Properties 256
    5.3.2 An Illustration: Horses and Mules on U.S. Farms ... 258
    5.3.3 An Illustration: Steam Locomotives on U.S.
          Railroads 260
5.4 Hyperbolically Variable Carrying Capacity 262
    5.4.1 Linearly Changing Crowding Coefficient 262
    5.4.2 The Growth Curves and Its Properties 263
    5.4.3 An Illustration: Growth Rates
          of Wheat Plant Components 263
5.5 Sinusoidally Variable Carrying Capacity 267
    5.5.1 Cyclic Variations in Growth
          and Transfer Phenomena 267
    5.5.2 The Growth Curve and Its Properties 267
    5.5.3 Phase Plane Display 270
    5.5.4 Exponentially Changing Carrying Capacity 275
    5.5.5 An Illustration: Railway Mileage
          in the United States 277
5.6 Power Law Logistic with a Power Law Logistically Variable
    Carrying Capacity 279
    5.6.1 The Power Law Logistic 279
    5.6.2 The Differential Equation and Its Solution 279
    5.6.3 An Illustration: Population of the United States 281

6 Phenomena with Time Delays 287
    6.0.1 Introduction 287
    6.0.2 Types and Features of Delay Equations 288
6.1 Discrete Time Delay in the Exponential Equation 289
    6.1.1 The Delay Differential Equation and Its Solution ... 289
    6.1.2 Roots of the Characteristic Equation 292
    6.1.3 Behavior of the Solutions 294
    6.1.4 An Illustration: Tinbergen's Shipbuilding Cycle 295
6.2 Discrete Time Delay in the Logistic Equation 297
    6.2.1 Introduction to the Delay Differential Equation 297
    6.2.2 Solution to the Discrete Delay Logistic Function 298
    6.2.3 Numerical Example 300
    6.2.4 An Illustration: Nicholson's Blowflies 302
6.3 Distributed Time Delay: Delay Integral
    in the Crowding Term 304
    6.3.1 The Integro-differential Equation 304
    6.3.2 Solution to the Integro-differential Equation 307
    6.3.3 An Illustration: Growth and Decline
          of the Populations of Northeast
          and East North Central American Cities 310

6.4   Distributed Time Delay: Delay Integral
in a Pollution Term 317
    6.4.1   The Integro-differential Equation and Its Solution .. 317
    6.4.2   An Approximate Sech-squared Solution 319
    6.4.3   Numerical Examples 320
    6.4.4   An Illustration: Growth and Self- Contamination
of Bacteria 322

7   **Phenomena with Spatial Diffusion** 327
    7.0.1   Introduction 327
    7.0.2   The Diffusion Equation 327
7.1   Diffusion from Instantaneous Sources 329
    7.1.1   Plane Source, Line Source and Point Source 329
    7.1.2   Rectilinear Diffusion with Convection 331
    7.1.3   An Illustration: Dispersion in Pipelines 332
    7.1.4   Radial Diffusion with Exponential Growth 335
    7.1.5   An Illustration: Biological Dispersion 336
7.2   Diffusion from Continuous Source 341
    7.2.1   Rectilinear Diffusion with Constant Boundary
Condition 341
    7.2.2   An Illustration: Bacterial Motility 343
    7.2.3   Rectilinear Diffusion with Variable Boundary
Condition 344
    7.2.4   An Illustration: Temperature Distribution
in the Soil 345
    7.2.5   Radial Diffusion 346
    7.2.6   An Illustration: Unsteady Fluid Flow in an Aquifer . 349
    7.2.7   Rectilinear Diffusion with Convection 352
7.3   Diffusion with Reaction in a Finite Region 354
    7.3.1   Dimensional Analysis 354
    7.3.2   Exponential Growth in a Finite Region 356
    7.3.3   Power Law Exponential Growth in a Finite Region . 361
    7.3.4   Logistic Growth in a Finite Region 362
    7.3.5   An Illustration: Zone of Regulated Fishing 366
7.4   Diffusion with Convection and Reaction 367
    7.4.1   The Differential Equation and Its Solution 367
    7.4.2   Exponential Growth with Convection and Diffusion . 371
    7.4.3   Exponential Decay with Convection and Diffusion .. 372
    7.4.4   Convection and Diffusion with Interphase Transfer .. 373
    7.4.5   An Illustration: Chemical Solute Removal
by Adsorption 378
7.5   Diffusion with Confined Exponential Growth 379
    7.5.1   Rectilinear Diffusion 379
    7.5.2   An Illustration: Heat Transfer from a River 380

7.5.3   Radial Diffusion 383
7.5.4   An Illustration: Population in Cities 385
7.5.5   Temporal-Spatial Diffusion 390
7.5.6   An Illustration: Population of London 395
7.6   Diffusion with Logistic Growth 402
7.6.1   Traveling Wave Solutions 402
7.6.2   A Power Law Traveling Wave Solution 408
7.6.3   An Illustration: Diffusion of Tractor Utilization 410
7.6.4   An Illustration: Adoption of Hybrid Corn Revisited . 414

8   Conclusion 419

References 429

Author Index 445

Subject Index 451

1. Introduction

This is a book about how fast and to what extent things grow, transfer and diffuse. For example

- trees, people, cities and federal debts grow;
- technology, heat, information and diseases are transferred or transmitted;
- innovations, insect colonies, groundwater contaminants and epidemics are diffused or dispersed or spread.

Sometimes growth, transfer and diffusion all occur simultaneously in a phenomenon. For example, as an epidemic spreads through a geographical region, the number of infected persons grows as the disease is transferred from those infected to persons who are susceptible. We are not going to be concerned about the fundamental reasons as to why these growth and diffusion processes take place. These are questions raised, studied and answered by cellular biologists, physical chemists, mechanical engineers, economic geographers, urban planners, rural sociologists and epidemiologists.

What we shall be concerned about is the methodology involved in utilizing various mathematical frameworks for the analysis and display of observed measurements and statistical data and how the resulting information can be employed to interpret and predict.

These frameworks are based on a surprisingly small number of basic principles and concepts. For example, according to a proposition set forth by Malthus, the human population of a nation grows exponentially, at least for a while. According to a model devised by Verhulst, the rate of transfer of a technology is proportional to the product of the numbers of haves and havenots. And according to a principle established by Fick, the rate of diffusion of a chemical in a liquid depends on the spatial gradient of the concentration of the chemical.

The history of growth and diffusion is itself an intriguing phenomenon of growth and diffusion. From this perspective, the origin of our time coordinate was about 200 years ago. In the closing years of the 18th century and the early decades of the 19th, a number of substantial advances were made in the study of growth phenomena. Especially noteworthy are three persons who made quite remarkable contributions to what we now term mathematical demography and mathematical ecology

- Thomas R. Malthus (1798). An essay on the principle of population.
- Benjamin Gompertz (1825). On the nature of the function expressive of the law of human mortality.
- Pierre F. Verhulst (1838). Notice sur la loi que la population suit dans son accroissement.

As we shall see, a specific growth rate is defined as $dN/N dt$ where N is, for example, a population at time t. In present-day vernacular, we would say that the specific growth rate for the Malthus relationship is a constant, a_0, which we term the intrinsic growth coefficient. The specific growth rate for the Gompertz equation decreases exponentially with time, i.e., as $a_0 \exp(-kt)$. And the specific growth rate for the Verhulst model is $a_0(1 - N/N_*)$ where N_* is the value of N after a long time; we shall call N_* the asymptotic value or carrying capacity. We note that the Gompertz and Verhulst relationships are generalizations of the Malthus equation.

Though numerous other investigators made many contributions to the overall body of knowledge of growth and diffusion processes during the latter half of the 19th century, surprisingly little real progress was made until around 1920.

Then, as Scudo and Ziegler (1978) describe it, the golden age of theoretical ecology began. During the period from about 1920 to 1940, a great many substantial advances were made by a number of remarkable people: Thompson (1917) On growth and form; Lotka (1924) Principles of physical biology; Volterra (1926) Variations and fluctuations in the numbers of coexisting animal species; Pearl (1927) The growth of populations; Gause (1934) The struggle for existence; and Kostitzin (1937) Biologie mathématique.

In the wake of these contributions came the works of Fisher (1937) The wave of advance of advantageous genes; Kolmogorov, Petrovsky and Piskounov (1937) Study of the diffusion equation with growth of the quantity of matter; Feller (1940) On the logistic law of growth and its empirical verifications in biology; and many others.

Throughout the 200 years of development of quantitative analyses of growth and diffusion phenomena, we cannot fail to observe the major roles played by mathematicians: Carslaw, Feller, Kolmogorov, Volterra and numerous others. Their contributions have made it feasible to describe these phenomena in a language understood by investigators of many diverse fields.

There are numerous excellent references for those interested in delving into the history of growth and diffusion processes. Some of these are the following

- Cole (1957). Sketches of General and Comparative Demography.
- Smith and Keyfitz (1977). Mathematical Demography.
- Scudo and Ziegler (1978). The Golden Age of Theoretical Ecology: 1923–1940.
- Oliveria-Pinto and Conolly (1982). Applicable Mathematics for Non-Physical Phenomena.

No review of the history of our subject would be complete without reference to Scudo (1971) and Scudo and Ziegler (1976) who present, respectively, biographical sketches of the lives of Volterra and of Kostitzin. Finally and especially noteworthy is the very readable book by Kingsland (1985): Modeling Nature – Episodes in the History of Population Ecology.

A conclusion to an introduction. It is abundantly clear that advances in growth-transfer-diffusion phenomena are being made at an increasing pace by a growing number of people in many disciplines. By whatever set of indices we might use to measure the growth of such activity and progress, it seems safe to say that we have not yet arrived at the inflection point of the growth curve of this activity.

As we begin a third century of endeavor – with horizons containing things like multi-species interactions, discrete systems analysis, deterministic and stochastic time dependent growth parameters, delay and integral equations, non-linear dynamics, chaos and fractals, and goodness knows what else – the inflection point may still be a long way off.

2. Some Basic Frameworks

We commence our analysis of growth phenomena with the introduction of three basic equations which will provide the main components of most of the mathematical frameworks we shall be devising.

These three fundamental equations are the following:

1. The exponential function. This is the most basic and simplest growth equation in our inventory. It is also the most ancient. In essence, this is the growth relationship considered by Malthus (1798), the English clergyman, in his early essays on the principles of population growth.

The exponential function has the following definition

$$\frac{dN}{dt} = aN \quad \text{or} \quad \frac{1}{N}\frac{dN}{dt} = a \quad , \qquad (2.0.1)$$

in which N is the magnitude of a growing quantity, t is the time, and a is the so-called intrinsic growth coefficient.

The quantity dN/dt is termed the growth rate and the quantity $dN/N\,dt$ is called the specific growth rate. Graphical displays of Eq. 2.0.1 are presented in Fig. 2.0.1. In the figure, the growth rate and specific growth rate are shown as ordinates; the quantity N is the abscissa.

2. The logistic distribution. As we shall see, the logistic distribution is an extremely important one in our collection of growth functions. Like the exponential function, it also stems from early times. It was formulated originally by Verhulst (1838), a Belgian mathematician, in his pioneering work on limits to population growth.

The logistic distribution is defined as follows

$$\frac{dN}{dt} = aN\left(1 - \frac{N}{N_*}\right) \quad \text{or} \quad \frac{1}{N}\frac{dN}{dt} = a\left(1 - \frac{N}{N_*}\right) \quad , \qquad (2.0.2)$$

where N_* is what we shall be calling the "carrying capacity". We note that if N_* is infinitely large, the logistic distribution reduces to the exponential function.

There are numerous plausibility arguments for postulating the form of Eq. 2.0.2. These arguments range from simple collision models in gas kinetics, to probability analyses involving contacts between infectives and susceptibles in

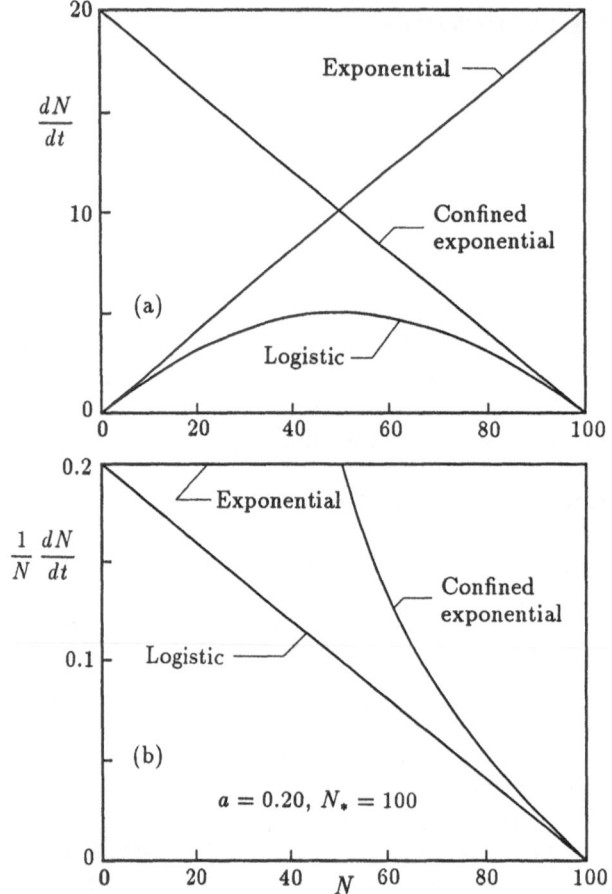

Fig. 2.0.1 Plots of (a) growth rate and (b) specific growth rate for the exponential function, logistic distribution, and confined exponential distribution

epidemics, to a rather cut-and-dried truncation of a Taylor series to fabricate Eq. 2.0.2. It is probably sufficient to simply make the statement, in a comparison of Eqs. 2.0.1 and 2.0.2, that the effect of a finite carrying capacity, N_*, is most logically and easily handled by a specific growth rate which decreases in direct proportion to the quantity, N/N_*. The relationships of Eqs. 2.0.2 are also shown in Fig. 2.0.1.

3. The confined exponential distribution. This growth function has the following definition

$$\frac{dN}{dt} = a(N_* - N) \quad \text{or} \quad \frac{1}{N}\frac{dN}{dt} = a\left(\frac{N_*}{N} - 1\right) . \tag{2.0.3}$$

It should be pointed out that this equation, or one quite similar, is sometimes identified as the exponential distribution or modified exponential distribution. For two reasons we shall call it the confined exponential distribution. First, we shall frequently interchange the words function and distribution in our descriptions of cumulative and density relationships. Inclusion of the adjective "confined" will avoid confusion with regard to the first of our three equations, the exponential function. Second, as we soon establish, the solution of Eq. 2.0.3 requires N to approach N_* for large t. Accordingly, with this vernacular we have an easy to visualize exponential growth curve which is bounded by or confined to a specified finite region. Again, the relationships of Eqs. 2.0.3 are displayed in Fig. 2.0.1.

In the plot of growth rate, dN/dt versus N, we note a parabolic shape for the logistic distribution, flanked by a linearly increasing exponential and a linearly decreasing confined exponential. The growth rate of the logistic is a maximum when the value of N is one-half of N_*.

In the display of specific growth rate, dN/Ndt versus N, we observe that the exponential maintains a constant value. By definition, the specific growth rate of the logistic decreases linearly with N. For the confined exponential, the specific growth rate is infinite when $N = 0$ and descends to zero when $N = N_*$.

In the analyses which follow, frequently we shall be looking at growth rate relationships similar to those of Fig. 2.0.1. By doing so, we can learn a good deal about the behavior of the particular mathematical framework involved as well as the specific phenomenon being examined.

2.1 Exponential Function

2.1.1 The Exponential Function and Its Properties

A logical place to begin our consideration of the subject of mathematical frameworks for growth phenomena is to examine the simplest case: the exponential function.

This case, as indicated in Chapter 2, corresponds to the differential equation

$$\frac{dN}{dt} = aN \quad , \tag{2.1.1}$$

where a is the growth coefficient; we assume it to be a constant. This equation says that the rate at which the quantity N changes with time is directly proportional to the amount of the quantity present at any instant. The solution of Eq. 2.1.1 is of the form, $N = N(t)$. To obtain the complete solution, it is necessary to impose a so-called initial condition. For this, we say that $N = N_0$ when $t = 0$, i.e., $N(0) = N_0$. Accordingly, the solution to Eq. 2.1.1 is

$$N = N_0 e^{at} \quad , \tag{2.1.2}$$

which is sometimes written in the form: $N = N_0 \exp(at)$. This expression is frequently identified as the Malthusian growth equation. However, from here on we will simply call it the exponential function or exponential equation.

It can be established that the infinite series expansion of the exponential function, $\exp(z)$, is given by

$$e^z = 1 + \frac{1}{1!}z + \frac{1}{2!}z^2 + \frac{1}{3!}z^3 + \dots \quad . \tag{2.1.3}$$

Substituting $z = 1$, into this expression gives the value: $e = 2.71828$.

Taking the natural logarithm of both sides of Eq. 2.1.2 we obtain

$$\log_e N = \log_e N_0 + at \quad . \tag{2.1.4}$$

In Fig. 2.1.1(a), a plot is presented of Eq. 2.1.2, using $N_0 = 10$ and $a = 0.15$. In Fig. 2.1.1(b), a semi-logarithmic plot is shown. In this plot, as Eq. 2.1.4 indicates, a straight line relationship of the form $y = k_0 + k_1 x$, is produced.

The first derivative with respect to time of Eq. 2.1.2 gives the *slope* of the exponential equation at any time, t. Likewise, the second derivative indicates the *curvature* of the equation. The integral with respect to time of Eq. 2.1.2 measures the *area* under the curve from $t = -\infty$ to time, t. These relationships are listed below

$$\frac{dN}{dt} = aN_0 e^{at} \tag{2.1.5}$$

$$\frac{d^2 N}{dt^2} = a^2 N_0 e^{at} \tag{2.1.6}$$

$$\int_{-\infty}^{t} N \, dt = \frac{1}{a} N_0 e^{at} \tag{2.1.7}$$

The geometrical interpretations of these three equations are shown in Fig. 2.1.1(a).

Before going further with our analysis, it may be helpful to introduce the matter of dimension and units.

In the ensuing sections, N will be referred to, on occasion, by the entirely versatile term, "item", regardless of the particular growth quantity or process involved. As we do in the case of the fundamental dimensions of force (F), mass (M), length (L), and time (T), the quantity N will be assigned the dimension: (N). In some of the topics and examples that follow, it may be possible to express the dimension (N) in terms of the above-indicated fundamental dimensions.

A particular "item" might be the height of a plant or the area of a leaf if we are considering a biological growth phenomenon. In this case, the dimensions of N are, respectively, length, L, or area, L^2. The units of N could be, in turn, meters (m) or centimeters squared (cm^2). Another "item" might be

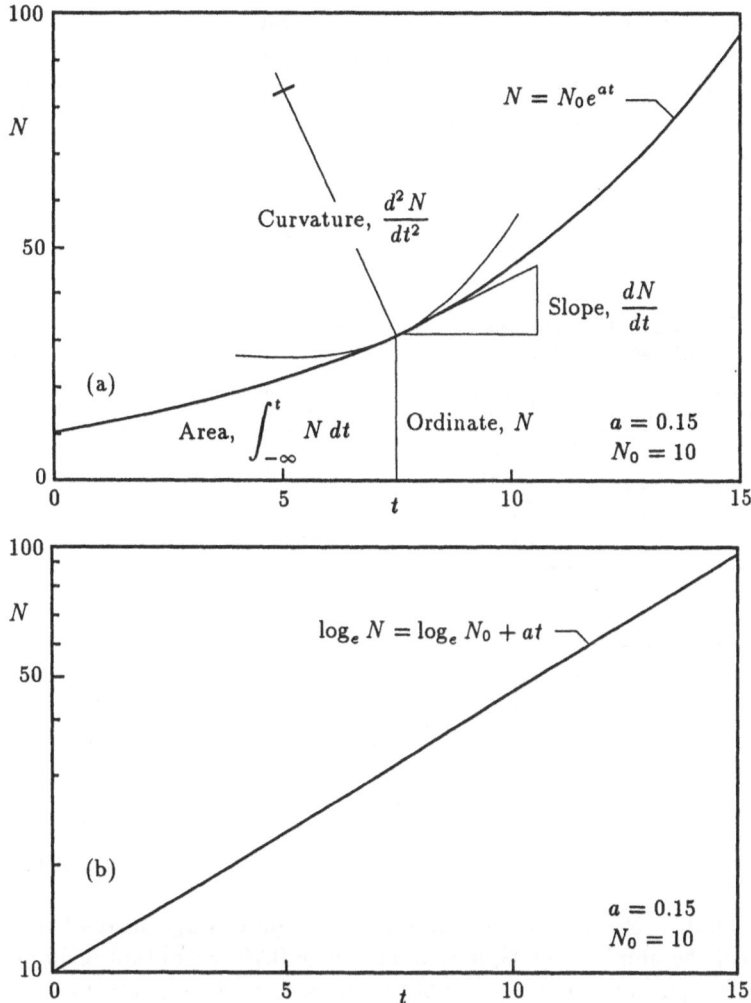

Fig. 2.1.1 (a) The exponential function and its main features. (b) Semi-logarithmic plot of the exponential function

the weight of grain produced per unit planted area. In this instance, the dimensions of N are weight per unit area (W/L^2); the units would probably be grams/meter^2 or tons/hectare.

Other "items", represented analytically and dimensionally by N, could be millions of people in a population problem, number of tractors in a technology transfer problem, solute concentration in a mass transfer problem, and so on.

In the example presented above, shown in Fig. 2.1.1, suppose that the "item" represented by N is the height of a plant. In this case, the dimension of N is length, L, and the unit is centimeters; so $N_0 = 10$ centimeters. The dimension of t is time, T, and its unit is weeks. Since the quantity at is dimensionless, then $a = 0.15$ per week (or 1/week or week^{-1}).

In Fig. 2.1.1(a), the ordinate, slope, curvature, and contained area of the exponential growth curve are shown for $t = 7.5$ weeks. Table 2.1.1 presents the numerical values of these growth characteristics, computed from Eqs. 2.1.5 through 2.1.7. The table also lists the various dimensions and units of these characteristics.

Table **2.1.1** Summary of characteristics of growth curve of Fig. 2.1.1. $N_0 = 10$ centimeters, $a = 0.15$ week^{-1}, $t = 7.5$ weeks

Characteristic	Symbol	Dimensions	Value	Units
Ordinate	N	L	30.802	centimeter
Slope	dN/dt	LT^{-1}	4.620	centimeter/week
Curvature	d^2N/dt^2	LT^{-2}	0.693	centimeter/week^2
Area	$\int_{-\infty}^{t} N\,dt$	LT	205.348	centimeter week

2.1.2 Doubling Times

If a quantity is growing exponentially, we can easily determine a parameter called the "doubling time". As implied, the question is: how long does it take for an item, increasing exponentially, to double in magnitude? Designating this time as t_2, letting $N = 2N_0$ in Eq. 2.1.2, and solving for t_2, we get

$$t_2 = \frac{1}{a}\log_e 2 = \frac{1}{a}(0.693) \quad . \tag{2.1.8}$$

If the growth coefficient, a, is expressed as a percentage instead of a decimal and if the number, 0.693, is rounded off to 0.70, we obtain a simple relationship for quickly estimating doubling times: $t_2 = 70/a$ (percent).

In the example presented above, with $a = 0.15$ week^{-1}, we obtain $t_2 = 4.62$ weeks. The "tripling time" is $t_3 = (1/a)\log_e 3$, and so on.

At this point, it is worthwhile to give a preliminary signal about the limitations of exponential growth. Proceeding as in the above paragraph, the time required for a 100-fold increase in N, would be $t_{100} = (1/a)\log_e 100 = 30.70$ weeks. That is, as Eq. 2.1.2 gives directly, when $t = 30.70$ weeks, the height of the plant, $N = 1000$ centimeters $= 10$ meters.

In the real world, this plant height is unlikely. So evidently the problem is not this simple. The exponential equation may accurately describe the growth for "small" values of time but after a certain period other factors become involved in the phenomenon which retard and limit the growth. We consider this matter in the following illustration.

2.1.3 An Illustration: Population of the United States

As an example of a phenomenon displaying exponential growth, at least over a certain period of time, we examine the growth of the population of the United States. In Table 2.1.2, the population, N, is listed for each decade commencing with the year 1790.

Table 2.1.2 Population of the United States, by decades, 1790 to 1980

Year	t [years]	N [millions]	Year	t [years]	N [millions]
1790	0	3.929	1890	100	62.980
1800	10	5.308	1900	110	76.212
1810	20	7.240	1910	120	92.228
1820	30	9.638	1920	130	106.021
1830	40	12.861	1930	140	123.203
1840	50	17.064	1940	150	132.165
1850	60	23.192	1950	160	151.326
1860	70	31.443	1960	170	179.323
1870	80	38.558	1970	180	203.302
1880	90	50.189	1980	190	226.546

The data of Table 2.1.2 are shown in Fig. 2.1.2 as arithmetic and semi-logarithmic plots. We disregard the solid lines in these figures for a moment.

In the semi-logarithmic plot of Eq. 2.1.2(b), it is observed that the points corresponding to the first seven or eight decades (say, 1790 through 1870) appear to fall on a straight line. Referring to Eq. 2.1.4, we surmise that during that 80-year period, the population of the U.S. was indeed growing exponentially. Although in this case it would be possible to visually construct a straight line through the first eight or nine points, a more accurate method is to determine the line of "best fit" by the "method of least squares". This method will be described in Section 2.2 in our examination of the logistic distribution.

A least squares computation involving the first eight data points of Table 2.1.2 gives the following results: $N_0 = 3.956$ million and $a = 0.0295\,\text{year}^{-1}$. With this value of the growth coefficient, the doubling time, $t_2 = 23.50$ years. The time required to increase the population by a factor of eight, $t_8 = (1/a)\log_e 8 = 70.49$ years. On this basis, in the year $1790 + 70 = 1860$ the approximate population would be $N = 8N_0 = 31.648$; this result agrees closely with the value $N = 31.443$ shown in Table 2.1.2.

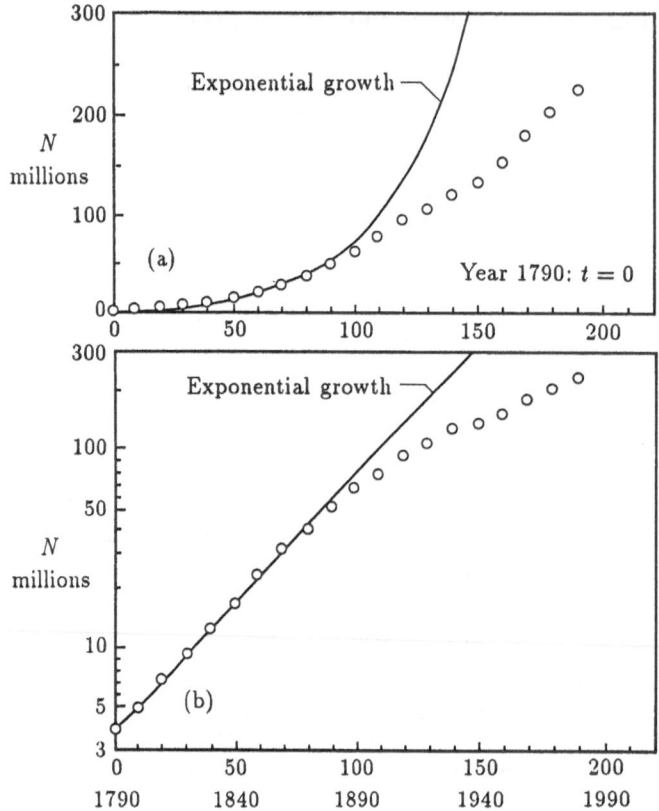

Fig. 2.1.2 Population of the United States, 1790 to 1980; (a) arithmetic plot and (b) semi-logarithmic plot

Substituting the above indicated values of a and N_0 into Eqs. 2.1.2 and 2.1.4 provides the solid lines shown in Fig. 2.1.2.

Now suppose we project this exponential growth for larger values of time. From Eq. 2.1.2 the following numbers result

Year	Time	Computed	Observed
1790	0	3.956	3.929
1840	50	17.292	17.064
1890	100	75.583	62.980
1940	150	330.377	132.165
1980	190	1075.169	226.546

It is seen that by 1890 ($t = 100$) the calculated value, based on exponential growth, is already in serious disagreement with the observed value. The computed results for 1940 and 1980 are ridiculous. This discrepancy is seen very clearly in Figs. 2.1.2(a) and (b).

From the above, we conclude that shortly after the Civil War in America (i.e., around 1865), the population of the nation was no longer increasing at an exponential rate. At that time, certain "crowding" effects began to be felt. No longer was there an infinite "carrying capacity" nor a demographic setting for unlimited growth. This matter of decelerated growth, reasonably well described by the Verhulst equation, or logistic equation, is examined in Section 2.2.

2.1.4 Exponential Function with Migration

A modest generalization can be given to the differential equation for exponential growth, Eq. 2.1.1, by writing

$$\frac{dN}{dt} = aN + s - h \quad , \tag{2.1.9}$$

where s is the number of items immigrating (or stocked) and h is the number of items emigrating (or harvested) per unit time. The solution to Eq. 2.1.9 with $N(0) = N_0$, is

$$N = N_0 e^{at} + \frac{s - h}{a}(e^{at} - 1) \quad . \tag{2.1.10}$$

If $s = h$, this equation reduces to Eq. 2.1.2.

As an example, suppose that at time, $t = 0$, there are $N_0 = 1000$ fish in a large lake in which there is an unlimited food supply and no predators. The intrinsic growth coefficient of the particular species of fish is, $a = 0.25\,\text{week}^{-1}$. It is emphasized that a is already the net growth coefficient, i.e., the birth rate minus the death rate. Also, with $s = h = 0$, the doubling time is, $t_2 = (1/a)\log_e 2 = 2.773$ weeks.

Suppose that 250 fish are added to the lake each week. Then the "stocking" (immigration) rate is $s = 250$ fish/week and $h = 0$. From Eq. 2.1.10, the total number of fish in the lake, $N = N(t)$, is computed; the result is shown as the extreme left-hand curve of Fig. 2.1.3. The two adjacent curves correspond to $s = 100$ and $s = 0$ respectively. When $s > 0$, the total number of fish increases very rapidly because of the combined effects of stocking and exponential growth of the fish population.

Now instead of adding fish to the lake, suppose that a certain number are removed each week. In other words, the lake is being harvested. Curves corresponding to "harvesting" (emigration) rates ranging from $h = 100$ to $h = 250$ fish/week (with $s = 0$, of course) are also shown in Fig. 2.1.3.

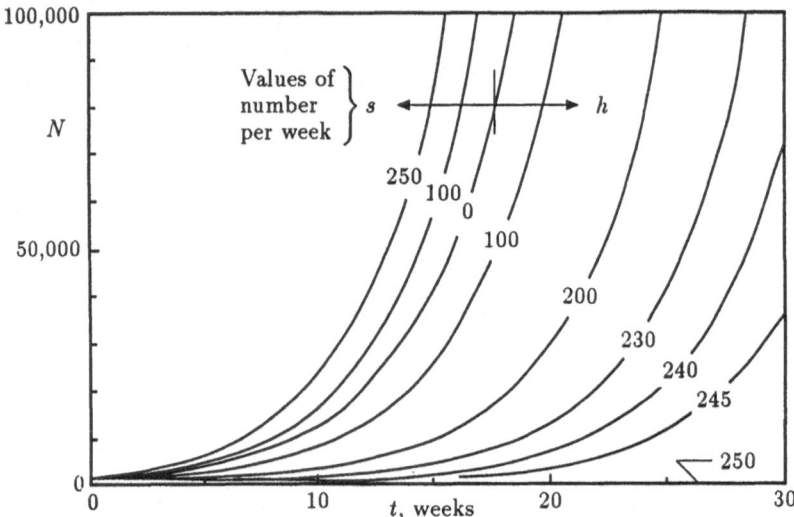

Fig. 2.1.3 Exponential growth of the fish population in a lake with stocking s and harvesting h

It is noted that no increase in N occurs when the harvesting rate, $h = aN_0$; this result follows immediately from Eq. 2.1.10. It is observed also that for any harvesting rate less than this amount, the value of N nevertheless becomes infinite eventually. However, if the harvesting rate, h, is larger than aN_0, there is an extinction time, t_e, resulting in $N = 0$

$$t_e = \frac{1}{a} \log_e \left(\frac{1}{1 - aN_0/h} \right) \quad . \tag{2.1.11}$$

In our example, if the harvesting rate $h = 300$ fish/week, then all the fish will be gone from the lake after $t_e = 7.167$ weeks.

For people who do not fish but like saving money. Suppose a young person (age 25, say) deposits \$1000 in a bank and thereafter, for the next 20 years, elects to save a weekly amount of \$25. The interest rate on the savings account is 5.2 percent annually or 0.001 per week. We have: $N_0 = \$1000$, $a = 0.001$, $s = \$25$ and $t = 20 \times 52 = 1040$ weeks. Substituting these numbers into Eq. 2.1.10 yields $N = \$48\,560$ of which \$2830 is the value of the original \$1000 and \$45\,730 is the value of the weekly deposits.

The person (now age 45) terminates the savings scheme and decides to withdraw a certain fixed amount each week; the interest rate is still 5.2 percent. The question is: how much can be withdrawn weekly before the total saved amount, $N_0 = \$48\,560$, is entirely gone. We solve Eq. 2.1.11 for h and then substitute desired values of extinction time, t_e. If $t_e = 1040$ weeks (20 years) then $h = \$75.11$ per week; if $t_e = 1560$ weeks (30 years) then $h = \$61.48$; and if $t_e = \infty$ then $h = \$48.56$.

We bear in mind that this example is based on so-called "instantaneous compounding". However, daily, weekly, or even monthly compounding periods do not greatly alter the above-indicated amounts.

2.1.5 Power Law Exponential Function

The exponential function can be given another kind of generalization by writing

$$\frac{dN}{dt} = aN^r \qquad (2.1.12)$$

where r is a constant. To avoid awkward dimensions and also to obtain a simpler answer, this differential equation is written in the dimensionless form

$$\frac{dW}{dT} = W^r \qquad (2.1.13)$$

where $W = N/N_0$ and $T = at$. This expression might be called a "power law" exponential equation since W is raised to the r-th power. The solution is

$$W = [1 + (1 - r)T]^{1/(1-r)} \qquad (2.1.14)$$

or

$$\frac{N}{N_0} = [1 + (1 - r)at]^{1/(1-r)} \quad . \qquad (2.1.15)$$

We now examine several cases starting with $r = 0$. Substituting this value into Eq. 2.1.15 gives the following linear relationship

$$\frac{N}{N_0} = 1 + at \qquad (2.1.16)$$

a result obtained immediately from Eq. 2.1.13.

Suppose that $r = 1/2$. Then from Eq. 2.1.15 we obtain a quadratic growth equation

$$\frac{N}{N_0} = (1 + \tfrac{1}{2}at)^2 \quad , \qquad (2.1.17)$$

and when $r = 3/4$, a quartic equation results

$$\frac{N}{N_0} = (1 + \tfrac{1}{4}at)^4 \quad . \qquad (2.1.18)$$

Next, we take $r = 1$. Substitution of this value into Eq. 2.1.15 produces an indeterminate answer. So we use the relationship

$$\lim_{m \to \infty} \left(1 + \frac{z}{m}\right)^m = e^z \qquad (2.1.19)$$

to establish the fact that, when $r = 1$, we again get the simple exponential solution

$$\frac{N}{N_0} = e^{at} \qquad\qquad (2.1.20)$$

Finally, an interesting case of "hyperbolic growth" arises if $r = 2$ is substituted into Eq. 2.1.15. The answer is

$$\frac{N}{N_0} = \frac{1}{1 - at} \ . \qquad\qquad (2.1.21)$$

In Fig. 2.1.4, curves are plotted for various values of r using $a = 0.020$.

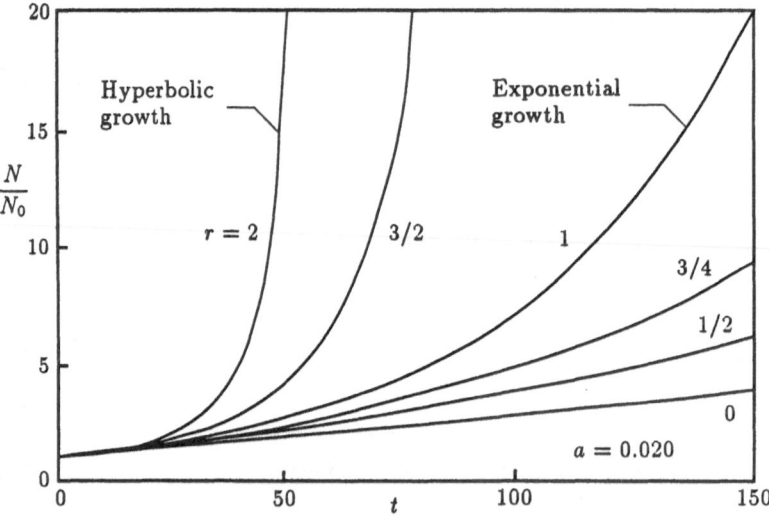

Fig. 2.1.4 Power law exponential growth for various values of the parameter r

2.1.6 An Illustration: Population of the World

It is observed in Fig. 2.1.4 that when $r = 2$, N becomes infinite at a definite value of t. From Eq. 2.1.21, this critical value is $t_e = 1/a$. This $r = 2$ case is essentially the one considered by Keyfitz (1968) in a dire example concerning world population explosion.

In Table 2.1.3, the population of the world is shown for various times commencing with the year 1650. We establish the time origin, $t = 0$, at that year.

Table **2.1.3** Population of the world, 1650–1990

Year	t [years]	Population N [billions]	Year	t [years]	Population N [billions]
1650	0	0.510	1965	315	3.354
1700	50	0.625	1970	320	3.696
1750	100	0.710	1975	325	4.066
1800	150	0.910	1980	330	4.432
1850	200	1.130	1985	335	4.822
1900	250	1.600	1990	340	5.318
1950	300	2.525			
1960	310	3.307			

The data of the table are displayed in Fig. 2.1.5. This figure illustrates the well-known fact that the world's population increased very slowly for many centuries. Indeed, not until the beginning of the present century did the population start to rise markedly and only after the mid-twentieth century mark did the world's population show really alarming increases.

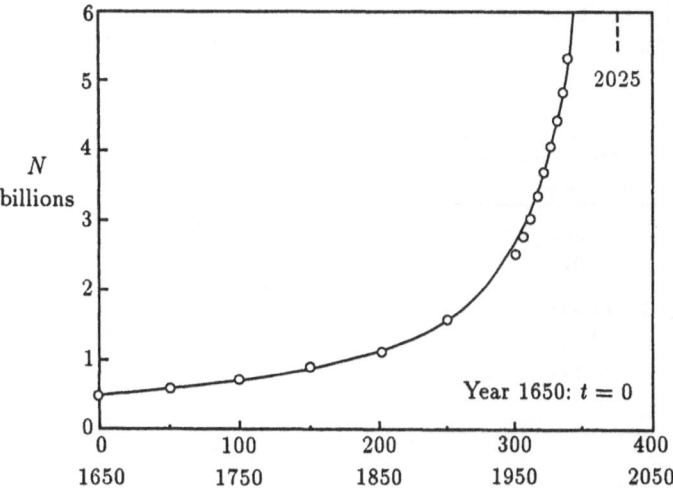

Fig. **2.1.5** Population of the world since 1650 and the corresponding hyperbolic growth curve

To follow Keyfitz, we assume that the growth of the world's population follows the inverse hyperbolic relationship indicated by Eq. 2.1.21. This equation can be written in the form

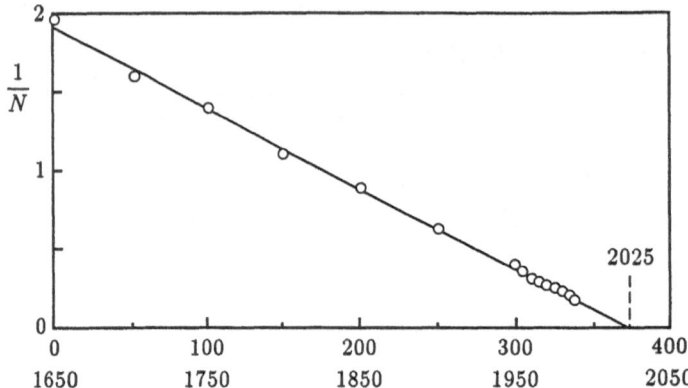

Fig. 2.1.6 A plot to determine the parameters of world population growth

$$\frac{1}{N} = \frac{1}{N_0} - \frac{a}{N_0}t \quad . \tag{2.1.22}$$

Hence, if the quantity, $1/N$, is plotted against time, t, a straight line relationship should be obtained if the assumption is correct. Such a plot is shown in Fig. 2.1.6. A least squares computation gives the values $N_0 = 0.525 \times 10^9$ and $a = 0.00267\,year^{-1}$ with a correlation coefficient of 0.9991. The solid lines of Figs. 2.1.5 and 2.1.6 are based on these results.

Using these values of a and N_0 in Eq. 2.1.21, we determine the following projected magnitudes of the world's population

Year	t [years]	Population N [billions]
1990	340	5.694
2000	350	8.015
2010	360	13.531
2020	370	43.388
2024	374	369.718
2024.53	374.53	∞

If the denominator of Eq. 2.1.21 is set equal to zero, N becomes infinite. The "explosion time" is $t_e = 374.53$ years, which corresponds approximately to the year $1650+375 = 2025$. Of course, nothing like this "infinite population explosion" will really take place. However, as Keyfitz states: "The minimum conclusion to be drawn is that the world population will turn from its present trend during the next 75 years, if indeed the trend is the hyperbola." This same problem was examined and the same conclusion reached by von Foerster et al (1960). These investigators devised a "coalition growth model" which incorporated the presumption that as human population increases so does its tendency to function as a single entity or coalition. This tendency evidently enhances standards of living and productivities; an important presumed consequence is that death rates decrease faster than birth rates. In other words, the net growth coefficient, a, increases with N.

Accordingly, in the view of von Foerster

$$\frac{dN}{dt} = a(N)N = aN^{1/k}N = aN^{1+1/k} \quad . \tag{2.1.23}$$

If we set $r = 1 + 1/k$ and $k = 1$, we again obtain Eq. 2.1.21.

On this basis, von Foerster and his colleagues forecast that November 13, 2026 will be "doomsday": the day on which the world's population will become infinite.

To get around this alarming and depressing situation, we shall examine a "modified coalition growth model" in Section 3.4.6 This model, devised by Austin and Brewer (1971), produces the more acceptable forecast that the world's population will be only 38 billion by the year 2060 and that it will never exceed 50 billion.

Even these magnitudes are undoubtedly very much on the high side. Never mind. As we shall see, the modified coalition model nicely combines the explosive feature of power law exponential growth with the stabilizing feature of logistic growth. The numerical parameters in the model can be easily changed to yield more likely projections of the world's population.

2.1.7 Combinations of Exponential Functions

There are a great many mathematical formulations in the physical, biological, and social sciences in which the exponential function or various combinations of it make an appearance. A simple example occurs in the kinetics of consecutive chemical reactions.

Following Moelwyn-Hughes (1961), we consider the following two consecutive first order irreversible reactions

$$A \xrightarrow{k_1} B \xrightarrow{k_2} C \quad , \tag{2.1.24}$$

where k_1 and k_2 are reaction rate coefficients. This relationship states that reactant A is converted into an intermediate product B which, in turn, is converted into a final product C. For example, Winkler and Hinshelwood (1935), in their studies of thermal decomposition of acetone, indicate the following reactions

$$(CH_3)_2CO \xrightarrow{k_1} CH_2 : CO + CH_4$$

$$CH_2 : CO \xrightarrow{k_2} \tfrac{1}{2}CH_4 + CO \quad . \tag{2.1.25}$$

The acetone (reactant A) produces the intermediate product, ketene (B), which subsequently yields the final product, methene (C). These investigators observed that the concentration of ketene attained a maximum value during an experiment and then decreased asymptotically to zero.

Let x, y, and z be the concentrations of A, B, and C and assume that first order irreversible reactions take place.

Accordingly, we obtain the following differential equations and impose the indicated initial conditions

$$\frac{dx}{dt} = -k_1 x; \qquad x(0) = x_0 \tag{2.1.26}$$

$$\frac{dy}{dt} = -k_1 x - k_2 y; \qquad y(0) = y_0 \tag{2.1.27}$$

$$\frac{dz}{dt} = -k_2 y; \qquad z(0) = z_0 \tag{2.1.28}$$

In most chemical reactions the initial values, y_0 and z_0, would be zero. However, for our purpose we will take these values to be positive constants; they can always be set to zero if we like. From the principle of conservation of mass we have

$$x + y + z = x_0 + y_0 + z_0 = r_0 \quad . \tag{2.1.29}$$

Before solving Eqs. 2.1.26–28, we digress for a moment. The first of these equations states that the rate at which a quantity decreases is proportional to the quantity present at any instant. This is the equation for radioactive decay. As in the case of "doubling times", it is easy to establish that the "half-life time" of a radioactively decaying substance is $t_{1/2} = \log_e 2/k_1$.

In Section 2.8 we examine a mathematical model concerning epidemics with a so-called SIR model which describes the interactions among the susceptible (S) to a disease, the infectives (I), and the removed or recovered (R). The SIR model is rather complicated. However, for now, in a simple linear model, suppose that x corresponds to S, y to I, and z to R. Then, as we will see shortly from Eq. 2.1.26, the number of susceptibles (x or S) decreases with time t, as this group increasingly becomes infected. From Eq. 2.1.27, we anticipate that the number of infectives (y or I) increases until $dy/dt = 0$ (i.e. until $y = k_1 x/k_2$) and thereafter decreases as this group recovers. Finally, from Eq. 2.1.28, the number of recovered (z or R) increases continuously right from the start.

In the matter of technology transfer, we again use our simplified (x, y, z) model. Selecting the example given by Sharif and Ramanathan (1982), the number of persons undecided about acquiring a black-and-white television (x: uncommitted) decreases as more and more buy the sets. The number of persons acquiring black-and-white televisions (y: adoptors of the technology) initially increases but then decreases because the group decides to purchase color TVs. Finally, the number of persons buying color TVs (z: rejectors of the earlier technology) increases from the outset.

2.1.8 Solutions and Properties of the Equations

We return to our problem consisting of the three linear differential equations. We easily obtain the following solution to Eq. 2.1.26

$$x = x_0 e^{-k_1 t} \quad . \tag{2.1.30}$$

This equation possesses no maximum point nor inflection point. Substituting this result into Eq. 2.1.27 yields

$$\frac{dy}{dt} + k_2 y = k_1 x_0 e^{-k_1 t} \quad . \tag{2.1.31}$$

This is a first order linear differential equation of the form

$$\frac{dy}{dt} + P(t)y = Q(t) \quad . \tag{2.1.32}$$

The general solution of such an equation is

$$y = c e^{-\int P(t)\,dt} + e^{-\int P(t)\,dt} \int Q(t) e^{\int P(t)\,dt}\, dt \quad . \tag{2.1.33}$$

The initial condition, $y(0)$, enables one to determine the value of the constant, c. From these relationships, it is not difficult to obtain the solution to Eq. 2.1.31

$$y = y_0 e^{-k_2 t} + \frac{k_1 x_0}{k_2 - k_1} \left(e^{-k_1 t} - e^{-k_2 t} \right) \quad . \tag{2.1.34}$$

If we take the first derivative of this expression and set it equal to zero (i.e., slope $= dy/dt = 0$), we obtain the time, t_m, corresponding to the maximum value of $y(t)$

$$t_{m,y} = \frac{1}{k_2 - k_1} \log_e \left[\frac{k_2}{k_1} \left(1 - \frac{(k_2 - k_1)y_0}{k_1 x_0} \right) \right] \quad . \tag{2.1.35}$$

Taking the second derivative of Eq. 2.1.34 and setting it to zero (i.e., curvature $= d^2 y/dt^2 = 0$), we determine the time, t_i, corresponding to the inflection point of $y(t)$

$$t_{i,y} = \frac{1}{k_2 - k_1} \log_e \left[\left(\frac{k_2}{k_1} \right)^2 \left(1 - \frac{(k_2 - k_1)y_0}{k_1 x_0} \right) \right] \quad . \tag{2.1.36}$$

We note that if the initial condition $y_0 = 0$, then Eqs. 2.1.34–36 are simplified considerably.

Finally, we substitute Eq. 2.1.34 into Eq. 2.1.28 to obtain

$$z = r_0 - y_0 e^{-k_2 t} - \frac{x_0}{k_2 - k_1} \left(k_2 e^{-k_1 t} - k_1 e^{-k_2 t} \right) \quad . \tag{2.1.37}$$

This equation has no maximum or minimum point for positive values of t. However, it does have an inflection point located at

$$t_{i,z} = \frac{1}{k_2 - k_1} \log_e \left[\frac{k_2}{k_1} \left(1 - \frac{(k_2 - k_1)y_0}{k_1 x_0} \right) \right] . \tag{2.1.38}$$

We note from this result, and Eq. 2.1.35, that the times of the maximum point of $y(t)$ and the inflection point of $z(t)$ are the same. Again, the expressions are simplified if $y_0 = 0$ and $z_0 = 0$.

From the preceding results it is not difficult to confirm that $x + y + z = x_0 + y_0 + z_0 = r_0$.

It is worthwhile to look at some limiting cases. First, suppose that $k_1 \gg k_2$. From Eqs. 2.1.34 and 2.1.37 we obtain

$$y = y_0 + x_0(1 - e^{-k_1 t}) \tag{2.1.39}$$

$$z = r_0 - (x_0 + y_0)e^{-k_2 t} . \tag{2.1.40}$$

On the other hand, if $k_2 \gg k_1$, then we get

$$y = y_0 e^{-k_2 t} \tag{2.1.41}$$

$$z = r_0 - x_0 e^{-k_1 t} - y_0 e^{-k_2 t} . \tag{2.1.42}$$

In our chemical kinetics problem, we can take $y_0 = z_0 = 0$. In this case, if $k_1 \gg k_2$, Eq. 2.1.40 reduces to $z = x_0[1 - \exp(-k_2 t)]$. On the other hand, if $k_2 \gg k_1$, Eq. 2.4.42 gives $z = x_0[1 - \exp(-k_1 t)]$. It is observed in both of these limiting cases that the creation of the final product, z, is controlled by the reaction involving the smaller reaction rate coefficient.

In the special case in which $k_1 = k_2 = k$, the preceding solutions become indeterminate. Consequently, it is necessary to return to the original differential equations, Eqs. 2.1.26–28, to obtain the following answers

$$x = x_0 e^{-kt} \tag{2.1.43}$$

$$y = (y_0 + k x_0 t)e^{-kt} \tag{2.1.44}$$

$$z = r_0 - [y_0 + x_0(1 + kt)]e^{-kt} . \tag{2.1.45}$$

The maximum and inflection points for $y(t)$ are

$$t_{m,y} = \frac{x_0 - y_0}{k x_0} \quad ; \quad t_{i,y} = \frac{2x_0 - y_0}{k x_0} . \tag{2.1.46}$$

As before, $z(t)$ has no maximum point; its inflection point is located at

$$t_{i,z} = \frac{x_0 - y_0}{k x_0} . \tag{2.1.47}$$

The results of some numerical calculations are shown in Fig. 2.1.7. In these plots, $x_0 = 100$, $y_0 = 0$, and $z_0 = 0$. The quantity λ is equal to k_2/k_1.

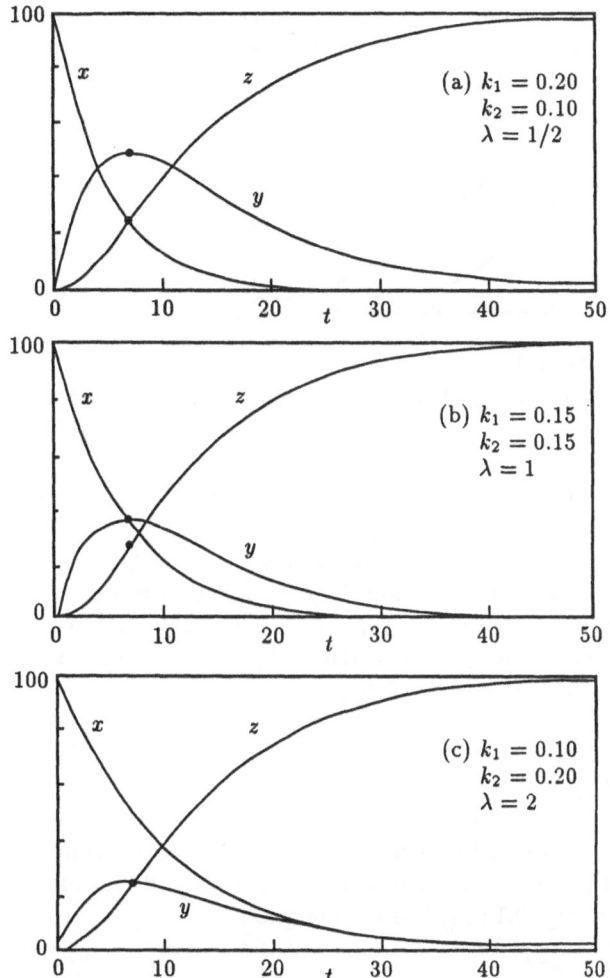

Fig. 2.1.7 Concentrations of reactants of consecutive chemical reactions. $\lambda = k_2/k_1$; $x_0 = 100$, $y_0 = z_0 = 0$

In Fig. 2.1.8, two curves are shown, again with $y_0 = z_0 = 0$. One curve is a dimensionless plot of the maximum value, y_m/x_0 as a function of $\lambda = k_2/k_1$. When $\lambda = 1$, $y_m/x_0 = 1/e$. For small and large values of λ, y_m/x_0 approaches unity and zero, respectively.

The other curve of Fig. 2.1.8 is a dimensionless display of the inflection point, z_i/x_0, as a function of λ. When $\lambda = 1$, $z_i/x_0 = 1-2/e$. For small and large values of λ, z_i/x_0 approaches zero.

In addition to their precise definitions in the above described problem in chemical kinetics, the three functions, $x(t)$, $y(t)$, and $z(t)$, can be useful

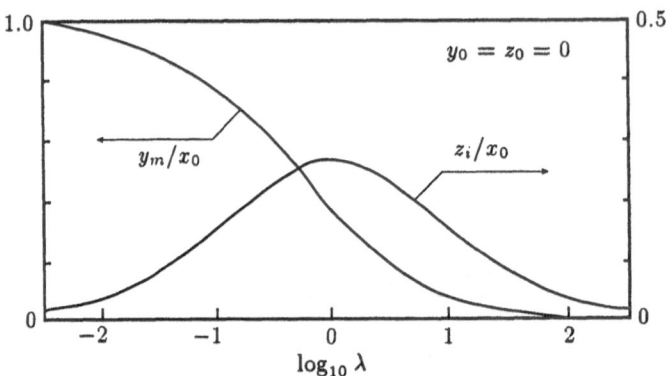

Fig. 2.1.8 Plots of maximum point of $y(t)$ and inflection point of $z(t)$ in consecutive chemical reactions

as mathematical frameworks for analyzing and interpreting other kinds of growth and transfer phenomena.

This point is illustrated in Fig. 2.1.9. The abscissa of the figure is the dimensionless quantity, $T = kt$. The ordinate, U, is a dimensionless quantity normalized by using suitable values of x_0, y_0, and z_0. Although we have not yet examined it, the well-known logistic equation is shown as Curve a in Fig. 2.1.9 to serve as a comparison. It has the definition

$$U = \frac{1}{1 + [(1/U_0) - 1]e^{-T}} \cdot \tag{2.1.48}$$

We select the initial value $U_0 = 0.10$.

Curve b is obtained from Eq. 2.1.45 with $y_0 = 0$. This curve is essentially the cumulative distribution of a gamma function which we will consider in Section 3.5. As plotted in Fig. 2.1.8, it has the equation

$$U = 1 - (1 - U_0)(1 + T)e^{-T} \quad . \tag{2.1.49}$$

Curve c is derived from either Eq. 2.1.40 or Eq. 2.1.42 with $y_0 = 0$. This produces the so-called confined exponential equation which we examine in Section 2.3. It has the definition

$$U = 1 - (1 - U_0)e^{-T} \quad . \tag{2.1.50}$$

We observe that the equations of all three of the curves in Fig. 2.1.9 contain the negative exponential function.

2.1.9 An Illustration: Oxygen Distribution in a River

We consider the steady flow of a river whose average velocity is u_0. At any point s along the river the concentration of dissolved oxygen (DO) is C (mg/l). At the same point, the concentration of biochemical oxygen demand

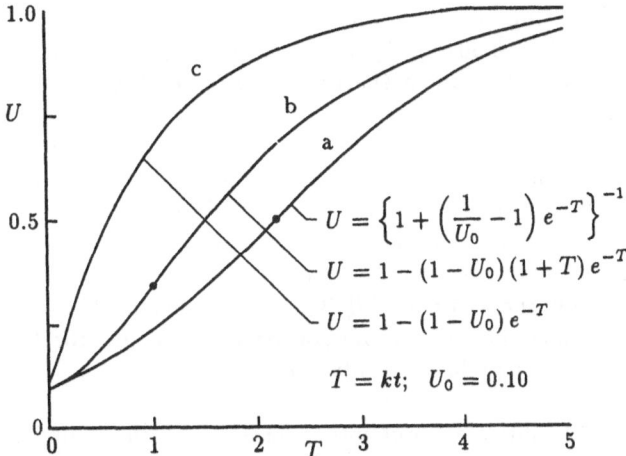

Fig. 2.1.9 Comparison of growth equations: (a) logistic function; (b) cumulative gamma function; (c) confined exponential function

(BOD) is L (mg/l). We assume that the only source of oxygen is the atmosphere. The temperature, atmospheric pressure, and water salinity establish the equilibrium concentration of oxygen, C_s (mg/l).

A mass balance on an elemental volume of the river provides the following differential equations

$$u_0 \frac{dL}{ds} = -K_1 L \tag{2.1.51}$$

$$u_0 \frac{dD}{ds} = K_1 L - K_2 D \quad , \tag{2.1.52}$$

where s is the linear distance along the river and $D = (C_s - C)$. According to Camp (1963), these expressions are known as the Streeter-Phelps equations or "oxygen sag" equations; they were devised years ago in connection with studies of pollution in the Ohio River. The constants, K_1 and K_2, are termed the deoxygenation coefficient and the reoxygenation or reaeration coefficient, respectively.

It is observed that Eqs. 2.1.51 and 2.1.52 are identical to Eqs. 2.1.26 and 2.1.27 with the equivalences $x : L$ and $y : D$ and the substitution $s = u_0 t$. Accordingly, we write down the solutions to the BOD–DO problem immediately from Eqs. 2.1.30 and 2.1.34. They are

$$L = L_0 e^{-K_1 s / u_0} \tag{2.1.53}$$

$$C = C_s - (C_s - C_0)e^{-K_2 s / u_0} - \frac{K_1 L_0}{K_2 - K_1}\left(e^{-K_1 s / u_0} - e^{-K_2 s / u_0}\right) . \tag{2.1.54}$$

The minimum point and inflection point for $C(t)$ are given by Eqs. 2.1.35 and 2.1.36.

We substitute the following numbers into our illustration: $u_0 = 25$ km/day, $K_1 = 0.25\,\mathrm{day}^{-1}$, $K_2 = 0.50\,\mathrm{day}^{-1}$, and $C_s = 9.2$ mg/l ($T = 20°\mathrm{C}$). The concentration of dissolved oxygen in the river at $s = 0$ is $C_0 = C_s = 9.2\,\mathrm{mg/l}$. We consider the following two cases.

Case 1. A heavily polluted waste enters the river at $s = 0$. At that point, the BOD concentration of the waste-river mixture is $L_0 = 36.8$ mg/l. Results of computations utilizing the previous equations are shown in Fig. 2.1.10. It is seen that the minimum DO concentration, $C_m = 0.0\,\mathrm{mg/l}$, occurs at a point $s = 69.3$ km downstream from the waste outfall. The entire reach of the river, from $s = 14.1\,\mathrm{km}$ to $s = 203.0$ km, has a DO concentration below the 5.0 mg/l value required by the state pollution control authority.

Case 2. Waste treatment facilities are put into operation which results in a reduction of BOD concentration at $s = 0$ to the value, $L_0 = 16.8$. As indicated in Fig. 2.1.11, the minimum DO concentration, $C_m = 5.0$ mg/l, occurs again at $s = 69.3\,\mathrm{km}$. In this case, the entire river has a dissolved oxygen concentration of 5.0 mg/l or more.

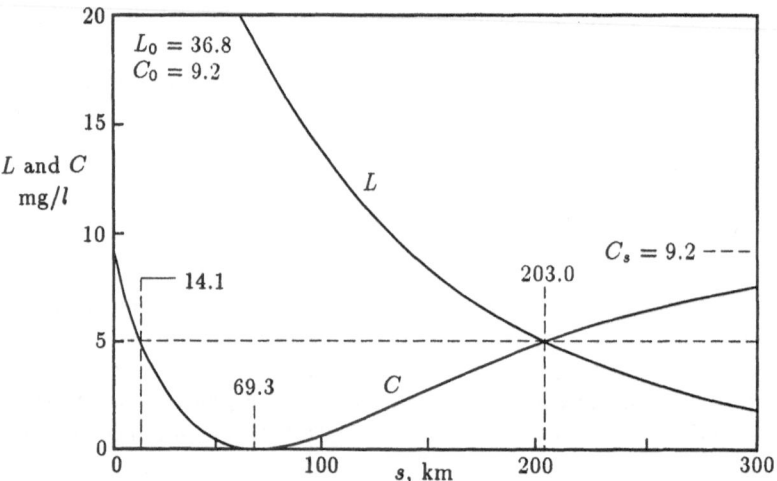

Fig. 2.1.10 Oxygen sag curve in a river. $u_0 = 25\,\mathrm{km/day}$, $K_1 = 0.25\,\mathrm{l/day}$, $K_2 = 0.50\,\mathrm{l/day}$, $C_s = 9.2\,\mathrm{mg/l}$, $L_0 = 36.8\,\mathrm{mg/l}$, $C_{\min} = 0.0\,\mathrm{mg/l}$

As we observed, our earlier dependent variable x corresponds to L, the biochemical oxygen demand, and y corresponds to D, the dissolved oxygen deficit. Accordingly, our third dependent variable, z, with $z_0 = 0$, corresponds to $(L_0 - L) + (C - C_0)$. Consequently, z represents the amount of oxygen

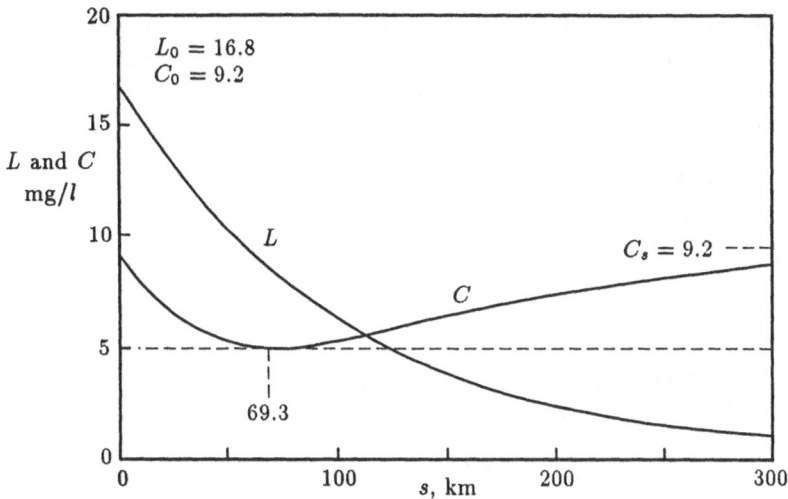

Fig. 2.1.11 Oxygen sag curve in a river. $u_0 = 25\,\text{km/day}$, $K_1 = 0.25\,\text{l/day}$, $K_2 = 0.50\,\text{l/day}$, $C_s = 9.2\,\text{mg/l}$, $L_0 = 16.8\,\text{mg/l}$, $C_{\text{min}} = 5.0\,\text{mg/l}$

available at time t, after having stabilized a portion of the organic material in the water plus the amount of oxygen available in excess of that originally present. As $t \to \infty$, $z_* = L_0 + C_s - C_0$. This simply says that after a long time, z_* represents the amount of oxygen necessary to stabilize all of the organic material plus the amount of oxygen needed to overcome the original oxygen deficit.

2.2 Logistic Distribution

2.2.1 The Differential Equation and Its Solution

In this chapter we introduce three basic distribution functions utilized in the analysis of growth phenomena, diffusion, and technology transfer. In Section 2.1 the simplest of these relationships, the exponential function, was examined in some detail. In the present and ensuing sections, other distribution functions are studied. Solutions to their differential equations are obtained and some of the more important properties of these solutions are presented. We begin with the logistic distribution. For this case, we have the following differential equation

$$\frac{dN}{dt} = aN - bN^2 \quad , \tag{2.2.1}$$

where a is the growth coefficient and b is a so-called "crowding coefficient".

The presence of the N^2 term indicates that this is a non-linear differential equation. Ordinarily it is not easy to obtain simple exact solutions for such equations; however, this case is a fortunate exception. There are several ways to obtain the solution to Eq. 2.2.1.

Method 1. Table of Integrals. The most straightforward method to solve Eq. 2.2.1 is simply to use a table of integrals. We rewrite the equation in the following form

$$\frac{dN}{aN - bN^2} = dt \quad .$$
(2.2.2)

Clearly, the problem is to determine the value of N as a function of time, t, i.e., $N = N(t)$. To obtain a complete solution to the problem we need to know the value of N at some time, t. Conveniently, we take $N = N_0$ when $t = 0$, i.e., $N(0) = N_0$. Writing Eq. 2.2.2 in integral form with this initial condition appearing as the lower limits we obtain

$$\int_{N_0}^{N} \frac{dN}{aN - bN^2} = \int_{0}^{t} dt \quad .$$
(2.2.3)

From a table of integrals and with some algebraic manipulation, the following intermediate answer is obtained

$$N = \left[\frac{1}{N_*} + \left(\frac{1}{N_0} - \frac{1}{N_*} \right) e^{-at} \right]^{-1} \quad ,$$
(2.2.4)

where $N_* = a/b$ is the so-called carrying capacity or equilibrium value. If the carrying capacity is infinitely large, i.e., $N_* = \infty$, then Eq. 2.2.4 reduces to the exponential case given by Eq. 2.1.2.

$$N = N_0 e^{at} \quad .$$
(2.2.5)

A slight modification of Eq. 2.2.4 gives the following final form of the logistic distribution

$$N = N_* \left[1 + \left(\frac{N_*}{N_0} - 1 \right) e^{-at} \right]^{-1} \quad .$$
(2.2.6)

In this analysis the growth coefficient, a, is assumed to be a positive constant. Accordingly, as time, t, approaches infinity, the second term in the denominator of Eq. 2.2.6 approaches zero and the value of N becomes N_*. Thus, N_* is simply the asymptotic value of N for very large t.

Nothing has been said about the relative values of N_0 and N_*. Consequently, N_0, the initial value of N, may be larger or smaller than N_*. Either way, the magnitude of N approaches N_* as $t \to \infty$. However, from here on we will assume that N_0 is less than N_*.

Method 2. Partial Fractions. We rewrite Eq. 2.2.2 in the following form

$$\frac{dN}{N(1 - N/N_*)} = a\,dt \quad . \tag{2.2.7}$$

The denominator of the left-hand member of this equation can be expressed in the following partial fraction form

$$\frac{1}{N(1 - N/N_*)} = \frac{k_1}{N} + \frac{k_2}{(1 + N/N_*)} \quad , \tag{2.2.8}$$

where k_1 and k_2 are constants. Constructing the same common denominator for both sides of this expression and equating the coefficients of like powers of N gives $k_1 = 1$ and $k_2 = 1/N_*$. So the equation becomes

$$\int_{N_0}^{N} \frac{dN}{N} + \frac{1}{N_*} \int_{N_0}^{N} \frac{dN}{(1 - N/N_*)} = a \int_0^t dt \quad . \tag{2.2.9}$$

A table of integrals is again needed unless one recalls that the integral of dN/N is $\log_e N$. In any event, solving Eq. 2.2.9 again yields Eq. 2.2.6.

Method 3. First Order Differential Equation. In fact, Eq. 2.2.1 is a special and simple case of the so-called Riccati differential equation which, in its general form, is

$$\frac{dy}{dx} + Q(x)y + R(x)y^2 = P(x) \quad . \tag{2.2.10}$$

Comparing this equation to Eq. 2.2.1, it is seen that $Q(x) = -a$, $R(x) = b$, and $P(x) = 0$. Making the substitution, $N = 1/C$, in Eq. 2.2.1, we obtain

$$\frac{dC}{dt} + aC = b \quad . \tag{2.2.11}$$

This is a first order linear differential equation whose solution is

$$C = ke^{-\int a\,dt} + e^{-\int a\,dt} \int be^{\int a\,dt}\,dt \quad , \tag{2.2.12}$$

where k is an integration constant. Since $N(0) = N_0$ and a and b are constants, the equation for the logistic distribution, Eq. 2.2.6, is again obtained.

2.2.2 Properties of the Logistic Distribution

To simplify the notation, Eq. 2.2.6 is written in the form

$$N = \frac{N_*}{1 + me^{-at}} \quad , \tag{2.2.13}$$

where $m = (N_*/N_0) - 1$. Before going further with the analysis, we look at a numerical example and a graphic display of the logistic distribution.

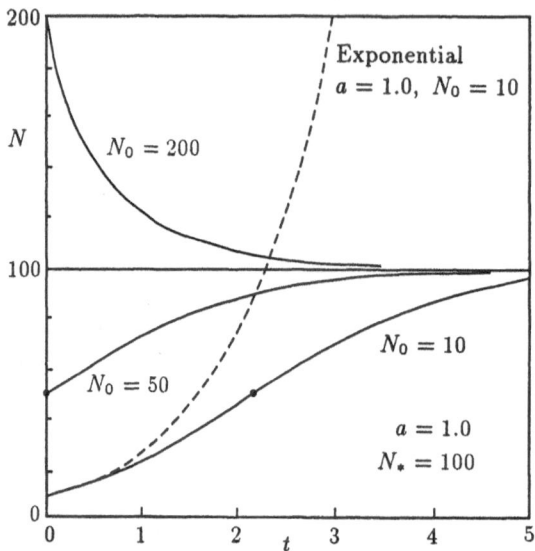

Fig. 2.2.1 Plots of the logistic equation for various values of N_0

In Fig. 2.2.1, three curves are shown for which $a = 1.0$ and $N_* = 100$. In the lowermost curve, $N_0 = 10$, and so $m = 9.0$. Substituting these values into Eq. 2.2.13 produces the familiar S-shaped curve associated with the logistic distribution. It is obvious that the slope of the curve is everywhere positive. The inflection point of the curve, which we will consider shortly, is located at $t_i = 2.197$, $N_i = 50$. This point is shown as the solid dot on the lower curve. When $t < t_i$, the curve is concave upward and when $t > t_i$, it is concave downward. For large values of time, the curve approaches the carrying capacity, $N_* = 100$.

The middle curve also has values of $a = 1.0$ and $N_* = 100$ but now $N_0 = 50$ and accordingly $m = 1.0$. In this case, the inflection point occurs at $t_i = 0$, $N_i = 50$. As the figure shows, the curve has a positive slope and is concave downward.

The uppermost curve is the case in which $N_0 = 200$, i.e., $N_0 > N_*$. In this case, the curve is concave upward, there is no inflection point, and the slope is negative. As mentioned we are not greatly interested in this case.

Finally, for comparison, the dashed curve shown in Fig. 2.2.1 is the exponential equation given by Eq. 2.2.5.

Returning to the analysis, we differentiate Eq. 2.2.13 with respect to time. The result is

$$n = \frac{dN}{dt} = amN_* \frac{e^{-at}}{(1 + me^{-at})^2} \; . \tag{2.2.14}$$

At this point, it is important to identify Eq. 2.2.13 as the cumulative distribution of the logistic equation. In turn, by definition, Eq. 2.2.14 is designated the density distribution, or frequency distribution, of the logistic. It should be clear that integrating Eq. 2.2.14 with respect to time takes us back to where we started. That is

$$N = \int_{-\infty}^{t} n(t)\, dt = \frac{N_*}{1 + me^{-at}} \quad . \tag{2.2.15}$$

With reference to Fig. 2.2.2 the following are some geometrical interpretations of the preceding two equations: (a) N is the ordinate of the cumulative distribution and, from Eq. 2.2.15, is also the area under the curve of the density distribution from $t = -\infty$ to $t = t$, and (b) n is the ordinate of the density distribution and, from 2.2.14, is also the slope of the cumulative distribution.

Setting the slope, dN/dt, equal to zero determines the values of t for which N has maximum and minimum values. Accordingly, if Eq. 2.2.14 is equated to zero we learn that N has a maximum value at $t = +\infty$, and a minimum at $t = -\infty$. This equation also says that as long as m is positive, i.e., $N_* > N_0$, the slope of the cumulative curve is positive

Differentiating Eq. 2.2.14 gives

$$\frac{dn}{dt} = \frac{d^2N}{dt^2} = a^2 m N_* \frac{e^{-at}(me^{-at} - 1)}{(1 + me^{-at})^3} \quad . \tag{2.2.16}$$

Setting this result equal to zero determines where the slope, dn/dt, of the density distribution is equal to zero; clearly, this identifies the maximum value of n. It also indicates where the curvature, d^2N/dt^2, of the cumulative distribution is zero ; this identifies the inflection point and thus the maximum slope of the cumulative curve. Accordingly, from Eq. 2.2.16 we obtain the following result

$$t_i = \frac{1}{a} \log_e m; \qquad m = \frac{N_*}{N_0} - 1 \quad , \tag{2.2.17}$$

where t_i is the time of the inflection point of the cumulative curve. Substituting this answer into Eqs. 2.2.13 and 2.2.14 gives the following result

$$Ni = \frac{1}{2} N_*; \qquad n_i = \left(\frac{dN}{dt}\right)_i = \frac{1}{4} a N_* \quad . \tag{2.2.18}$$

These expressions give, respectively, the values of the cumulative distribution, the density distribution, and the slope of the cumulative distribution at the inflection point of the cumulative curve.

It is useful to present the logistic distribution, Eq. 2.2.13, in an alternative form. First, from Eq. 2.2.17, we note that $m = \exp(at_i)$. Substituting this result into Eq. 2.2.13 yields

$$N = \frac{N_*}{1 + e^{-a(t-t_i)}} \quad . \tag{2.2.19}$$

We can rewrite this last equation in the following form

$$N = \frac{N_*}{2}\left(1 + \tanh\left[\tfrac{1}{2}a(t - t_i)\right]\right) \quad , \tag{2.2.20}$$

where tanh is the hyperbolic tangent. Likewise, Eq. 2.2.14 can be expressed as

$$n = \frac{aN_*}{4}\operatorname{sech}^2\left[\tfrac{1}{2}a(t - t_i)\right] \quad , \tag{2.2.21}$$

where sech is the hyperbolic secant.

We now differentiate Eq. 2.2.16 to obtain the second derivative of the density distribution, d^2n/dt^2, and the third derivative, d^3N/dt^3, of the cumulative distribution. Setting the answer equal to zero gives the inflection points of the density function and maximum curvature points of the cumulative function. The results are

$$t_c = \frac{1}{a}\log_e[(2 \pm \sqrt{3})m] \tag{2.2.22}$$

$$N_c = \frac{N_*}{2}\left(1 \pm \frac{1}{\sqrt{3}}\right) \tag{2.2.23}$$

$$n_c = \left(\frac{dN}{dt}\right)_c = \frac{1}{6}aN_* \tag{2.2.24}$$

$$\left(\frac{dn}{dt}\right)_c = \left(\frac{d^2N}{dt^2}\right)_c = \pm\frac{1}{6\sqrt{3}}a^2N_* \quad . \tag{2.2.25}$$

Finally, we determine the mean value or expectation, $\bar{t}$, and the variance, σ^2. By definition, the mean value is

$$\bar{t} = \frac{1}{N_*}\int_{-\infty}^{\infty} tn(t)\, dt \quad . \tag{2.2.26}$$

Substituting Eq. 2.2.14 into this equation yields the answer

$$\bar{t} = \frac{1}{a}\log_e m \quad , \tag{2.2.27}$$

which, not surprisingly, is the same as the expression for the inflection point time, t_i, given by Eq. 2.2.17.

The variance is defined as

$$\sigma^2 = \frac{1}{N_*}\int_{-\infty}^{\infty} (t - \bar{t})^2 n(t)\, dt \quad . \tag{2.2.28}$$

Using Eqs. 2.2.14 and 2.2.27 in this expression and carrying out the definite integration gives the answer

$$\sigma^2 = \frac{\pi^2}{3a^2} \quad . \tag{2.2.29}$$

The standard deviation, which by definition is the square root of the variance, is

$$\sigma = \frac{\pi}{\sqrt{3}a} \quad .$$ (2.2.30)

It is noted, from its origin in Eq. 2.2.28, that σ is measured from the coordinate of the mean value, $\bar{t}$.

The cumulative and density distributions, for the numerical example presented above and displayed as the lower-most curve of Fig. 2.2.1, are shown in Fig. 2.2.2. Various features of the distributions are indicated in the figure and their numerical values are presented in Table 2.2.1.

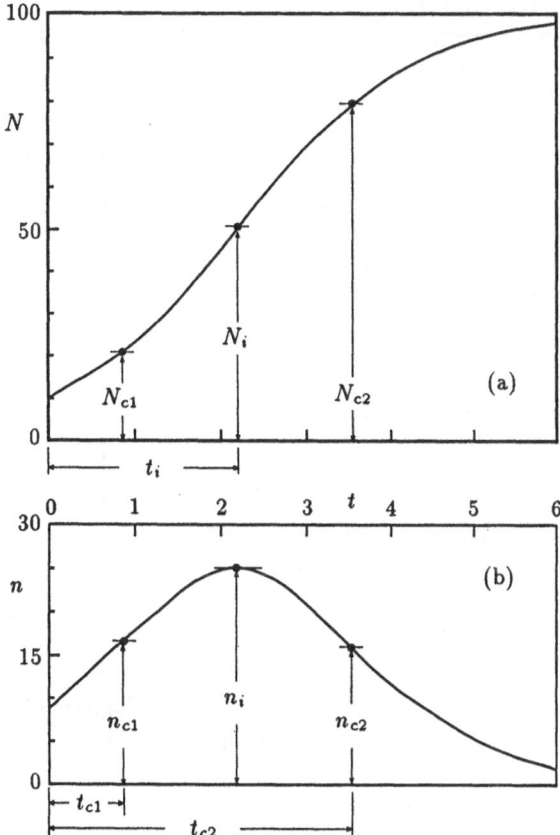

Fig. 2.2.2 Definition sketch for various critical (t, N) coordinates of the ordinary logistic equation; (a) cumulative distribution and (b) density distribution

Table **2.2.1** Various features of the logistic distribution with the following numerical values: $a = 1.0$, $N_0 = 10$, $N_* = 100$; see Fig.2.2.2

Symbol	Quantity	Value
t_i	value of t at cumulative inflection point	2.197
N_i	value of N at cumulative inflection point	50.0
$n_i = (dN/dt)_i$	value (maximum) of n at cumulative inflection point; value (maximum) of slope of cumulative curve at cumulative inflection point	25.0
t_c	values of t at density inflection points	0.880 3.514
N_c	values of N at density inflection points	21.13 78.87
$n_c = (dN/dt)_c$	values of n at density inflection points; values of slope of cumulative curve at density inflection points	16.67
$(dn/dt)_c = (d^2N/dt^2)_c$	values (maxima) of slope of density curve at density inflection points; values (maxima) of curvature of cumulative curve at density inflection points	± 9.62
$\bar{t}$	mean value of density distribution	2.197
σ	standard deviation of density distribution; measured from mean value of origin	1.814

The above-described properties, and other properties, of the logistic distribution are listed in Section 2.5 where the logistic is compared to the normal probability distribution.

2.2.3 An Illustration: Technology Substitution

Throughout history there have been almost countless examples in which one kind of technology has gradually replaced or substituted for another kind of technology. Gunpowder weapons replaced bows and arrows, steam-powered ships replaced sailing vessels, tractors replaced horses and mules, diesel locomotives replaced steam locomotives, jet engines replaced propellors in aircraft, color television replaced black-and-white, and so on.

This phenomenon of technology replacement or substitution is the topic of our next illustration. The subject has been examined by many investigators. Typical studies are those of Mansfield (1961), Ayres (1969), Fisher and Pry (1971), Blackman (1972) Lanford (1972), and Linstone and Sahel (1976). More recent are the contributions by Sharif (1983), Mahajan and Peterson (1985), Lal, Karmeshu and Kaicker (1988), and Bhargava (1989).

Our illustration is based on a study made by Fisher and Pry (1971) who examined 17 cases of technology substitution. These cases range from the replacement of the Bessemer process by the open-hearth process in steel production in the early part of the century to the relatively recent substitution of plastics for metal in automobile manufacture.

Table 2.2.2 Fraction of annual consumption of synthetic fibers to total fibers in the United States, 1930–1967. From Fisher and Pry (1971)

Year	t	U	$\left(\frac{1}{U} - 1\right)$	$\log_e \left(\frac{1}{U} - 1\right)$
1930	0	0.044	21.73	3.08
1935	5	0.079	11.66	2.46
1940	10	0.10	9.00	2.20
1945	15	0.14	6.14	1.81
1950	20	0.22	3.55	1.27
1955	25	0.28	2.57	0.94
1960	30	0.29	2.45	0.90
1965	35	0.43	1.33	0.29
1967	37	0.47	1.13	0.12

We go into detail concerning one of the other cases studied by Fisher and Pry: substitution of synthetic fibers for natural fibers (cotton, wool, silk, flax) in cloth production. The data corresponding to this case are shown in Table 2.2.2 which refer to the annual consumption of fibers in the United States commencing with the year 1930. The quantity, $U = N/N_*$, indicates the fraction of synthetic fiber to total fiber. Plots of the data are shown in Figs. 2.2.3(a).

We begin our analysis with the assumption that the logistic distribution, Eq. 2.2.6, provides an appropriate framework for description of the data. In this illustration the analysis is greatly simplified by knowing that N_* is 1.0 or 100 percent. So, with $U = N/N_*$, Eq. 2.2.6 becomes

$$U = \frac{1}{1 + (1/U_0 - 1)e^{-at}} \quad , \tag{2.2.31}$$

which can be re-written as follows

$$\log_e\left(\frac{1}{U} - 1\right) = \log_e\left(\frac{1}{U_0} - 1\right) - at \quad . \tag{2.2.32}$$

This expression is in the form: $y = k_0 + k_1 x$.

In our illustration of the next section we examine the method of least squares as a way to determine the values of k_0 and k_1. Briefly, this method enables us to compute the "best" values of the intercept and slope of the straight line expressed by Eq. 2.2.32 and shown in Fig. 2.2.3(b). Such a computation yields the values: $(1/U_0 - 1) = 18.94$, $U_0 = 0.050$, and $a = 0.0757$. Substitutions of these numerical values into Eqs. 2.2.31 and 2.2.32 yields the curves shown in Figs. 2.2.3(a) and (b).

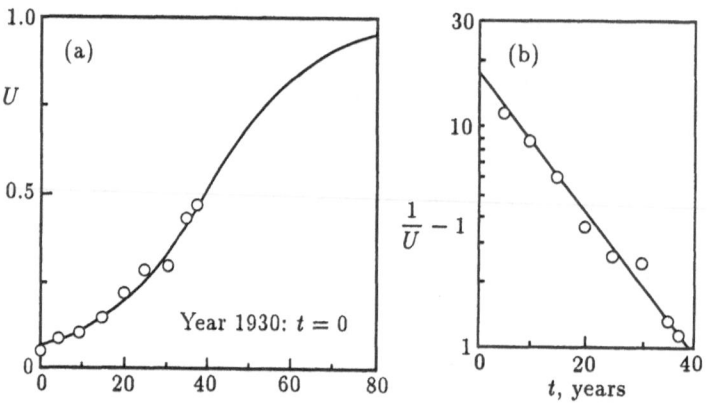

Fig. 2.2.3 Technology substitution of synthetic fibers for natural fibers; (a) cumulative substitution by years and (b) plot to determine parameters U_0 and a. Data of Fisher and Pry (1971)

From Eq. 2.2.17, the inflection point time, $t_i = 38.8$. Since $t = 0$ corresponds to 1930, t_i corresponds to 1968.8. At this value of time, the substitution fraction $U = 0.50$, i.e., half of the total consumption is synthetic fiber and half is natural fiber. We re-write Eq. 2.2.31 in the form

$$U = \frac{1}{1 + e^{-a(t-t_i)}} \quad . \tag{2.2.33}$$

A "takeover time", Δt, is defined by Fisher and Pry as the time required for the technology substitute to increase from $U = 0.10$ at t_1 to $U = 0.90$ at t_2. Letting $\Delta t = t_2 - t_1$, we obtain from Eq. 2.2.33 the relationship: $\Delta t = \log_e 81/a$. With $a = 0.0757$, the takeover time, $\Delta t = 58.0$ years. With this relationship, Eq. 2.2.33 becomes

$$U = \frac{1}{1 + e^{-2.197\phi}}, \quad \text{where} \quad \phi = 2(t - t_i)/\Delta t . \tag{2.2.34}$$

In the preceding equation we note that when $\phi = -1$, 0, and $+1$, we obtain, respectively, $U = 0.10, 0.50$, and 0.90 as we require. There is a reason for this "normalization" of the time coordinate. In their study, Fisher and Pry determined the values of the transfer coefficient, a, the inflection point time, t_i, and the takeover time, Δt, for 17 kinds of technology substitutions. Expressing all of their results in the framework of Eq. 2.2.34, they were able to obtain a single "universal" plot for the entire ensemble of 17 cases. Their quite remarkable final result is shown in Fig. 2.2.4.

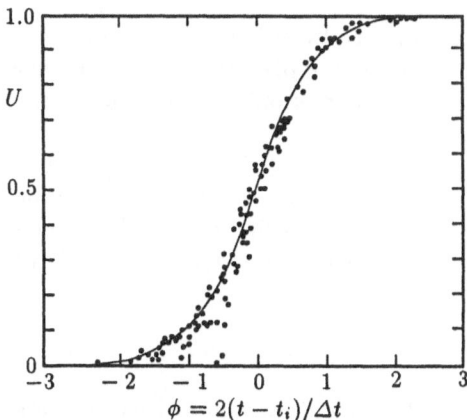

Fig. 2.2.4 Universal plot with normalized time for 17 cases of technology substitution. From Fisher and Pry (1971)

We conclude that the ordinary logistic equation can provide an appropriate framework for the analysis and display of data concerning technology substitution.

Three comments. Our illustration concerning the substitution of synthetic fibers for natural fibers has utilized the logistic distribution as the vehicle for analysis. This same case of fiber substitution, provided by Fisher and Pry, has been analyzed by Stapleton (1976) using the normal probability distribution. In Section 2.5 we examine the normal distribution and show that, in many ways, it is virtually the same as the logistic. This is nothing new. Comparisons of the two distribution functions were made long ago by Winsor (1932b), Feller (1940), Becking (1948) and others.

Secondly, the effects of time delay on innovation diffusion and technology transfer have been examined by Lal, Karmeshu and Kaicker (1988). We study the problem of discrete and distributed time delays in Chapter 6.

Finally, Bhargava (1989) analyzes the topic of technology substitutuion utilizing the framework of the so-called Lotka-Volterra equations. These equations include the much studied prey-predator problem; Bhargava considers

the old technology to be the prey and the new the predator. We take a brief
look at this kind of "two-species" phenomenon in Section 5.5.

2.2.4 An Illustration: Diffusion of Improved Pasture Technology in Uruguay

We now consider another example of the use of the logistic function in technology transfer. This illustration is based on a study made by Jarvis (1981)
of the diffusion of technology concerning fertilized grass-legume pastures in
Uruguay.

Jarvis examines the growth of the number of ranchers who adopted this
technology as well as the increase of the total area planted in which the new
technology was utilized. In our illustration, we look only at the first feature,
i.e., the growth of the number of ranchers.

The data obtained by Jarvis are listed in Table 2.2.3 and are displayed
in Fig. 2.2.5. The upper figure shows the cumulative number of adopting
ranchers, N, commencing at $t = 0$ in 1961. The lower figure gives the number
of adopting ranchers, n , during each year over the period 1961 to 1976. We
disregard the smooth solid curves until later.

Table 2.2.3 Growth of the number of ranchers adopting improved pasture technology in Uruguay. From Jarvis (1981)

Year	t [years]	Newly adopting ranchers, n	Cumulative adopting ranchers, N
1961	0	141	141
1962	1	120	261
1963	2	136	397
1964	3	300	697
1965	4	247	944
1966	5	501	1 455
1967	6	615	12 060
1968	7	1187	13 247
1969	8	2037	15 284
1970	9	1815	17 999
1971	10	2455	19 554
1972	11	1911	11 465
1973	12	2302	13 767
1974	13	911	14 678
1975	14	320	14 998
1976	15	475	15 473

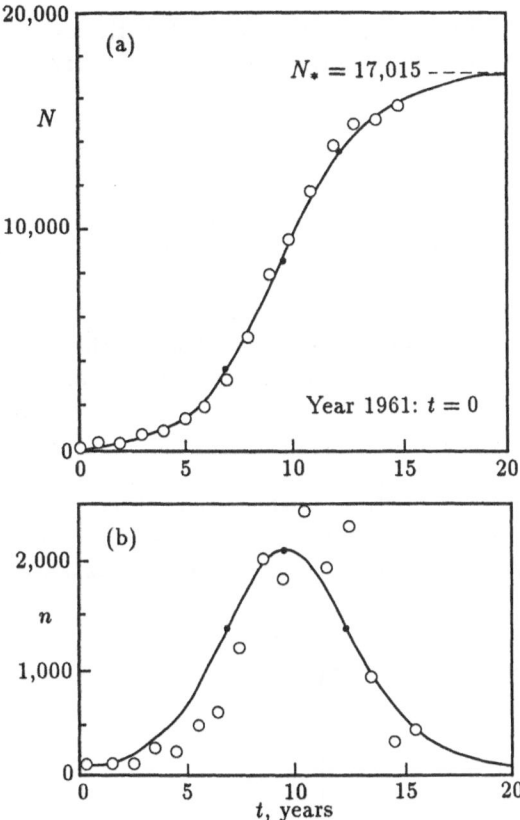

Fig. 2.2.5 Number of ranchers adopting new pasture technology in Uruguay; (a) cumulative distribution; (b) density distribution. Data of Jarvis (1981)

The assumption is made that the diffusion of this technology, represented by the number of adopting ranchers, can be quantitatively described by the logistic function. So again we select Eq. 2.2.6 as the framework for our analysis

$$N = N_* \left[1 + \left(\frac{N_*}{N_0} - 1 \right) e^{-at} \right]^{-1} . \tag{2.2.35}$$

Our problem is one of "curve-fitting" the data of Table 2.2.3 to Eq. 2.2.35. In other words, we need to determine the numerical values of the three parameters, a, N_0, and N_*.

There are numerous methods and techniques for computing these coefficients. In many problems, the curve-fitting process is greatly simplified by the fact that we know, for one reason or another, that N_* is equal to, say, 1.0 or 100 percent. This was the case in our preceding illustration involving substitution of synthetic fibers.

However, in the present illustration, the ultimate number of adopting ranchers, N_*, is not known. Also unknown is the value of the transfer coefficient, a. We do know that the observed initial value is $N_0 = 141$. However, in a statistical analysis of data, it is certainly unwise to use a single observation as the initial value for the mathematical problem.

A straightforward method to determine the values of a and N_* is to write the original differential equation, Eq. 2.2.1, in the following finite difference form

$$\frac{1}{N} \frac{\Delta N}{\Delta t} = a - b\overline{N} \quad . \tag{2.2.36}$$

In this equation, ΔN, is the change of N in time interval, Δt. If we take $\Delta t = 1$ year, then $\Delta N/\Delta t$ is simply n, as listed in Table 2.2.3. For the values of $\overline{N}$ in Eq. 2.2.36, we use the average value of N, at the beginning and end of each interval.

The functional form of Eq. 2.2.36 is the linear equation, $y = k_0 + k_1 x$. Hence, if we plot, on arithmetically scaled paper, $(\Delta N/\Delta t)/\overline{N}$ versus $\overline{N}$, the value of a is given by the intercept $(\overline{N} = 0)$; the value of b is obtained from the slope of the straight line correlating the variables of Eq. 2.2.36. The data listed in Table 2.2.3 are computed in the form indicated by Eq. 2.2.36; the results are shown in Fig. 2.2.6.

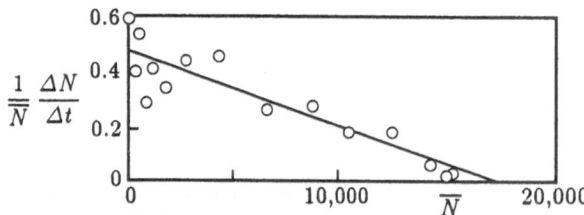

Fig. 2.2.6 Ranchers adopting new pasture technology in Uruguay. Plot to determine values of the growth coefficient a and the carrying capacity N_*.

To obtain the "best" linear relationship between the two variables of Eq. 2.2.36, we employ the method of least squares. This methodology begins with an equation of the form

$$\delta = y - (k_0 + k_1 x) \tag{2.2.37}$$

where δ is the difference between an observed ordinate, y, and its calculated value expressed by the quantity in parentheses.

To avoid any confusion regarding positive and negative quantities computed from Eq. 2.2.37, we take the square of both sides of the equation and then calculate the sum corresponding to all data points. This procedure gives the expression

$$S = \sum \delta^2 = \sum [y - (k_0 + k_1 x)]^2 \quad . \tag{2.2.38}$$

We want to determine the values of k_0 and k_1 which provide the smallest or "least square" value of S. To determine this minimum we calculate the partial derivatives of S with respect to k_0 and k_1 and set the resulting two expressions equal to zero. This yields

$$k_0 M + k_1 \sum x = \sum y$$

$$k_0 \sum x + k_1 \sum x^2 = \sum xy \quad , \tag{2.2.39}$$

in which M is the number of data points. Solving for k_0 and k_1 we obtain

$$k_0 = \frac{\sum y \sum x^2 - \sum x \sum xy}{M \sum x^2 - (\sum x)^2} \; ; \qquad k_1 = \frac{M \sum xy - \sum x \sum y}{M \sum x^2 - (\sum x)^2} \quad . \tag{2.2.40}$$

For our illustration, the indicated computations produce the results: $k_0 = 0.490$ and $k_1 = -0.00002883$. The correlation coefficient is $r = 0.8989$. Accordingly, $a = 0.490$, $b = 0.00002883$, and $N_* = a/b = 17\,015$.

The remaining parameter to determine is N_0. An easy way to compute its value is to obtain an expression for N_0 from the logistic equation, Eq. 2.2.35, substitute the respective values of (t, N) into the expression, and then calculate the average value of N_0. This method is not too satisfactory if some of the N-values, for large t, happen to be larger than N_*. However, this method is satisfactory in the present example; the result is $N_0 = 151.5$.

An alternative method for determination of N_0 is the one we utilized in the synthetic fiber illustration. Since we now know that $N_* = 17\,015$, we easily compute $U = N/N_*$ for each data point. We then express the logistic in the form given by Eq. 2.2.32 and carry out another least squares computation. With this procedure we obtain, for our present illustration, $N_0 = 148.7$ and $a = 0.491$; these values are very close to the above indicated values.

Substituting $a = 0.490$, $N_0 = 150$, and $N_* = 17\,015$ into Eqs. 2.2.13 and 2.2.14 produces, respectively, the cumulative and density distribution curves shown in Fig. 2.2.5. The various critical (t, N) points and other quantities are computed from Eqs. 2.2.17 and 2.2.18 and Eqs. 2.2.22 through 2.2.30. The results of computations are presented in Table 2.2.4 The critical points are identified by small dots in Figs. 2.2.5.

Table 2.2.4 Critical (t, N) coordinates for the illustration: Diffusion of improved pasture technology in Uruguay. $a = 0.490$, $N_0 = 150$, $N_* = 17\,015$, $m = 112.433$

$t_i = 9.64$	$N_i = 8508$	$n_i = (dN/dt)_i = 2084$
$t_c = 6.95$ and 12.32		$N_c = 3596$ and $13\,419$
$n_c = (dN/dt)_c = 1390$		$(dn/dt)_c = (d^2 N/dt^2)_c = \pm 393$
$\bar{t} = 9.64$	$\sigma^2 = 13.70$	$\sigma = 3.70$

On the basis of these numerical results, we make the following observations:

- Commencing with about 150 ranchers in 1961, the technology of improved pastures was diffused at such a rate that approximately 8500 ranchers had adopted the technology 9.64 years later.

- At that time, in 1970, the technology was being adopted at the maximum rate of around 2100 ranchers per year.

- The 8500 ranchers who had adopted the technology by 1970 represented about one-half of the number who would ultimately accept the improved technology, viz., approximately 17 000.

- About 90 percent (15 300 ranchers) of the ultimate number would adopt the technology 14.1 years (1975) after the start, around 95 percent (16 165 ranchers) would adopt 15.7 years later (1977), and approximately 99 percent (16 845 ranchers) would be utilizing the technology 19.0 years (1980) after it was introduced.

At this point, we digress briefly to consider some features of the "kinematics" of a growth phenomenon or transfer process. Suppose the quantity, N, is regarded as a distance or length. Indeed if we were examining the growth of a plant, it is likely that that is precisely what we are considering. On this basis, the rate of change of N with respect to time, dN/dt, represents a velocity and d^2N/dt^2 an acceleration.

From Eq. 2.2.16 and the ensuing paragraph, it is recalled that setting the second derivative, the acceleration, equal to zero determined when the first derivative, the velocity, is a maximum. This analysis produced, in Eqs. 2.2.17 and 2.2.18, expressions for t_i, N_i, and $(dN/dt)_i$. Specifically, from these equations, we determined the location of the inflection point of $N(t)$ and the maximum velocity.

By the same token, setting the third derivative, d^3N/dt^3, equal to zero established when the second derivative, the acceleration, is a maximum. This analysis led to Eqs. 2.2.22 to 2.2.25 which specified the values of t_c, N_c, $(dN/dt)_c$, and $(d^2N/dt^2)_c$. From this information, we identified the maximum curvature points of $N(t)$ and the maximum acceleration.

It is noted from Eq. 2.2.22 and Fig. 2.2.2 that there are two maximum curvature points. One of these, (t_{c1}, N_{c1}), occurs before the inflection point (t_i, N_i). This is the point of maximum positive acceleration; the $N(t)$ curve is concave upward. The other, (t_{c2}, N_{c2}), occurs after the inflection point. This is the point of maximum negative acceleration (deceleration); the $N(t)$ curve is concave downward.

In his analysis of innovation diffusion, Rogers (1983) indicates that the S-shaped diffusion curve "takes off" at about the 10 to 25 percent adoption level. Accordingly, to be bit more precise, the maximum acceleration point could be identified as the "take off" or "up-turning" point. If so, the maximum deceleration point could be labelled the "settling down" or "down-turning" point. Whatever vernacular is used, the precise magnitudes of these critical points are given by Eqs. 2.2.22 to 2.2.25.

In the illustration concerning improved pastures, the "take off" time, determined from Eq. 2.2.22, occurred at $t_{c1} = 6.95$ years (1968). At that time, the number of adopters, computed from Eq. 2.2.23, was $N_{c1} = N_* = 3596$.

It was established, in Eq. 2.2.27, that the mean value of the density distribution is given by $t = (1/a)\log_e m$. As Eq. 2.2.17 indicates, this is also the time, t_i, at which $N = N_*/2$. In other words, at this value of t, 50 percent of the ultimate number of adopters have already done so. From Eq. 2.2.19, the following equation is easily obtained to calculate the times for other adoption percentages

$$t = t_i - \frac{1}{a}\log_e\left(\frac{N_*}{N} - 1\right) \quad . \tag{2.2.41}$$

For the pasture illustration, these times, and the corresponding percentages and numbers of adopters, are listed in Table 2.2.5.

Table 2.2.5 Years for the indicated percentages and numbers of ranchers to adopt improved pasture technology ($t = 0$ in 1961)

N/N_*	N	t	N/N_*	N	t	N/N_*	N	t
1	170	0.26	50	8508	9.64	75	12761	11.88
5	851	3.63				90	15314	14.12
10	1702	5.15				95	16164	15.65
25	4254	7.39				99	16845	19.01

In his study of innovation transfer, Rogers categorizes an adopter population according to how quickly, or how tardily, an innovation is adopted. He defines the categories by the number of standard deviations, σ, before and after the mean time, t. His categorization is based on the normal probability distribution, not the logistic. However, as we will see in Section 2.5, the differences between the two distributions are slight. Incidentally, unlike the case of the normal distribution, the standard deviation points of the logistic equation do not coincide with the inflection points of the density distribution. This and other features are examined in Section 2.5 where these two important distribution functions are compared.

From Eqs. 2.2.19 and 2.2.30, we obtain the following expression to determine the values of N/N_* at the standard deviation points

$$\frac{N}{N_*} = \frac{1}{1 + \exp(\pm j\pi\sqrt{3})} \quad ; \quad j = 0, 1, 2 \quad . \tag{2.2.42}$$

Table 2.2.6 shows the categories proposed by Rogers and the corresponding percentages in each category. The numbers of ranchers adopting the improved pasture technology are also shown.

Table **2.2.6** Categories of adopters, corresponding percentages, and numbers of ranchers adopting improved pasture technology. Based on the logistic distribution function

Adopter category	Category limits From	To	Adopter percentage	Number of adopters
Innovators	$-\infty$	-2σ	2.59	440
Early adopters	-2σ	$-\sigma$	11.43	1945
Early majority	$-\sigma$	0	35.98	6122
Late majority	0	$+\sigma$	35.98	6122
Early laggards	$+\sigma$	$+2\sigma$	11.43	1945
Late laggards	$+2\sigma$	$+\infty$	2.59	440

2.2.5 An Illustration: Growth of Prime Mover Horsepower in the United States

Commencing in 1850 periodic surveys have been made by the federal government to determine the total horsepower of all prime movers in the United States. According to the Department of Commerce: "Prime movers are mechanical engines and turbines which originally convert fuels or force into work and power." The total horsepower of all prime movers in the United States in 1985 is categorized in Table 2.2.7.

Table **2.2.7** Total horsepower of all prime movers in the United States by category, 1985. Reference: Statistical Abstract of the United States 1987, U.S. Department of Commerce, Government Printing Office, Washington, D.C.

Category	N [million HP]	Category	N [million HP]
Automotive	30 792	Farms	358
Factories	65	Power plants	912
Mines	47	Aircraft	268
Railroads	58		
Merchant ships	29	Total	32 529

As indicated, record keeping began in 1850. In that year, the total horsepower, N, was 8.5 million. By 1870, the amount had approximately doubled to 16.9 million and at the turn of the century, it had increased to 63.95 million. We commence our analysis with the year 1900 ($t = 0$). Table 2.2.8 lists the data for the period 1900 through 1985.

Table **2.2.8** Growth of total horsepower of all prime movers of the United States 1900–1985. Reference: Historical Statistics of the United States, Colonial Times to 1970; Statistical Abstract of the United States 1987, Department of Commerce, Government Printing Office, Washington, D.C.

Year	t	N [million HP]	Year	t	N [million HP]
1900	0	64	1960	60	11 008
1910	10	139	1965	65	15 096
1920	20	435	1970	70	20 408
1930	30	1664	1975	75	25 100
1940	40	2773	1980	80	28 992
1950	50	4754	1985	85	32 529
1955	55	7158			

A plot of the data contained in Table 2.2.8 is shown in Fig. 2.2.7. Our task is to fit a growth curve to the indicated points. The illustration is somewhat remarkable in that there is a 500-fold increase in the value of N over the specified time range.

Later on, in our consideration of the power law logistic and hyperlogistic equations, we examine growth expressions of the form

$$\frac{dU}{dt} = aU^r(1 - U^s) \quad , \tag{2.2.43}$$

in which $U = N/N_*$ and r and s are non-negative constants. This equation is a five parameter relationship; that is, there are five parameters which must be determined: a, r, s, N_*, and the initial value, N_0. If $r = s = 1$, Eq. 2.2.37 reduces to the three-parameter ordinary logistic equation.

For our present illustration, we will assume that $s = 1$ but, at this point, we are not certain that $r = 1$. For example, in Section 2.1.6, in our illustration concerning the growth of the world's population, we had $r = 2$.

For small values of U, Eq. 2.2.43 becomes a power law exponential

$$\frac{dN}{dt} = kN^r \quad , \tag{2.2.44}$$

where k is a constant. Expressing this relationship in a finite difference form and taking logarithms we obtain

$$\log_e\left(\frac{\Delta N}{\Delta t}\right) = \log_e k + r\log_e \overline{N} \quad , \tag{2.2.45}$$

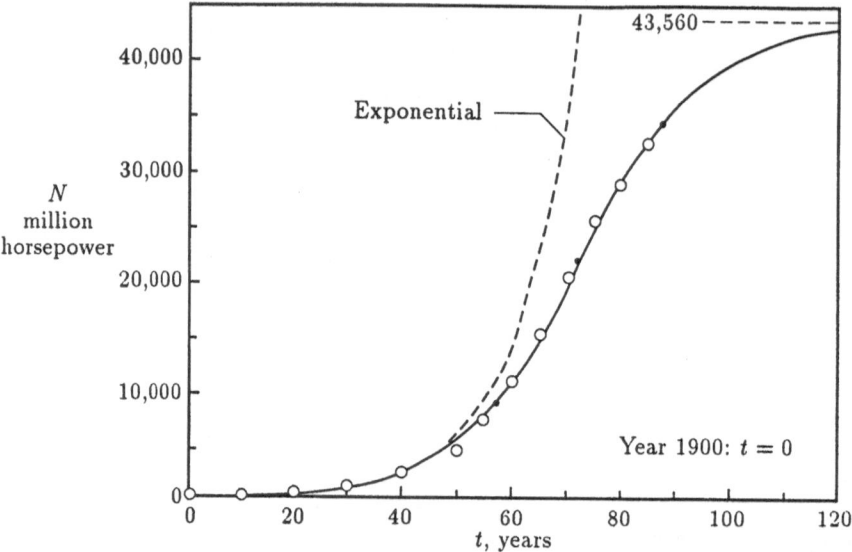

Fig. 2.2.7 Total horsepower of all prime movers in the United States, 1900–1985

in which $\overline{N}$ is the average value of N in an interval Δt. It is clear that a log-log plot of Eq. 2.2.45 should produce a straight line whose slope is r. A least squares calculation, utilizing the data of Table 2.2.8 corresponding to small values of N (1900 through 1960, say), yields the correlation shown in Fig. 2.2.8. The computation gives $r = 0.949$; the magnitude of k does not concern us. Consequently, we take $r = 1$ and so

$$\frac{dN}{dt} = aN \left(1 - \frac{N}{N_*} \right) \quad . \tag{2.2.46}$$

It was mentioned earlier that there are numerous computation procedures to determine the numerical values of the three parameters. We now examine a few of the methods to calculate approximate or starting values. These methods and values suffice for our purpose. References dealing with more accurate techniques are given later on.

A. Finite Difference Method. This is the method we employed in our illustration concerning improved pasture technology in Uruguay. The framework for data analysis is the differential equation, not the integrated solution. The method is straightforward though in some cases, there may be sizeable variations of $\Delta N / \Delta t$, especially when N is small. In addition, there may be some ambiguity in the computation of N_0 when N is large.

Writing Eq. 2.2.46 as a finite difference expression

$$\frac{1}{\overline{N}} \frac{\Delta N}{\Delta t} = a - \frac{a}{N_*} \overline{N} \tag{2.2.47}$$

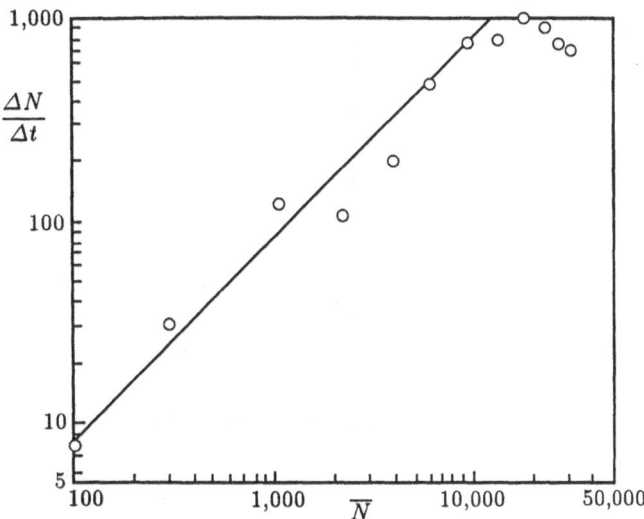

Fig. 2.2.8 Horsepower of prime movers in the United States. Plot to establish the approximate value $r = 1$

and utilizing the data of Table 2.2.8 in the form stipulated by Eq. 2.2.47, yields the plot displayed in Fig. 2.2.9. A least squares computation gives: $a = 0.0871$ and $N_* = 43\,560$. We substitute these values into the integrated logistic equation and solve for N_0 using all values of N. The resulting average value is $N_0 = 80.7$.

With the magnitudes of the three parameters established we calculate the coordinates of the inflection point and the maximum curvature points. Our results are listed below.

$$a = 0.0871; \qquad N_* = 43\,560; \qquad N_0 = 80.7$$

$$t_i = 71.21; \qquad N_i = 21\,780; \qquad n_i = 948.5$$

$$t_c = 57.09 \text{ and } 87.33; \qquad N_c = 9205 \text{ and } 34\,355$$

$$n_c = (dN/dT)_c = 632.3; \qquad (dn/dt)_c = (d^2 N/dt^2)_c = \pm 31.80.$$

The solid curve shown in Fig. 2.2.7 is based on these numerical values; the solid dots identify the critical points. For comparison, exponential growth is identified as the dashed line.

Based on these results we project a value, $N = 38\,300$ for 1995. By the end of the century, the total horsepower of all prime movers in America will be approximately $40\,000$ million.

B. Point Matching Method. This method is the oldest, the simplest, and probably the least reliable method for "curve-fitting" the logistic equation.

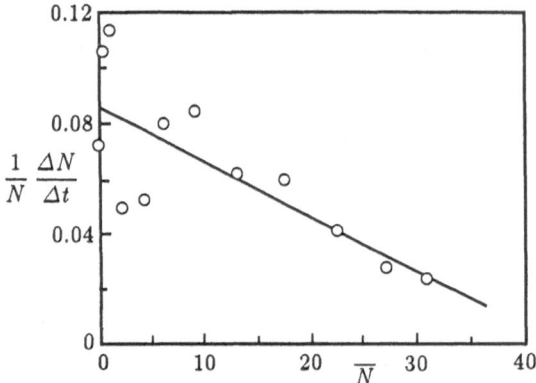

Fig. 2.2.9 Horsepower of prime movers in the United States. Plot to determine values of the growth coefficient a and carrying capacity N_*.

It was utilized many years ago by Pearl and Reed (1920), Lotka (1924), Yule (1925), and other early investigators including, most likely, Verhulst (1838). Nevertheless, the method is suggested by Chakravarti et al (1967) and Draper and Smith (1981) as a method for acquiring initial estimates of the parameters.

The concept is simple. Since the values of three parameters are un-known, we "point match" at three (t, N) coordinates to obtain three equations and then solve for the unknowns. There are numerous ramifications of this method; some these are described by Nair (1954), Nelder (1961), Oliver (1964), and Ashton (1972).

We write the logistic equation in the following form

$$N = \frac{N_*}{1 + \exp(c - at)} \quad , \tag{2.2.48}$$

where $c = \log_e[(N_*/N_0) - 1]$. Substituting values of t and N (with subscripts, 1, 2 and 3) and carrying out the algebra yields

$$N_* = \frac{N_2^2(N_1 + N_3) - 2N_1 N_2 N_3}{N_2^2 - N_1 N_3} \tag{2.2.49}$$

$$a = \frac{1}{t_3 - t_1} \log_e \left[\frac{N_3}{N_1} \left(\frac{N_* - N_1}{N_* - N_3} \right) \right] \tag{2.2.50}$$

$$c = \frac{1}{t_3 - t_1} \left[t_3 \log_e \left(\frac{N_* - N_1}{N_1} \right) - t_1 \log_e \left(\frac{N_* - N_3}{N_3} \right) \right] \quad . \tag{2.2.51}$$

From the last equation, the initial value, N_0, is determined from the relationship: $N_0 = N_*/(1 + e^c)$.

An inherent weakness in this method is the arbitrariness involved in selecting the three matching points There are no rules. It is suggested that the two time intervals be equal and that the two end points encompass most of the time range of the data.

For example, in our illustration we take the following (t, N) points: 1: (0, 64), 2: (40, 2773), and 3: (80, 28 922). Substituting these values into Eqs. 2.2.49 through 2.2.51 gives: $N_* = 36\,420$, $a = 0.0962$, $c = 6.342$, and $N_0 = 64.0$.

C. Quadratic Least Squares. This method involves a series of reiterated calculations to determine the value of the carrying capacity, N_*; subsequently, the magnitudes of the growth coefficient, a, and initial value, N_0, are easily computed.

We begin by writing the logistic equation in the form

$$\log_e \left(\frac{N_*}{N} - 1 \right) = \log_e \left(\frac{N_*}{N_0} - 1 \right) - at \quad . \tag{2.2.52}$$

Next, we select a value of N_*, compute the data according to Eq. 2.2.52, and then prepare a semi-logarithmic plot of the results. If we had made the proper selection of N_* initially then, according to Eq. 2.2.52, we would obtain a straight line. Problem: how do we know we have a straight line?

At this point it is helpful to use some numbers. Suppose we select, for our prime mover illustration, the quite absurd value, $N_* = 100\,000$. The resulting plot of the computed data is shown as the upper curve in Fig. 2.2.10. Obviously, this is not a straight line; it is distinctly concave upward. At the other extreme, we select an equally absurd $N_* = 33\,000$ and repeat the process. The result is shown in Fig. 2.2.10 as the lower curve. Again, it is not a straight line; this time it is concave downward.

To resolve this matter, we write Eq. 2.2.52 as follows

$$y = k_0 + k_1 x + k_2 x^2 \quad , \tag{2.2.53}$$

in which y is the left-hand member of Eq. 2.2.52, k_0 is the first term in the right-hand member, $k_1 = -a$, k_2 is another constant, and t is replaced by x.

The first and second derivatives of Eq. 2.2.53 represent the slope and curvature, respectively. So we have

$$\frac{dy}{dx} = k_1 + k_2 x; \qquad \frac{d^2 y}{dx^2} = 2k_2 \quad . \tag{2.2.54}$$

If k_2 is positive, then the curvature is positive and the parabola or quadratic expression of Eq. 2.2.53 is concave upward. If k_2 is negative, then the curvature is negative and Eq. 2.2.53 is concave downward. If $k_2 = 0$ then we have a straight line. So, it is clear that we require k_2 to be zero.

This is accomplished by following a procedure similar to that which led to Eq. 2.2.40 for linear least squares analysis. We start with an expression of the form

$$\delta = y - (k_0 + k_1 x + k_2 x^2) \quad , \tag{2.2.55}$$

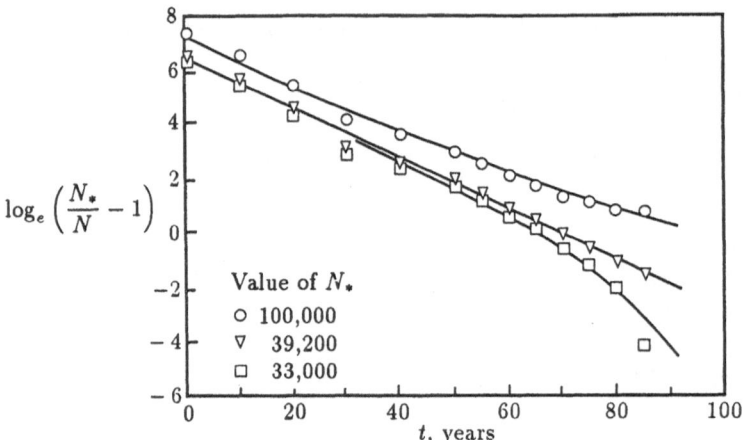

Fig. 2.2.10 Horsepower of prime movers in the United States. Plot to determine value of the carrying capacity N_* based on quadratic least squares analysis

in which δ is the difference between an observed ordinate, y, and its computed value given by the quantity in the parentheses.

Next, in order to avoid difficulty involving positive and negative differences, we consider the square of both members of Eq. 2.2.55 and then take the sum of these expressions for all the data points. This yields

$$S = \sum \delta^2 = \sum [y - (k_0 + k_1 x + k_2 x^2)]^2 \quad . \tag{2.2.56}$$

We want to determine the values of k_0, k_1 and k_2 which will minimize the quantity S. Accordingly, if we take the partial derivatives of S with respect to each of these three parameters and equate each to zero, we obtain

$$k_0 M + k_1 \sum x + k_2 \sum x^2 = \sum y$$

$$k_0 \sum x + k_1 \sum x^2 + k_2 \sum x^3 = \sum xy \tag{2.2.57}$$

$$k_0 \sum x^2 + k_1 \sum x^3 + k_2 \sum x^4 = \sum x^2 y$$

where M is the number of data points. From these three equations we could easily calculate the values of k_0, k_1, and k_2.

However, our present task is simply to require that k_2 be zero. From Eqs. 2.2.57 it is not difficult to establish that the criterion for $k_2 = 0$ is

$$\left[\left(\sum x \right) \left(\sum x^3 \right) - \left(\sum x^2 \right)^2 \right] + \left[\left(\sum x \right) \left(\sum x^2 \right) - M \left(\sum x^3 \right) \right] \sum xy$$

$$+ \left[M \left(\sum x^2 \right) - \left(\sum x \right)^2 \right] \sum x^2 y = 0 \quad . \tag{2.2.58}$$

We construct a computation procedure based on this expression. We select a value of N_*, carry out the calculations indicated by Eq. 2.2.58, and acquire a positive or negative number. If it is positive, the curvature is concave upward and so we reduce the value of N_*. This process is continued until Eq. 2.2.58 is satisfied, i.e., $k_2 = 0$.

The computation involving the final and correct selection of N_*, producing $k_2 = 0$, is itself a linear least squares computation which provides the desired values of k_0 and k_1. So the problem of acquiring reasonably accurate values of the parameters is solved.

In our illustration concerning total horsepower of prime movers, the above outlined procedure yields: $N_* = 39\,200$, $a = 0.0921$, and $N_0 = 67.3$.

D. Maximum Correlation Coefficient Method. Finally, a straight-forward method for the determination of approximate values of the parameters is simply to substitute values of N_* into Eq. 2.2.52 and carry out linear least squares computations. The criterion for identifying the proper value of N_* is that value which yields the largest correlation coefficient.

In our horsepower illustration, this procedure gives: $N_* = 41\,000$, $a = 0.0905$, and $N_0 = 70.0$.

The results obtained from the four above-outlined computation methods are summarized in Table 2.2.9.

Table 2.2.9 Summary of results. Values of parameters for logistic growth of horsepower of prime movers in the United States

Method	N_*	a	N_0
A. Finite difference	43 560	0.0871	80.7
B. Point matching	36 420	0.0962	64.0
C. Quadratic least squares	39 200	0.0921	67.3
D. Maximum correlation	41 000	0.0905	70.0

It has been stressed that the preceding methods for calculation of parameters are simply rapid ways to obtain initial estimates. Numerous more complicated methods have been devised for determining more accurate parameter values or handling more difficult problems. These include: generalized least squares method, maximum likelihood estimation, Gauss–Newton procedure, and minimum logit chi-squared analysis.

Many references are available which deal with these topics including the works of Berkson (1953), Cox (1970), Dobson (1983), Draper and Smith (1981), McCullagh and Nelder (1983), and Ratkowsky (1983).

Back to horsepower. From our finite difference method computation, we project a total of about 40 000 million prime mover horsepower in America by the year 2000. On this basis it will have increased, during the 20th century, by an amount 625 times larger than the 64 million horsepower recorded for the year 1900.

During the same period, the nation's population will have increased by a factor of about 3.6. Thus, per capita horsepower will have grown nearly 175 times. It is noted that since 1960, nearly 95 percent of prime mover horsepower has been categorized as automotive: cars, trucks, buses and motorcycles.

2.3 Confined Exponential Distribution

2.3.1 Scope of Applications of the Distribution

Without doubt, the most important single mathematical function in growth, diffusion and transfer phenomena is the exponential function. We examined this function in considerable detail in Section 2.1. As we now see, one very important type of growth framework in which the exponential function again appears is the so-called "confined" exponential distribution. This relationship is extremely useful in the physical sciences and engineering in phenomena involving heat transfer and mass transfer. Kreith (1958) and Bird, Stewart and Lightfoot (1960) consider these aspects at length.

The confined exponential equation represents an important analytical framework in technology transfer. Sharif (1983) describes the utilization of this equation in innovation diffusion models.

It serves as the basis of a source process in diffusion theories and models in geography. This subject has been studied by Brown (1981), Ralston (1980) and numerous other geographers, regional scientists and planners. Bartholomew (1981, 1982) utilizes the confined exponential equation in various mathematical models arising in the social sciences.

This distribution has also found application in agriculture. Referring to it as the Mitscherlich equation, Batschelet (1979) and Goldsworthy (1974) have utilized the confined exponential equation in connection with the use of fertilizers to increase crop yields. Valentine (1985) employed the Mitscherlich equation in his models of tree growth.

In Section 2.2, the logistic function was considered at length. With regard to the matter of technology transfer, we can say that the logistic equation provides the basis for a personal, pure interaction mechanism for transfer. By the same token, we say that the confined exponential equation gives the basis for an impersonal, pure source mechanism for transfer. This topic is considered further in Section 2.4 where these two important equations are combined.

2.3.2 The Differential Equation with Constant Coefficients

The differential equation for the confined exponential distribution is

$$\frac{dN}{dt} = a_*(N_* - N) \quad , \tag{2.3.1}$$

in which a_* is a growth or transfer coefficient and N_* is the carrying capacity or equilibrium value. For now, a_* and N_* are assumed to be constants. With the initial condition, $N(0) = N_0$, the solution to Eq. 2.3.1 is

$$N = N_* - (N_* - N_0)e^{-a_* t} \quad , \tag{2.3.2}$$

which we label the confined exponential equation.

Before going further with this result, suppose that $N_* = 0$ in Eqs. 2.3.1 and 2.3.2. This substitution yields

$$N = N_0 e^{-a_* t} \quad . \tag{2.3.3}$$

In this case, as t approaches infinity, N approaches zero. In contrast to the exponential growth described by Eq. 2.1.2, we now have an exponential decay. Instead of a "doubling time", we have a "half-life time", $t_{1/2} = (1/a_*)\log_e 2$.

The first derivative of Eq. 2.3.2 gives the slope of the confined exponential equation

$$\frac{dN}{dt} = a_*(N_* - N_0)e^{-a_* t} \quad . \tag{2.3.4}$$

If $N_* > N_0$, the slope is positive; otherwise it is negative. The second derivative gives the curvature

$$\frac{d^2 N}{dt^2} = -a_*(N_* - N_0)e^{-a_* t} \quad . \tag{2.3.5}$$

If $N_* > N_0$, the curvature is negative (i.e., the curve is concave downward); otherwise it is positive (concave upward). The curve has no inflection point and the maximum slope and curvature occur at $t = 0$. The value of N approaches N_* as $t \to \infty$. We note that this simple type of exponential growth is always "confined" to a region bounded by N_*. In Fig. 2.3.1, curves are presented of the confined exponential distribution for various values of a_*.

2.3.3 The Differential Equation with Variable Coefficients

We return to Eq. 2.3.1 and examine it in its more generalized form

$$\frac{dN}{dt} = a_*(t)[N_*(t) - N] \quad , \tag{2.3.6}$$

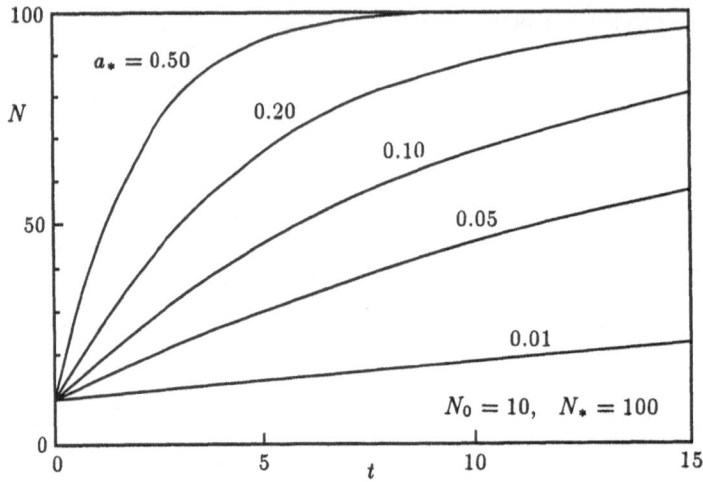

Fig. 2.3.1 Confined exponential curves for various values of the growth or transfer coefficient a_*

where now $a_*(t)$ and $N_*(t)$ are functions of time; as before, the initial condition is $N(0) = N_0$.

Writing Eq. 2.3.6 in the form

$$\frac{dN}{dt} + a_*(t)N = a_*(t)N_*(t) \quad , \tag{2.3.7}$$

we see that we have a first order, linear, complete (non-homogeneous) differential equation with time-dependent coefficients. The solution to this equation is

$$N = k\mathrm{e}^{-\int a_*\,dt} + \mathrm{e}^{-\int a_*\,dt}\int a_*N_*\mathrm{e}^{\int a_*\,dt}\,dt \tag{2.3.8}$$

where k is an integration constant. However, this form is not convenient. For every substitution of prescribed $a_*(t)$ or $N_*(t)$, we must go to the trouble of imposing the initial condition, $N(0) = N_0$, to evaluate k.

A better answer is obtained as follows. Let $N(t) = N_1(t)N_2(t)$ where $N_1(t)$ is the general solution of the reduced (homogeneous) equation. Then writing

$$\frac{dN_1}{dt} + a_*(t)N_1 = 0 \tag{2.3.9}$$

we obtain

$$N_1(t) = N_{1,0}\mathrm{e}^{-\int_0^t a_*\,dt} \quad . \tag{2.3.10}$$

Now since

$$N(t) = N_1(t)N_2(t) \tag{2.3.11}$$

then

$$\frac{dN}{dt} = N_1 \frac{dN_2}{dt} + N_2 \frac{dN_1}{dt} \quad . \tag{2.3.12}$$

Substitution of these expressions into Eq. 2.3.7 and utilizing Eqs. 2.3.9 and 2.3.10 yields

$$N_2(t) = N_{2,0} + \frac{1}{N_{1,0}} \int_0^t a_* N_* e^{\int_0^t a_* \, dt} dt \quad . \tag{2.3.13}$$

Accordingly, the generalized solution to Eq. 2.3.6 is

$$N(t) = N_0 \exp\left(-\int_0^t a_*(\xi)d\xi\right)$$

$$\times \left[1 + \frac{1}{N_0} \int_0^t a_*(\xi)N_*(\xi) \exp\left(\int_0^\xi a_*(\eta)d\eta\right) d\xi\right] \quad . \tag{2.3.14}$$

The quantities ξ and η are variables of integration.

In the following sections, we consider two main categories of problems involving the confined exponential equation: (1) $a_* = a_*(t)$, $N_* = N_{*0} =$ constant and (2) $a_* = a_{*0} =$ constant, $N_* = N_*(t)$.

2.3.4 Variable Transfer Coefficient

Suppose that the growth coefficient is a specified function of time, $a_* = a_*(t)$, and that the carrying capacity or equilibrium value of N is a constant, $N_*(t) = N_{*0}$. In this case, Eq. 2.3.14 becomes

$$N = N_{*0} - (N_{*0} - N_0) \exp\left[-\int_0^t a_*(\xi)d\xi\right] \quad . \tag{2.3.15}$$

Several cases corresponding to prescribed values of $a_*(t)$ are now examined.

Case A.1. Constant a_*. As the first and simplest case, suppose that $a_*(t) = a_{*0} =$ constant. If this definition of the transfer coefficient is substituted into Eq. 2.3.15, we again obtain the ordinary confined exponential equation

$$N = N_{*0} - (N_{*0} - N_0)e^{-a_{*0}t} \tag{2.3.16}$$

as presented in Eq. 2.3.2 and shown in Fig. 2.3.1. Incidentally, in all the numerical examples which follow in this section, we shall use the values: $a_{*0} = 0.20$, $N_0 = 10$, and $N_{*0} = 100$.

Case A.2. Linearly Variable a_*. Next we consider the case in which $a_*(t)$ varies linearly with time

$$a_*(t) = a_{*0}(1 + ct) \quad , \tag{2.3.17}$$

where c is a positive or negative constant. Substituting Eq. 2.3.17 into Eq. 2.3.15 gives

$$N = N_{*0} - (N_{*0} - N_0) \exp[-a_{*0}(t + \tfrac{1}{2}ct^2)] \quad . \tag{2.3.18}$$

This equation is plotted in Fig. 2.3.2 for several values of c.

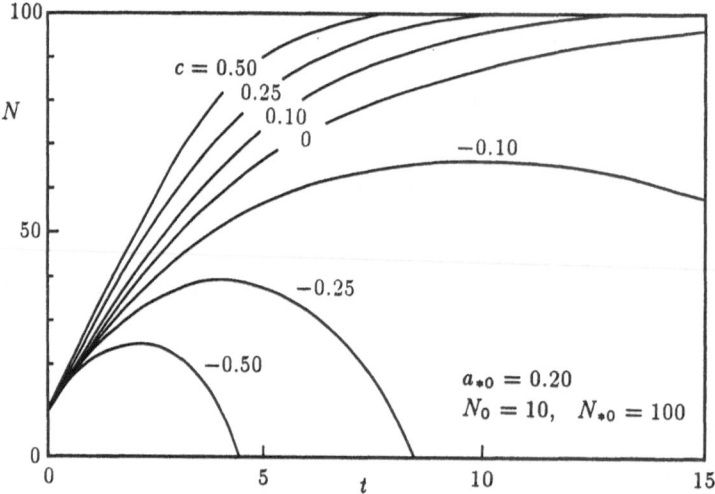

Fig. 2.3.2 Confined exponential curves for linearly variable transfer coefficients

Differentiating Eq. 2.3.18 and setting the result equal to zero indicates that N has a maximum value at

$$t_m = -\frac{1}{c} \quad ; \qquad N_m = N_{*0} - (N_{*0} - N_0)e^{a_{*0}/2c} \tag{2.3.19}$$

as long as c is negative. It is easily established that for values of t larger than t_m the magnitude of N decreases to the initial value, N_0, when $t = -2/c$ and N becomes zero at

$$t_e = -\frac{1}{c}\left[1 + \sqrt{1 + \frac{2c}{a_{*0}}\log_e\left(1 - \frac{N_0}{N_{*0}}\right)}\right] \quad . \tag{2.3.20}$$

It is emphasized that Eqs. 2.3.19 and 2.3.20 apply only for negative c. As Eq. 2.3.18 indicates, the value of N asymptotically approaches N_{*0} for non-negative c.

An example of the above analysis concerns the phenomenon of evaporation from a water surface due to wind. Sutton (1953) indicates that this kind of evaporation can be described by the Dalton equation

$$E = k(e_s - e) \quad , \tag{2.3.21}$$

in which E is the rate of evaporation, e_s is the saturated vapor pressure at the water surface, e is the vapor pressure of the air, and k is a mass transfer coefficient which is proportional to the wind velocity.

With respect to Eq. 2.3.6, the analogy is that e corresponds to N and k corresponds to $a_*(t)$. With caution that the analogy not be carried too far, if the mass transfer coefficient changes linearly over a period of time, then the preceding analysis describes the evaporation process.

Case A.3. Exponentially Variable a_*. Suppose that the transfer coefficient varies in the following manner

$$a_*(t) = a_{*0}e^{ct} \tag{2.3.22}$$

in which c may be positive or negative. In this case we obtain the solution

$$N = N_{*0} - (N_{*0} - N_0)\exp\left[-\frac{a_{*0}}{c}\left(e^{ct} - 1\right)\right] \quad . \tag{2.3.23}$$

Plots of this equation are shown in Fig. 2.3.3. For large values of t, $N = N_{*0}$ for positive c and $N = N_{*0} - (N_{*0} - N_0)\exp(a_{*0}/c)$ for negative c.

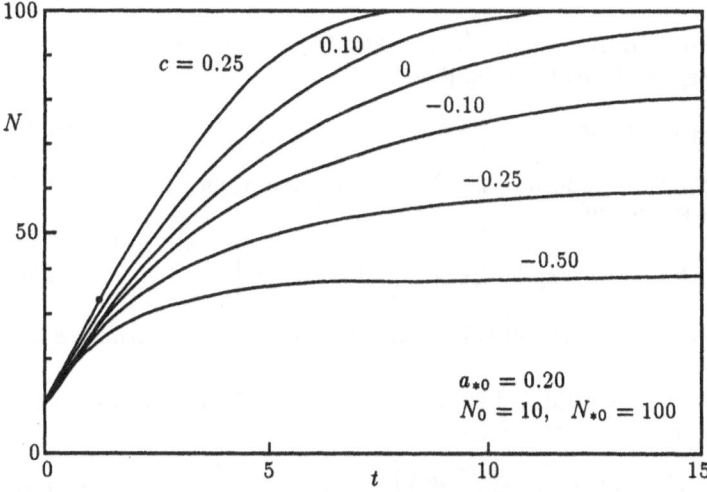

Fig. 2.3.3 Confined exponential curves for exponentially variable transfer coefficients

The form of Eq. 2.3.23 is similar to the Gompertz relationship we consider in Section 3.1. From Eq. 2.3.23 we determine that an inflection point occurs at

$$t_i = \frac{1}{c} \log_e \frac{c}{a_{*0}} \quad ; \qquad N_i = N_{*0} - (N_{*0} - N_0)e^{(a_{*0}-c)/c} \tag{2.3.24}$$

as long as $c > a_{*0}$. As shown in Fig. 2.3.3, for the curve corresponding to $c = 0.25$, an inflection point occurs at $t = 0.89$, $N = 26.32$.

By way of example, suppose that an organization launches an effort to promote the adoption of an innovation in a community. Imagine that the transfer and adoption mechanisms concerning the innovation are due entirely to the information source provided by the organization; that is, there is no interacting information transfer process between the adopters and potential adopters about the innovation. Accordingly, we surmise that the confined exponential equation provides the suitable framework for analysis.

On the basis of the preceding analysis, if the organization exponentially increases its efforts to transfer the innovation (positive c), the effective, or virtual, information transfer coefficient will be increased as a consequence. As illustrated in Fig. 2.3.3, this means that the, say, 95 percent adoption level will be attained in a shorter period of time than would have been the case with a constant level of effort ($c = 0$). On the other hand, if the organization exponentially decreases its efforts (negative c), not only will the effective transfer coefficient be decreased, more importantly the ultimate adoption level will be something less than 100 percent. This point is brought out in the figure.

Case A.4. Sinusoidally Variable a_*. Finally, we examine the case in which the transfer coefficient is described by the equation

$$a_*(t) = a_{*0} + a_m \sin \omega t \quad , \tag{2.3.25}$$

in which a_m is the amplitude, $\omega = 2\pi/T$ is the frequency, and T is the period. The solution to Eq. 2.3.25 is

$$N = N_{*0} - (N_{*0} - N_0) \exp\left[-a_{*0}t - \frac{a_m}{\omega}(1 - \cos \omega t)\right] \quad . \tag{2.3.26}$$

In Fig. 2.3.4, curves are shown of this equation with $T = 5$ for various values of a_m.

From the derivative of Eq. 2.3.26 we establish that N has maximum and minimum values at

$$t_m = T\left[\frac{j}{2} + (-1)^{j+1} \frac{1}{2\pi} \arcsin\left(\frac{a_{*0}}{a_m}\right)\right] \quad , \tag{2.3.27}$$

where odd values of j correspond to the maxima and even values to the minima. It is seen from Eq. 2.3.27 that maxima and minima occur only when $(a_{*0}/a_m) < 1$.

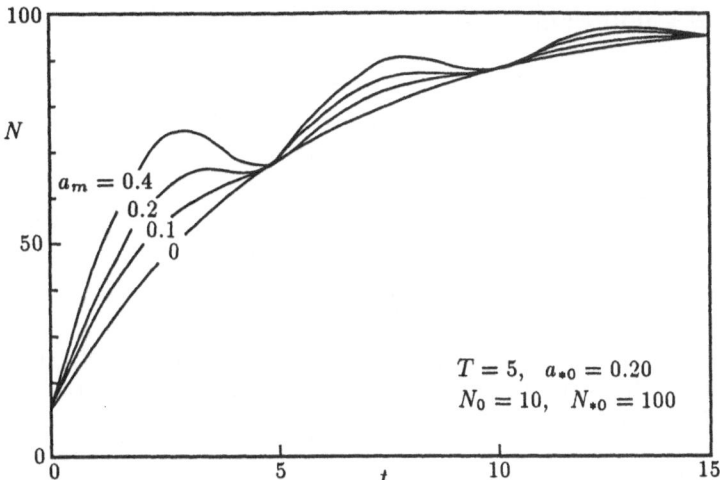

Fig. 2.3.4 Confined exponential curves for sinusoidally variable transfer coefficients

We return to the example presented above concerning the organization involved in transferring an innovation in a community of potential adopters. This time, suppose that the organization schedules its promotion efforts in a cyclic pattern with monthly or yearly frequency. That is, the intensity of its efforts, scaled by the information transfer coefficient, a_*, is described by Eq. 2.3.25.

Under these circumstances, the number of adopters, N, when $a_m > 0$, is larger than it would be if $a_m = 0$, at a particular intermediate time in the cycle, e.g., $t = T/2 = 2.5$. This is clear from Fig. 2.3.4. However, during the ensuing portion of the $a_*(t)$ cycle, as adoption efforts decline, the rate of adoption also declines. In fact, for values of a_m larger than a_{*0}, the number of adopters actually decreases; this is apparent in the figure. However, the outcome is that after a full cycle, or any number of full cycles, the number of adopters is the same regardless of whether a_m is larger than or equal to zero.

The preceding remarks are based on the assumption that the phase angle $\phi = 0$ in the more general equation, $a_*(t) = a_{*0} + a_m \sin(\omega t + \phi)$. If $\phi = \pi$, for example, the adoption efforts by the organization descend during the first portion of the cycle and subsequently increase. However, after a full cycle, or an arbitrary number of full cycles, the outcome is as it was before: the value of N is the same regardless of the magnitude of a_m. This matter concerning the inclusion of the phase angle ϕ in the definition of a_* is considered in Section 4.4.

2.3.5 Variable Equilibrium Value

Next we consider the second category of problems involving the confined exponential equation: $a_* = a_{*0} =$ constant and $N_* = N_*(t)$. For this category, the general solution given by Eq. 2.3.14 reduces to

$$N = \exp(-a_{*0}t) \left[N_0 + a_{*0} \int_0^t N_*(\xi) \exp(a_{*0}\xi)d\xi \right] \quad . \tag{2.3.28}$$

As before, we examine several specific forms of $N_*(t)$.

Case B.1. Constant N_*. Beginning with the easiest case, let $N_*(t) = N_{*0} =$ constant. As we expect, Eq. 2.3.28 immediately reduces to the ordinary confined exponential equation

$$N = N_{*0} - (N_{*0} - N_0)e^{-a_{*0}t} \quad . \tag{2.3.29}$$

Again, in the numerical examples presented in the remainder of the section, the following values are used: $a_{*0} = 0.20$, $N_0 = 10$, and $N_{*0} = 100$.

In the preceding paragraph, the imposition of $N_*(t) = N_{*0} =$ constant represents a "step function" that applies over the time period $0 < t < \infty$; Eq. 2.3.29 is the "transient response" to this imposed initial condition. Now, suppose that instead of a step function, we have a "rectangular pulse function". That is, the step function of magnitude, N_{*0}, does not continue indefinitely but instead extends over a finite time interval, t_*, and then immediately drops back to zero. In this case we obtain

$$N = N_{*0}e^{-a_{*0}(t-t_*)} - (N_{*0} - N_0)e^{-a_{*0}t} \tag{2.3.30}$$

which applies for $t > t_*$. For $t < t_*$, Eq. 2.3.29, of course, describes the magnitude of N. In Fig. 2.3.5, curves are shown of $N(t)$ for values of $t_* = 1$, 5, and 10.

Case B.2. Linearly Variable N_*. Suppose that $N_*(t)$ is a linear function of time; that is

$$N_*(t) = N_{*0}(1 + ct) \tag{2.3.31}$$

where c is positive or negative. Substituting this relationship into Eq. 2.3.28 gives

$$N = N_{*0}\left[1 + c\left(t - \frac{1}{a_{*0}}\right)\right] - \left[N_{*0}\left(1 - \frac{c}{a_{*0}}\right) - N_0\right]e^{-a_{*0}t} \quad . \tag{2.3.32}$$

This equation is plotted in Fig. 2.3.6 for three values of c. The dashed lines in the figure show $N_* = N_{*0}(1 + ct)$.

From Eq. 2.3.32 we determine that N has a maximum value at

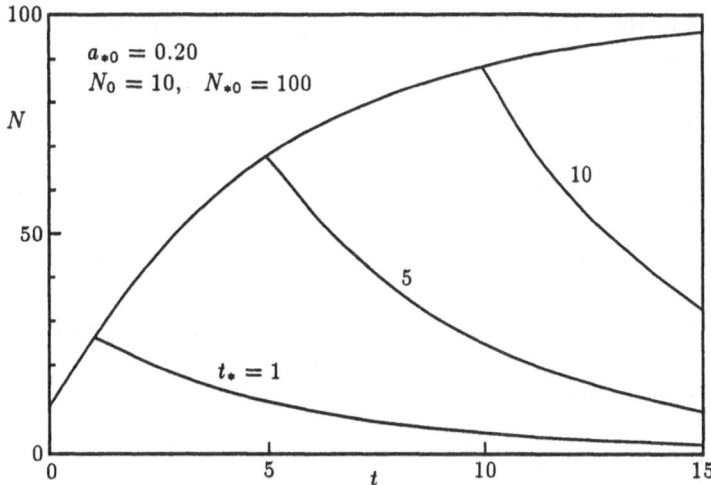

Fig. 2.3.5 Confined exponential curves for rectangular pulse functions

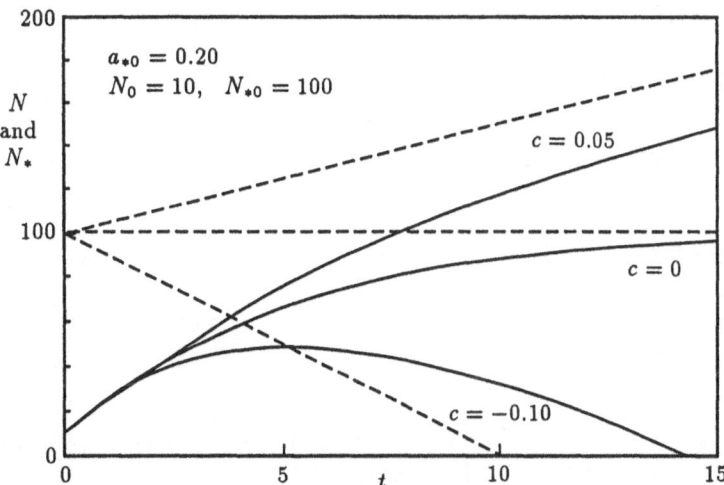

Fig. 2.3.6 Confined exponential curves for linearly variable carrying capacities or equilibrium values. Solid lines (—) are the growth curves, N. Dashed lines (- - -) are the equilibrium value curves, N_*

$$t_{\mathrm{m}} = \frac{1}{a_{*0}} \log_e \left[1 - \frac{a_{*0}}{c} \left(1 - \frac{N_0}{N_{*0}} \right) \right] \quad , \tag{2.3.33}$$

but only when c is negative. This result is illustrated in Fig. 2.3.6. The corresponding value of N is

$$N_{\mathrm{m}} = N_{*0} \left\{ 1 + \frac{c}{a_{*0}} \log_e \left[1 - \frac{a_{*0}}{c} \left(1 - \frac{N_0}{N_{*0}} \right) \right] \right\} \quad . \tag{2.3.34}$$

It is not difficult to show that the two curves, $N(t)$ and $N_*(t)$, intersect at the maximum point.

The gap between the two curves, $N_* - N$, is obtained by subtracting Eq. 2.3.32 from Eq. 2.3.31. The result is

$$N_* - N = \frac{N_{*0}c}{a_{*0}} + \left[N_{*0}\left(1 - \frac{c}{a_{*0}}\right) - N_0 \right] e^{-a_{*0}t} \quad . \tag{2.3.35}$$

For large values of time, the gap has the constant value

$$N_* - N = \frac{N_{*0}c}{a_{*0}} \tag{2.3.36}$$

where, again, c may be positive or negative.

Case B.3. Exponentially Variable N_*. Consider the case in which $N_*(t)$ behaves exponentially, that is

$$N_*(t) = N_{*0}e^{ct} \tag{2.3.37}$$

where again c is positive or negative. In this case we obtain

$$N = N_{*0}\left(\frac{a_{*0}}{a_{*0} + c}\right) e^{ct} - \left[N_{*0}\left(\frac{a_{*0}}{a_{*0} + c}\right) - N_0 \right] e^{-a_{*0}t} \quad . \tag{2.3.38}$$

This relationship is shown in Fig. 2.3.7 for several values of c. A maximum value of N occurs when

$$t_m = \frac{1}{a_{*0} + c} \log_e \left[-\frac{a_{*0}}{c}\left(1 - \frac{N_0}{N_{*0}}\right) + \frac{N_0}{N_{*0}} \right] \tag{2.3.39}$$

as long as c is negative.

The gap, $N_* - N$, is

$$N_* - N = N_{*0}\left(\frac{c}{a_{*0} + c}\right) e^{ct} + \left[N_{*0}\left(\frac{a_{*0}}{a_{*0} + c}\right) - N_0 \right] e^{-a_{*0}t} \quad . \tag{2.3.40}$$

As t approaches infinity, the second term in this equation goes to zero. With respect to the first term, if c is positive, the gap becomes infinitely large as $t \to \infty$; if c is negative, the gap eventually vanishes.

By way of illustration, we look briefly at the results of an analysis by Sharif and Ramanathan (1981). These investigators studied a number of innovation diffusion models with time-dependent potential adopter populations. They examined the logistic equation, the confined exponential equation, and a combination of these two equations, which we consider in Section 2.4. For each of these they imposed the following forms for $N_*(t)$: (1) constant, (2) linear, (3) exponential, (4) confined exponential, and (5) logistic. Indeed, Sharif and Ramanathan present the solutions given by Eqs. 2.3.32 and 2.3.38.

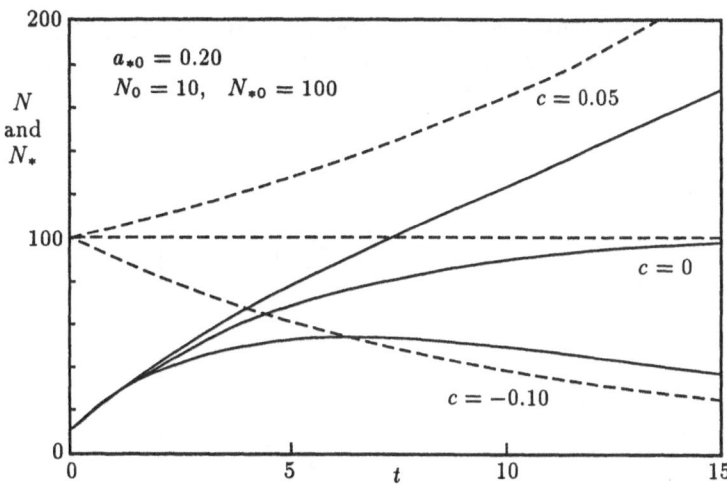

Fig. 2.3.7 Confined exponential curves for exponentially variable equilibrium values

They carried out a number of case studies on the rate of adoption of innovations, including one involving use of fluoridated water in the United States. As a "community innovation", as contrasted with a "consumer innovation", they demonstrated that the adoption mechanism of this particular innovation is best described by the confined exponential equation. The exponentially increasing form for expressing the potential adopter population, $N_*(t) = N_{*0} \exp(ct)$, gave quite satisfactory agreement with observed data.

Case B.4. Sinusoidally Variable N_*. As our final case, suppose that $N_*(t)$ is given by

$$N_*(t) = N_{*0} + N_m \sin \omega t \qquad (2.3.41)$$

in which N_m is the amplitude, $\omega = 2\pi/T$, is the frequency, and T is the period. In this case the solution is

$$N = N_{*0} + \left(\frac{\theta}{\theta^2 + 1}\right) N_m (\theta \sin \omega t - \cos \omega t)$$

$$- \left[N_{*0} - N_0 - \left(\frac{\theta}{\theta^2 + 1}\right) N_m\right] e^{-a_{*0} t} \quad , \qquad (2.3.42)$$

where $\theta = a_{*0}/\omega$. For large values of time, the last term in this equation vanishes and the steady periodic solution is

$$N = N_{*0} + \left(\frac{\theta}{\theta^2 + 1}\right) N_m (\theta \sin \omega t - \cos \omega t) \quad . \qquad (2.3.43)$$

We recall that the ordinary confined exponential solution, Eq. 2.3.29, is termed the "transient response" due to a step function $N_*(t)$. By the same token, Eq. 2.3.42 is the "frequency response" due to a sinusoidal $N_*(t)$.

In Fig. 2.3.8 curves are presented which show $N_*(t)$ and $N(t)$, for $N_m = 50$ and $T = 5$. It is observed that the two curves intersect at the points of maximum and minimum N.

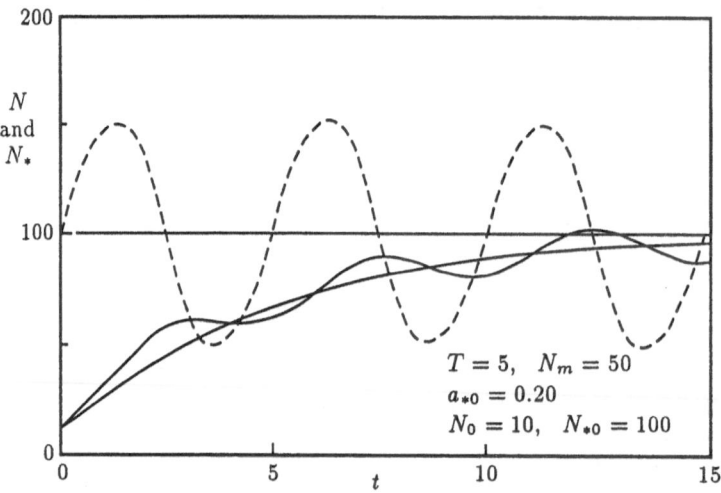

Fig. 2.3.8 Confined exponential curves for sinusoidally variable equilibrium values

Finally, as an example, we look at a problem in heat transfer. Newton's law of cooling states that

$$q = hA(\theta_* - \theta) \tag{2.3.44}$$

in which q is the rate of heat transferred, in joules/second, across an area A. The temperature of a solid or liquid, from or into which heat is being transferred, is θ. The temperature of an adjacent fluid, the so-called ambient temperature, is θ_*. The quantity h is the heat transfer coefficient whose units are joules/sec m^2 $^\circ$C. In general, the magnitude of h depends on the velocity of the ambient fluid. There is extensive literature on the subject of heat transfer; a comprehensive reference is that of Rohsenow and Hartnett (1973).

For our example, the analogy is that θ and h correspond to N and a_{*0}, respectively. Now, suppose that the temperature of air over a well-mixed body of water is described by Eq. 2.3.41. The heat transfer coefficient, h, is assumed to have a reasonably constant value. Then Eq. 2.3.42 describes the temperature changes in the water due to changes of the air temperature. For example, Eq. 2.3.42 might represent the initial warming and subsequent cycling of the temperature of a solar pond due to diurnal changes of the ambient temperature. Alternatively, this result might describe the varying temperature in a shallow lake due to annual ambient temperature cycles.

2.3.6 An Illustration: Oxygen Transfer Across a Water Surface

In the formulation of water quality models for rivers and lakes it is usually necessary to make a so-called oxygen balance on the body of water. To carry out this balance, one needs to know or be able to estimate, the rate at which oxygen is transferred from the atmosphere to the lake or river. The rate of oxygen transfer may depend on the velocity of the water, or on the wind velocity, wave action, rainfall, or other mechanisms producing turbulence at the water surface.

Oxygen transfer rates can be determined from laboratory experiments. In such experiments, the initial oxygen concentration in a tank of water is reduced to a low value by the addition of a chemical, e.g., sodium sulfite. Then the oxygen concentration in the water is allowed to increase due to the effects of the test variable, e.g., wind velocity or mechanical mixing. Measurement of the dissolved oxygen concentration at time intervals during the experiment allows one to determine the oxygen transfer coefficient or reaeration coefficient.

In this illustration, the "item" we are considering to be N is the dissolved oxygen concentration, C, in milligrams per liter. The differential equation is

$$\frac{dC}{dt} = K_2(C_s - C) \pm S \tag{2.3.45}$$

where C_s is the equilibrium concentration of oxygen in water and K_2 is the reaeration coefficient. The quantity S represents oxygen sources (e.g., photosynthesis) or oxygen sinks (respiration).

With $S = 0$, the solution to Eq. 2.3.45 is

$$C = C_s - (C_s - C_0)e^{-K_2 t} \quad . \tag{2.3.46}$$

This equation can be written

$$\log_e \left(\frac{C_s}{C_s - C} \right) = \log_e \left(\frac{C_s}{C_s - C_0} \right) + K_2 t \tag{2.3.47}$$

which is of the form $y = k_0 + k_1 x$. A semi-logarithmic plot of the variables of Eq. 2.3.47 will yield a straight line if the analytical model is correct.

The results of a typical reaeration experiment are listed in Table 2.3.1. The equilibrium concentration at the test temperature is $C_s = 7.00\,\text{mg/l}$. Values of the oxygen concentration, C, for the indicated values of time, t, are shown in Fig. 2.3.9. The numerical values of the quantity, $C_s/(C_s - C)$, appearing in Eq. 2.3.47 and listed in Table 2.3.1, are plotted against time, t, in Fig. 2.3.10.

Table 2.3.1 Experimental data to determine the reaeration coefficient

t min	C mg/l	$C_s/(C_s - C)$	$\log_e[C_s/(C_s - C)]$
0	0.36	1.054	0.0528
1	1.54	1.282	0.2485
2	2.55	1.573	0.4530
3	3.37	1.928	0.6567
4	3.98	2.318	0.8407
5	4.56	2.869	1.0539
6	5.03	3.553	1.2679
7	5.32	4.167	1.4271
8	5.67	5.263	1.6607
9	5.88	6.250	1.8326
10	6.13	8.046	2.0852
11	6.22	8.974	2.1944
12	6.42	12.069	2.4906
13	6.49	13.725	2.6193
14	6.59	17.073	2.8375
15	6.68	21.875	3.0853

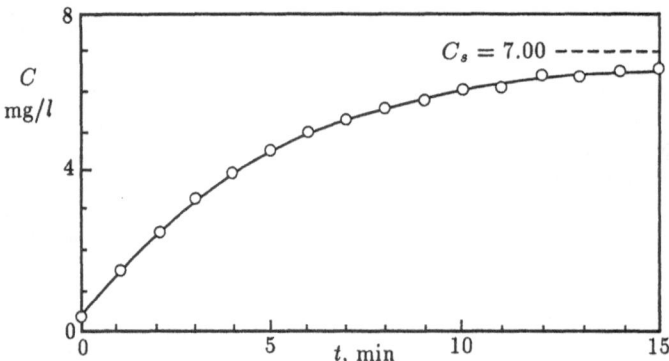

Fig. 2.3.9 Concentration of dissolved oxygen versus time as an example of confined exponential growth

A least squares analysis of the variables of Eq. 2.3.47 gives the results: $K_2 = 0.200\,\text{min}^{-1}$ and $C_0 = 0.34$ mg/l. Substitution of these numerical values into Eqs. 2.3.46 and 2.3.47, yields the solid curves shown in Figs. 2.3.9 and 2.3.10.

We conclude that the confined exponential function provides a suitable framework for a quantitative description of oxygen transfer.

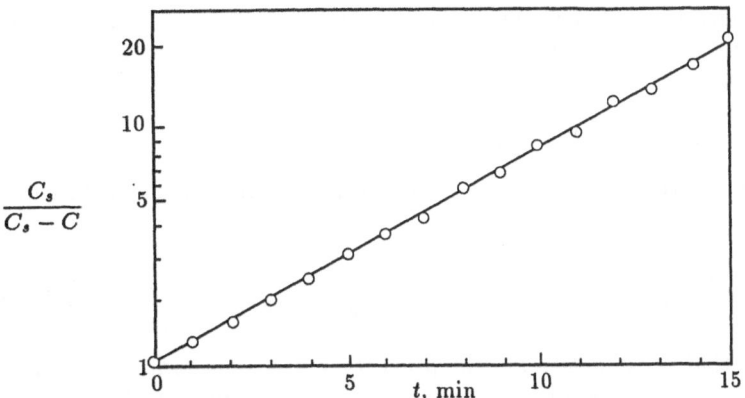

Fig. 2.3.10 Semi-logarithmic plot of dissolved oxygen concentration data to determine the reaeration coefficient K_2

2.3.7 An Illustration: Biochemical Oxygen Demand

In studies concerning the quality of water in rivers, lakes and estuaries, there are numerous indices which collectively describe the overall quality. Without question the two most important indices are the dissolved oxygen (DO), which we considered above, and biochemical oxygen demand (BOD). This quantity is defined as the amount of oxygen utilized by microorganisms for the stabilization of organic material in a wastewater.

The BOD is determined from laboratory tests by measuring the DO concentration each day, over a period of several days, from samples incubated at constant temperature, normally $20\,^\circ$C.

It is usually assumed that the exertion of BOD is described by the first order equation

$$\frac{dL}{dt} = K_1(L_* - L) \tag{2.3.48}$$

in which $L = L(t)$ is the BOD, L_* is the ultimate BOD, and K_1 is the deoxygenation coefficient. If the initial value is $L(0) = L_0$, the solution to Eq. 2.3.48 is

$$L = L_* - (L_* - L_0)e^{-K_1 t} \quad . \tag{2.3.49}$$

Assuming that $L_0 = 0$, we have

$$L = L_*(1 - e^{-K_1 t}) \quad . \tag{2.3.50}$$

The problem is to determine the numerical values of L_* and K_1 from experimental data. A simple method to accomplish this has been devised by Thomas (1950) based on the approximation described below.

We consider the following quantity, S_1, and make an expansion of the negative exponential through the fourth term of the series. This yields

$$S_1 = 1 - e^{-z} = z\left(1 - \frac{1}{2}z + \frac{1}{6}z^2 - \frac{1}{24}z^3\right) \quad . \tag{2.3.51}$$

Next we consider the quantity S_2, and expand this quantity through the third term. This gives

$$S_2 = \left(1 + \frac{1}{6}z\right)^{-3} = \left(1 - \frac{1}{2}z + \frac{1}{6}z^2 - \frac{5}{108}z^3\right) \quad . \tag{2.3.52}$$

It is seen that the quantities in the brackets in the right-hand members of Eqs. 2.3.51 and 2.3.52 are identical except for the last terms. In these terms, the coefficients differ only slightly: 1/24 (0.04167) versus 5/108 (0.0463).

Accordingly, with this approximation, we re-write Eq. 2.3.50 in the form

$$L = L_*(K_1 t)(1 + \frac{1}{6}K_1 t)^{-3} \tag{2.3.53}$$

which can be expressed as follows

$$\left(\frac{t}{L}\right)^{1/3} = \frac{1}{(K_1 L_*)^{1/3}} + \frac{K_1^{2/3}}{6L_*^{1/3}}t \quad . \tag{2.3.54}$$

If we plot $(t/L)^{1/3}$ versus t, we can expect a linear relationship with the indicated intercept and slope.

A numerical example illustrates the method. In Table 2.3.2, measured values of L(BOD) are shown for a number of days following collection and incubation of the water samples. Shown also in the table are the computed values of $(t/L)^{1/3}$. A graphical display of L versus t is given in Fig. 2.3.11.

A plot of the coordinates stipulated by Eq. 2.3.54 is presented in Fig. 2.3.12. A least squares calculation gives $K_1 = 0.168\,\mathrm{day}^{-1}$ and $L_* = 176$ mg/l. Substituting these values into Eq. 2.3.50 yields the solid curve shown in Fig. 2.3.11.

Table **2.3.2** Laboratory measurements of biochemical oxygen demand, L

t days	L mg/l	$(t/L)^{1/3}$	t days	L mg/l	$(t/L)^{1/3}$
0.5	14	0.330	5	102	0.366
1	30	0.322	6	109	0.381
2	48	0.347	8	131	0.394
3	71	0.349	10	139	0.416
4	82	0.366	12	152	0.429

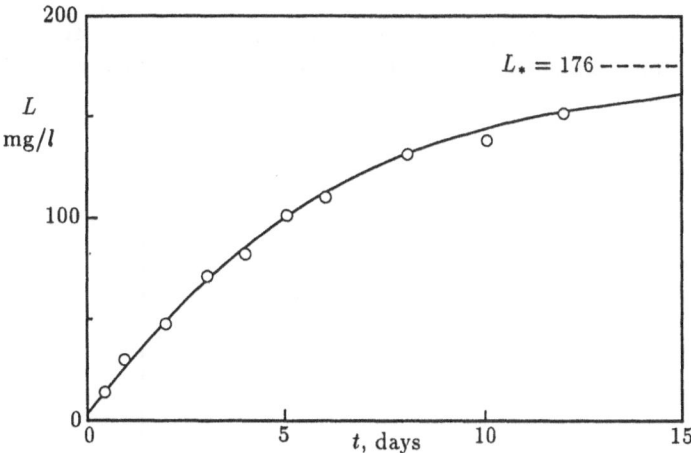

Fig. 2.3.11 Measured values of biochemical oxygen demand (BOD) versus time, t

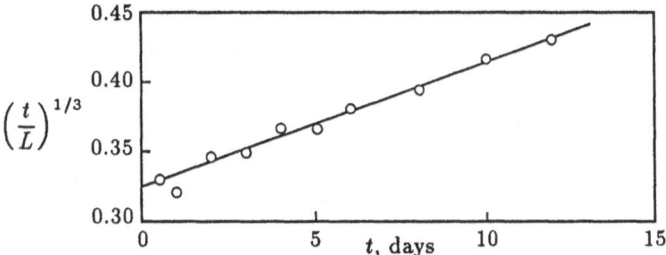

Fig. 2.3.12 Plot to determine values of the deoxygenation coefficient, K_1, and the ultimate BOD, L_*

2.3.8 An Illustration: Growth of Humans

It is not surprising that many studies have been carried out over a period of many years concerning the growth of humans. One of the earliest was the work of Quetelet published in 1835. The results of that study, and other early studies, are summarized in the unabridged version of Thompson's classic "On Growth and Form" (1917).

Many of the advances made by researchers in recent times on the subject of human growth are described in a review by Zeger and Harlow (1987). The book by Bogin (1988) gives extensive coverage to topics of growth and growth rates of humans.

One of the most comprehensive studies on the subject of human growth is that of Stoudt et al. (1960). They determined the average heights and weights of American Caucasian males and females from birth to age 85. Their data indicate that the average male attains his maximum height at age 25 and maximum weight at age 45. The average female acquires her maximum height at age 18 and maximum weight at age 55.

In Fig. 2.3.13, a graphical display is presented of their data for the average height, H, and average weight, W, for males and females over the age range $t = 0$ to $t = 32$ years.

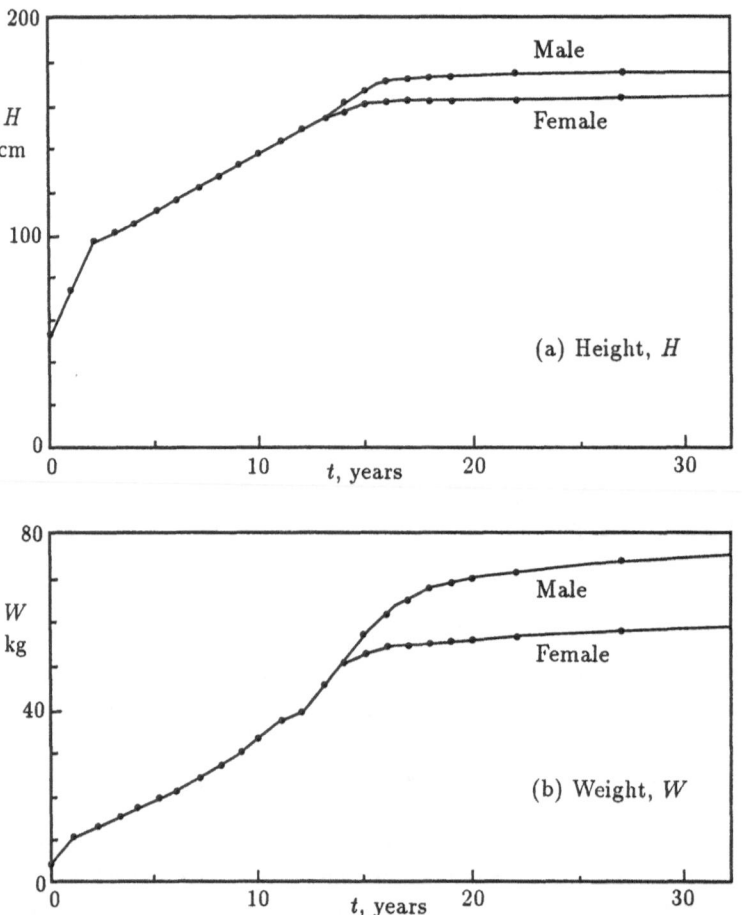

Fig. 2.3.13 Growth rates of American Caucasian males and females: (a) height H and (b) weight W. Data of Stoudt, Damon, and McFarland (1960)

It is apparent in the plots of Fig. 2.3.13 that the curvatures of the growth functions, $H(t)$ and $W(t)$, are negative for small values of t. That is, the plots are concave downward near $t = 0$. This is certainly the most distinguishing characteristic of the confined exponential curve and for this reason it is selected as the framework for our analysis.

We observe in the plots that both height and weight begin to taper off at age 16 to 18. We note also that the weight curve, in particular, is concave upward in the age range from approximately $t = 2$ to $t = 15$. This feature is markedly different from that of the confined exponential equation with a constant growth coefficient.

For our illustration we select the relationship involving weight, W, and rate of weight increase, $w = dW/dt$, of male children in the age range $t = 0$ to 18. It is apparent that the growth coefficient, a_*, is not constant but, rather, a function of time. We assume that we can describe $a_*(t)$ as follows

$$a_*(t) = a_{*0} + a_{m1} \sin \omega t + a_{m2} \cos \omega t \quad , \tag{2.3.55}$$

in which $\omega = 2\pi/T$ is a frequency and T is a period of oscillation. Two trigonometric functions are utilized in this equation to provide flexibility in fitting the growth curve to observed data.

Substituting Eq. 2.3.55 into Eq. 2.3.15 gives the solution

$$w = W_{*0} - (W_{*0} - W_0) \exp \left(-\left[a_{*0}t + \frac{a_{m1}}{\omega}(1 - \cos \omega t) + \frac{a_{m2}}{\omega} \sin \omega t \right] \right) . \tag{2.3.56}$$

The first derivative of this expression yields

$$W = (W_{*0} - W_0) \exp \left(-\left[a_{*0}t + \frac{a_{m1}}{\omega}(1 - \cos \omega t) + \frac{a_{m2}}{\omega} \sin \omega t \right] \right)$$

$$\times (a_{*0} + a_{m1} \sin \omega t + a_{m2} \cos \omega t) \quad . \tag{2.3.57}$$

The data of Stoudt et al. (1960) for the weights, $W(t)$, and weight rate of growth, $w(t)$, of male children for the age range $t = 0$ to 18 are shown in Table 2.3.3 and Fig. 2.3.14.

Table 2.3.3 Mean weights and weight increments of American Caucasian male children, age 0–18 years. From Stoudt et al. (1960)

Age, t [years]	Weight, W [kg]	Increment, w [kg/yr]	Age, t [years]	Weight, W [kg]	Increment, w [kg/yr]
0	3.4	–	9	29.9	2.7
0.5	7.9	9.0	10	33.1	3.2
1	10.4	5.0	11	37.2	4.1
2	12.7	2.3	12	39.5	2.3
3	14.5	1.8	13	44.9	5.4
4	16.8	2.3	14	51.3	6.4
5	19.1	2.3	15	58.1	6.8
6	21.3	2.2	16	62.1	4.0
7	24.5	3.2	17	64.9	2.8
8	27.2	2.7	18	67.6	2.7

We take $W(0) = W_0 = 3.4\,\mathrm{kg}$ and $T = 18\,\mathrm{years}$; hence $\omega = 2/T = 0.349\,\mathrm{yr}^{-1}$. In addition, the following values are determined for the various parameters

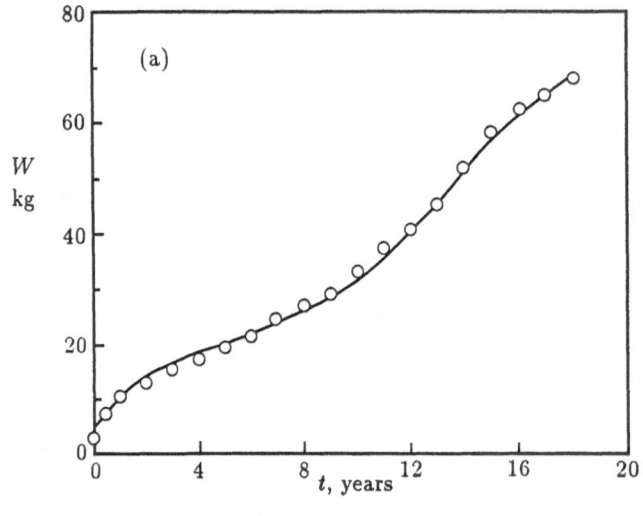

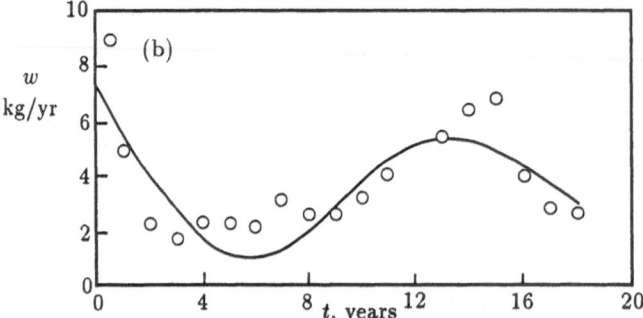

Fig. 2.3.14 Comparison of measured and computed (a) weight and (b) weight rate of increase of American Caucasian male children

$$a_{*0} = 0.048; \qquad N_{*0} = 114.6; \qquad a_{m1} = -0.033; \qquad a_{m2} = +0.016$$

Substitution of these numbers into Eq. 2.3.56 and 2.3.57 produces the solid lines shown in Fig. 2.3.14.

It is seen that the model provides a fairly good description of the weight curve, $W(t)$. Modification of the trigonometric components of $a_*(t)$ would make it even better. Indeed, a harmonic analysis involving the sum of terms like

$$a_*(t) = a_{*0} + \sum_{n=1} a_{mn} \sin(n\omega t + \phi_n) \qquad (2.3.58)$$

would make agreement as precise as we please. The price we would pay, of course, would be the introduction of additional parameters.

Numerous other mathematical models of human growth have been presented. Preece and Baines (1978) also utilize the differential equation for confined exponential growth. For the time-dependent growth coefficient they employ a pair of logistic functions. Hauspie et al (1980) determined that the Preece-Baines model provides an accurate description of the heights of Belgian girls. Other investigators have employed various modifications of the logistic and Gompertz equations as frameworks for expressing human growth. These various growth models are discussed by Falkner and Tanner (1978) and by Bogin (1988).

One of the most adaptable frameworks for description of growth phenomena is that presented by Brody (1945). The concept and method are simple and straightforward. Instead of seeking a single continuous function to describe growth over an entire range of interest, Brody proposed the use of "piece-wise" functions properly matched at appropriate junctions.

For example, in the above illustration concerning weight increase of male children we could have utilized the following scheme

Age range	Growth function
0–2	confined exponential
2–15	exponential
15–45	confined exponential
45–90	another appropriate function

We will illustrate this approach and method in Section 2.4.9 with an example concerning the growth of population of the state of California.

2.3.9 An Illustration: Public Interest in a News Event

We conclude our consideration of the confined exponential distribution with an illustration familiar to all of us: the growth and decline of public interest in a major news event. We have all had numerous experiences relating to news reporting of some major local, national or international crisis which increased rapidly during a few days, reached a peak, and then gradually died out. The public's interest in the particular event evidently followed the same pattern.

We include this illustration here since the confined exponential describes the externally influenced, impersonal mechanism of information transfer characterized by the television, radio, and newspaper media.

We begin by introducing the so-called gamma function. Later on, in Section 3.5.5, we examine this function in more detail. For the present, suppose that the carrying capacity is described by the following simple form of this important and useful function

$$N_*(t) = N_{*0}(ct)e^{-ct} \quad , \tag{2.3.59}$$

where c is a positive constant. Differentiating this equation it is easy to determine the following maximum and inflection points of N_*

$$t_{\mathrm{m}} = 1/c; \qquad N_{*\mathrm{m}} = N_{*0}/e \tag{2.3.60}$$

and

$$t_{\mathrm{i}} = 2/c; \qquad N_{*\mathrm{i}} = 2N_{*0}/e^2 \quad . \tag{2.3.61}$$

If we substitute Eq. 2.3.59 into the general solution, Eq. 2.3.28, and employ the initial condition, $N(0) = 0$, we obtain the equation of the growth curve

$$N = N_{*0}\frac{a_{*0}c}{(a_{*0} - c)^2}\left\{[(a_{*0} - c)t - 1]e^{-ct} + e^{-a_{*0}t}\right\} \quad . \tag{2.3.62}$$

Setting the first derivative equal to zero gives

$$(a_{*0} - c)t_{\mathrm{m}} + \log_e\left[1 - \frac{c}{a_{*0}}(a_{*0} - c)t_{\mathrm{m}}\right] = 0 \quad , \tag{2.3.63}$$

which, by trial-and-error, provides the value of t_{m} corresponding to the maximum value of N.

In the special case in which $c = a_{*0}$ it is not difficult to establish that

$$N = \tfrac{1}{2}N_{*0}a_{*0}^2 t^2 e^{-a_{*0}t} \quad . \tag{2.3.64}$$

This equation has maximum and inflection points at

$$t_{\mathrm{m}} = \frac{2}{a_{*0}}; \qquad t_{\mathrm{i}} = \frac{2}{a_{*0}}\left(1 \pm \frac{1}{\sqrt{2}}\right) \quad . \tag{2.3.65}$$

Two numerical examples of these analytical results are shown in Fig. 2.3.15. In both examples, $a_{*0} = 0.20$, $N_0 = 0$, and $N_{*0} = 100$. The left-hand plot corresponds to $c = 0.10$; the right-hand plot, with a shift of time scale, refers to $c = a_{*0} = 0.20$. We note, in this example involving the gamma function, that both the carrying capacity curve and the growth curve start from zero, attain maxima, and subsequently decrease asymptotically to zero.

The preceding analysis provides the background we need for our illustration. This is a fabricated and hence idealized example; in fact, it is based on the value, $c = 0.30$. However, it should illustrate the point.

Carrying capacity curve. For the function N_*, we determine and tabulate the number of minutes devoted each day by our several local television stations relating to a particular major news event.

Growth curve. For the function N, we ascertain and record the percentage of persons responding in the affirmative to the telephoned question: "at the present time, are you keenly interested in the events concerning . . . ?"

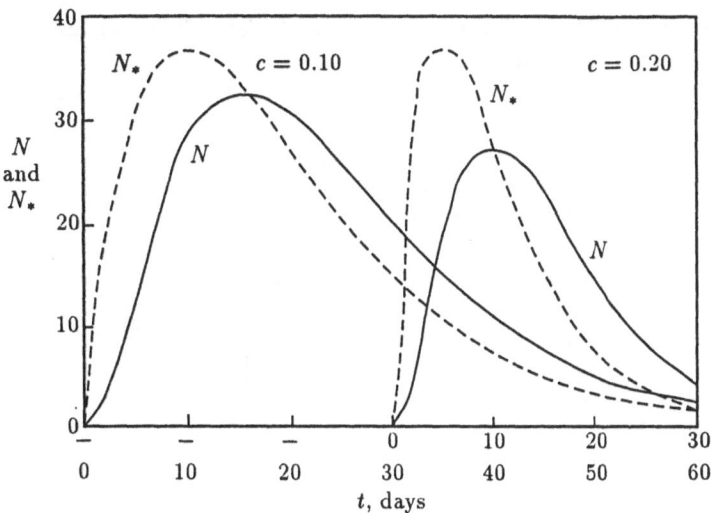

Fig. 2.3.15 Growth curves corresponding to a gamma function carrying capacity or equilibrium value

The units of N and N_* need not be the same, though with different units we do lose some mathematical properties (e.g., the carrying capacity curve does not pass through the maximum point of the growth curve). Our mythical data are listed in Table 2.3.4 and are plotted in Fig. 2.3.16.

Table 2.3.4 Minutes of daily television coverage of a major news event (N_*) and public interest in the event (U)

t [days]	N_* [min/day]	U [percent]	t [days]	N_* [min/day]	U [percent]
0	0	0	8	131	23.2
1	133	2.3	10	90	21.5
2	198	7.1	12	59	18.4
3	220	12.2	15	30	13.2
4	217	16.6	20	9	6.5
5	201	19.9	25	3	2.9
6	179	22.0	30	1	1.2

The numerical values of c and N_{*0} are easily obtained by rewriting Eq. 2.3.59 in the form

$$\log_e \frac{N_*}{t} = \log_e(N_{*0}c) - ct \quad . \tag{2.3.66}$$

A least squares computation of the $N_*(t)$ data of Table 2.3.4 yields $c = 0.30\,\mathrm{day}^{-1}$ and $N_{*0} = 600\,\mathrm{minutes}$.

From Fig. 2.3.16 we observe that $U = N/N_*$ is a maximum when $t = 7.5$ or 8.0 days (the more precise value is $t = 7.63$ days). Knowing that $c = 0.30$, iterated solutions of Eq. 2.3.63 give $a_{*0} = 0.20\,\mathrm{day}^{-1}$. With the numerical values of the constants now available, the solid curves of Fig. 2.3.16 are constructed from Eqs. 2.3.59 and 2.3.62.

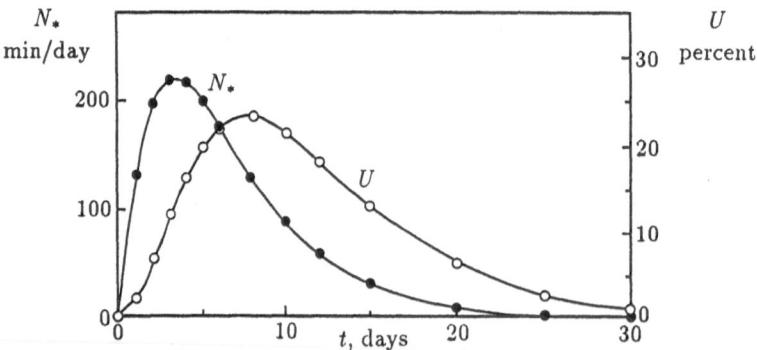

Fig. 2.3.16 Minutes of daily television coverage, N_*, and public interest, U, in a particular news event

The results displayed in Fig. 2.3.16 are certainly compatible with our own experiences. Public interest in a particular news event grows and declines, with a certain time lag, according to the rise and fall of information received from the impersonal sources. Other mathematical forms of $N_*(t)$, of course, would yield other responses in $N(t)$.

2.4 Combination of the Logistic Distribution and the Confined Exponential Distribution

2.4.1 Comparison of the Two Distributions

In Section 2.3, we considered the confined exponential distribution, commencing with the basic differential equation

$$\frac{dN}{dt} = a_*(N_* - N) \quad .$$ (2.4.1)

With the assumption that a_* and N_* are constants, and using the initial condition, $N(0) = N_0$, the following solution was obtained

$$N = N_* - (N_* - N_0)e^{-a_* t} \quad .$$
(2.4.2)

Likewise, in Section 2.2, we examined the logistic distribution, starting with its differential equation

$$\frac{dN}{dt} = aN \left(1 - \frac{N}{N_*} \right)$$
(2.4.3)

and acquired the answer

$$N = \frac{N_*}{1 + [(N*/N_0) - 1]e^{-at}} \quad .$$
(2.4.4)

We now make comparisons of these two basic distributions from several points of view. Then the two will be combined to obtain a generalized answer which reduces to the confined exponential equation as one limiting case and the logistic equation as the other.

2.4.2 Phenomena in Industrial Technology Transfer

First we look at these two equations as descriptions of phenomena in industrial technology transfer. Sharif and Ramanathan (1981) refer to the confined exponential case as the Coleman model and indicate the following assumptions in this model

- The population of potential adopters of the innovation is limited and remains constant over time.
- All members of the population eventually adopt.
- The diffusion process proceeds from a constant source, independent of the number of adopters.
- The impact of this constant and impersonal source on all non-adopters is the same.

These assumptions provide the basis of the differential equation, Eq. 2.4.1, and its solution, Eq. 2.4.2.

Sharif and Ramanathan describe the logistic case as the Dodd model and list the following assumptions for this model

- The population of potential adopters of the innovation is limited and remains constant over time.
- All members of the population eventually adopt.
- All adopters are imitators and adopt only after observing others using the innovation.

- The adoption rate is dependent not only on the number who have adopted but also on the proportion of the maximum number of adopters not yet realized.
- The probability of any one pair of adopters meeting is the same as that of any other pair meeting.

These assumptions lead to the differential equation, Eq. 2.4.3, and its answer, Eq. 2.4.4.

As we shall do in Section 2.4.6, Sharif and Ramanathan combined these two cases to obtain the differential equation and solution for the generalized case.

2.4.3 Phenomena in Social Innovation Diffusion

In their book concerning mathematical modeling in geography, Wilson and Kirkby (1980) present the following equation

$$\frac{dN}{dt} = \lambda N^r (N_* - N) \quad . \tag{2.4.5}$$

If we set $\lambda = a_*$ and $r = 0$ in this equation we generate the confined exponential equation. Alternatively, if we set $\lambda = a/N_*$ and $r = 1$, we produce the logistic. Accordingly, Eq. 2.4.5 is a somewhat generalized relationship. However, to establish the form we want for the present purpose, we set Eq. 2.4.5 aside until we get to Section 3.4.

Instead, we write the generalized differential equation as simply the sum of the differential equations for the confined exponential and the logistic. That is

$$\frac{dN}{dt} = a_*(N_* - N) + aN\left(1 - \frac{N}{N_*}\right) \quad . \tag{2.4.6}$$

This is the approach taken by Sharif and Ramanathan (1981). It is also the approach utilized by Ralston (1980) in his work on the dynamics of communication and by Haynes, Mahajan and White (1977) in their analysis of innovation diffusion. Bartholomew (1981) presents Eq. 2.4.6 in connection with a generalized treatment of diffusion of news, innovations and rumors and the spread of epidemics.

Lekvall and Wahlbin (1973) examined the subject of external and internal influence in the communication of an innovation. Their approach was the same. In presenting Eq. 2.4.6, they rationalized that "external" influences on innovation diffusion are described by the confined exponential equation and "internal" influences by the logistic.

With respect to Eq. 2.4.6, it is seen that if $a = 0$, the equation reduces to the confined exponential distribution. We term this the pure source, externally influenced model. Alternatively, if $a_* = 0$, the equation becomes the logistic distribution. We identify this as the pure interaction, internally influenced model. As we shall see in Section 2.4.6, when neither parameter is zero there is a blend of source diffusion with interaction diffusion.

2.4.4 Phenomena in Chemical Reaction Kinetics

One of the fundamental principles in physical chemistry is the rate law of chemical reactions. This law states that the rate of a chemical reaction is proportional to the concentrations of the reactants. That is

$$-\frac{dA}{dt} = kABC\ldots \tag{2.4.7}$$

in which k is the rate constant and A, B, $C\ldots$, are the concentrations of the reactants expressed in, say, moles per liter.

If the reaction is one in which the rate of reaction depends only on the concentration of a single reactant then

$$-\frac{dA}{dt} = k_1 A \quad . \tag{2.4.8}$$

We term this a first order reaction. If the initial concentration of the reactant is A_0, then the solution to Eq. 2.4.8 is $A = A_0 \exp(-k_1 t)$. Substituting our own symbols, $A = (N_* - N)$ and $A_0 = (N_* - N_0)$, we obtain

$$\frac{dN}{dt} = k_1 (N_* - N) \tag{2.4.9}$$

and

$$N = N_* - (N_* - N_0)e^{-k_1 t} \quad . \tag{2.4.10}$$

These expressions are the same as Eqs. 2.4.1 and 2.4.2. We term this first order case a monomolecular reaction.

Next suppose that the chemical reaction depends on the concentrations of two reactants. In this case we have

$$-\frac{dA}{dt} = k_2 AB \quad , \tag{2.4.11}$$

with a similar expression for dB/dt. This is termed a second order reaction. If the initial concentrations of the reactants are A_0 and B_0 and letting N be the concentration of the products, Eq. 2.4.11 becomes

$$\frac{dN}{dt} = k_2 (A_0 - N)(B_0 - N) \tag{2.4.12}$$

with a specified value of $N(0)$, the solution to Eq. 2.4.12 is easily obtained.

The solution to Eq. 2.4.12 1s not carried out largely because we are primarily interested in another type of second order reaction: autocatalysis. In such a reaction, one of the reactants is converted to a product of a kind which catalyzes its own further formation. Thus, the rate of increase of the concentration of the product is proportional to the concentration of the reactant and the product. To quote Atkins (1982): "In an autocatalytic reaction, the products contribute to the rate of the forward reaction, and so the rate is expected to increase, possibly very rapidly, as products are generated, and then to stop suddenly when the reactants are exhausted."

In this case, the rate equation becomes

$$\frac{dN}{dt} = k_2 A N \qquad\qquad (2.4.13)$$

where N is the product concentration. As before, we take $A = (N_* - N)$; accordingly Eq. 2.4.13 becomes

$$\frac{dN}{dt} = k_2 N (N_* - N) \qquad\qquad (2.4.14)$$

and with $N(0) = N_0$

$$N = \frac{N_*}{1 + [(N_*/N_0) - 1]\,e^{-k_2 N_* t}} \qquad\qquad (2.4.15)$$

These last two equations are identical to Eqs. 2.4.3 and 2.4.4, respectively, with a replaced by $k_2 N_*$. We term this special second order case an autocatalytic reaction. Occasionally one sees the logistic equation referred to as the autocatalytic equation.

A background note: In the early years of the century, a physiologist–biochemist named Robertson (1923) became a strong advocate of a theory that the autocatalytic equation provided the complete quantitative description of all plant and animal growth phenomena. In her very lucid book, Kingsland (1985) describes the Robertson episode and many other highlights relating to the history of population ecology.

2.4.5 Phenomena in the Psychology of Learning

Over a period of many years, psychologists have devoted much attention to the development of theories and models concerning the phenomena of learning and forgetting. An extensive review of early work on mathematical models of learning is given by Lewis (1966).

In recent years, psychologists have made advances in learning theory by utilizing stochastic approaches to the problem in contrast to the deterministic considerations given earlier. One of the first to utilize stochastic methods to describe the learning process was Estes (1950). Even so, relatively simple deterministic models continue to provide suitable frameworks for presentation and analysis of experimental data.

Over the years, numerous deterministic learning models have been proposed by psychologists investigating the phenomenon. The two models most thoroughly examined and accepted are the following

- The linear-operator model devised by Bush and Mosteller (1955). This model assumes that the improvement in performance induced by an event is proportional to the amount of improvement still possible.

Expressed in mathematical terminology, this assumption leads to the confined exponential equation.

- The beta response-strength model proposed by Luce (1959). This model assumes that the improvement in performance induced by an event is proportional to the product of the improvement still possible and the amount already achieved.

In quantitative terms, this assumption yields logistic equation.

In summary, the analogies described above are listed in Table 2.4.1.

Table 2.4.1 Analogies among industrial technology transfer, social innovation diffusion, chemical reaction kinetics, and the psychology of learning

Phenomenon	Confined exponential	Ordinary logistic
Differential equation	$dN/dt = a_*(N_* - N)$	$dN/dt = aN(1 - N/N_*)$
Solution	$N = N_* - (N_* - N_0)e^{-a_*t}$	$N = \dfrac{N_*}{1 + \left(\frac{N_*}{N_0} - 1\right)e^{-at}}$
Industrial technology transfer	Concentrated, impersonal information source, Coleman model	Distributed, personal information source, Dodd model
Social innovation diffusion	Externally influenced pure source model, $a = 0$	Internally influenced pure interaction model, $a_* = 0$
Chemical reaction kinetics	First order mono-molecular reaction	Second order, auto-catalytic reaction
Psychology of learning	Linear operator model of Bush and Mosteller	Beta response-strength model of Luce

2.4.6 The Differential Equation, Its Solution and Properties

To proceed with the analysis, we combine the differential equations for the confined exponential and the logistic as follows

$$\frac{dN}{dt} = a_*(N_* - N) + aN\left(1 - \frac{N}{N_*}\right) \quad . \tag{2.4.16}$$

We write this equation in dimensionless form by letting $U = N/N_*$, $T = at$ and $w = a/(a + a_*)$. If $w = 0$ (i.e., $a = 0$), this expression reduces to the confined exponential; if $w = 1$ (i.e., $a_* = 0$), it becomes the logistic. Accordingly,

$$\frac{dU}{dt} = \left(\frac{1-w}{w} + U\right)(1 - U) \quad . \tag{2.4.17}$$

This dimensionless equation relates the growth rate, dU/dT, to U for values of w between zero and one. Several curves of Eq. 2.4.17 are shown in Fig. 2.4.1. The lower-most curve, corresponding to $w = 1$, is the symmetrical parabola of the ordinary logistic equation. From Eq. 2.4.17, we establish that the maximum value, $(dU/dT)_m = 1/(4w^2)$ occurs at $U_m = 1 - (1/2w)$. The dashed line in Fig. 2.4.1 passes through the maximum points.

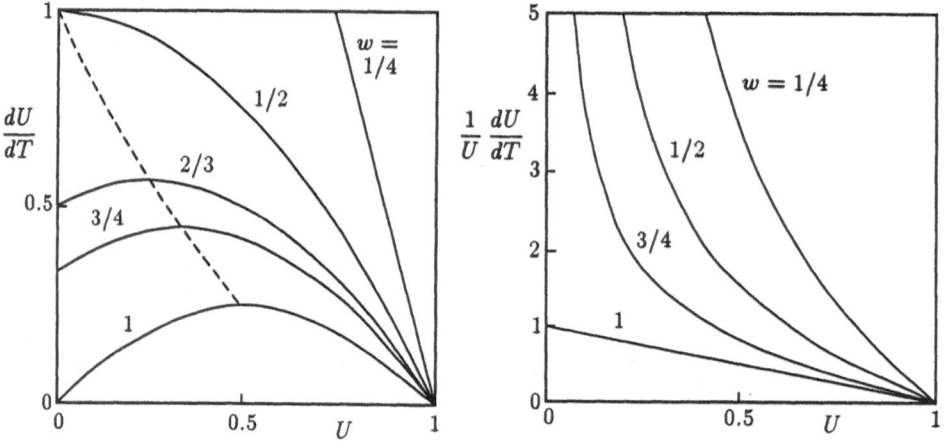

Fig. 2.4.1 (left) Dimensionless growth rates for combination of confined exponential and logistic distributions

Fig. 2.4.2 (right) Dimensionless specific growth rates for combination of confined exponential and logistic distributions

Dividing Eq. 2.4.17 by U gives

$$\frac{1}{U}\frac{dU}{dt} = \frac{1}{U}\left(\frac{1-w}{w} + U\right)(1 - U) \quad . \tag{2.4.18}$$

This equation gives the specific growth rate, $dU/U\,dT$, as a function of U. Fig. 2.4.2 displays this relationship for various values of w. Again, the lower-most curve shows the linear relationship of the logistic.

The case $w = 1/2$ is interesting. In this instance, Eqs. 2.4.17 and 2.4.18 become

$$\frac{dU}{dt} = 1 - U^2 \qquad\qquad (2.4.19)$$

and

$$\frac{1}{U}\frac{dU}{dt} = \frac{1}{U} - U \quad . \qquad\qquad (2.4.20)$$

As Eq. 2.4.19 and Fig. 2.4.1 indicate, the growth rate for this case, is a maximum when $U = 0$. Since $w = 1/2$, then $a_* = a$, and we can say, with an eye on Table 2.4.1, that the phenomenon is equally an "externally influenced, pure source model" and an "internally influenced, pure interaction model".

We return to the basic differential equation, Eq. 2.4.16, but digress for a moment. In this equation, if the substitution $N = 1/C$ is made and if a_* and a trade places, we obtain

$$\frac{dC}{dt} = a_*(C_* - C) + aC\left(1 - \frac{C}{C_*}\right) \qquad\qquad (2.4.21)$$

which is exactly the same as Eq. 2.4.16. Thus, there is an interesting "reciprocal" relationship between the confined exponential and logistic equations. This relationship exists even if a_*, a, and N_* are functions of time.

The most direct method to solve Eq.2.4.16 is to use a table of integrals, with an integrand of the type $(c_0 + c_1 x + c_2 x^2)^{-1}$. Another method of solution is to write Eq. 2.4.16 in the form

$$\frac{dN}{dt} + (a_* - a)N + \frac{a}{N_*}N^2 = a_*N_* \quad . \qquad\qquad (2.4.22)$$

This is a Ricatti differential equation. Making the substitution

$$N = \frac{N_*}{C}\frac{dC}{dt} \quad , \qquad\qquad (2.4.23)$$

this equation becomes

$$\frac{d^2C}{dt^2} + (a_* - a)\frac{dC}{dt} - aa_*C = 0 \qquad\qquad (2.4.24)$$

which is a second order linear homogeneous differential equation with constant coefficients. With the usual initial condition, $N(0) = N_0$, we obtain the solution

$$N = N_*\frac{1 - (ma_*/a)e^{-(a+a_*)t}}{1 + me^{-(a+a_*)t}} \quad , \qquad\qquad (2.4.25)$$

where

$$m = \frac{\left(\frac{N_*}{N_0} - 1\right)}{\left(\frac{N_*}{N_0}\frac{a_*}{a} + 1\right)} \quad . \tag{2.4.26}$$

The slope of the combined logistic-confined exponential distribution or, equivalently, the ordinate of the corresponding density distribution, is

$$n = \frac{dN}{dt} = N_* m a \left(1 + \frac{a_*}{a}\right)^2 \frac{e^{-(a+a_*)t}}{\left(1 + m e^{-(a+a_*)t}\right)^2} \quad . \tag{2.4.27}$$

From the derivative of this expression we obtain the inflection point

$$t_i = \frac{1}{a + a_*} \log_e m; \quad N_i = \tfrac{1}{2} N_* \left(1 - \frac{a}{a_*}\right) \quad . \tag{2.4.28}$$

For a numerical example of these results, we select the following values: $N_0 = 10$, $N_* = 100$, and $(a + a_*) = 0.20$; we let $w = 0, 1/4, 1/2, 3/4$, and 1. Results of computations are shown in Fig. 2.4.3. Numerical values of the main parameters of the example are listed in Table 2.4.2. The solid dots in Fig. 2.4.3 identify inflection points.

Table 2.4.2 A numerical example of the combined logistic–confined exponential equation. $N_0 = 10$, $N_* = 100$, $a + a_* = 0.20$

w	a	a_*	m	t_i	N_i	Remarks
0	0.00	0.20	0	$-\infty$	$-\infty$	Conf. exponential
1/4	0.05	0.15	0.290	-6.184	-100.00	
1/2	0.10	0.10	0.818	-1.003	0.00	
3/4	0.15	0.05	2.077	3.655	33.33	
1	0.20	0.00	9.000	10.990	50.00	Ordinary logistic

We return briefly to the special case $w = 1/2$. From Eq. 2.4.25 or, more simply, from a direct integration of Eq. 2.4.19, we get

$$U = \frac{1 - m_{1/2} e^{-2T}}{1 + m_{1/2} e^{-2T}} \quad , \tag{2.4.29}$$

where

$$m_{1/2} = \frac{N_* - N_0}{N_* + N_0} \quad . \tag{2.4.30}$$

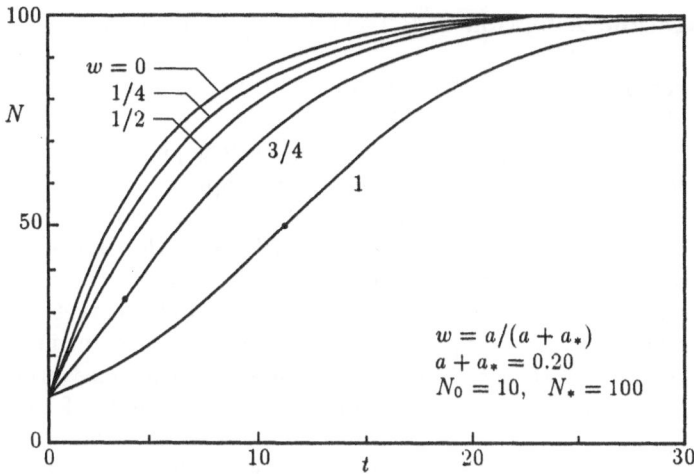

Fig. 2.4.3 Growth curves for combination of confined exponential and logistic equations for various values of the parameter w

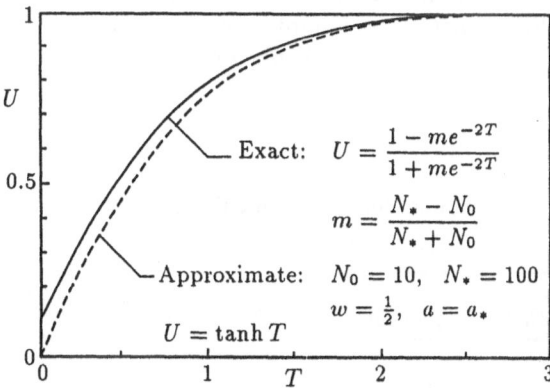

Fig. 2.4.4 Exact and approximate growth curves for the case $w = 1/2$

With $N_0 = 10$ and $N_* = 100$, the exact relationship of Eq. 2.4.29 is plotted in Fig. 2.4.4. If N_* is very large with respect to N_0, then $m_{1/2} = 1$ and Eq. 2.4.29 reduces to the simple relationship

$$U = \tanh T \quad . \tag{2.4.31}$$

This approximate relationship is also shown in Fig. 2.4.4 Thus, in the case in which $w = 1/2$ (i.e., $a = a_*$), the growth curve or technology diffusion equation is approximately and simply described by Eq. 2.4.31.

2.4.7 An Illustration: Adoption of a Tornado Warning Device

Throughout midwest America, tornadoes are highly likely to occur during the summer months. Almost invariably they do a great deal of damage and not infrequently they cause numerous injuries and fatalities. The physical characteristics of tornadoes include very high wind velocities, rapid and substantial decrease of atmospheric pressure, considerable rainfall and occasionally hail, a dark green coloring of the skies, and an unnatural stillness in the immediate environment shortly before the tornado strikes.

Some years ago, a manufacturer produced and successfully marketed a "tornado warning device". This compact device contained an instrument to determine the direction and velocity of the wind, a barometer for measuring pressure, a rain gauge, and a photoelectric cell to indicate changes of light intensity.

After an initial period of planning, the device was widely advertised for a prolonged period on the radio and in the newspapers, especially in the rural areas. Subsequently, in one county of a midwestern state, surveys were taken each month to determine the number of households who had purchased the device. The cumulative number sold since the start of the survey was then converted to a percentage of the estimated total number of potential adopters. The results of the monthly surveys are listed in Table 2.4.3 and are plotted in Fig. 2.4.5.

Table 2.4.3 Adoption of a tornado warning device. Cumulative percentage adoption by months

t [months]	N [percent]	t [months]	N [percent]
0	5.1	–	–
1	16.5	7	82.4
2	30.1	8	88.0
3	42.9	9	91.4
4	55.7	10	94.4
5	66.2	11	95.8
6	75.6	12	97.4

In this case, it can be assumed that the adoption of the innovation was due to two factors: (a) the constant and impersonal transmission of information about the device through advertising media and (b) the exchanges of information and opinions about the device among those who had already acquired it and those who had not yet done so. Accordingly, the information

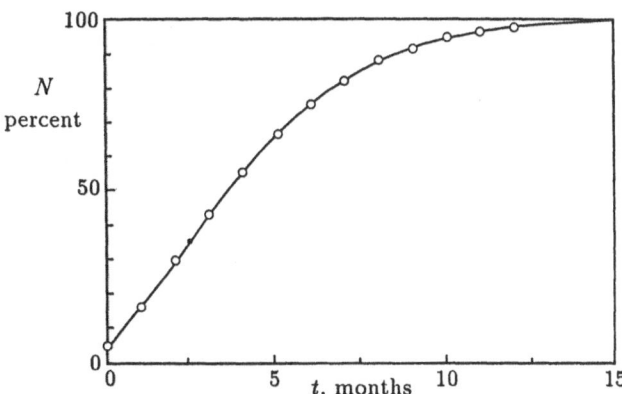

Fig. 2.4.5 Rate of adoption of a tornado warning device

diffusion process was apparently a blend of a pure source model ($w = 0$) and a pure interaction model ($w = 1$).

We begin our analysis of the problem by determining the various coefficients of Eq. 2.4.25. In this case, we know fortunately that the value of N_* is 100 percent; this greatly simplifies the problem. So it is necessary to determine the values of a, a_*, and N_0; this analysis is quite straightforward.

We write the basic differential equation, Eq. 2.4.16, in the finite difference form

$$\frac{1}{\overline{N}(N_* - \overline{N})}\frac{\Delta N}{\Delta t} = \frac{a}{N_*} + a_*\frac{1}{\overline{N}} \tag{2.4.32}$$

which is in the form, $y = k_0 + k_1 x$. So if the quantity in the left-hand member is plotted against the quantity $1/\overline{N}$, a linear correlation should result. Plotting the data of Table 2.4.3 in the form needed for this correlation yields Fig. 2.4.6. A least squares analysis gives $a = 0.306$ and $a_* = 0.096$; the value of $w = a/(a + a_*) = 0.761$.

The remaining coefficient to determine is N_0. This is done by solving Eq. 2.4.25 for m and then substituting the tabulated values of (t, N). The average value of m, computed in this fashion, is $m = 2.571$. Finally, from Eq. 2.4.26 we obtain $N_0 = 5.42$.

Substituting these numerical values into Eq. 2.25 produces the curve shown in Fig. 2.4.5. The inflection point, computed from Eq. 2.4.28, is $t_i = 2.35$ and $N_i = 34.30$.

The main conclusion of this illustration is that the technology diffusion process involved is a blend of the effects due to an external impersonal information source and an internal personal information source. This conclusion is based on the fact that $w = a/(a + a_*) = 0.761$.

Since the comparative values of the two transfer coefficients are $a_* = 0.096$ and $a = 0.306$, we conclude that the major factor affecting adoption is an internally influenced, interacting communication mechanism.

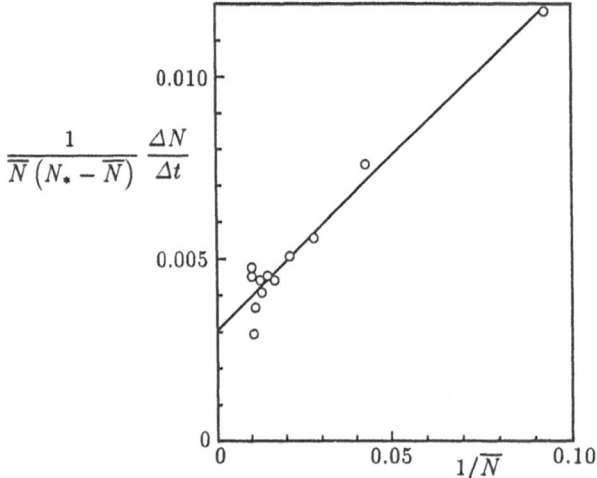

$$\frac{1}{\overline{N}\,(N_* - \overline{N})}\;\frac{\Delta N}{\Delta t}$$

Fig. 2.4.6 Adoption of a tornado warning device. Plot to determine a and a_*.

2.4.8 Combination of the Exponential Function and the Confined Exponential Distribution

Closely related to the subjects examined above is a topic on which we conclude this section: a combination of the exponential function and the confined exponential distribution. This time we will not simply add the two distributions as we did for the confined exponential and logistic distributions.

Instead we suppose that the exponential function is valid over a period of time during which the growth rate is increasing. We call this Stage 1. In the ensuing Stage 2, during which the growth rate is decreasing, we assume that the confined exponential distribution is applicable. This is the approach taken by Brody (1945) in his studies of growth phenomena, particularly of livestock.

Accordingly, we write the following equations

Stage 1 $\dfrac{dN}{dt} = aN$ (2.4.33)

$N = N_0 e^{at}$ (2.4.34)

and

Stage 2 $\dfrac{dN}{dt} = a_*(N_* - N)$ (2.4.35)

$N = N_* - (N_* - N_i)e^{-a_*(t-t_i)}$ (2.4.36)

The initial conditions are: (1) $N = N_0$, $t = 0$ and (2) $N = N_i$, $t = t_i$.

In order to "match" the distributions, we require that the ordinate, N_i, and the slope, $(dN/dt)_i$, be the same at the junction of the two curves. The slope of the exponential distribution (Stage 1) is always positive and the slope of the confined exponential is always negative. Consequently, the junction itself is an inflection point.

Utilizing the above equations, we obtain the following expressions for the coordinates of the inflection point

$$t_i = \frac{1}{a} \log_e \frac{N_i}{N_0} \quad ; \quad N_i = \left(\frac{a_*}{a + a_*}\right) N_* \quad . \tag{2.4.37}$$

A numerical example illustrates the method. We use the values: $a = 0.20$, $N_0 = 10$, $N_* = 100$, and values of $a/a_* = 0$, 1/2, 1, 2, and 5. Results of computations employing the preceding equations are shown in Fig. 2.4.7. For comparison, the ordinary logistic distribution is shown by the dashed line.

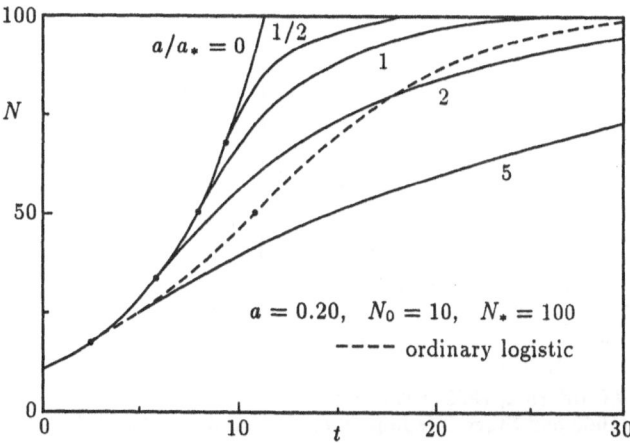

Fig. 2.4.7 Combination of the exponential function and the confined exponential distribution

The inflection points are identified by the solid dots. We note that the exponential curve itself is the locus of these points. This wide range of inflection point ordinates, from N_0 to N_*, is an attractive feature of this particular combination of distributions.

2.4.9 An Illustration: Population of California

As an illustration of this approach we consider the following. The population of the state of California, commencing with the year 1860, is listed in Table 2.4.4. The data of the table are shown in the arithmetic plot of Fig. 2.4.8 and the semi-logarithmic plot of Fig. 2.4.9.

Table 2.4.4 Population of California, 1860 to 1990. Reference: The Universal Almanac 1990, Andrews and McMeel, Kansas City, Mo., 1989

Year	t [years]	N [millions]	Year	t [years]	N [millions]
1860	0	0.380	1930	70	5.677
1870	10	0.560	1940	80	6.907
1880	20	0.865	1950	90	10.586
1890	30	1.214	1960	100	15.717
1900	40	1.485	1970	110	19.971
1910	50	2.738	1980	120	23.668
1920	60	3.427	1990	130	29.126

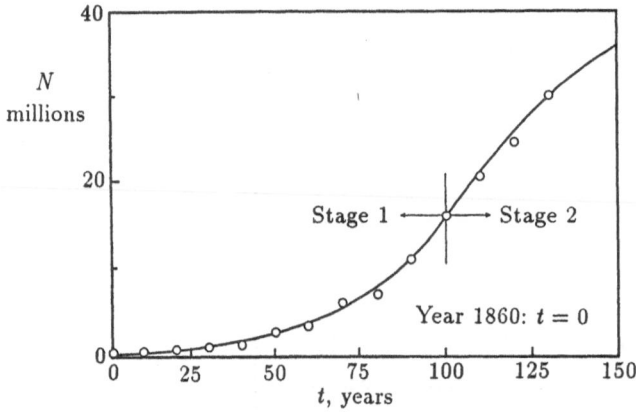

Fig. 2.4.8 Population of California, 1860–1990. Arithmetic plot shows Stage 1 (exponential growth) 1860–1960 and Stage 2 (confined exponential growth) 1960–1990

From these plots, especially that of Fig. 2.4.9, it appears that the population of California grew at an exponential rate during the 100-year period from 1860 to 1960. On this basis, we select the year 1960 as the junction between Stage 1 exponential growth and Stage 2 confined exponential growth. That is, $t_i = 100$.

From Eq. 2.4.34 we obtain

$$\log_e N = \log_e N_0 + at \quad . \tag{2.4.38}$$

A least squares analysis involving the first 11 entries of Table 2.4.4 gives $a = 0.0368$ and $N_0 = 0.390$. These results, substituted into Eq. 2.4.34, yield the solid line of Stage 1 shown in the figures. At the inflection point, $t_i = 100$, we determine that $N_i = 15.462$.

For Stage 2 we proceed as follows. Rewriting Eq. 2.4.36 gives

$$\log_e(N_* - N) = \log_e(N_* - N_i) - a_*(t - t_i) \quad . \tag{2.4.39}$$

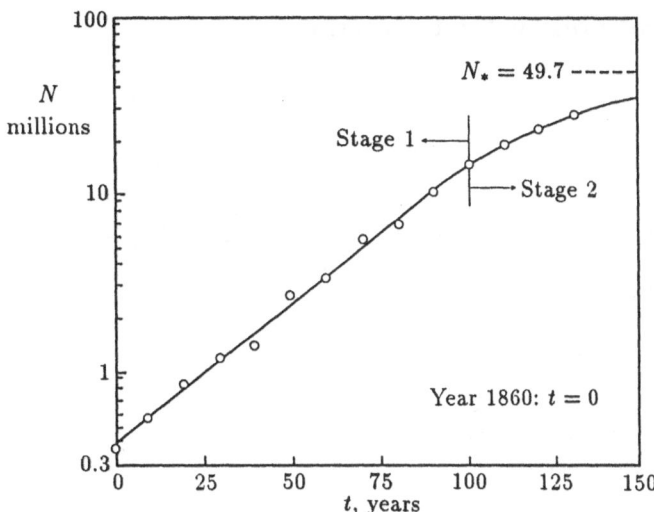

Fig. 2.4.9 Population of California, 1860 to 1990. Semi-logarithmic plot. Calculated carrying capacity, N_*, is 49.7 million

Since Eq. 2.4.37 provides

$$(N_* - N_i) = aN_i/a_* \quad , \tag{2.4.40}$$

we obtain

$$\log_e(N_* - N) = \log_e(aN_i/a_*) - a_*(t - t_i) \quad , \tag{2.4.41}$$

which is a linear equation. Least squares computation involving the last four entries of Table 2.4.4 provides the values of a_* and N_*. An iterated solution featuring assumed values of N_* is necessary. In this case, the solution criterion is that the magnitude of a_* computed from the intercept value of Eq. 2.4.41 must be the same as that calculated from the slope value. This procedure yields $a_* = 0.0162$ and $N_* = 49.7$. Substituting these quantities into Eq. 2.4.36 produces the solid line of Stage 2 displayed in Figs. 2.4.8 and 2.4.9.

Whether or not the asymptotic population of California will be 49.7 million is somewhat beside the point. It seems to be a reasonable figure.

The more important matter is to illustrate the simple methodology involved in this type of combination of distributions. As we have seen, the concept is simple, the approach is straightforward, and determination of the values of the various parameters is not difficult.

2.5 Normal Probability Distribution

2.5.1 The Normal Probability Function and Its Features

Our next topic involves one of the most important distribution functions of all: the Gaussian or normal probability function. As we shall see, the normal distribution is quite similar to the logistic in many ways. The cumulative distribution of the normal probability function has the following definition

$$U(t) = \frac{1}{\sigma\sqrt{2\pi}} \int_{-\infty}^{t} e^{-\frac{1}{2}\left(\frac{\zeta-m}{\sigma}\right)^2} d\zeta \quad , \tag{2.5.1}$$

in which m is the mean and σ^2 is the variance. Making the substitution $z = (\zeta - m)/\sigma$, Eq. 2.5.1 becomes

$$U(T) = \frac{1}{\sqrt{2\pi}} \int_{-\infty}^{T} e^{-\frac{1}{2}z^2} dz \quad , \tag{2.5.2}$$

where $T = (t - m)/\sigma$. With the additional substitutions, $U = N/N_*$, $\sigma = 1/a_n$, and $m = t_i$, Eq. 2.5.2 becomes

$$N = \frac{N_*}{\sqrt{2\pi}} \int_{-\infty}^{a_n(t-t_i)} e^{-\frac{1}{2}z^2} dz \quad , \tag{2.5.3}$$

in which a_n is the growth coefficient and t_i is the time corresponding to the inflection point. The density or frequency distribution is

$$n = \frac{dN}{dt} = \frac{a_n N_*}{\sqrt{2\pi}} e^{-\frac{1}{2}[a_n(t-t_i)]^2} \tag{2.5.4}$$

and the derivative of this equation is

$$\frac{dn}{dt} = \frac{d^2 N}{dt^2} = \frac{a_n^2 N_*}{\sqrt{2\pi}} (t - t_i) e^{-\frac{1}{2}[a_n(t-t_i)]^2} \tag{2.5.5}$$

Setting this result equal to zero establishes that the time $t = t_i$ indeed corresponds to the inflection point of the cumulative curve and also to the maximum value of the density curve. From Eqs. 2.5.3 and 2.5.4 we obtain

$$N_i = \tfrac{1}{2}N_* \quad ; \quad n_i = \left(\frac{dN}{dt}\right)_i = \frac{a_n N_*}{\sqrt{2\pi}} \quad . \tag{2.5.6}$$

Differentiation of Eq. 2.5.5 gives the second derivative of the density function, d^2n/dt^2, and the third derivative of the cumulative function, d^3N/dt^3.

If this result is equated to zero, we obtain the inflection points of the density curve and the points of maximum curvature of the cumulative. The result is

$$t_c = t_i \pm 1/a_n \quad .$$ (2.5.7)

Substituting this answer into Eqs. 2.5.3 to 2.5.5 gives

$$N_c = \frac{N_*}{\sqrt{2\pi}} \int_{-\infty}^{\pm 1} e^{-\frac{1}{2}z^2} dz = \begin{cases} 0.1587 N_* \\ 0.8413 N_* \end{cases}$$ (2.5.8)

$$n_c = \left(\frac{dN}{dt}\right)_c = \frac{a_n N_*}{\sqrt{2\pi e}}$$ (2.5.9)

$$\left(\frac{dn}{dt}\right)_c = \left(\frac{d^2 N}{dt^2}\right)_c = \pm \frac{a_n N_*}{\sqrt{2\pi e}} \quad .$$ (2.5.10)

The mean value, $\bar{t}$, and standard deviation, σ, are obtained from Eqs. 2.2.26 and 2.2.28. The answers are

$$\bar{t} = t_i \quad ; \qquad \sigma = \pm \frac{1}{a_n} \quad .$$ (2.5.11)

For a numerical example and graphical display, we take $a_n = 1.0$, $N_0 = 10$, and $N_* = 100$. These are the same numbers used in Section 2.2 for the example involving the logistic distribution. To determine the value of t_i, we substitute the initial condition, $N(0) = N_0$ into Eq. 2.5.3 and obtain

$$N_0 = \frac{N_*}{\sqrt{2\pi}} \int_{-\infty}^{-a_n t_i} e^{-\frac{1}{2}z^2} dz \quad .$$ (2.5.12)

From this relationship, with N_0 known, the magnitude of the upper limit, $-a_n t_i$, is determined. Conveniently, Brandt (1976) provides tables which give direct determination of this inversion. In our example, $t_i = 1.282$.

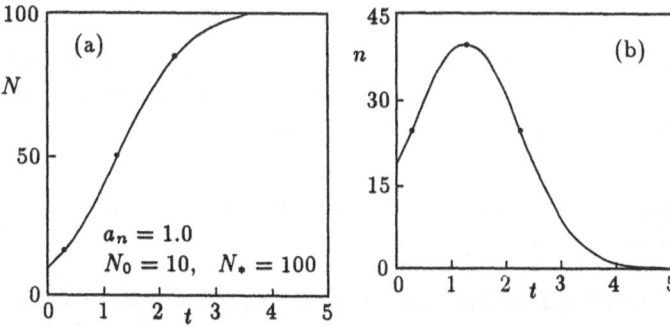

Fig. 2.5.1 Plots of the normal probability function; (a) cumulative distribution and (b) density distribution

Table 2.5.1 Various features of the normal probability distribution with the following numerical values: $a_n = 1.0$, $N_0 = 10$, $N_* = 100$. See Fig. 2.5.1

Symbol	Quantity	Value
t_i	Value of t at cumulative inflection point	1.282
N_i	Value of N at cumulative inflection point	50.00
$n_i = (dN/dt)_i$	Value (maximum) of n at cumulative inflection point; value (maximum) of slope of cumulative curve at cumulative inflection point	39.89
t_c	Values of t at density inflection points	0.282 0.282
N_c	Values of N at density inflection points	15.87 84.13
$n_c = (dn/dt)_c$	Values of n at density inflection points; values of slope of cumulative curve at density inflection points	24.19
$(dn/dt)_c = (d^2 N/dt^2)_c$	Values (maxima) of slope of density curve at density inflection points; values (maxima) of curvature of cumulative curve at density inflection points	24.19
$\bar{t}$ σ	Mean value of density distribution Standard deviation of density distribution; measured from mean value origin	1.282 1.000

The cumulative and density distributions of the normal probability function are shown in Fig. 2.5.1. The dots identify the critical points. Definitions and numerical values of the various features are listed in Table 2.5.1.

2.5.2 Relationship Between the Normal Probability Function and the Error Function

An important mathematical function, closely related to the normal probability distribution, is the so-called error function. It has the following definition

$$\text{erf}\, T = \frac{2}{\sqrt{\pi}} \int_0^T e^{-z^2} dz \quad . \tag{2.5.13}$$

Comparing this expression to Eq. 2.5.2 we establish that

$$\text{erf}\, T = 2U(T\sqrt{2}) - 1 \tag{2.5.14}$$

or inversely

$$U(T) = \tfrac{1}{2}\left\{1 + \text{erf}\left(\frac{1}{\sqrt{2}}T\right)\right\} \quad . \tag{2.5.15}$$

The error function appears in many problems in the theory of heat conduction and molecular diffusion. These problems are examined extensively by Carslaw and Jaeger (1959) and Crank (1975). We shall see a lot of the error function when we get to Chapter 7.

2.5.3 Approximate Expressions for the Normal Function and Its Inverse

Unlike the relatively simple and easily manipulated expression for the logistic distribution, e.g., Eq. 2.2.19, the corresponding expression for the normal probability distribution, Eq. 2.5.3, is relatively complicated and awkward, at least for some kinds of analysis and computation. This is primarily because it is not possible to obtain an exact explicit equation for the inverse function, $t = t(N)$, as we can easily do in the case of the logistic.

However, the following relatively simple and surprisingly accurate approximation for the normal function is given by Abramowitz and Stegun (1965)

$$U(T) = \tfrac{1}{2} + \tfrac{1}{2}\left(1 - e^{-2T^2/\pi}\right)^{1/2} \quad . \tag{2.5.16}$$

From this result, the following inverse relationship is obtained

$$T(U) = \left(\frac{\pi}{2}\log_e \frac{1}{4U(1-U)}\right)^{1/2} \quad . \tag{2.5.17}$$

In Table 2.5.2, comparisons are made of exact and approximate values of $U(T)$ and $T(U)$ utilizing these two equations.

Table 2.5.2 Comparison of exact and approximate values of the normal probability function and its inverse

T	$U(T)$ approx	$U(T)$ exact	U	$T(U)$ approx	$T(U)$ exact
0.0	0.5000	0.5000	0.5	0.0000	0.0000
0.5	0.6918	0.6915	0.6	0.2532	0.2535
1.0	0.8431	0.8413	0.7	0.5233	0.5242
1.5	0.9363	0.9332	0.8	0.8373	0.8413
2.0	0.9800	0.9772	0.9	1.2668	1.2821
2.5	0.9953	0.9938	1.0	∞	∞

2.5.4 Comparison of the Logistic and Normal Probability Distributions

Quite frequently, in articles concerning growth phenomena, innovation diffusion and technology transfer, the author of the article quite properly describes a cumulative distribution curve as an S-shaped or sigmoid curve. If the author endeavors to quantify such a description, nearly always the logistic equation is indicated. Likewise, if it is mentioned at all, the density distribution curve is generally described as a "bell-shaped" curve; in this case, the author usually makes reference to the normal probability distribution. Thus, in many instances the logistic equation is presented as the framework to express, for example, the cumulative number of adopters of an innovation, followed by a categorization of types of adopters based on the density distribution of the normal equation.

The fact that there is not always entire consistency in these presentations is not of paramount importance. What is important to mention is that many authors have utilized both of these two well-known distribution functions to describe growth and transfer processes. Indeed, numerous investigators have introduced several other functions, with similar features and characteristics, for these purposes. A number of these other functions are examined in Chapter 3.

Covering a span of many years, the following works are notable examples of endeavors along these lines: Winsor (1932b), Feller (1940), Berkson (1944), Becking (1948) and Prentice (1976). In a classic paper on symmetric growth curves, Winsor compared the logistic, normal probability, arctangent, and Pearson Type VII distributions. In an early publication concerning experimental verifications in biology, Feller compared the logistic, arctangent exponential, and error functions. Berkson made a comparison of "logits" (logistic units) and "probits" (normal probability units). Becking carried out a comprehensive analysis of the logistic and normal distributions. Prentice devised a generalized parametric model which reduces to the logistic distribution, the normal probability distribution, and several other distributions for specified values of various parameters. We examine his generalized equation in Section 3.3.

Our comparison of the logistic and normal probability distributions commences with the definitions

$$U = \frac{N}{N_*} \quad ; \quad u = \frac{dU}{dT} \quad ; \quad T = \left\{ \begin{matrix} a \\ a_n \end{matrix} \right\} (t - t_i) \quad , \qquad (2.5.18)$$

in which the upper quantity in the expression for T refers to the logistic equation and the lower to the normal equation. Accordingly, we have the following equations

$$\underset{\text{Logistic}}{} \qquad \underset{\text{Normal}}{}$$

$$U = \frac{1}{1+e^{-T}} \quad ; \quad U = \frac{1}{\sqrt{2\pi}} \int_{-\infty}^{T} e^{-\frac{1}{2}z^2} dz \qquad (2.5.19)$$

and

$$u = \frac{e^{-T}}{(1+e^{-T})^2} \quad ; \quad u = \frac{1}{\sqrt{2\pi}} e^{-\frac{1}{2}T^2} \qquad (2.5.20)$$

The two expressions of Eq. 2.5.20, provide "differential equations" for the two distribution functions in the form: $dU/dT = f(T)$.

The logistic distribution can be expressed in terms of a "differential equation" of the type $dU/dT = g(U)$. From Eq. 2.2.1 we have

$$\frac{dU}{dT} = U(1-U) \quad ; \quad \frac{1}{U}\frac{dU}{dT} = (1-U) \qquad (2.5.21)$$

which describe the growth rate and specific growth rate, respectively.

It is not possible to write an exact equation of the form $dU/dT = g(U)$, for the normal probability equation. An approximate relationship, can be obtained from Eq. 2.5.17. The result is

$$\frac{dU}{dT} = \left(\frac{2^{\pi-1}}{\pi}\right)^{1/2} [U(1-U)]^{\pi/4} \quad . \qquad (2.5.22)$$

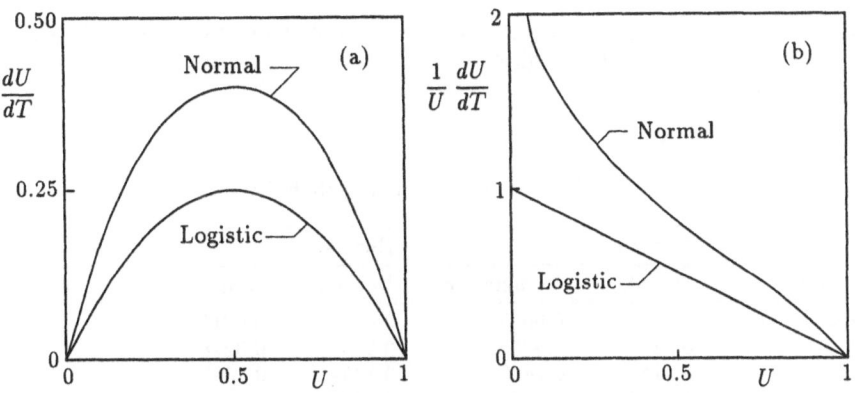

Fig. 2.5.2 Comparison of the logistic and normal probability distributions; (a) growth rates and (b) specific growth rates

Plots of the growth rates and specific growth rates for the logistic and normal probability functions are given in Fig. 2.5.2. Curves showing the cumulative and density distributions are presented in Fig. 2.5.3. In these displays, the growth coefficients a and a_n are numerically equal.

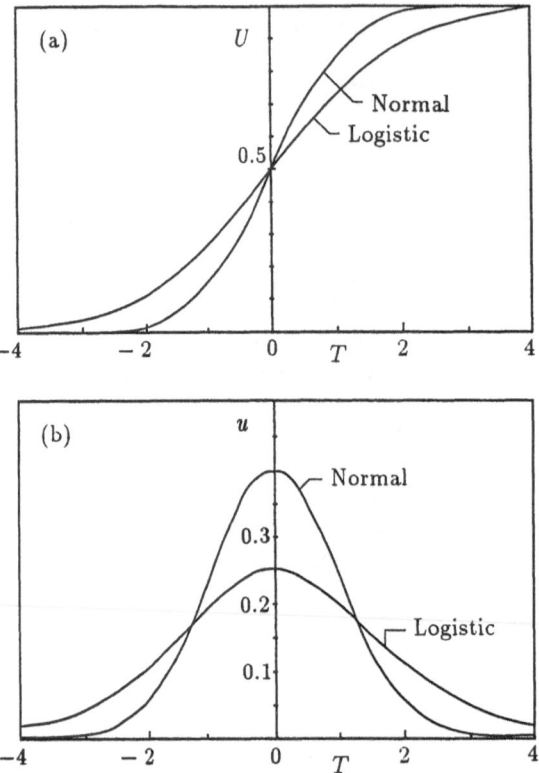

Fig. 2.5.3 Comparison of the logistic and normal probability distributions for equal values of growth or transfer coefficients, a and a_n; (a) cumulative distributions and (b) density distributions

Table 2.5.3 Comparison of logistic and normal probability distributions

Interval	Logistic		Normal	
	Density	Cumulative	Density	Cumulative
$-\infty$ to -3σ	0.0043	0.0043	0.0013	0.0013
-3σ to -2σ	0.0216	0.0259	0.0215	0.0228
-2σ to $-\sigma$	0.1143	0.1402	0.1359	0.1587
$-\sigma$ to 0	0.3598	0.5000	0.3413	0.5000
0 to σ	0.3598	0.8598	0.3413	0.8413
σ to 2σ	0.1143	0.9741	0.1359	0.9772
2σ to 3σ	0.0216	0.9957	0.0215	0.9987
3σ to ∞	0.0043	1.0000	0.0013	1.0000

From the preceding analysis, for the normal distribution the standard deviation $\sigma = 1$. For the logistic distribution the standard deviation $\sigma = \pi/\sqrt{3}$. In Table 2.5.3, a comparison is made of areas under the density distribution curves and the ordinates of the cumulative distribution curves at integer intervals of the respective standard deviations.

Another comparison is presented in Table 2.5.4 which shows the values of the dimensionless time, T, for the two distributions at various values of U.

Table 2.5.4 Comparison of values of dimensionless time T for the logistic and normal distributions at various values of U

U	Logistic	Normal
0.01	−4.595	−2.327
0.05	−2.944	−1.645
0.10	−2.197	−1.282
0.25	−1.099	−0.674
0.50	0	0
0.75	1.099	0.674
0.90	2.197	1.282
0.95	2.944	1.645
0.99	4.595	2.327

In Figs. 2.5.3, the fact that the two growth coefficients, a and a_n, are numerically equal camouflages the close similarity of the logistic and normal distributions. To make this similarity more apparent, we now require that the slopes of the two cumulative curves be the same at $\tau = (t - t_i) = 0$. This is equivalent to equating the magnitudes of the two density curves at $\tau = 0$. Accordingly, from Eqs. 2.5.20

$$\frac{dU}{d\tau} = \frac{ae^{-a\tau}}{(1 + e^{-a\tau})^2} \tag{2.5.23}$$

for the logistic distribution and

$$\frac{dU}{d\tau} = \frac{a_n}{\sqrt{2\pi}} e^{-\frac{1}{2}(a_n \tau)^2} \tag{2.5.24}$$

for the normal. Substituting $\tau = 0$ in both equations and equating, we obtain

$$\frac{a_n}{a} = \frac{1}{2}\sqrt{\frac{\pi}{2}} \cdot \tag{2.5.25}$$

On this basis, Eqs. 2.5.19 and 2.5.20 take on the forms

Logistic   Normal

$$U = \frac{1}{1 + e^{-T}} \quad ; \quad U = \frac{1}{\sqrt{2\pi}} \int_{-\infty}^{(\sqrt{\pi/2})T/2} e^{-\frac{1}{2}z^2} dz \qquad (2.5.26)$$

$$U = \frac{e^{-T}}{(1 + e^{-T})^2} \quad ; \quad U = \tfrac{1}{4}e^{-\pi T^2/16} \quad . \qquad (2.5.27)$$

These relationships are shown in Fig. 2.5.4. It is seen that the growth coefficient scaling relationship of Eq. 2.5.25 produces a result which brings out the very close similarity of the two distributions. It also signals the need to be cautious when comparing numerical values of growth and transfer coefficients computed from different distribution functions.

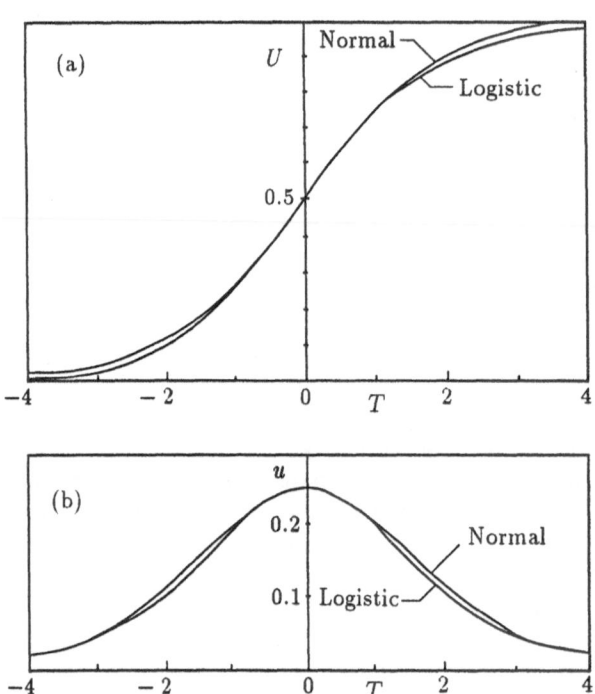

Fig. 2.5.4 Comparison of the logistic and normal probability distributions with equal slopes of the cumulative curve at $T = 0$; (a) cumulative distributions and (b) density distributions

In a comparison of the logistic and the normal probability distributions, it is appropriate to introduce the terms: "logit" and "probit".

Logit. The term "logit" (logistic unit) was introduced by Berkson (1944) to provide a convenient expression for the inversion of the logistic equation. We write

$$U = \Psi[a(t - t_i)] \quad , \tag{2.5.28}$$

where, for the logistic distribution

$$\Psi = \frac{1}{1 + e^{-a(t - t_i)}} \quad . \tag{2.5.29}$$

Then, by definition, the logit is

$$\text{logit } U = \Psi^{-1}(U) = a(t - t_i) \quad . \tag{2.5.30}$$

In the case of the logistic distribution, the inversion is mathematically easy. From the previous two equations we get

$$\text{logit } U = \log_e \left(\frac{U}{1 - U} \right) = a(t - t_i) \quad . \tag{2.5.31}$$

Thus, the logit transformation provides a linear relationship between U and t. It also stretches the range of the dependent variable U from $(0, 1)$ to $(-\infty, \infty)$. Finally, it offers an interesting ratio of "success-to-failure", $U/(1 - U)$.

Probit. The term "probit" (probability unit) was invented before the "logit"; it was introduced by Bliss (1935). For this quantity we write

$$U = \Phi[a_n(t - t_i)] \quad , \tag{2.5.32}$$

in which, for the normal probability distribution

$$\Phi = \frac{1}{\sqrt{2\pi}} \int_{-\infty}^{a_n(t - t_i)} e^{-\frac{1}{2}z^2} \, dz \quad . \tag{2.5.33}$$

Hence, the definition of the probit is

$$\text{probit } U = \Phi^{-1}(U) = a_n(t - t_i) \quad . \tag{2.5.34}$$

Unlike the case of the logistic equation, it is not possible to analytically express the inverse of the normal probability equation in simple terms. As we saw, Eq 2.5.17 provides a reasonably accurate approximation. The inverse is tabulated by Abramowitz and Stegun (1965) and by Brandt (1976).

By whatever means the inversion is obtained, the probit transformation provides a linear relationship between U and t. Indeed a direct plotting of time data, or whatever the independent variable may be, versus U on probability-scaled graph paper yields a straight line. The probits on such a plot consist of intervals of constant length measured along the probability axis.

The subjects of probits and logits have been examined by numerous investigators. The books by Finney (1971) and by Ashton (1972) provide comprehensive coverage of the matter.

To conclude the comparison of the logistic and normal equations, a summary of the main definitions and features of the two distributions is presented in Table 2.5.5.

Table 2.5.5 Comparison of the logistic and normal probability distributions. $U = N/N_*$; $u = dU/dT$; $T = \left\{ {a \atop a_n} \right\}(t - t_i)$

Distribution and feature	Logistic (coefficient a)	Normal (coefficient a_n)
Cumulative	$U = \dfrac{1}{1 + e^{-T}}$ or $U = \frac{1}{2}(1 + \tanh\dfrac{T}{2})$	$U = \dfrac{1}{\sqrt{2\pi}} \displaystyle\int_{-\infty}^{T} e^{-\frac{1}{2}z^2} dz$
Density	$u = \dfrac{e^{-T}}{(1 + e^{-T})^2}$ or $u = \frac{1}{4}\operatorname{sech}^2\dfrac{T}{2}$	$u = \dfrac{1}{\sqrt{2\pi}}e^{-\frac{1}{2}T^2}$
$\dfrac{dU}{dT} = f(T)$ $\dfrac{dU}{dT} = g(U)$	$\dfrac{dU}{dT} = \dfrac{e^{-T}}{(1 + e^{-T})^2}$ $\dfrac{dU}{dT} = U(1 - U)$	$\dfrac{dU}{dT} = \dfrac{1}{\sqrt{2\pi}}e^{-\frac{1}{2}T^2}$ not possible; see Eq. 2.5.22 for approximation
Inflection point of cumulative distribution	$T_i = 0; \quad U_i = \frac{1}{2}$ $U_i = \left(\dfrac{dU}{dT}\right)_i = \frac{1}{4}$	$T_i = 0; \quad U_i = \frac{1}{2}$ $u_i = \left(\dfrac{dU}{dT}\right)_i = \dfrac{1}{\sqrt{2\pi}}$
Inflection point of density distribution	$T_c = \log_e(2 \pm \sqrt{3})$ or $T_c = \pm 2\operatorname{arcsinh}\dfrac{1}{\sqrt{2}}$ $U_c = \frac{1}{2}(1 \pm \frac{1}{\sqrt{3}})$ $u_c = \left(\dfrac{dU}{dT}\right)_c = \frac{1}{6}$ $\left(\dfrac{du}{dT}\right)_c = \pm\dfrac{1}{6\sqrt{3}}$	$T_c = \pm 1$ $U_c = \dfrac{1}{\sqrt{2\pi}} \displaystyle\int_{-\infty}^{\pm 1} e^{-\frac{1}{2}z^2} dz$ $u_c = \left(\dfrac{dU}{dT}\right)_c = \dfrac{1}{\sqrt{2\pi e}}$ $\left(\dfrac{du}{dT}\right)_c = \pm\dfrac{1}{\sqrt{2\pi e}}$
Mean value Standard deviation	$\overline{T} = T_i = 0$ $\sigma = \dfrac{\pi}{\sqrt{3}}$ $\operatorname{logit}U = \Psi^{-1}(U) = T$ $\operatorname{logit}U = \log_e\left(\dfrac{U}{1-U}\right)$	$\overline{T} = T_i = 0$ $\sigma = 1$ $\operatorname{probit}U = \Phi^{-1}(U) = T$ not possible; see Eq. 2.5.17 for approximation

2.5.5 An Illustration: Adoption of Herbicides by Mexican Barley Farmers

Over the years, studies have been carried out at the International Maize and Wheat Improvement Center (CIMMYT) in Mexico on the rate and sequence of adoption of improved cereal technologies. One such study, reported by Byerlee and Hesse (1982), dealt with the adoption rates of improved varieties of seeds, herbicides, and fertilizers, as well as the adoption rates of various types of mechanical components. The field research was carried out, commencing in 1975, in wet and dry barley producing zones in central Mexico.

To illustrate the preceding analysis, the CIMMYT data on the adoptions of herbicides in the wet zone are presented in Table 2.5.6. The authors based their study on the assumption that $N_* = 100$ percent. The table includes the numerical values of the logits and probits needed to determine the growth coefficients and inflection point times. The data are displayed in Fig. 2.5.5.

Table 2.5.6 Adoption of herbicides by barley farmers in Central Mexico. Year 1966 corresponds to $t = 0$. From Byerlee and Hesse (1982)

Year	t [years]	$U = N/N_*$	$U/(1-U)$	logit U	probit U
1966.1	0.1	0.09	0.099	−2.314	−1.341
1968.4	2.4	0.12	0.136	−1.992	−1.175
1969.1	3.1	0.33	0.493	−0.908	−0.439
1970.6	4.6	0.38	0.613	−0.490	−0.305
1972.5	6.5	0.40	0.667	−0.405	−0.253
1974.5	8.5	0.59	1.439	0.346	0.227
1975.1	9.1	0.64	1.776	0.574	0.358
1976.1	10.1	0.71	2.451	0.896	0.553
1977.0	11.0	0.80	4.000	1.386	0.878

Least squares analyses of the data shown in Table 2.5.6 give the following results

	Logistic	Normal
Correlation coefficient	$r = 0.9699$	$r = 0.9721$
Transfer coefficient	$a = 0.320$	$a_n = 0.192$
Inflection point time	$t_i = 7.08$	$t_i = 7.03$

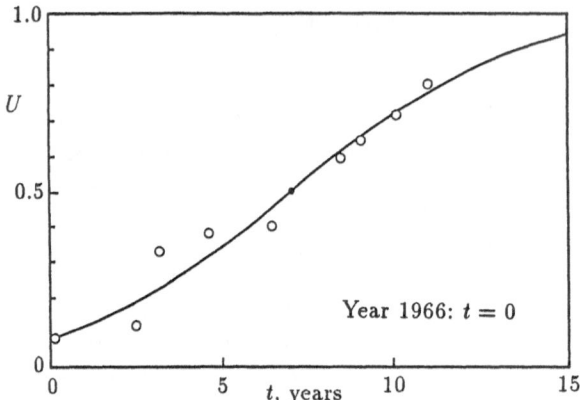

Fig. 2.5.5 Adoption of herbicides by barley farmers in Central Mexico. Data of Byerlee and Hesse (1982)

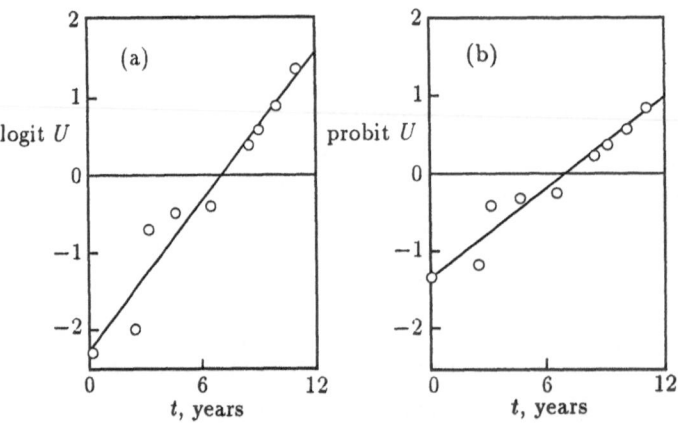

Fig. 2.5.6 Plot for determination of transfer coefficients, a and a_n, and inflection point time, t_i. (a) logit U vs. t and (b) probit U vs. t

The solid lines shown in Figs. 2.5.6 (a) and (b) are based on the above values of a, a_n, and t_i. We note that the two different definitions and values of the transfer coefficients, a and a_n, describe the data equally well. We also note that the ratio, $a_n/a = 0.192/0.320 = 0.600$, is close to the value given by Eq. 2.5.25, viz, $(\sqrt{\pi/2})/2 = 0.627$. As Eqs. 2.5.30 and 2.5.34 indicate, the values of a and a_n are established by the slopes of logit U and probit U versus t, respectively. So one would expect the measured ratio a_n/a, to be approximately equal to the computed ratio.

For the logistic distribution, the initial value, N_0, is computed from Eq. 2.2.17; the result: $N_0 = 0.094$. For the normal distribution, Eq. 2.5.12 provides the relationship to calculate N_0; in this case, $N_0 = 0.089$.

The cumulative distribution curves $U = U(t)$ were computed using Eqs. 2.5.19 and the above indicated numerical values. The solid curve shown in Fig. 2.5.5 is based on the normal probability function. The curve computed from the logistic equation is almost indistinguishable from that produced by the normal equation.

2.6 Power Law Logistic Distribution

2.6.1 The Differential Equation and Its Features

We have seen in our examinations of the logistic and normal probability distributions that the inflection point of the cumulative curve always occurs at one-half the asymptotic or equilibrium value. This is certainly the case in many actual growth and transfer phenomena; the examples given in Sections 2.2 and 2.5 illustrate this point.

However, in general, a growth curve may inflect at quite a different value of N. For example, in our consideration of the Gompertz distribution of Section 3.1, we shall see that the inflection point occurs at $N_i = N_*/e$. Likewise, in Sections 3.2 and 3.3 it will be seen that inflection points and, more importantly, the shape of the growth curves themselves, of the Weibull and extreme value distributions, are substantially different from those of the normal probability and logistic equations.

In the present section, we consider a functional form of $N(t)$ which gives us more flexibility, in specifying the shape of the growth curve. We term this function a "power law logistic" distribution because, as we shall see, the quantity, N/N_*, in the growth rate equation is raised to a power; that is, it carries an exponent, s.

The analysis begins with the following equation proposed by Bertalanffy (1968) for describing animal metabolism and growth

$$\frac{dN}{dt} = aN^\xi - bN^\eta \quad ,$$

(2.6.1)

where a, b, ξ, and η are constants. In Section 3.4, a general solution to this equation is presented. For the present, we set $\xi = 1$, and write Eq. 2.6.1 in the form

$$\frac{dN}{dt} = aN \left[1 - \left(\frac{N}{N_*} \right)^s \right] \quad ,$$

(2.6.2)

where s is a constant. Clearly, if $s = 1$ the differential equation for the ordinary logistic is recovered.

This expression has been studied by numerous investigators. Gilpin and Ayala (1973) emphasize the important feature of this equation that allows non-symmetrical growth patterns with maximum growth rates occurring at values greater or less than $N_*/2$. Turner et al. (1976) present a growth equation of which Eq. 2.6.2 is a special case. We examine relationships similar to their more general expression in Section 3.4.

As before, if $U = N/N_*$ and $T = at$, Eq. 2.6.2 becomes

$$\frac{dU}{dT} = U(1 - U^s) \qquad (2.6.3)$$

and

$$\frac{1}{U}\frac{dU}{dT} = (1 - U^s) \quad . \qquad (2.6.4)$$

Plots of Eqs. 2.6.3 and 2.6.4 are shown in Fig. 2.6.1. For the case, $s = 1$, the symmetrical parabola for the growth rate and straight line for the specific growth rate are noted.

It is observed in Fig. 2.6.1(a) that as s increases, the maximum growth rate occurs at increasingly larger values of U. As we establish in Eq. 2.6.12, at these maximum points

$$U_i = \frac{1}{(1+s)^{1/s}} \quad ; \qquad \left(\frac{dU}{dT}\right)_i = \frac{s}{(1+s)^{(1+s)/s}} \quad . \qquad (2.6.5)$$

We can show that when $s = 0$, the maximum growth rate occurs at $U_i = 1/e$.

In Fig. 2.6.1(b) it is noted that when $s \gg 1$, the specific growth rate is almost unity over a wide range of U; this means that growth is rapid (in fact, exponential) during most of the growth period and then quickly slows down. On the other hand, if $s \ll 1$, the specific growth rate decreases rapidly with increasing U and then takes on a nearly constant small value.

2.6.2 The Growth Curve and Its Properties

We rewrite Eq. 2.6.2 in the form

$$\frac{dN}{dt} = aN - \frac{a}{N_*^s}N^{s+1} \quad . \qquad (2.6.6)$$

This expression is a simple case of the Bernoulli differential equation of the generalized form

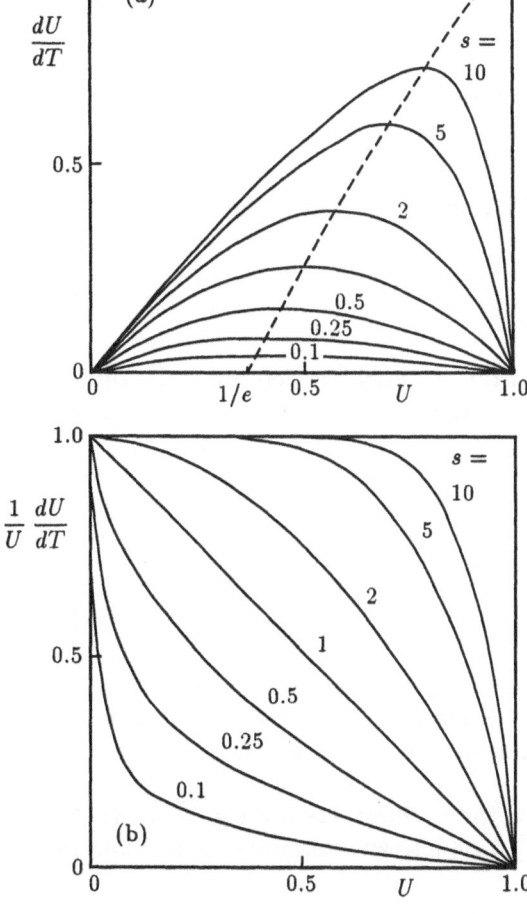

Fig. 2.6.1 Power law logistic distribution; (a) growth rate and (b) specific growth rate for various values of s

$$\frac{dy}{dx} = f(x)y + g(x)y^p \quad . \tag{2.6.7}$$

The first step in solving this type of non-linear differential equation is to convert it to a linear equation with the substitution

$$N = C^{-1/s} \quad . \tag{2.6.8}$$

Accordingly, Eq. 2.6.6 becomes

$$\frac{dC}{dt} + asC = \frac{as}{N_*^s} \tag{2.6.9}$$

which is the same as Eq. 2.2.11. So the solution, using Eq. 2.2.12 and the initial condition $N(0) = N_0$, is

$$N = \frac{N_*}{\left\{1 + \left[\left(\frac{N_*}{N_0}\right)^s - 1\right] e^{-ast}\right\}^{1/s}} \, . \tag{2.6.10}$$

As we would expect, when $s = 1$ this answer reduces to the ordinary logistic. The first derivative of Eq. 2.6.10 gives the density function

$$n = am N_* \frac{e^{-ast}}{(1 + me^{-ast})^{(1+s)/s}} \tag{2.6.11}$$

where $m = (N_*/N_0)^s - 1$. From this equation we determine the time corresponding to the inflection point of the cumulative curve and the maximum point of the density curve

$$t_i = \frac{1}{as} \log_e(m/s) \quad ; \qquad N_i = \frac{N_*}{(1 + s)^{1/s}} \quad ;$$

$$n_i = \left(\frac{dN}{dt}\right)_i = \frac{as N_*}{(1 + s)^{(1+s)/s}} \, . \tag{2.6.12}$$

We note that if $s = 1$, then $N_i = N_*/2$. However, if $s = 2$, for example, then $N_i = 0.577 N_*$ and if $s = 0.5$, then $N_i = 0.444 N_*$. So, by bringing the parameter s into the analysis, the inflection point need not necessarily occur at the 50 percent point.

As we did in Section 2.2 for the ordinary logistic equation, we put the expression for N, given by Eq. 2.6.10, into an alternative form. Some algebra yields

$$N = \frac{N_*}{2^{1/s}} \left(1 + \tanh\frac{\phi}{2}\right)^{1/s} \tag{2.6.13}$$

where $\phi = as(t - t_i) - \log_e s$. Differentiating Eq. 2.6.13 gives

$$n = \frac{a N_*}{2^{(1+s)/s}} \frac{\operatorname{sech}^2(\phi/2)}{[1 + \tanh(\phi/2)]^{(s-1)/s}} \, . \tag{2.6.14}$$

These two expressions are the generalized hyperbolic forms of Eqs. 2.2.20 and 2.2.21 corresponding to the ordinary logistic ($s = 1$) case.

The results of a numerical example are presented in Fig. 2.6.2. The cumulative distribution is shown in Fig. 2.6.2(a). The dashed line indicates the locus of the inflection points. From the relationships of Eq. 2.6.12, we determine that if $N_*/N_0 \gg 1$, the minimum value of t_i occurs when $s = e$.

For $s \to \infty$, the growth is exponential. As $s \to 0$, the asymptotic value of $N_i = N_*/e$.

The density distribution is displayed in Fig. 2.6.2(b) for the same range of s. The density function for the ordinary logistic ($s = 1$) is symmetrical about the cumulative inflection point. The curves are negatively skewed for $s > 1$ and positively skewed for $s < 1$. We examine "skewness" in some detail in Section 3.3 and 3.5.

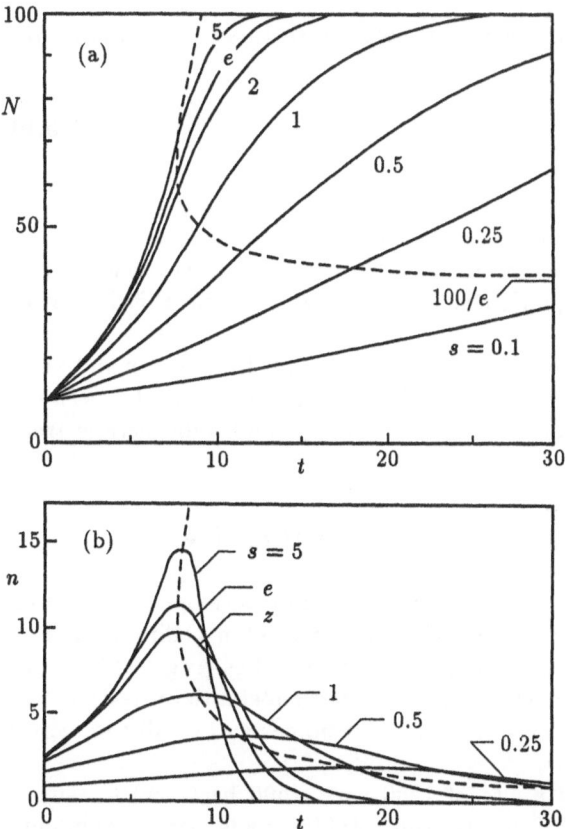

Fig. 2.6.2 Numerical example of power law logistic. (a) Cumulative distribution and (b) density distribution. $a = 0.25$, $N_0 = 10$, $N_* = 100$.

2.6.3 The Richards Function

Closely related to the growth function defined by the differential equation, Eq. 2.6.2, and its solution, Eq. 2.6.10, is the so-called Richards function. This function is given defined by the expression

$$\frac{dN}{dt} = \frac{a'}{s}N\left[1 - \left(\frac{N}{N_*}\right)^s\right] \quad , \tag{2.6.15}$$

where a'/s replaces the quantity a. The solution to this equation is termed the Richards function. That is

$$N = \frac{N_*}{(1 + me^{-a't})^{1/s}} \quad , \tag{2.6.16}$$

where, as before, $m = (N_*/N_0)^s - 1$.

Several special cases ensue from this equation. If $s = -1$, we acquire the confined exponential equation and if $s = 1$, the ordinary logistic is generated. If $s = 0$, the Gompertz equation is obtained; we examine this important function in Section 3 .1.

The Richards function has been utilized extensively by Causton and Venus (1981) in connection with the biometry of plant growth. It has also been employed by Usher (1980) in his work concerning optimal radiotherapy treatment of cancer.

2.6.4 An Illustration: Sale of Development Property

In the early 1980s, there was a vigorous program of development of an extensive arid region in the south central part of Southern California due, in large measure, to the continual expansion of greater Los Angeles. Large areas of land were acquired by developers and, after substantial investments in infrastructure, were offered for sale as sites for homes and small commercial enterprises.

One developer prepared a site of 1250 acres (506 hectares) for such a purpose. He utilized about 100 acres of this total for roads, parks, community club house and grounds, and the like. After considerable advertising and intensive promotion, the entire property was put on sale in parcels of land ranging from one to three acres.

The subsequent schedule of weekly sales is summarized in Table 2.6.1. The "n" column (density distribution) is the record of sales in acres per week; the "N" column (cumulative distribution) indicates the total number of acres sold including the original 100 acres of infrastructure. A display of the information listed in Table 2.6.1 is shown in Fig. 2.6.3.

Table 2.6.1 Sale of development property. $N_0 = 100$; $N_* = 1250$

t [weeks]	n [acres/week]	N [acres]	t [weeks]	n [acres/week]	N [acres]
0	0	100	8	85	495
1	27	127	9	104	599
2	22	149	10	145	744
3	28	177	11	139	883
4	55	232	12	141	1024
5	40	272	13	132	1156
6	50	322	14	58	1214
7	88	410	15	19	1233

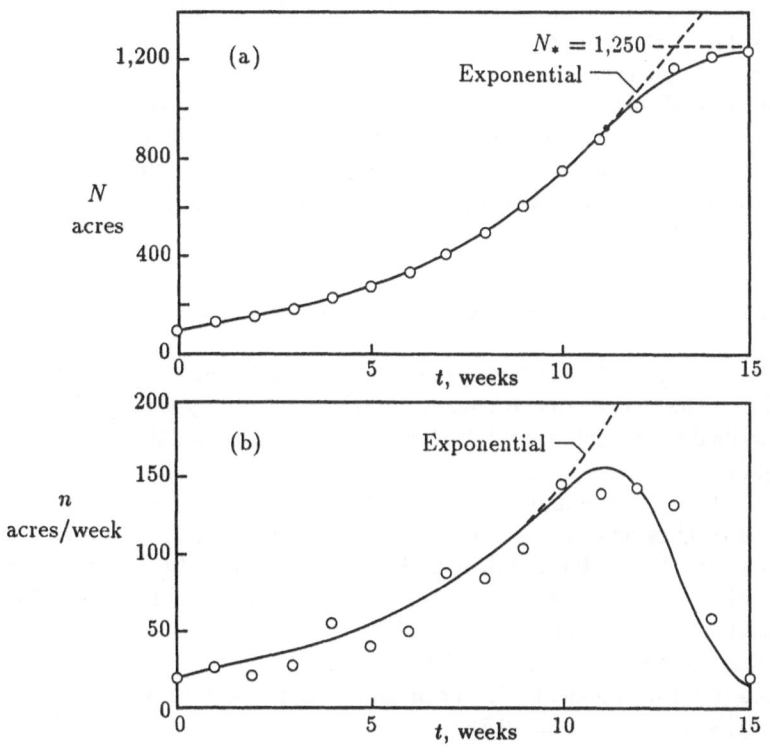

Fig. 2.6.3 Sale of development property showing (a) cumulative number of acres sold and (b) number of acres sold each week

From the appearance of the plotted points of Fig. 2.6.3(a), we suspect that the curve is not well described by an ordinary logistic ($s = 1$). The ordinate of the inflection point, N_i, appears to be somewhat larger than

$N_*/2$; accordingly, with reference to Fig. 2.6.2(a), s is evidently greater than unity.

With regard to Eq. 2.6.10, with $N_* = 1250$, $N_0 = 100$, and $s > 1$, the quantity, $(N_*/N_0)^s \gg 1$. So, for small values of t, Eq. 2.6.10 reduces to the simple exponential function, $N = N_0 \exp(at)$. A least squares computation of $\log_e N$ versus t, for the first 12 points of the table yields: $a = 0.198\,\text{week}^{-1}$ and $N_0 = 101.2$ acres.

There remains the task of determining the value of s. A reasonably accurate approximation is obtained by estimating, from Fig. 2.6.3(a), that the value of N at the inflection point is $N_i = 900$. From Eq. 2.6.12, a trial and error solution gives $s = 5.85$. Substituting this value of s, along with $a = 0.198$, $N_0 = 101.2$, and $N_* = 1250$, into Eq. 2.6.10, we compute the values of N for each of the 16 points. Subsequently, we determine, for each point, the difference between the observed value, N_{obs}, and the computed value, N_{com}. From this we calculate the quantity $\sum(N_{obs} - N_{com})^2$. We stipulate that the minimum value of this quantity is the criterion for the correct value of s. A few iterations of this type yields $s = 6.20$.

Substitution of these numerical values into Eqs. 2.6.12 gives $t_i = 11.21$ weeks, $N_i = 909.7$ acres, and $n_i = (dN/dt)_i = 155.1$ acres/week. Finally, from Eqs. 2.6.10 and 2.6.11 we determine the smooth curves for $N(t)$ and $n(t)$ shown in Fig. 2.6.3. For comparison, curves corresponding to an exponential rate of sales are shown.

2.6.5 An Illustration: Growth of Pine Trees in New Zealand

In his study of the growth of radiata pine trees in New Zealand, Garcia (1983) concluded that the Bertalanffy–Richards model, expressed by Eqs. 2.6.15 and 2.6.16, described the height–age relationship of the trees better than did other growth equations.

Two sets of data were examined by Garcia. One set consisted of 91 sample plots with a total of 543 measurements of radiata pine in Kaingaroa Forest in New Zealand. These data, shown in Fig 2.6.4(a), are labeled "range of measurements". The other set was comprised of 58 sample plots with 247 measurements of radiata pine in Southland Conservancy also in New Zealand. These data are presented in Fig. 2.6.4(b).

The values of the exponent s for both sets of Garcia's data were negative (respectively, $s = -0.621$, $s = -0.783$).

Rewriting Eq. 2.6.16 with $c = -s$ gives

$$N = N_* \left\{ 1 + \left[\left(\frac{N_0}{N_*} \right)^c - 1 \right] e^{-a't} \right\}^{1/c} . \tag{2.6.17}$$

From Eqs. 2.6.12, with $a' = as$ and $s = -c$, the inflection point coordinates are

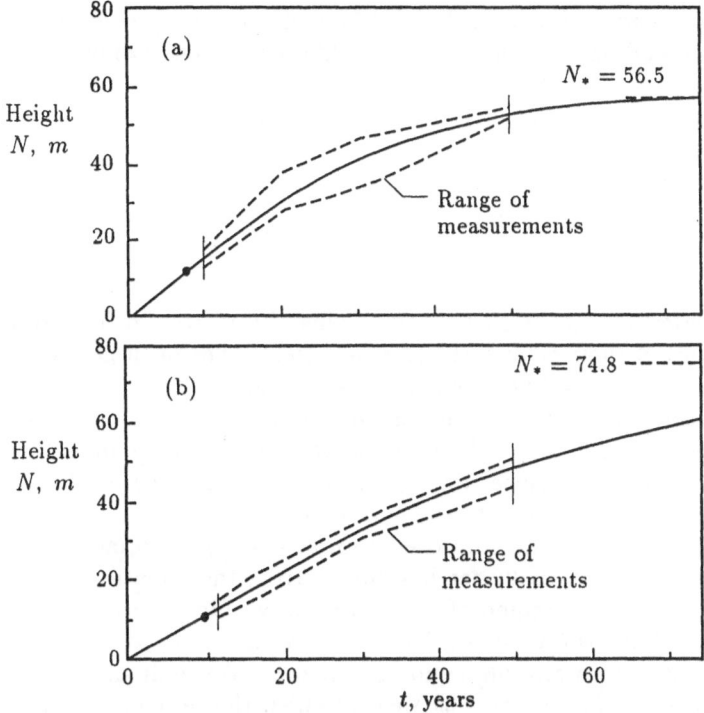

Fig. 2.6.4 Heights and ages of radiata pine trees. (a) Kaingaroa Forest, New Zealand and (b) Southland Conservancy, New Zealand. Data of Garcia (1983)

$$t_i = \frac{1}{a'} \log_e \left\{ \frac{1}{c} \left[1 - \left(\frac{N_0}{N_*} \right)^c \right] \right\}$$

$$N_i = N_*(1-c)^{1/c} \quad ; \qquad \left(\frac{dN}{dt} \right)_i = a' N_*(1-c)^{(1-c)/c} \quad . \qquad (2.6.18)$$

From the field measurements and analysis of Garcia, and from the above equations, the following numerical values are obtained

	Fig. 2.6.4(a)	Fig. 2.6.4(b)
N_*	56.5 m	74.8 m
N_0	0.0	0.0
a	0.057	0.025
c	0.621	0.783
t_i	8.5 yrs	9.8 yrs
$(dN/dt)_i$	1.78 m/yr	1.22 m/y
N_i	11.8 m	10.6 m

The calculated curves, based on Eq. 2.6.17 and these numerical values, are shown as the solid lines in Fig. 2.6.4. The inflection points are indicated by the solid dots.

It is apparent that this model provides an adequate description of the growth of the pine trees.

2.6.6 An Illustration: Adoption of Hybrid Corn in the United States

In midwest America during the 1930s, new strains of corn were being developed and, because of their greater stamina and yields over open-pollinated varieties, began to be utilized by farmers in the region.

This hybrid corn innovation commenced in the heart of the Corn Belt – the states of Iowa and Illinois – where its adoption was slow at first, then more rapid, and finally, in typical logistic growth fashion, leveled off after virtually all farmers had switched to the new variety.

Concurrently, the use of this improved corn strain began to increase in neighboring states and eventually far beyond. Clearly, the adoption of this innovation was not only a "temporal" phenomenon within a particular geographical setting but also a "spatial" one in that the innovation spread throughout the entire corn-growing region of America. We shall analyze the spatial aspects later on. For the present, we continue with the temporal problem.

An extensive study of this phenomenon of hybrid corn adoption was carried out by Griliches (1957, 1960). He determined the proportions of corn acreage planted with hybrid seed to the total acreage available in 31 states over the period from 1933 to 1958 . One of the states was Alabama; we use that as our illustration.

In his study, Griliches endeavored to establish a relationship between the rate of adoption of this new technology and the "profitability" involved in such adoption. He assumed that all of the hybrid corn adoption patterns followed the trend function described by the ordinary logistic equation. This equation, of course, contains the growth coefficient, a, which Griliches utilized to measure the rate of adoption.

Certainly the logistic equation is the simplest trend function to use and was undoubtedly adequate in most of the cases in the hybrid corn study. However, to improve some of the correlations, Braun (1982) suggested that a combination of the confined exponential equation (to account for the effects of "external influence") and the logistic (effects of "internal influence") might have been a more suitable analytical framework. Along the same lines, Dixon (1980) indicated that the Gompertz equation would have been more appropriate for analysis of some of the data.

For our illustration we utilize the power law logistic defined by the differential equation, Eq. 2.6.2, and its solution, Eq. 2.6.10 The data cited by

Griliches (1960) for Alabama are listed in Table 2.6.2 and displayed in Fig. 2.6.5.

Table **2.6.2** Fraction of corn acreage planted with hybrid seed in Alabama: Years 1944 through 1958. From Griliches (1960)

Year	t	U	Year	t	U
1944	9	0.02	1952	17	0.39
1946	11	0.03	1953	18	0.50
1947	12	0.05	1954	19	0.57
1948	13	0.10	1955	20	0.65
1949	14	0.15	1956	21	0.72
1950	15	0.17	1957	22	0.80
1951	16	0.26	1958	23	0.83

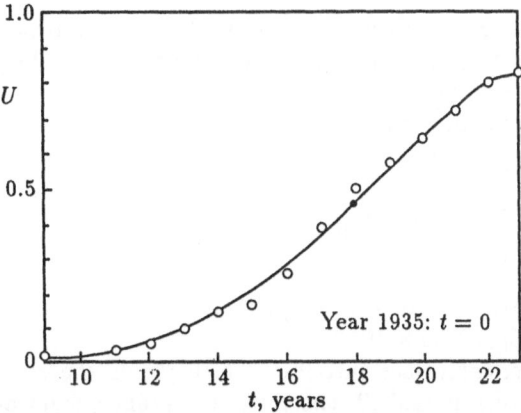

Fig. 2.6.5 Adoption of hybrid corn in Alabama. Data of Griliches (1960)

In our analysis of the data it is necessary to determine the numerical values of the three parameters appearing in Eq. 2.6.10: a, s, and U_0. The simplest and most straightforward approach is to initially utilize the value of the growth coefficient given by Griliches, $a = 0.51$. Then, expressing Eq. 2.6.3 in finite difference form we obtain

$$\log_e \left(1 - \frac{1}{\overline{U}}\frac{\Delta U}{\Delta T}\right) = s \log_e \overline{U} \quad . \tag{2.6.19}$$

Substituting the appropriate numbers provided by Table 2.6.2 into this equation yields the plot shown in Fig. 2.6.6(a). A least squares computation gives $s = 0.616$.

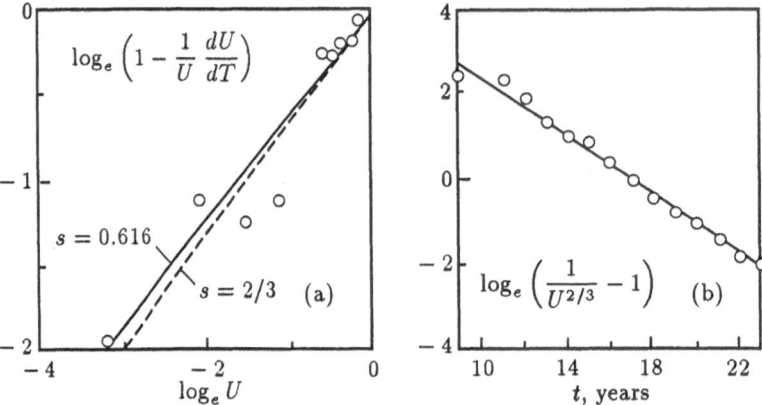

Fig. 2.6.6 Adoption of hybrid corn in Alabama. Plots to determine parameters of power law logistic; (a) value of s and (b) values of a and U_0

For reasons we will take up in Section 7.6, involving both spatial diffusion and logistic growth, we set $s = 2/3$. The line corresponding to this slightly altered values of s is shown in Fig. 2.6.6(a).

Finally, we write Eq. 2.6.10 in the form

$$\log_e \left(\frac{1}{U^{2/3}} - 1 \right) = \log_e \left(\frac{1}{U_0^{2/3}} - 1 \right) - \frac{2}{3} at \quad . \tag{2.6.20}$$

A plot of this relationship is presented in Fig. 2.6.6(b). A least squares calculation gives: $a = 0.522$ and $U_0 = 1.50 \times 10^{-4}$.

Substitution of these numerical values into Eq. 2.6.10 yields the solid curve shown in Fig. 2.6.5. From Eqs. 2.6.12, the inflection point occurs at $t_i = 18.05$, $U_i = 0.465$; the solid dot identifies this point in the figure. The 50 percent time occurs soon after: $t_{50} = 18.40$. On this basis, we compute that by mid-1953, half of all the corn acreage in Alabama had been switched to the hybrid variety. Calculated adoption rose to $U = 0.917$ by 1960, when data collection was discontinued, and to $U = 0.99$ by early 1962.

We return to the main theme of this extensive study by Griliches. He determined the values of the growth coefficient, a, for all 31 states; the values ranged from $a = 1.02$ in Iowa to $a = 0.36$ in Tennessee. He was then able to establish a good correlation between the growth coefficient and the variables: yield per acre (Y) and average number of acres per farm (A) devoted to corn production. With these latter variables reflecting income to the farmers, he thus established a positive relationship between profitability and the rate of adoption.

2.7 Logistic Growth with Migration

2.7.1 Immigration and Emigration

In Section 2.1 we examined the exponential equation and a number of its ramifications. In particular, we considered the phenomenon of exponential growth with constant immigration (stocking) or constant emigration (harvesting).

An appropriate next topic is the problem of logistic growth with constant immigration or emigration. Examples of immigration are the planting of trees in a forest and the movement of people into a geographical area over a period of time. Examples of emigration are the extraction of fish from the ocean and the harvesting of agricultural products.

The subject of logistic growth with migration has been examined by numerous investigators including Brauer and Sanchez (1975), Clark (1976, 1981) and Hallam (1986).

We start with the ordinary logistic equation with migration, study it to some extent, and then consider a problem involving the harvesting of fish from a fish cultivation pond. We conclude the section with an illustration involving an endangered species: the sandhill crane.

2.7.2 Logistic Growth with Constant Stocking

The differential equation for logistic growth with constant immigration (or stocking, depositing, injection) is

$$\frac{dN}{dt} = aN - bN^2 + s \quad , \tag{2.7.1}$$

where s is the stocking rate expressed in items of N per unit time. It is easy to establish that this equation is essentially the same as Eq. 2.4.16 which describes the linear combination of logistic growth and confined exponential growth.

Letting $U = N/N_*$, $T = at$, and $S = s/aN_*$, the dimensionless form of Eq. 2.7.1 is

$$\frac{dU}{dT} = U(1 - U) + S \quad . \tag{2.7.2}$$

The initial condition is $U(0) = U_0$. We obtain the following solution, expressed in terms of the hyperbolic tangent instead of exponential functions as in Section 2.4.6

$$U = \frac{U_0 - \left(\frac{1-2U_0-\lambda^2}{2\lambda}\right)\tanh\left(\frac{\lambda T}{2}\right)}{1 - \left(\frac{1-2U_0}{\lambda}\right)\tanh\left(\frac{\lambda T}{2}\right)} \tag{2.7.3}$$

in which $\lambda = \sqrt{1 + 4S}$. If we set $dU/dT = 0$ in Eq. 2.7.2, we determine that the asymptotic value of U is

$$U_\infty = \tfrac{1}{2}(1 + \lambda) \ . \tag{2.7.4}$$

The second derivative of Eq. 2.7.3 yields the inflection point

$$T_i = \frac{2\phi}{\lambda} \quad ; \qquad U_i = \tfrac{1}{2} \quad ; \qquad \left(\frac{dU}{dT}\right)_i = \frac{\lambda^2}{4} \quad , \tag{2.7.5}$$

where

$$\phi = \text{arctanh}\left(\frac{1 - 2U_0}{\lambda}\right) \ . \tag{2.7.6}$$

We note that $U_i = 1/2$ regardless of the value of λ.

The maximum curvature points are obtained from the third derivative of Eq. 2.7.3

$$T_c = \frac{2}{\lambda}\left(\phi \pm \text{arctanh}\frac{1}{\sqrt{3}}\right) \quad ; \qquad U_c = \frac{1}{2}\left(1 \pm \frac{\lambda}{\sqrt{3}}\right) \quad ; \tag{2.7.7}$$

$$\left(\frac{dU}{dT}\right)_c = \frac{\lambda^2}{6} \quad ; \qquad \left(\frac{d^2U}{dT^2}\right)_c = \frac{\lambda^3}{6\sqrt{3}} \quad . \tag{2.7.8}$$

A numerical example is shown in Fig 2.7.1. The value of S ranges from $S = 0$ (ordinary logistic) to $S = 2.0$; the initial value is $U_0 = 0.1$. As we would expect, magnitudes of U increase with T for all values of S.

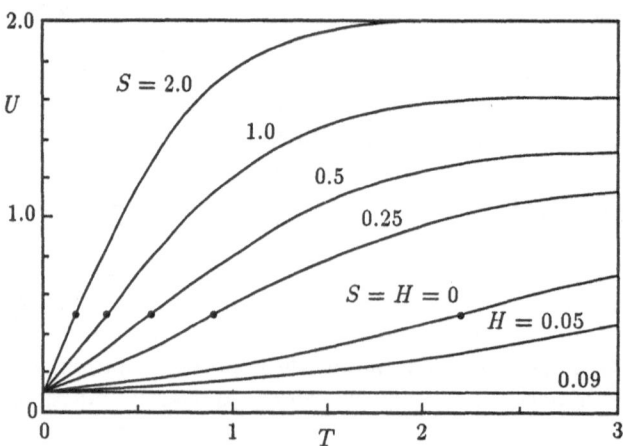

Fig. 2.7.1 Curves of logistic growth with various rates of stocking(S) and small rates of harvesting (H)

2.7.3 Logistic Growth with Constant Harvesting

Now instead of logistic growth with constant immigration, we consider logistic growth with emigration (or harvesting, withdrawing, extraction). In this case, the differential equation is

$$\frac{dN}{dt} = aN - bN^2 - h \quad , \tag{2.7.9}$$

in which h is the harvesting rate. In dimensionless terms this equation becomes

$$\frac{dU}{dT} = U(1-U) - H \quad . \tag{2.7.10}$$

where $H = h/aN_*$. In this case we obtain the same answer as Eq. 2.7.3

$$U = \frac{U_0 - \left(\frac{1-2U_0-\lambda^2}{2\lambda}\right)\tanh\left(\frac{\lambda T}{2}\right)}{1 - \left(\frac{1-2U_0}{\lambda}\right)\tanh\left(\frac{\lambda T}{2}\right)} \tag{2.7.11}$$

where, now, $\lambda = \sqrt{1-4H}$.

The appearance of the minus sign in this definition of makes the phenomenon of harvesting somewhat more complicated, though more interesting, than the phenomenon of stocking. There are several cases to consider.

Case 1: $H < 1/4$. In this case we observe in Fig. 2.7.1 that even though there is harvesting, the magnitude of U nevertheless may increase with increasing T and approaches the asymptotic value $U = (1+\lambda)/2$, as in the case of stocking.

The reason for this continued increase in U is clear enough when we look at Eq. 2.7.10. The magnitude of dU/dT is positive as long as the effect of exponential growth (the U term) is greater than the combined effects of crowding (the U^2 term) and harvesting (the H term). Clearly, when H reaches a certain critical value, the magnitude of dU/dT becomes zero and thereafter, with increasing H, becomes negative.

To be more specific, the slope of the growth curve at $T = 0$ is

$$\left(\frac{dU}{dT}\right)_0 = U_0(1-U_0) - H \quad . \tag{2.7.12}$$

If $H < U_0(1-U_0)$, then the initial slope is positive and remains so with increasing T. This case is illustrated by the $H = 0.05$ curve in Fig. 2.7.1.

If $H = U_0(1-U_0)$, the slope of the growth curve is zero. This is shown as the $H = 0.09$ curve in the figure.

If $H > U_0(1 - U_0)$, the initial slope of the growth curve is negative and remains so. This is the case for the curves labeled $H = 0.10$ through $H = 0.40$ in Fig. 2.7.2.

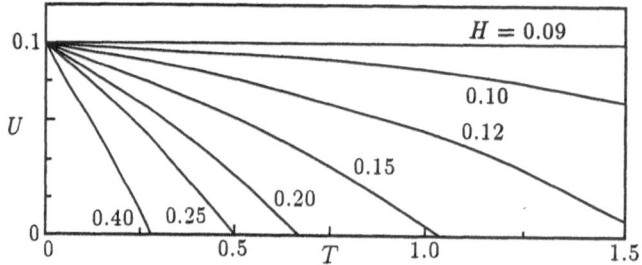

Fig. 2.7.2 Curves of logistic growth with large rates of harvesting (H)

In these three regimes for which $H < 1/4$, growth computations utilize Eq. 2.7.11. When $H > U_0(1 - U_0)$ an "extinction time", T_e, exists whose value is established by setting $U = 0$. This is determined by equating the numerator of Eq. 2.7.11 to zero. The result is

$$T_e = \frac{2}{\lambda}\text{arctanh}\left(\frac{2U_0\lambda}{1 - 2U_0 - \lambda^2}\right) . \tag{2.7.13}$$

Case 2: $H = 1/4$. If the dimensionless harvesting rate $H = 1/4$, the differential equation, Eq. 2.7.10 has the solution

$$U = \frac{1}{2}\left[1 - \frac{1 - 2U_0}{1 - (1 - 2U_0)T/2}\right] . \tag{2.7.14}$$

In this case the extinction time is simply

$$T_e = \frac{4U_0}{1 - 2U_0} . \tag{2.7.15}$$

The curve $H = 0.25$ in Fig. 2.7.2 corresponds to this special situation.

Case 3: $H > 1/4$. Finally, we suppose that $H > 1/4$. In this instance,

$$U = \frac{U_0 - \left(\frac{1 - 2U_0 - \mu^2}{2\mu}\right)\tan\left(\frac{\mu T}{2}\right)}{1 - \left(\frac{1 - 2U_0}{\mu}\right)\tan\left(\frac{\mu T}{2}\right)} \tag{2.7.16}$$

in which $\mu = \sqrt{4H - 1}$. Incidentally, instead of solving the differential equation all over again we can obtain Eq. 2.7.16 directly from Eq. 2.7.11. Noting that $\lambda^2 = -\mu^2$, we substitute $\lambda = i\mu$ (where $i = \sqrt{-1}$) into Eq. 2.7.11. Paying close attention to the rules for converting a hyperbolic function with an imaginary argument to a trigonometric function with a real argument, we get Eq. 2.7.16. In this regard, we have the relationship, $\tanh(iz) = i\tan(z)$.

The extinction time T_e when $H > 1/4$ is

$$T_e = \frac{2}{\mu}\operatorname{arctanh}\left(\frac{2U_0\mu}{1 - 2U_0 + \mu^2}\right) \quad . \tag{2.7.17}$$

This case is illustrated by the curve $H = 0.40$ in Fig. 2.7.2.

2.7.4 Logistic Growth with Variable Harvesting

In the preceding section, we stipulated a constant rate of harvesting, h. We can generalize Eq. 2.7.9 somewhat by writing

$$\frac{dN}{dt} = aN - bN^2 - h(N) \quad . \tag{2.7.18}$$

The last term on the right-hand side indicates that the harvesting rate depends on the instantaneous value of N. The function $h(N)$ can be of whatever form we wish.

As an example, suppose that

$$h(N) = h_0 + h_1 N + h_2 N^2 \quad . \tag{2.7.19}$$

In this case, Eq. 2.7.18 becomes

$$\frac{dN}{dt} = (a - h_1)N - (b + h_2)N^2 - h_0 \quad . \tag{2.7.20}$$

Letting $T = (a - h_1)t$, $N_* = (a - h_1)/(b + h_2)$, $H = h_0/(a - h_1)N_*$, and $U = N/N_*$, Eq. 2.7.20 reduces to Eq. 2.7.10. Consequently, the solution to the rather complicated-looking "parabolic harvesting function", expressed by Eq. 2.7.19, is given by either Eq. 2.7.11 or Eq. 2.7.16 with the indicated modifications in the definitions of the growth coefficient, crowding coefficient, and carrying capacity.

The parabolic harvesting function of Eq. 2.7.19 has the convenient feature of allowing a constant harvesting rate ($h_1 = h_2 = 0$), a linear harvesting rate ($h_2 = 0$), or a parabolic harvesting rate without any additional complication in the mathematics.

2.7.5 An Illustration: Fish Harvesting

As an illustration of the above analysis, suppose we have a very large pond which we intend to use for fish cultivation and harvesting. We establish that the growth coefficient of the fish species is $a = 0.25$ /week and the carrying capacity of the pond is $N_* = 25\,000$. The crowding coefficient is $b = a/N_* = 0.000010$ /fish-week. The pond is initially stocked with $N_0 = 1000$ fish.

At $t = 0$, ordinary logistic growth commences as shown in Fig. 2.7.3. The inflection point occurs at $t_i = (1/a)\log_e[(N_*/N_0) - 1] = 12.71$ weeks and $N_i = N_*/2 = 12\,500$.

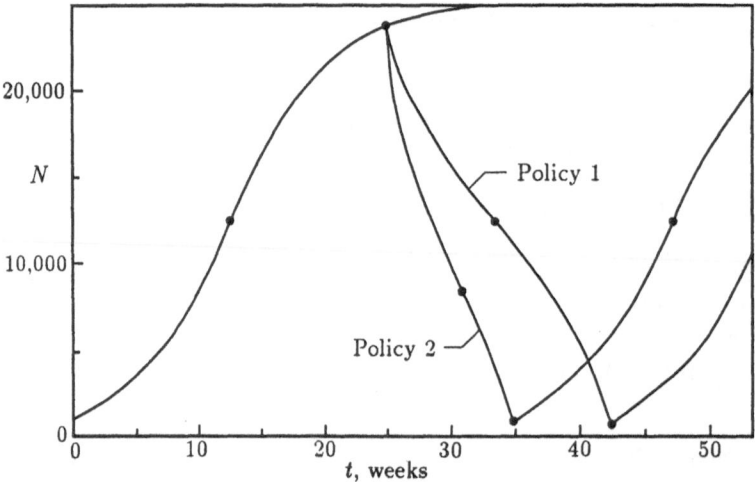

Fig. 2.7.3 Ordinary logistic growth followed by constant harvesting (Policy 1) and parabolic harvesting (Policy 2)

We examine the following harvesting policies.

Harvesting Policy 1. When $N = 0.95N_* = 23\,750$ fish, which occurs at $t_0 = 24.50$ weeks, we commence harvesting at the constant rate, $h_0 = 2500$ fish/week until we reach $N = 1000$. At that time, harvesting is ceased and ordinary logistic growth is again resumed for another cultivation cycle. Accordingly, we have the following information for the harvesting computation

$$h_0 = 2500 \quad ; \quad a = 0.25 \quad ; \quad b = 1.0 \times 10^{-5} \quad ;$$

$$t_0 = 24.50 \quad ; \quad N_0 = 23\,750 \quad ; \quad N_* = 25\,000 \quad .$$

Since $H = 0.40$, we utilize Eq. 2.7.16 to calculate the curve identified as Policy 1 in Fig. 2.7.3. From Eq. 2.7.17 the extinction time $t_e = 42.79$ weeks measured from the original start.

Fish harvesting is ceased when $N = 1000$ at $t = 42.37$ weeks after a harvesting time of 17.87 weeks. The total number of fish harvested is $M = 17.87 \times 2500 = 44\,675$ under this policy.

Harvesting Policy 2. As in policy 1, when $N = 0.95 N_* = 23\,750$ at $t_0 = 24.50$ weeks, harvesting is commenced. However, under policy 2 we utilize the parabolic harvesting rate described by Eq. 2.7.19. The following numerical values are employed

$$h_0 = 2500 \quad ; \quad h_1 = 0.050 \quad ; \quad h_2 = 2.0 \times 10^{-6} \quad ;$$

$$t_0 = 24.50 \quad ; \quad N_0 = 23\,750 \quad ; \quad a' = a - h_1 = 0.20 \quad ;$$

$$b' = b + h_2 = 1.2 \times 10^5 \quad ; \quad N'_* = (a - h_1)/(b + h_2) = 16\,667 \quad .$$

We again use Eq. 2.7.16 to compute the curve; this is shown as Policy 2 in Fig. 2.7.3. The extinction time is $t_e = 35.33$ weeks. As before, fish harvesting is terminated when $N = 1000$; this occurs at $t = 34.92$.

It is not difficult to show, using Eq. 2.7.19, that the average harvesting rate in this case is $h_m = 3510$ fish/week. Accordingly, under policy 2, the total amount harvested during the cycle is $M = 10.42 \times 3510 = 36\,575$. This compares with $M = 44\,675$ with policy 1. However, considering the duration of the total cycle, cultivation plus harvesting, the average weekly production under policy 1 is $m = 44\,675/42.37 = 1054$. Under policy 2, $m = 36\,575/34.92 = 1047$. We note that, in this case, the average weekly yield is approximately the same under the two policies. Other numerical results would be obtained, of course, by using other values of the constants in Eq. 2.7.19.

This example serves as an introduction to the many kinds of questions involving economics that arise in connection with fish cultivation and harvesting. These problems in "bioeconomics" have been examined by Clark (1976, 1981), Conrad (1986), and Conrad and Clark (1987).

2.7.6 An Illustration: The Sandhill Crane

We now consider a problem not greatly different from the fish harvesting example. This problem concerns the sandhill crane, an endangered species in the view of Miller and Botkin (1974).

These investigators indicate that the equilibrium level or carrying capacity of sandhill cranes in the United States is approximately 194 600. They also indicate that the net growth coefficient of the species is about 0.087 per year.

Starting with an initial value of $N_0 = N_* = 194\,600$, Miller and Botkin calculate the extents to which the total crane population would be reduced as a result of different rates of annual harvest by hunting. They use values of h ranging from 2000 to 12 000 cranes harvested per year.

Case 1: $H < 1/4$. We recall that $H = h/aN_*$. In the present problem, with $N_0 = N_*$, we have $U_0 = 1$. Accordingly, Eq. 2.7.11 reduces to

$$U = \frac{1 + \left(\frac{1+\lambda^2}{2\lambda}\right)\tanh\left(\frac{\lambda T}{2}\right)}{1 + \frac{1}{\lambda}\tanh\left(\frac{\lambda T}{2}\right)} \quad , \tag{2.7.21}$$

where $\lambda = \sqrt{1 - 4H}$. From this equation we obtain the asymptotic value

$$U_\infty = \tfrac{1}{2}(1 + \lambda) \quad . \tag{2.7.22}$$

In this case, the sandhill crane population is reduced to a new and lower equilibrium value but the birds continue to survive.

Case 2: $H = 1/4$. For this special case, taking $U_0 = 1$, Eq. 2.7.14 becomes

$$U = \frac{1}{2}\left(\frac{4+T}{2+T}\right) \quad , \tag{2.7.23}$$

and, as $T \to \infty$, we have $U_\infty = 1/2$.

Case 3: $H > 1/4$. For this final case, with $U_0 = 1$, Eq. 2.7.16 reduces to

$$U = \frac{1 - \left(\frac{1-\mu^2}{2\mu}\right)\tan\left(\frac{\mu T}{2}\right)}{1 + \frac{1}{\mu}\tan\left(\frac{\mu T}{2}\right)} \quad . \tag{2.7.24}$$

In this range of H, an inflection point occurs at

$$T_i = \frac{2}{\mu}\arctan\left(\frac{1}{\mu}\right) \quad ; \qquad U_i = \frac{1}{2} \quad ; \qquad \left(\frac{dU}{dT}\right)_i = -\frac{\mu}{4} \quad . \tag{2.7.25}$$

Most importantly, an extinction time exists. From Eq. 2.7.17 we obtain

$$T_e = \frac{2}{\mu}\arctan\left(\frac{2\mu}{\mu^2 - 1}\right) \quad . \tag{2.7.26}$$

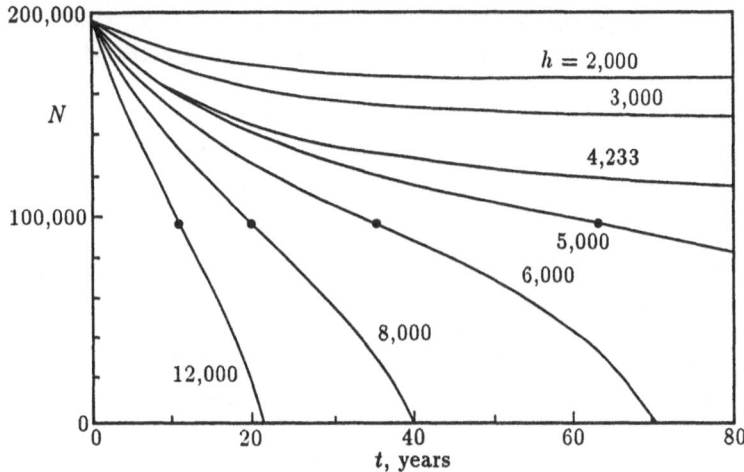

Fig. 2.7.4 Reduction in the number of sandhill cranes due to harvesting by hunting. From Miller and Botkin (1974)

Utilizing the equations presented above, the curves shown in Fig. 2.7.4 are constructed. The curve labelled $h = 4233$ corresponds to the value $H = 1/4$. These curves, almost identical to those of Miller and Botkin, show the sandhill crane population for various rates of harvesting.

We note in the figure that a sustained annual harvest rate of 12 000, for example, would reduce the crane population to one-half its initial value of 194 600 in 10.8 years and to extinction of the cranes after 21.6 years. Brauer and Sanchez (1975) discuss this sandhill crane problem raised by Miller and Botkin.

2.8 Epidemics and Technology Transfer

2.8.1 Simple and General Epidemics

For our next topic we consider the phenomenon of epidemics. We shall see that mathematical frameworks of epidemiology are similar to those which describe certain kinds of growth phenomena and technology transfer.

The mathematics of epidemiology has been examined by many researchers over a long period of time. The most important early studies of the subject are those of Ross (1911) and of Kermack and McKendrick (1927). The classic book by Bailey (1975) presents a comprehensive treatment of mathematical epidemiology from both the deterministic and stochastic points of view. The contributions of Bartholomew (1982), Dietz (1967), Hoppensteadt (1975), Mollison (1977), Murray (1989), Waltman (1974) and others have greatly enlarged our knowledge of the subject. Several studies, including those of

Mansfield and Hensley (1960) and Sharif and Ramanathan (1982), relate the mathematical aspects of epidemiology to those of technology transfer and innovation diffusion.

It is customary to identify an epidemic as either a simple epidemic or a general epidemic; their defining components are shown in Table 2.8.1.

Table **2.8.1** Simple and general epidemics: their components and technology transfer equivalents

Type of epidemic	Epidemic component	Mathematical symbol	Technology transfer equivalent
simple	susceptible	S	potential adopter
	infective	I	adopter
general	susceptible	S	potential adopter
	infective	I	adopter
	removed	R	rejector

In a simple epidemic (SI model) there are two components or categories of persons: those who have either already acquired the disease (infectives) and those who are going to acquire it (susceptibles). In the vernacular of technology transfer these categories correspond, respectively, to actual adopters and to potential adopters.

In the general epidemic (SIR model) there is a third category: those who have either recovered or died from the disease or have been isolated or are immune. Collectively, this removed category is the technology transfer equivalent of the category who had adopted the technology but subsequently rejected it. In a more complicated general epidemic model, a fourth category is sometimes included: those who have been exposed to the disease but are not yet infectious. We will not consider this SEIR model.

2.8.2 The Differential Equations and Phase Plane Display

We consider a community in which there is a constant population, N_*, which initially is composed of S_0 susceptibles, I_0 infectives, and R_0 removed. For $t > 0$, as the distribution among the three components continuously changes, the total population remains constant. Accordingly

$$N_* = S + I + R = S_0 + I_0 + R_0 \quad . \tag{2.8.1}$$

The following differential equations provide the basis of our analysis

$$\frac{dS}{dt} = -\beta SI \quad , \tag{2.8.2}$$

$$\frac{dI}{dt} = \beta SI - \gamma I \quad , \tag{2.8.3}$$

$$\frac{dR}{dt} = \gamma I \quad , \tag{2.8.4}$$

where β (the infection rate) and γ (the removal rate) are non-negative constants.

These three equations are about as simple as we can make them and still have a fairly realistic model. The first equation assumes that the rate of decrease of susceptibles is proportional to the number of contacts per unit time between the susceptibles and the infectives. Its precise form, βSI, is similar to the crowding term, bN^2, in the single species growth phenomenon leading to the ordinary logistic equation. Since S, I, and β are non-negative quantities, it is clear from Eq. 2.8.2 that dS/dt is always a decreasing function.

The second equation states that the rate at which the number of infectives increases or decreases is equal to the difference between the rates at which the number of susceptibles decreases and the number of removed increases. The same result is obtained by differentiating Eq. 2.8.1.

The third equation assumes that the rate at which the removed category is increased is directly proportional to the number of infectives. Clearly dR/dt is an increasing function.

Before seeking the solutions, $S(t)$, $I(t)$ and $R(t)$, we establish a few useful relationships. First we divide Eq. 2.8.2 by Eq. 2.8.4 to obtain

$$S = S_0 \exp\left[-\frac{1}{\rho}(R - R_0)\right] \quad , \tag{2.8.5}$$

where S_0 and R_0 are the initial values. The quantity $\rho = \gamma/\beta$ is termed the relative removal rate. It is easily established that the dimensions of the three constants are: $[\beta] = 1/NT$, $[\gamma] = 1/T$, and $[\rho] = N$. We remind ourselves that in any mathematical analysis it is advisable to check the dimensions of terms and groups from time to time.

We obtain additional useful information by writing Eq. 2.8.3 in the form

$$\frac{dI}{dt} = \beta(S - \rho)I \quad . \tag{2.8.6}$$

Now since S is a decreasing function it follows that $S(t) < S_0$. Accordingly, if $S_0 < \rho$ in the above equation then $(dI/dt)_0 < 0$. This means that the infection "dies out" and no epidemic occurs. Alternatively, if $S_0 > \rho$ then $(dI/dt)_0 > 0$ and the epidemic spreads. Thus, we observe that $S_0 = \rho$ represents a threshold number of susceptibles. Values of S_0 larger (smaller) than ρ assure the spreading (dissipation) of the epidemic.

Another helpful relationship is provided by dividing Eq. 2.8.3 by Eq. 2.8.2. This yields

$$\frac{dI}{dS} = \frac{\rho}{S} - 1 \quad , \tag{2.8.7}$$

which, when integrated, gives

$$I = S_0 + I_0 - S + \rho \log_e(S/S_0) \quad . \tag{2.8.8}$$

Setting Eq. 2.8.7 equal to zero indicates that when $S = \rho$, the number of infectives has the maximum value

$$I_m = S_0 + I_0 + \rho \log_e(\rho/eS_0) \quad . \tag{2.8.9}$$

We present the above results in graphical form; for simplicity, it is assumed that $R_0 = 0$. So Eq. 2.8.8 becomes

$$I = N_* - S + \rho \log_e(S/S_0) \quad . \tag{2.8.10}$$

This relationship is shown in the phase plane displays of Figs. 2.8.1 and 2.8.2 for values of $\rho = 20$ and 50, respectively, for various values of S_0. Incidentally, we shall use the term "phase plane" for graphical displays of this and other kinds of plots relating dependent variables of a problem.

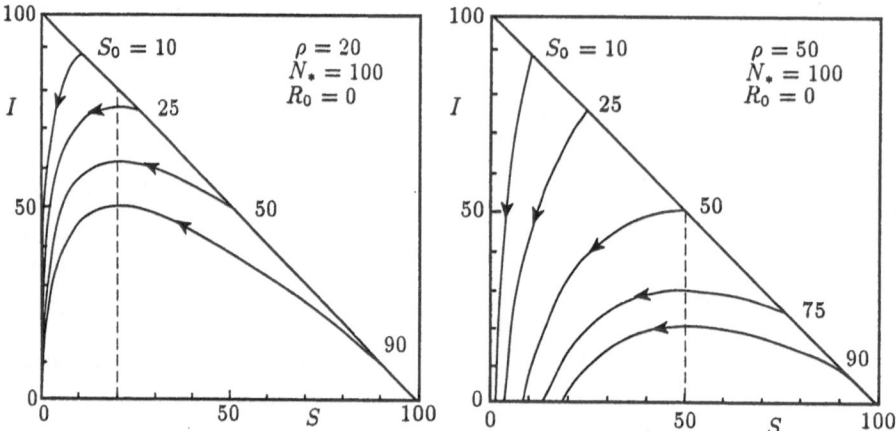

Fig. 2.8.1 (left) Phase plane display of infectives versus susceptibles; relative removal rate $\rho = 20$

Fig. 2.8.2 (right) Phase plane display of infectives versus susceptibles; relative removal rate $\rho = 50$

Now since $S(t) < S_0$, points on the curves of the two figures move with increasing t in the directions indicated by the arrows. As we saw, the maximum number of infectives occurs when $S = \rho$. The curves approach $(S_*, 0)$ as $t \to \infty$. The magnitude of this limiting value of S, obtained by setting $I = 0$ in Eq. 2 .8.10, is the root of the expression

$$S_* = N_* + \rho \log_e(S_*/S_0) \quad . \tag{2.8.11}$$

2.8.3 Solutions to the Differential Equations

If Eq. 2.8.8 is substituted into Eq. 2.8.2, we obtain

$$\frac{dS}{dt} = \beta S[S - (S_0 + I_0)] - \gamma S \log_e(S/S_0) \quad . \tag{2.8.12}$$

So, in principle, we have the solution, $S = S(t)$. Then, from Eq. 2.8.5, we also have $R = R(t)$. Finally, using either Eq. 2.8.1 or Eq. 2.8.8, we obtain $I(t)$. Accordingly, in theory, we know the time distributions of all three components of the epidemic.

Unfortunately it is not possible to obtain an exact solution to Eq. 2.8.12. We pursue the matter of an approximate solution in a moment. For now, we digress to obtain a solution to the simple epidemic problem.

Simple epidemic (SI model) In this case, there is no removed category and so $R = R_0 = 0$. Further, the removal rate coefficient, $\gamma = 0$. Accordingly, since $N_* = S + I = S_0 + I_0$, Eq. 2.8.3 becomes

$$\frac{dI}{dt} = \beta I(N_* - I) \quad , \tag{2.8.13}$$

which, of course, is the differential equation of the ordinary logistic with growth coefficient $a = \beta N_*$, crowding coefficient $b = \beta$, and carrying capacity N_*. The solution to Eq. 2.8.13 is

$$I = \frac{N_*}{1 + \left(\frac{N_*}{I_0} - 1\right) e^{-\beta N_* t}} \tag{2.8.14}$$

and so, $S(t) = N_* - I(t)$. Thus we see that the simple epidemic corresponds exactly to ordinary logistic growth. The infective category has its technology transfer equivalent in the adopter group and the susceptible category in the potential adopter.

General epidemic (SIR model) Returning to the considerably more complicated general epidemic problem, we utilize Eqs. 2.8.1, 2.8.4 and 2.8.5 to obtain

$$\frac{dR}{dt} = \gamma \left(N_* - R - S_0 e^{-\frac{1}{\rho}(R - R_0)} \right) \quad . \tag{2.8.15}$$

Anticipating an impossible exact integration of this equation, we expand the exponential quantity in a series involving the quadratic term. Simplifying, we obtain

$$\frac{dR}{dt} = \gamma(N_* - S_e) + \gamma\left(\frac{S_e}{\rho} - 1\right)R - \frac{\gamma S_e}{2\rho^2}R^2 \quad , \tag{2.8.16}$$

where $S_e = S_0 \exp(R_0/\rho)$. Seeing another logistic-like differential equation, we write Eq. 2.8.16 in the form

$$\frac{dN}{dt} = r + aN - bN^2 \quad , \tag{2.8.17}$$

where, for the moment, N replaces R. The definitions of r, a, and b are clear enough from Eq. 2.8.16. The r-term in this equation represents a constant immigration (stocking: $r = s$) or a constant emigration (harvesting: $r = -h$). Writing Eq. 2.8.17 in definite integral form we have

$$\int_{N_0}^{N} \frac{dN}{r + aN - bN^2} = \int_0^t dt \quad , \tag{2.8.18}$$

which, after integration and some algebra, becomes

$$N = \frac{1}{2b}[a + \sigma\tanh(\tfrac{1}{2}\sigma t - \phi)] \quad , \tag{2.8.19}$$

where

$$\sigma = \sqrt{a^2 + 4br} \tag{2.8.20}$$

and

$$\phi = \text{arctanh}\left(\frac{a - 2bN_0}{\sigma}\right) \quad . \tag{2.8.21}$$

The cumulative distribution function, $N(t)$, is given by Eq. 2.8.19. The density distribution is

$$n = \frac{\sigma^2}{4b}\text{sech}^2(\tfrac{1}{2}\sigma t - \phi) \quad . \tag{2.8.22}$$

From this equation we determine the inflection point of the cumulative curve and, of course, the maximum point of the density curve. The results are

$$t_i = \frac{2\phi}{\sigma} \quad ; \qquad N_i = \frac{a}{2b} \quad ; \qquad n_i = \frac{\sigma^2}{4b} \quad . \tag{2.8.23}$$

The value of N as $t \to \infty$ is

$$N_* = \frac{1}{2b}(a + \sigma) \quad . \tag{2.8.24}$$

Returning to the epidemic problem, we substitute the equivalent values of R, a, and b into the preceding equations, with R replacing N, to obtain

$$R = \frac{\rho^2}{S_e} \left[\left(\frac{S_e}{\rho} - 1 \right) + \alpha \tanh(\tfrac{1}{2}\alpha\gamma t - \phi) \right] \tag{2.8.25}$$

where $S_e = S_0 \exp(R_0/\rho)$ and

$$\alpha = \sqrt{\left(\frac{S_e}{\rho} - 1 \right)^2 + \frac{2S_e(N_* - S_e)}{\rho^2}} \quad , \tag{2.8.26}$$

$$\phi = \operatorname{arctanh} \left[\frac{1}{\alpha} \left(\frac{S_e}{\rho} - 1 - \frac{S_e R_0}{\rho^2} \right) \right] \quad , \tag{2.8.27}$$

$$t_i = \frac{2\phi}{\alpha\gamma} \quad ; \qquad R_i = \frac{\rho^2}{S_e} \left(\frac{S_e}{\rho} - 1 \right) \quad , \tag{2.8.28}$$

$$\left(\frac{dR}{dt} \right)_i = \frac{\gamma\alpha^2\rho^2}{2S_e} \quad ; \qquad R_* = \frac{\rho^2}{S_e} \left[\left(\frac{S_e}{\rho} - 1 \right) + \alpha \right] \quad . \tag{2.8.29}$$

The infective function, $I(t)$, is easily obtained from Eqs. 2.8.4 and 2.8.22. The answer is

$$I = \frac{\alpha^2\rho^2}{2S_e} \operatorname{sech}^2(\tfrac{1}{2}\alpha\gamma t - \phi) \quad . \tag{2.8.30}$$

The coordinates of the maximum infective point are

$$t_i = \frac{2\phi}{\alpha\gamma} \quad ; \qquad I_m = \frac{\alpha^2\rho^2}{2S_e} \quad . \tag{2.8.31}$$

To complete the problem, we determine the susceptible function, $S(t)$, from Eq. 2.8.1. Accordingly, to the extent that the series approximation of the exponential term is sufficiently accurate, we now know the time distributions of each of the three components of the epidemic.

We look at a couple of numerical examples; values of various constants and computed quantities are listed in Table 2.8.2. Substituting the constants indicated in the table into the preceding equations yields the curves shown in Fig. 2.8.3 for Example 1 and in Fig. 2.8.4 for Example 2.

Table 2.8.2 Design constants and computed quantities for Examples 1 and 2 (Figs. 2.8.3 and 2.8.4)

Parameter		Example 1 Fig. 2.8.3	Example 2 Fig. 2.8.4
(a) Design constants			
	N_*	100	100
	S_0	90	66.67
	I_0	10	33.33
	R_0	0	0
	S_e	90	66.67
	ρ	55	66.67
	γ	0.1	0.1
	α	1.0	1.0
	ϕ	0.752	0.0
(b) Computed quantities			
	t_i	15.04	0.0
	R_i	21.39	0.0
	$(dR/dt)_i$	1.68	3.33
	R_*	55.0	66.67
	I_*	0.0	0.0
	S_*	45.0	33.33
	I_m	16.81	33.33

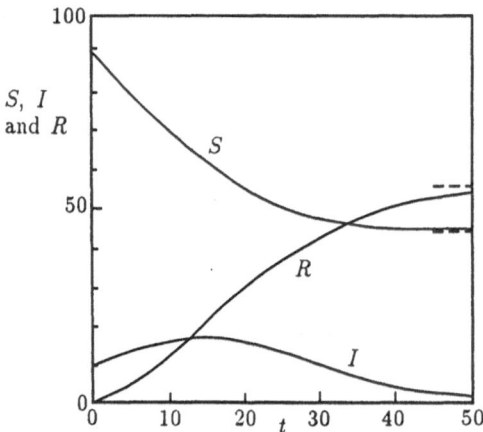

Fig. 2.8.3 Plots of $S(t)$, $I(t)$, and $R(t)$. Example 1. See Table 2.8.2

Incidentally, it is interesting to compare these plots to the curves, shown in Fig. 2.1.7, corresponding to an entirely linear epidemic model.

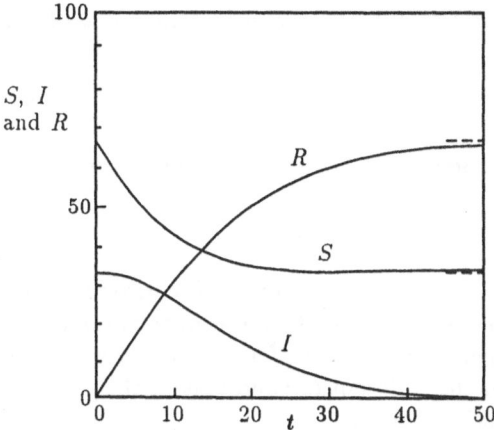

Fig. 2.8.4 Plots of $S(t)$, $I(t)$, and $R(t)$. Example 2. See Table 2.8.2

2.8.4 Logarithmic Form of the Solutions

The solution to the logistic equation

$$\frac{dN}{dt} = r + aN - bN^2 \qquad (2.8.32)$$

can be expressed either in terms of logarithmic functions or in terms of hyperbolic functions. We have already obtained the hyperbolic-form solutions for the general epidemic problem; these are expressed by Eqs. 2.8.19 and 2.8.22. We now present them in the more familiar logarithmic or, equivalently, exponential form.

To do so we could utilize the identity

$$\text{arctanh} z = \tfrac{1}{2} \log_e \frac{1+z}{1-z} \quad . \qquad (2.8.33)$$

However, it is probably easier to return to Eq. 2.8.18 and select the logarithmic-type solution which most tables of integrals give as an alternative answer. By either route we obtain the following results

$$N = \frac{1}{2b} \left[\frac{(\sigma+a) - m(\sigma-a)e^{-\sigma t}}{1 + me^{-\sigma t}} \right] \qquad (2.8.34)$$

and

$$n = \frac{m\sigma^2}{b} \frac{e^{-\sigma t}}{(1 + me^{-\sigma t})^2} \quad , \qquad (2.8.35)$$

where

$$\sigma = \sqrt{a^2 + 4br} \quad ; \qquad m = \frac{(\sigma+a) - 2bN_0}{(\sigma-a) + 2bN_0} \quad . \qquad (2.8.36)$$

The asymptotic value is $N_* = (a + \sigma)/2b$. At the inflection point of the cumulative distribution

$$t_i = \frac{1}{\sigma} \log_e m \quad ; \qquad N_i = \frac{a}{2b} \quad ; \qquad n_i = \frac{\sigma^2}{4b} \quad . \tag{2.8.37}$$

As we would expect, these are the same answers as those given by Eq. 2.8.23.

2.8.5 A Technology Transfer Analogy

In Section 2.4, we examined the linear combination of a concentrated impersonal diffusion source (Coleman model) and a distributed personal diffusion source (Dodd model) in connection with the transfer of a technology. For this purpose, we arbitrarily added the differential equations of the confined exponential and the ordinary logistic to obtain

$$\frac{dN}{dt} = a_*(N_* - N) + aN(1 - N/N_*) \quad . \tag{2.8.38}$$

This equation can be rewritten in the form

$$\frac{dN}{dt} = a_* N_* + (a - a_*)N - \frac{a}{N_*} N^2 \quad , \tag{2.8.39}$$

which is mathematically identical to Eq. 2.8.16 (for the removed category function in the general epidemic problem) and to Eq. 2.8.17 (for the constant migration-logistic growth problem). So we have obtained a somewhat generalized set of solutions. Indeed the answers given by Eqs. 2.8.34 through 2.8.37 are essentially the same as those expressed by Eqs. 2.4.25 through 2.4.28.

By way of conclusion, Table 2.8.3 presents a comparison of the various coefficients involved in these three equivalent phenomena. For this purpose, the migration-logistic equation

$$\frac{dN}{dt} = r + aN - bN^2 \tag{2.8.40}$$

serves as the reference differential equation and Eq. 2.8.19, or equivalently, Eq. 2.8.34, provides the reference solution.

Table 2.8.3 Analogous quantities among phenomena of migration-logistic growth, general epidemics, and technology transfer

Constant migration with logistic growth	Removal category in general epidemic	Coleman–Dodd technology transfer
$N(t)$	$R(t)$	$N(t)$
r	$\gamma(N_* - S_e)$	$a_* N_*$
a	$\gamma[(S_e/\rho) - 1]$	$(a - a_*)$
b	$\gamma S_e/2\rho^2$	a/N_*

2.8.6 An Illustration: Bombay Plague of 1905-1906

In their early studies of epidemics, Kermack and McKendrick (1927) presented data on the number of deaths resulting from a rat-spread plague in
Bombay during the period December 1905 to July 1906. These data, displayed
in the density distribution curve of Fig. 2.8.5, correspond to the presentation
made in the authors' publication. The cumulative curve, shown in Fig. 2.8.6
was prepared from the density data.

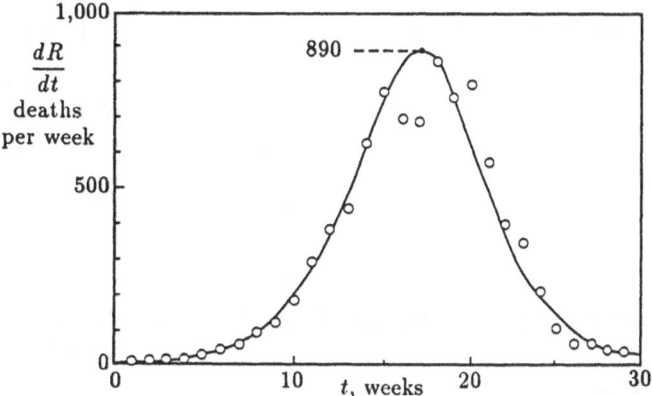

Fig. 2.8.5 Density distribution curves for the Bombay plague of 1905–1906

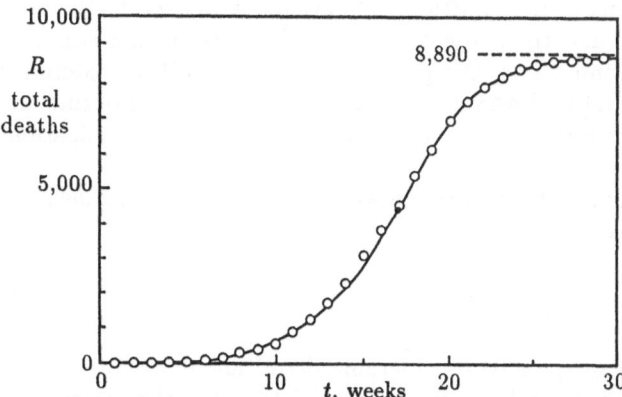

Fig. 2.8.6 Cumulative distribution curve of the Bombay plague of 1905–1906

Kermack and McKendrick give the following equation for the density distribution

$$\frac{dR}{dt} = 890\,\mathrm{sech}^2(0.20t - 3.4) \quad . \tag{2.8.41}$$

From Eqs. 2.8.4 and 2.8.30 we have

$$\frac{dR}{dt} = \frac{\gamma \alpha^2 \rho^2}{2S_e} \operatorname{sech}^2(\tfrac{1}{2}\alpha\gamma t - \phi) \quad , \tag{2.8.42}$$

and from Eq. 2.8.25

$$R = \frac{\rho^2}{S_e}\left[\left(\frac{S_e}{\rho} - 1\right) + \alpha \tanh(\tfrac{1}{2}\alpha\gamma t - \phi)\right] \quad . \tag{2.8.43}$$

From the numerical values given by Eq. 2.8.41, the cumulative curve becomes

$$R = 4440 + 4450\tanh(0.20t - 3.4) \quad . \tag{2.8.44}$$

The solid curves shown in the figures were computed from Eqs. 2.8.41 and 2.8.44. The cumulative curve inflection point coordinates are $t_i = 17.0$ weeks, $R_i = 4440$ deaths. The maximum number of deaths occurred during the seventeenth week at the rate of 890 deaths per week. The asymptotic value is $R_* = 8890$ deaths.

2.9 Some Modifications of the Logistic Distribution

2.9.1 Use of Taylor Series

We now examine another category of growth functions which we characterize as simply "modifications of the logistic distribution". To generate this category of functions we raise the subject of time delays. Since this is a very important subject in connection with growth processes, it will be examined at some length in Chapter 6. However, we now bring in the matter of time delays simply to establish a mechanism for manufacturing these modifications of the logistic.

The differential equation for logistic growth with a discrete time delay, τ, in the crowding term is

$$\frac{dN(t)}{dt} = aN(t)\left(1 - \frac{N(t-\tau)}{N_*}\right) \quad . \tag{2.9.1}$$

As indicated, later on we go into considerable detail about this and other equations involving discrete and distributed time delays. For the moment we present Eq. 2.9.1 only to provide a kind of "generating function" for what we are after.

Applying Taylor series

$$f(x + x_0) = f(x) + \frac{x_0}{1!}\frac{df(x)}{dx} + \frac{x_0^2}{2!}\frac{d^2 f(x)}{dx^2} + \cdots \tag{2.9.2}$$

to the quantity $N(t-\tau)$ yields

$$N(t - \tau) = N(t) - \tau \frac{dN(t)}{dt} + \frac{1}{2}\tau^2 \frac{d^2 N(t)}{dt^2} - \cdots \quad . \tag{2.9.3}$$

If we substitute this result into Eq. 2.9.1, we can produce a first order, a second order, or an n-th order differential equation, depending on how many terms of the series we decide to use.

2.9.2 First Order Differential Equations

We start with the simplest case. Suppose we utilize only the first two terms of the right-hand side of Eq. 2.9.3. Substitution into Eq.2.9.1 gives

$$\frac{dN}{dt} = \frac{aN\left(1 - \frac{N}{N_*}\right)}{1 - a\tau \frac{N}{N_*}} \quad . \tag{2.9.4}$$

As before, this equation is put into dimensionless form by letting $U = N/N_*$, $T = at$ and $\tau_* = a\tau$. Also, to dismiss the matter of time delays, we let $\tau_* = -c$. Accordingly Eq. 2.9.4 becomes

$$\frac{dU}{dT} = \frac{U(1 - U)}{1 + cU} \quad . \tag{2.9.5}$$

We note that, when $c = -1$, this equation reduces to the exponential function. When $c = 0$, it becomes the ordinary logistic.

Imposing the initial condition $U(0) = U_0$, and integrating Eq. 2.9.5 we obtain the solution

$$T = \log_e \left(\frac{U}{U_0}\right) \left(\frac{1 - U_0}{1 - U}\right)^{1+c} \tag{2.9.6}$$

or, alternatively,

$$\frac{(1 - U)^{1+c}}{U} = \frac{(1 - U_0)^{1+c}}{U_0} e^{-T} \quad . \tag{2.9.7}$$

The derivative of Eq. 2.9.5 provides the ordinate of the inflection point

$$U_i = \frac{1}{c}(\sqrt{1 + c} - 1) \quad . \tag{2.9.8}$$

Curves showing the growth rate and specific growth rate are given in Fig. 2.9.1 for various values of c. We note the similarity of these curves with those of the power law logistic shown in Fig. 2.6.1.

Plots of the modified logistic function, Eq. 2.9.6, are presented in Fig. 2.9.2 for the same range of c and $U_0 = 0.10$. The dashed line is the locus of the inflection points. From Eqs. 2.9.6 and 2.9.8 we establish that the minimum inflection point time is $T_i = 2.145$. Corresponding values are $U_i = 0.662$ and $c = -0.739$. The solid dot in the figure identifies this point. It is seen that this modified logistic permits a wide range of inflection point ordinates.

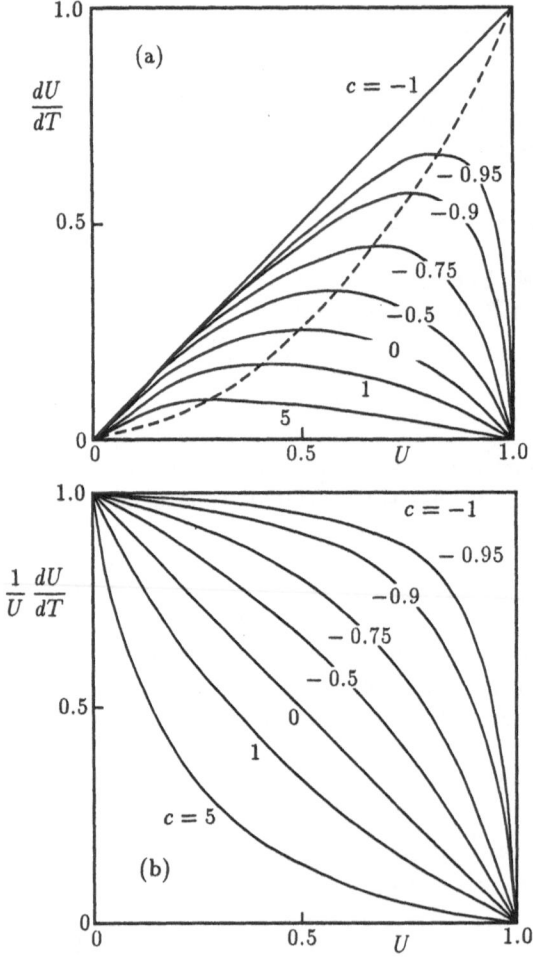

Fig. 2.9.1 Modified logistic distribution. (a) Growth rate and (b) specific growth rate for various values of c

2.9.3 Illustration: Growth of Water Fleas and Trees

Example 1. In his study of the population dynamics of the water flea *Daphnia magna*, Smith (1963) presents the following expression to describe the growth of the organisms

$$\frac{1}{M}\frac{dM}{dt} = r\left(\frac{K - M}{K + \left(\frac{r}{s}\right)M}\right) \quad , \tag{2.9.9}$$

in which M is the dry mass of the population of *D. magna* at time, t; K is the value of M at saturation; r is the rate of increase with unlimited food; and

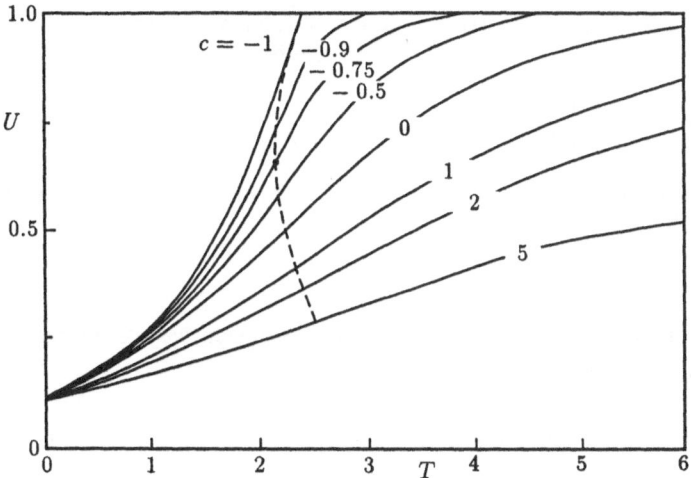

Fig. 2.9.2 Plots of modified logistic distribution for various values of c

s is the rate of replacement of mass in the population at saturation. Smith gives values of $r = 0.44$ mg/mg-day, $s = 0.127$ mg/mg-day, and $K = 15.0$ mg.

We note that Eq. 2.9.9 is identical to Eq. 2.9.5 with $U = M/K$, $T = rT$, and $c = r/s$.

In his experiments, Smith observed that his plots of specific growth curves, similar to Fig. 2.9.1(b), were substantially different from the linear relationship ($c = 0$) of the logistic function. In fact, his curves were distinctly concave upward. This is compatible with his observed value of $c = r/s = 0.44/0.127 = +3.46$.

Smith states that the ordinary logistic equation was inadequate for the problem in population dynamics he studied. He concludes that his experimental data are well described by the framework provided by Eq. 2.9.9. Discussions of Smith's work are presented by May (1974), Moss, Watson and Ollason (1982), and Pielou (1977).

Example 2. A tree-growth model, devised by Valentine (1985), utilizes this same mathematical framework. He presents the following differential equation

$$\frac{dH}{dt} = \frac{H(a'K - b' - bH)}{z + H}, \qquad (2.9.10)$$

where H is tree height, t is time, and the other quantities are constants. Valentine sets $m = a'K - b'$, $n = bz/m$, and $H_* = m/b$. Accordingly Eq. 2.9.10 becomes

$$\frac{dH}{dt} = \frac{mH}{z}\left(\frac{1 - \frac{H}{H_*}}{1 + \frac{1}{n}\frac{H}{H_*}}\right). \qquad (2.9.11)$$

Taking $a = m/z$, $T = at$, $U = H/H_*$, and $c = 1/n$, we obtain Eq. 2.9.5. Valentine presents his solution to Eq. 2.9.11 in the form

$$\frac{(m - bH)^{1+n}}{H^n} = \frac{(m - bH_0)^{1+n}}{H_0^n}e^{-bt} \tag{2.9.12}$$

which is essentially the same as Eq. 2.9.7. As Valentine points out, setting $n = 0$ in Eq 2.9.12 yields

$$H = H_* - (H_* - H_0)e^{-bt} \quad , \tag{2.9.13}$$

which is the confined exponential or Mitscherlich equation we considered in Section 2.3.

2.9.4 An Illustration: Sale of Development Property Revisited

We return to the illustration given in Section 2.6.4 concerning the sale of development property. We do so in order to make a comparison between our modified logistic and the power law logistic examined in the earlier section.

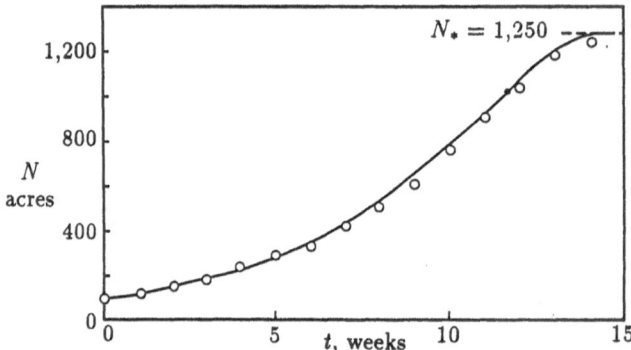

Fig. 2.9.3 Sale of development property. Compare with the power law logistic model of Fig. 2.6.3

The data presented in Table 2.6.1 are shown in Fig. 2.9.3. The solid line is based on Eq. 2.9.6 with the following values of the various constants

$$a = 0.204 \quad ; \quad N_0 = 101.2 \quad ; \quad N_* = 1250 \quad ; \quad c = -0.928 \quad .$$

The inflection point occurs at $t_i = 11.67$ weeks, $N_i = 985.5$ acres. For the power law logistic, the value was $t_i = 11.21$ weeks, $N_i = 909.7$ acres. For the power law, the value $s = 6.20$ was obtained. Accordingly, this value of s is equivalent to $c = -0.928$ of the modified logistic.

It is seen that these two growth functions are similar. The power law logistic is somewhat easier to handle mathematically since an explicit solution, $N = N(t)$, is possible. For the modified logistic we have to settle for an answer in the form $t = t(N)$.

2.9.5 Second Order Differential Equations

Next we examine a more complicated case by taking the first three terms, instead of two, of the Taylor series given by Eq. 2.9.3. Substituting into Eq. 2.9.1 and simplifying we obtain

$$\frac{d^2N}{dt^2} + \frac{2}{\tau^2}\left(\frac{1}{a}\frac{N_*}{N} - \tau\right)\frac{dN}{dt} + \frac{2}{\tau^2}N = \frac{2}{\tau^2}N_* \quad . \tag{2.9.14}$$

In dimensionless form with $\tau_* = a\tau$, this becomes

$$\frac{d^2U}{dT^2} + \frac{2}{\tau_*^2}\left(\frac{1}{U} - \tau_*\right)\frac{dU}{dT} + \frac{2}{\tau_*^2}U = \frac{2}{\tau_*^2} \tag{2.9.15}$$

which agrees with the result given by Cunningham (1954). If we set $\tau_* = 0$ in Eq. 2.9.15 we regain the differential equation for the ordinary logistic, $dU/dT = U(1 - U)$.

Incidentally, had we started with the confined exponential equation, instead of the logistic, the same procedure would have yielded

$$\frac{d^2U}{dT^2} + \frac{2}{\tau_*^2}(1 - \tau_*)\frac{dU}{dT} + \frac{2}{\tau_*^2}U = \frac{2}{\tau_*^2} \tag{2.9.16}$$

Both Eqs. 2.6.15 and 2.7.16 are second order differential equations. As we shall see, the exact solution of a linear equation, like Eq. 2.9.16, is easy to obtain. However, because of the appearance of the quantity, U, in the second term, Eq. 2.9. 15 is nonlinear; an exact solution is impossible. Techniques for solving equations of this type are given by Davis (1961), Jordan and Smith (1977) and other references on nonlinear differential equations.

Looking ahead, in Section 6.2 we examine the topic of time delay phenomena and we will see the computer solution of nonlinear Eq. 2.9.15. However, for the present we remind ourselves that we have utilized the discrete delay Eq. 2.9.1, along with Taylor series, only as a mechanism to generate differential equations.

Accordingly, at this point, we linearize Eq. 2.9.15 by simply writing

$$\frac{d^2U}{dT^2} + \alpha\frac{dU}{dT} + \beta U = \beta_* \quad , \tag{2.9.17}$$

where

$$\alpha = \beta \left(\frac{1}{U_*} - \sqrt{\frac{2}{\beta}} \right) \quad ; \quad \beta = \frac{2}{\tau_*^2} \quad . \tag{2.9.18}$$

Two comments. First, in going from Eq. 2.9.15 to Eq. 2.9.17, we have arbitrarily replaced β with another constant, β_*, on the right-hand side of Eq. 2.9.17 in order to make the equation somewhat more generalized. Second, we have also arbitrarily replaced the dependent variable, U, in the second term of Eq. 2.9.15 with an equivalent constant, U_*. Because of this substantial simplification, Eq. 2.9.17 is no longer a logistic-type equation. We note also, in Eq. 2.9.18, that α may be positive or negative.

We now have, in Eq. 2.9.17, a second order linear differential equation with constant coefficients. Its solution is not difficult. Utilizing the initial conditions $U(0) = U_0$ and $(dU/dT)_0 = U'(0) = U'_0$, and letting $\gamma = \beta_*/\beta$, we have the following solutions

Case 1. $\alpha^2 = 4\beta$

$$U = \gamma - \{(\gamma - U_0) + [\tfrac{1}{2}\alpha(\gamma - U_0) - U'_0]T\}e^{-\frac{1}{2}\alpha T} \quad . \tag{2.9.19}$$

Case 2. $\alpha^2 > 4\beta$; $p = \sqrt{\alpha^2 - 4\beta}$

$$U = \gamma + \frac{1}{p}e^{-\frac{1}{2}\alpha T}$$

$$\times \left\{ \left[U'_0 - \left(\frac{\alpha + p}{2} \right)(\gamma - U_0) \right] e^{\frac{1}{2}pT} - \left[U'_0 - \left(\frac{\alpha - p}{2} \right)(\gamma - U_0) \right] e^{-\frac{1}{2}pT} \right\} \quad . \tag{2.9.20}$$

Case 3. $\alpha^2 < 4\beta$; $q = \sqrt{4\beta - \alpha^2}$

$$U = \gamma - e^{-\frac{1}{2}\alpha T}\{(\gamma - U_0)\cos(\tfrac{1}{2}qT) + \frac{1}{q}[\alpha(\gamma - U_0) - 2U'_0]\sin(\tfrac{1}{2}qT)\} \quad .\tag{2.9.21}$$

Case 4. $\alpha = 0$

This is a special case of Case 3 which reduces Eq. 2.9.21 to the following

$$U = \gamma - \left[(\gamma - U_0)\cos(\sqrt{\beta}T) - \frac{U'_0}{\sqrt{\beta}}\sin(\sqrt{\beta}T) \right] \quad . \tag{2.9.22}$$

In this instance, oscillations produced by the cosine and sine terms are neither amplified nor attenuated.

By way of numerical example, suppose that the constant amplitude oscillation of Eq 2.9.22 occurs when $\tau_* = \pi/2$. In this case, from Eq. 2.9.18, we obtain $\beta = 8/\pi^2$ and, since $\alpha = 0$, we have $U_* = 2/\pi$.

Thus, we have established a value for the equivalent constant we introduced above; we use this value, $U_* = 2/\pi$, in the remainder of our example. Consequently, Eq. 2.9.18 becomes

$$\alpha = \beta \left(\frac{\pi}{2} - \tau_* \right) \quad ; \qquad \beta = \frac{2}{\tau_*^2} \quad . \tag{2.9.23}$$

It should be pointed out that normally α and β are entirely independent quantities. In our particular example, however, they are related by Eq. 2.9.23. Thus, if a value of τ_* is selected, then β and, in turn, α are specified.

In Fig. 2.9.4 a number of curves are shown based on Eqs. 2.9.19 to 2.9.22 with $\beta = \beta_*$. We take as initial conditions: $U_0 = 0.10$ and $U_0' = 0$. Incidentally, it is noted that the abscissa of Fig. 2.9.4 is $\theta = t/\tau$ instead of $T = at$. The two time parameters are related: $T = \tau_* \theta$. This scaling of the time coordinate is made to allow a direct comparison of the linearized solution of Eq. 2.9.15, displayed in Fig. 2.9.4, and its exact solution shown in Fig. 6.2.1.

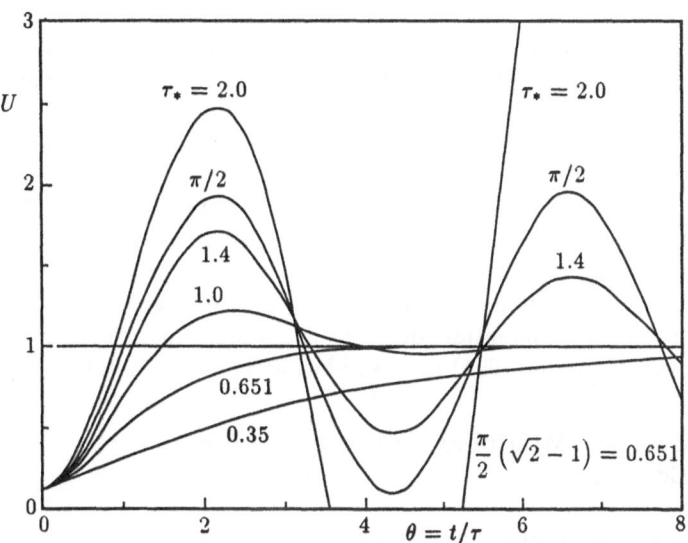

Fig. 2.9.4 Plot of solutions to second order linear differential equations for various values of $\tau_* = a\tau$. Compare with nonlinear plot of Fig. 6.2.1

In Eq. 2.9.21 the period of oscillation, T_{osc}, is

$$T_{\text{osc}} = \frac{2\pi}{\frac{1}{2}q} = \frac{4\pi}{\sqrt{4\beta - \alpha^2}} = \frac{2\pi\tau_*^2}{\sqrt{\tau_*^2 + \pi\tau_* - (\pi/4)^2}} \quad . \tag{2.9.24}$$

If we consider the limiting Case 1, for which $\alpha^2 = 4\beta$, then we have

$$\tau_* = \frac{\pi}{2}(\sqrt{2} - 1) = 0.651 \quad ; \qquad T_{\text{osc}} = \infty \quad . \tag{2.9.25}$$

The corresponding Eq. 2.9.19 is shown in Fig. 2.9.4 as the curve which asymptotically approaches $U = 1$ without oscillation.

If $\tau_* > \pi/2$, we see from Eq. 2.9.23 that α becomes negative. In this case, as Eq. 2.9.21 indicates, U oscillates with ever-increasing maximum and minimum amplitudes. Portions of the $\tau_* = 2.0$ curve are shown in Fig. 2.9.4. On the other hand, if $\tau_* < \pi/2$, the amplitudes decrease, as seen in the figure.

Second order ordinary differential equations arise in the study of many areas of the physical, biological, and social sciences.

In physics and engineering these equations make their appearance in topics concerning the mechanics of motion and vibration theory. They are also encountered in problems involving electrical circuits and control systems. Many references are available which examine the subject of differential equations in the physical sciences; representative are the books by Boyce and DiPrima (1986), Kreyszig (1983), Murphy (1960), Spiegel (1981), and Zwillinger (1989).

In the biological sciences, numerous publications are devoted to topics involving differential equations. Among these are the contributions of Batschelet (1979), Eason et al (1980), Jones and Sleeman (1983), and Segel (1980).

In the social sciences, differential equations have been utilized to a great extent in the field of economics. The historic text by Samuelson (1983) and the comprehensive books by Allen (1965), Gandolfo (1971), and Lancaster (1968) on mathematical economics are noteworthy. A lucid and concise introduction to the subject is presented by Clements (1984).

Indeed we now examine, as an illustration, a problem in economics featuring second order linear differential equations. This illustration is adapted from examples given by Allen (1965) and Burghes and Wood (1984).

2.9.6 An Illustration: Multiplier-Accelerator Model of a National Economy

We begin our analysis by writing the equation

$$z = c + x + g \quad , \tag{2.9.26}$$

in which z is the annual demand for goods and services of a nation, c is the consumption of these goods and services, x is investment, and g is the amount of government expenditure.

The assumption is made that

$$c = ky = (1 - s)y \quad , \tag{2.9.27}$$

in which y is the supply or output of goods and services, k is a "propensity to consume", and s is a "propensity to save". The quantity $1/s$ is termed the multiplier. We require that $0 < s < 1$.

It is further assumed that

$$\frac{dy}{dt} = \lambda(z - y) \quad , \tag{2.9.28}$$

where λ is the speed of response of supply to demand. Accordingly, Eq. 2.9.26 becomes

$$\frac{dy}{dt} + \lambda sy = \lambda(x + g) \quad . \tag{2.9.29}$$

Finally we assume that

$$\frac{dx}{dt} = \kappa \left(v\frac{dy}{dt} - x \right) \quad , \tag{2.9.30}$$

in which κ is the speed of response of investment to the rate of change of supply. The quantity v is termed the accelerator.

Solving Eq. 2.9.29 for x yields

$$x = \frac{1}{\lambda}\frac{dy}{dt} + sy - g \quad , \tag{2.9.31}$$

and taking the derivative of this expression gives

$$\frac{dx}{dt} = \frac{1}{\lambda}\frac{d^2y}{dt^2} + s\frac{dy}{dt} - \frac{dg}{dt} \quad . \tag{2.9.32}$$

Substituting Eqs. 2.9.31 and 2.9.32 into Eq. 2.9.30 yields the following expression

$$\frac{d^2y}{dt^2} + (\lambda s + \kappa - \kappa\lambda v)\frac{dy}{dt} + \kappa\lambda sy = \lambda(\kappa g + \frac{dg}{dt}) \quad . \tag{2.9.33}$$

Taking $g = g_0 = \text{constant}$ and letting

$$\alpha = (\lambda s + \kappa - \kappa\lambda v) \quad ; \qquad \beta = \kappa\lambda s \quad ;$$

$$\beta_* = \kappa\lambda g_0 \quad ; \qquad \gamma = \beta_*/\beta = g_0/s \tag{2.9.34}$$

provides the following second order linear differential equation

$$\frac{d^2y}{dt^2} + \alpha\frac{dy}{dt} + \beta y = \beta_* \quad . \tag{2.9.35}$$

This result is identical to Eq. 2.9.17 with y replacing U and t replacing T. Consequently the solutions given by Eqs. 2.9.19 through 2.9.22 are applicable.

As a numerical example suppose that

$$s = 0.58 \quad ; \qquad v = 1.0 \quad ; \qquad \kappa = 1.0 \quad ;$$

$$\lambda = 2.75 \quad ; \qquad y_0 = 4.0 \quad ; \qquad g_0 = 3.0 \quad .$$

Then

$$\alpha = (\lambda s + \kappa - \kappa\lambda v) = -0.155 \quad ; \qquad \beta = \kappa\lambda s = 1.595 \quad ;$$

$$q = \sqrt{4\beta - \alpha^2} = 2.521 \quad ; \qquad \gamma = g_0/s = 5.172 \quad .$$

Since $\alpha^2 < 4\beta$, we are dealing with Case 3 whose solution is given by Eq. 2.9.21. Also since α is negative, the amplitudes of the solution, $y = y(t)$, increase with increasing values of t. The period of oscillation is $t_{\text{osc}} = 2\pi/(q/2) = 4.985$.

Suppose that the initial conditions are $y(0) = y_0$ and $y'(0) = y'_0 = 0$. Then the solution for the supply equation is

$$y = \gamma - (\gamma - y_0)e^{-\frac{1}{2}\alpha t}[\cos(\tfrac{1}{2}qt) + \frac{\alpha}{q}\sin(\tfrac{1}{2}qt)] \quad . \tag{2.9.36}$$

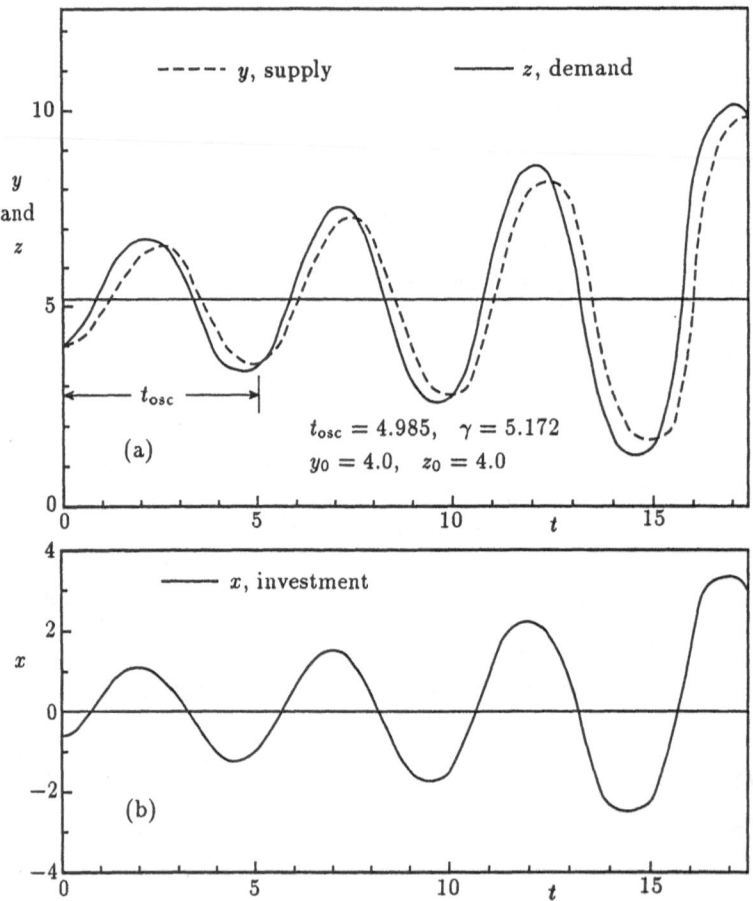

Fig. 2.9.5 Multiplier–accelerator model of a national economy. (a) Supply and demand curves; (b) investment curve

The derivative of this equation is

$$\frac{dy}{dt} = \tfrac{1}{2}(\gamma - y_0)e^{-\frac{1}{2}\alpha t}\left(q + \frac{\alpha^2}{q}\right)\sin(\tfrac{1}{2}qt) \quad . \tag{2.9.37}$$

Substituting Eqs. 2.9.36 and 2.9.37 into Eq. 2.9.28 yields the demand equation

$$z = \gamma - (\gamma - y_0)e^{-\frac{1}{2}\alpha t}\left[\cos(\tfrac{1}{2}qt) + \frac{1}{q}\left(\alpha - \frac{2\beta}{\lambda}\right)\sin(\tfrac{1}{2}qt)\right] \quad . \tag{2.9.38}$$

To complete the analysis we determine the investment equation. Substituting Eqs. 2.9.36 and 2.9.37 into Eq 2.9.31 gives

$$x = -s(\gamma - y_0)e^{-\frac{1}{2}\alpha t}\left[\cos(\tfrac{1}{2}qt) + \frac{1}{q}\left(\alpha - \frac{2\beta}{\lambda s}\right)\sin(\tfrac{1}{2}qt)\right] \quad . \tag{2.9.39}$$

Plots of the supply equation (y), demand equation (z), and investment equation (x) are presented in Fig. 2.9.5 utilizing the numerical values listed above.

A substantially more difficult problem would be that in which government expenditure, g, is not a constant but is instead a function of time.

For example, suppose that $g(t) = g_0 \sin \omega t$, where $\omega = 2\pi/t_g$; ω is the frequency and t_g is the period, not necessarily the same as t_{osc}. In this case, Eq. 2.9.33 becomes

$$\frac{d^2y}{dt^2} + \alpha\frac{dy}{dt} + \beta y = \beta_*\left(\sin \omega t + \frac{\omega}{\kappa}\cos \omega t\right) \quad , \tag{2.9.40}$$

where, again, $\beta_* = \kappa\lambda g_0$. This and other more complicated problems are considered by Allen (1965) and by the authors of the previously indicated references on differential equations.

3. Some Additional Frameworks

Up to here, we have examined rather closely several basic frameworks for the analysis and display of various types of growth and transfer processes.

We started with the simplest framework, the exponential function, and then proceeded to the more realistic and much more applicable logistic distribution. Considerable attention was given to the confined exponential distribution and to its linear combination with the logistic. The extremely important normal probability distribution was studied. We also took a look at the power law exponential and the power law logistic.

Judging from the wide range of topics of the illustrations of the preceding sections, we can conclude that these several basic frameworks are adequate for handling many kinds of practical problems.

Nevertheless, to enlarge our perspective and methodology we now consider some additional frameworks. We start with the very useful Gompertz and Weibull distributions and explore illustrations concerning tumor growth and technology substitution. This leads to an analysis of the so-called extreme value distribution with an example concerning dose response analysis of beetle mortality data. Then the quite generalized "hyperlogistic" distribution is introduced with illustrations featuring growth of the population of the world and growth of the public debt of the United States.

The total number of other distribution functions available to a researcher, though not endless, is certainly very long. We conclude the ensuing sections with an examination of some that seem most relevant for our purpose.

3.1 Gompertz Distribution

3.1.1 The Gompertz Distribution and Its Features

As we saw in Section 2.1, one of the consequences of simple Malthusian, or exponential, growth is that sooner or later, the magnitude, N, of the growing quantity becomes ridiculously large. How quickly it becomes enormously big depends on the value of the intrinsic growth coefficient, a, and the time, t, elapsed since it had the initial value, N_0. In any event, the exponential model finally becomes meaningless since nothing really ever goes off to infinity.

In an endeavor to construct a growth model more realistic than the exponential relationship, and at the same time establish a generalization of that simpler result, we devise the following Gompertz model.

We assume a growth phenomenon in which there is no inherent or natural constraint in the growth system or environment to retard the growth. That is, we assume an unlimited carrying capacity or, in other words, a crowding coefficient, b, equal to zero. Thus, as in the case of exponential growth, we commence with the differential equation

$$\frac{dN}{dt} = aN \quad . \tag{3.1.1}$$

At this point it is postulated that the growth coefficient, a, itself changes with time according to the relationship

$$\frac{da}{dt} = -ka \quad , \tag{3.1.2}$$

in which k is the decay coefficient of the growth coefficient; we assume that $k > 0$. With the initial condition $a(0) = a_0$, the solution to Eq. 3.1.2 is

$$a = a_0 e^{-kt} \quad . \tag{3.1.3}$$

Thus, the main feature of the Gompertz model is the incorporation of an exponentially decreasing growth coefficient.

Substituting Eq. 3.1.3 into Eq. 3.1.1 gives

$$\frac{dN}{dt} = a_0 e^{-kt} N \quad . \tag{3.1.4}$$

This equation is easily integrated. With the initial condition $N(0) = N_0$, we obtain

$$N = N_0 \exp\left[\frac{a_0}{k}\left(1 - e^{-kt}\right)\right] \quad . \tag{3.1.5}$$

This expression defines the Gompertz distribution function, or Gompertz growth equation. We would expect this equation to reduce to the case of exponential growth when $k = 0$. As it stands, the quantity in the brackets is indeterminate for this value of k. However, writing the exponential term in the parentheses of Eq. 3.1.5 in the form of a series, dividing by k and then letting $k = 0$, does indeed yield the exponential result, $N = N_0 \exp(a_0 t)$.

In Eq. 3.1.5, it is seen that as $t \to \infty$

$$N = N_* = N_0 e^{a_0/k} \quad . \tag{3.1.6}$$

Substituting this result into Eq. 3.1.5 gives an alternative form of the Gompertz equation

$$N = N_* \exp\left(-\frac{a_0}{k}e^{-kt}\right) \quad . \tag{3.1.7}$$

The quantity N_* is the value of N when t becomes very large, i.e., it is the asymptotic value of N. In this sense, it does represent a kind of carrying capacity and it can be interpreted that way. However, its appearance in the answer is not the result of a crowding coefficient, b. Rather, its presence is the mathematical outcome of the trade-off between an exponentially amplified growth (the $N = N_0 \exp(a_0 t)$ effect) and an exponentially retarded growth coefficient (the $a = a_0 \exp(-kt)$ effect).

The density function, $n(t)$, and the slope of the cumulative function dN/dt, of the Gompertz equation are obtained by differentiating Eq. 3.1.7. The result is

$$n = \frac{dN}{dt} = a_0 N_* \exp(-kt) \exp\left(-\frac{a_0}{k} e^{-kt}\right) \quad . \tag{3.1.8}$$

Utilizing Eq. 3.1.7 this expression gives

$$\frac{dN}{dt} = kN(\log_e N_* - \log_e N) \quad . \tag{3.1.9}$$

We observe that the specific growth rate, dN/Ndt, of the Gompertz function is given by the difference of the logarithms of N_* and N. Recall that the specific growth rate of the ordinary logistic is simply the difference between N_* and N.

The coordinates of the inflection point are

$$t_i = \frac{1}{k} \log_e \frac{a_0}{k} \quad ; \qquad N_i = \frac{1}{e} N_* \quad ; \qquad n_i = \frac{1}{e} kN_* \quad . \tag{3.1.10}$$

The original contributions of Gompertz are quite ancient. The main portion of his 1825 publication concerning "the law of human mortality" is reproduced by Smith and Keyfitz (1977); a brief summary of his mathematical analysis appears in Impagliazzo (1985). A number of years ago, Winsor (1932a) made an extensive comparison of the Gompertz and logistic distributions.

In Fig. 3.1.1, Gompertz growth curves are shown for various values of the decay coefficient k. For each of the curves, $a_0 = 0.50$ and $N_0 = 10$. Inflection points are indicated in the figure.

A comparison is made of exponential, logistic, and Gompertz growth curves in Fig. 3.1.2. The initial value, $N_0 = 10$, asymptotic value, $N_* = 100$, and growth coefficient, $a_0 = 0.50$, are the same for all three. From Eq. 3.1.6, the corresponding value of the decay coefficient for the Gompertz curve is $k = 0.217$. Inflection points of the logistic and Gompertz curves are shown.

3.1.2 An Illustration: Growth of Plant Leaves

In their book on the biometry of plant growth, Causton and Venus (1981) indicate that biologists have found the Gompertz function to be more appropriate than any other S-shaped function in quantitatively describing biological growth phenomena.

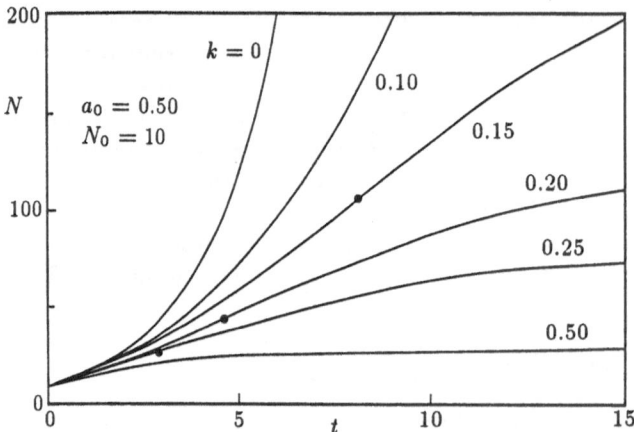

Fig. 3.1.1 Gompertz growth curves for various values of the decay coefficient k

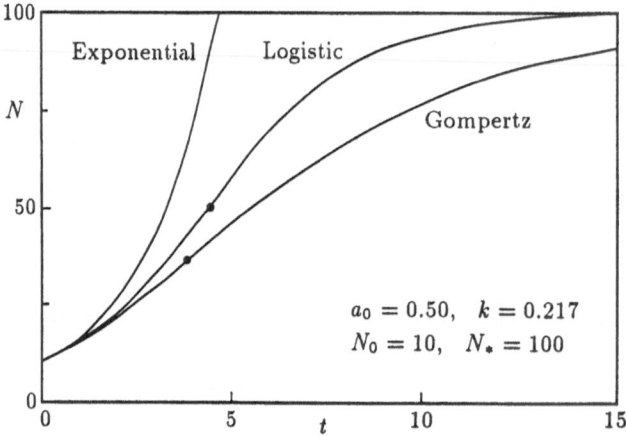

Fig. 3.1.2 Comparison of exponential, logistic, and Gompertz growth curves

In this regard, we examine the utilization of the Gompertz equation by considering the data shown in Table 3.1.1. In the table, the total area of the leaves of a plant, N (in cm^2), is shown as a function of time t (in days). These data are displayed in graphical form in Fig. 3.1.3; for the present, we disregard the smooth curve in the figure.

To fit the data of Table 3.1.1 to the Gompertz equation, it is necessary to determine the numerical values of three parameters: a_0, k, and either N_* or N_0; the latter two quantities are related by Eq. 3.1.6.

Table 3.1.1 Total area of the leaves of a plant

Time t [days]	N [cm^2]	Time t [days]	N [cm^2]
0	9.0	80	186.6
20	39.7	100	209.7
40	92.5	120	230.5
60	142.7	140	235.4

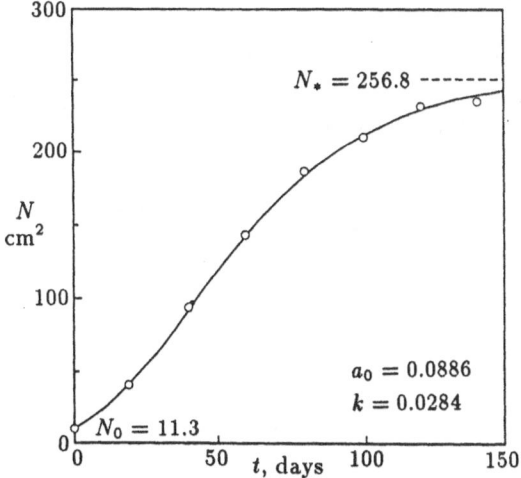

Fig. 3.1.3 Example of Gompertz growth equation. Observed and computed values of total area of leaves of a plant

As we have done several times before, we begin by writing Eq. 3.1.9 in finite difference form

$$\frac{1}{\overline{N}}\frac{\Delta N}{\Delta t} = k \log_e N_* - k \log_e \overline{N} \quad , \tag{3.1.11}$$

which is a linear relationship. Casting the data of the table into the form stipulated by Eq. 3.1.11 and carrying out a least squares computation gives $N_* = 256.8$ and $k = 0.02745$. This step of our analysis is displayed in Fig. 3.1.4.

Now knowing the values of N_* and k we solve Eq. 3.1.7 for a_0 and then substitute the eight pairs of (t, N) listed in Table 3.1.1 to compute a_0. The average value is $a_0 = 0.0882$.

Alternatively, we write Eq. 3.1.7 in the form

$$\log_e z = \log_e \left(\frac{a_0}{k}\right) - kt \quad , \tag{3.1.12}$$

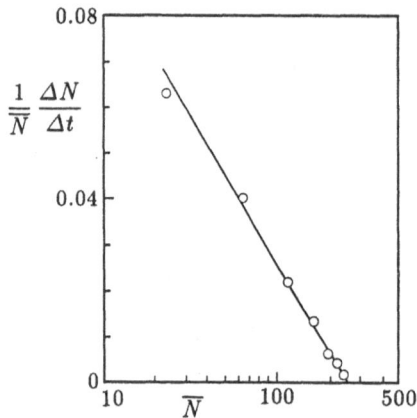

Fig. 3.1.4 Growth of plant leaves. Plot to determine values of the parameters N_* and k

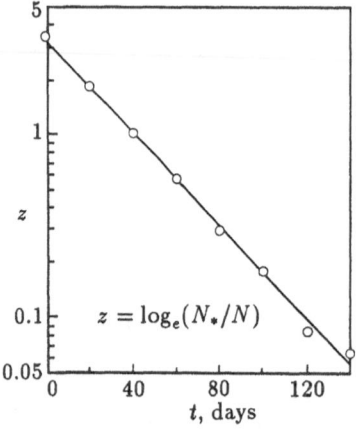

Fig. 3.1.5 Growth of plant leaves. Plot to determine values of the parameters a_0 and k

where $z = \log_e(N_*/N)$. Again we have a linear equation: $y = k_0 + k_1 x$. This method, illustrated in Fig. 3.1.5, yields $a_0 = 0.0890$ and $k = 0.02930$. These values are reasonably close to those calculated above.

Using $N_* = 256.8$ and the average values $a_0 = 0.0886$ and $k = 0.0284$, we obtain $N_0 = 11.3$ from Eq. 3.1.6. Substitution of these numbers into Eq. 3.1.7 yields the solid curve shown in Fig. 3.1.3. From Eq. 3.1.10, the inflection point coordinates are $t_i = 40.16$ days and $N_i = 94.5\,\text{cm}^2$. The maximum growth rate is $(dN/dt)_i = 2.68\,\text{cm}^2/\text{day}$.

It can be concluded that the Gompertz distribution provides a suitable framework for describing this particular botanical phenomenon.

Finally, in their book on mathematical models in agriculture, France and Thornley (1989) examine the Gompertz equation at length in connection with plant and animal growth. In addition, they present and analyze the so-called Chanter growth function, a hybrid of the logistic and Gompertz distributions.

3.1.3 An Illustration: Dynamics of Tumor Growth

A noteworthy study by Laird (1965) on the growth of a variety of transplanted and primary tumors in mice, rats, and rabbits involves the use of the Gompertz equation.

The study gave the results of 19 examples involving 12 different tumors. Observed data are shown in Fig. 3.1.6.

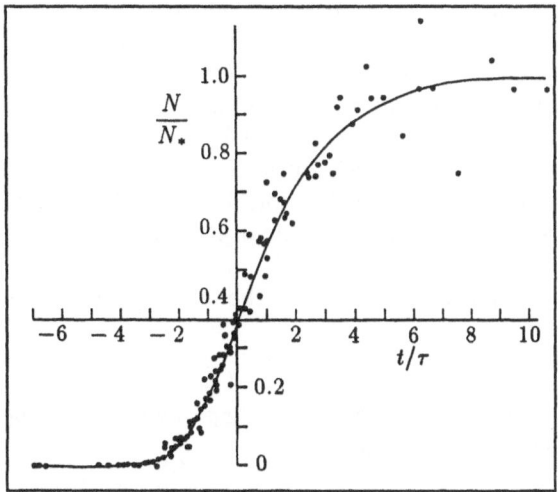

Fig. 3.1.6 A dimensionless plot of the growth of tumors based on the Gompertz distribution. From Laird (1965)

For each example, Laird plotted the size of the tumor, N, (expressed in grams or number of cells) versus time, t, (measured in hours or days). For each tumor the asymptotic value, N_*, was determined and the ordinate of the plot was then converted to the dimensionless quantity, N/N_*.

The time scale was established by determining the factor required to make the time for calculated doubling, preceding the inflection point, equal to unity. That is, the origin of the dimensionless time scale, $t/\tau = 0$, was set at the inflection point; the curve was forced to pass through the point $t/\tau = -1$, $N/N_* = 1/2e$. Incidentally, the time scale could have been normalized by "slope matching" at the inflection point. The final dimensionless plot would have been essentially the same.

The results of Laird's analysis are shown in Fig. 3.1.6. The solid curve shown in this figure is the Gompertz distribution, Eq. 3.1.7, with coordinates normalized as described.

3.2 Weibull Distribution

3.2.1 The Weibull Distribution and Its Features

Another useful and relatively simple distribution function is the Weibull distribution. It has found considerable application in renewal theory, which is the analysis of recurring events in phenomena involving the duration of life of components in a system or environment. This subject is covered extensively by Cox (1967). The Weibull distribution is utilized in problems involving failure rates, survival distribution times, and reliability theory. These and closely related topics are considered by McCullagh and Nelder (1983).

The Weibull distribution has also found application as a trend curve in technological forecasting; its simplicity and flexibility makes it very useful for this purpose. Sharif (1983) has included this distribution in his comprehensive monograph on technology transfer. Indeed in Section 3.2.2, an illustration is presented involving an example of technology substitution considered by Sharif.

The Weibull cumulative distribution function has the following definition

$$N = N_* \left(1 - e^{(t/b)^c}\right) \quad , \tag{3.2.1}$$

in which $b > 0$ is a scale parameter and $c > 0$ is a shape parameter. The corresponding density distribution function is

$$n = N_* \frac{c}{b} \left(\frac{t}{b}\right)^{c-1} e^{-(t/b)^c} \quad . \tag{3.2.2}$$

Defining $U = N/N_*$, $u = (bn)/N_*$, and $T = t/b$, Eqs. 3.2.1 and 3.2.2 are put into the following dimensionless forms

$$U = 1 - e^{-T^c} \quad ; \tag{3.2.3}$$

$$u = cT^{c-1}e^{-T^c} \quad . \tag{3.2.4}$$

These cumulative and density distribution functions, $U(T)$ and $u(T)$ are shown in Fig. 3.2.1 for several values of c. We note from Eq. 3.2.3 that when $T = 1$, the value of $U = 1 - (1/e)$ regardless of the value of c.

From Eq. 3.2.4 we obtain the inflection point of the cumulative distribution and the maximum value of the density distribution

$$T_i = \left(\frac{c-1}{c}\right)^{1/c} \quad ; \tag{3.2.5}$$

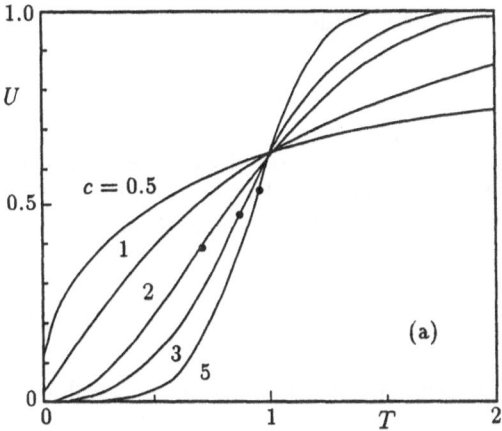

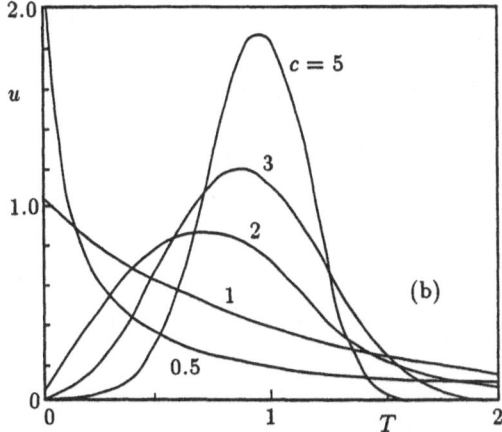

Fig. 3.2.1 Plots of the Weibull distribution for various values of c. (a) Cumulative distribution and (b) density distribution

$$U_i = 1 - e^{-(c-1)/c} \quad ; \tag{3.2.6}$$

$$u_i = c \left(\frac{c-1}{c} \right)^{(c-1)/c} e^{-(c-1)/c} \quad . \tag{3.2.7}$$

Several statistical parameters of the Weibull distribution are listed below.

The mode is defined as the value of the independent variable T at which the density distribution function has its maximum value. We have already determined this quantity; the answer, expressed by T_i, is given by Eq. 3.2.5.

As indicated in Eq. 2.2.26, the mean or expected value of T is defined as follows

$$\overline{T} = \int_{-\infty}^{\infty} T u(T) \, dT \quad . \tag{3.2.8}$$

Substitution of Eq. 3.2.4 into this equation yields

$$\overline{T} = \Gamma\left(\frac{c+1}{c}\right) \tag{3.2.9}$$

where $\Gamma(z)$ is the gamma function. This function is tabulated in Abramowitz and Stegun (1965) and other mathematical handbooks.

From Eq. 2.2.28, the variance of T has the following definition

$$\sigma^2 = \int_{-\infty}^{\infty} (T - \overline{T})^2 u(T)dT \quad . \tag{3.2.10}$$

If Eqs. 3.2.4 and 3.2.9 are substituted into this expression, the following answer is obtained

$$\sigma^2 = \Gamma\left(\frac{c+2}{c}\right) - \left[\Gamma\left(\frac{c+1}{c}\right)\right]^2 \quad . \tag{3.2.11}$$

The standard deviation is the square root of the variance.

Finally, the coefficient of variation is defined as the ratio of the standard deviation to the mean. From Eqs. 3.2.9 and 3.2.11 we get

$$\text{C.V.} = \frac{\sigma}{\overline{T}} = \left(\frac{\Gamma\left(\frac{c+2}{c}\right)}{[\Gamma\left(\frac{c+1}{c}\right)]^2} - 1\right)^{1/2} \quad . \tag{3.2.12}$$

In Table 3.2.1, numerical values of these statistical parameters are shown for various values of c.

Table 3.2.1 Numerical values of statistical parameters of the Weibull distribution for various values of c

c	T_i	$\overline{T}$	σ^2	σ	C.V.
1	0.000	1.000	1.000	1.000	1.000
2	0.707	0.886	0.215	0.464	0.524
3	0.874	0.893	0.106	0.326	0.365
4	0.931	0.906	0.065	0.255	0.281
5	0.956	0.918	0.044	0.210	0.229
6	0.970	0.928	0.032	0.179	0.193
7	0.078	0.935	0.025	0.158	0.169
8	0.983	0.942	0.019	0.138	0.146
9	0.987	0.947	0.016	0.126	0.133
10	0.990	0.951	0.014	0.118	0.124
∞	1.000	1.000	0.000	0.000	0.000

Unlike most of the other cumulative distribution functions we have considered, or will consider, the Weibull distribution has the initial condition $U(0) = 0$. That is, the curve passes through the origin. This characteristic is useful in growth or transfer processes in which we know or insist that $U(0) = 0$.

One very significant advantage of the Weibull distribution, in addition to its simplicity, is its flexibility. For example, the inflection point of the ordinary logistic distribution is required to be located at $U_i = 1/2$ and for the Gompertz distribution it must be at $U_i = 1/e$.

In contrast, as we establish from Eq. 3.2.6, we can make U_i as close to zero as we like by selecting values of c close to but larger than unity. On the other hand, we can make U_i as large as (but not larger than) $U_i = 1 - (1/e) = 0.632$ by selecting very large values of c.

If we desire $U_i = 1/2$, then we determine from Eq. 3.2.6 that $c = 1/(1 - \log_e 2) = 3.259$. For this value of c, the density distribution function is almost symmetrical about the inflection point time, T_i. For values of c greater than 3.259, the density curve is negatively skewed (the longer tail extends in the negative direction); for values of c less than 3.259, the density curve is positively skewed (the longer tail extends in the positive direction).

A number of curves are shown in Fig. 3.2.2 of the cumulative distribution for values of $b = 3$ and 5 and $c = 1, 2, 4, 6$, with $N_* = 100$. In addition, curves for the critical value, $c = 3.259$, are indicated. The small dots show inflection points.

3.2.2 An Illustration: Substitution of Diesel and Electric Locomotives for Steam Locomotives in the United States

During the entire period from the early years of the 1830s until immediately after World War II, practically all railroad traffic in the United States, freight and passenger alike, was handled by steam locomotives. In the years prior to the war, there had been some electrification of railway lines, particularly in the northeastern part of the country. Also, during the 1930s several railroads had put streamlined diesel-powered passenger trains into service.

Starting around 1945, virtually every railroad system in America began to replace its steam engines with diesel or electrically powered locomotives. During the ensuing 15 years, such replacement was essentially completed.

This particular example of technology substitution has been analyzed by Sharif and Islam (1980) who utilized the Weibull distribution as the framework for their analysis. The data presented by Sharif and Islam are listed in Table 3.2.2 and are shown in Fig. 3.2.3(a). The data represent the ratio of the number of diesel and electric locomotives to the total number of locomotives. Accordingly, $N_* = 1.0$. By a trial-and-error process, involving the maximization of the correlation coefficient as the solution criterion, the authors determined the base year ($t = 0$) to be 1918.

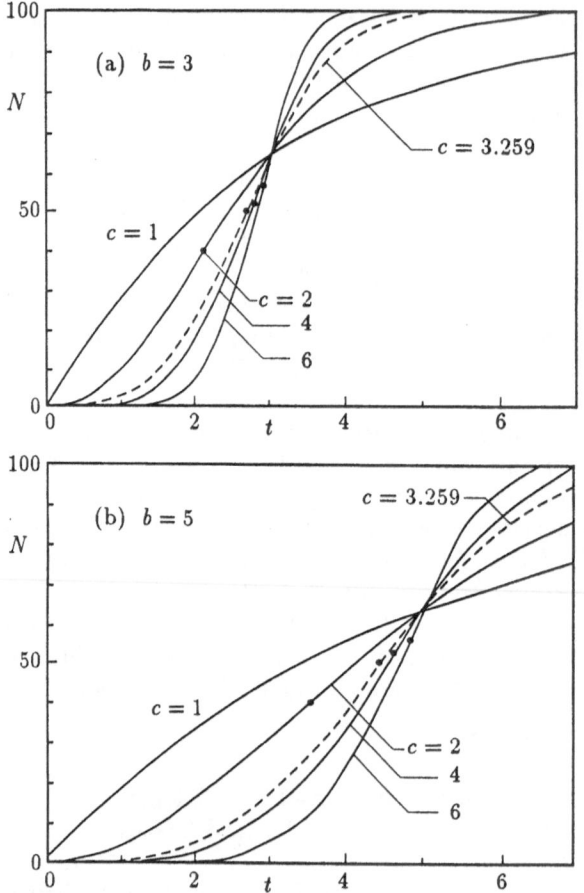

Fig. 3.2.2 Weibull cumulative distribution curves for various values of b and c

Table 3.2.2 Substitution of diesel and electric locomotives for steam locomotives in the United States. Fraction of diesel and electric to total, N/N_*. From Sharif and Islam (1980)

Year	t [years]	N/N_*	Year	t [years]	N/N_*
1918	0	0.00	1951	33	0.48
1945	27	0.11	1953	35	0.67
1947	29	0.18	1955	37	0.81
1949	31	0.30	1957	39	0.91

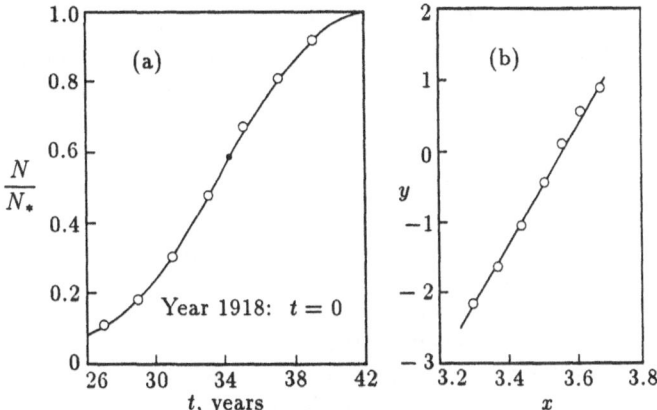

Fig. 3.2.3 Substitution of diesel and electric locomotives for steam locomotives in the United States. (a) Observed values and computed curve and (b) plot to determine values of b and c. Data of Sharif and Islam (1980)

One may write Eq. 3.2.1 in the following form

$$\log_e\left[\log_e\left(\frac{1}{1-N/N_*}\right)\right] = -c\log_e b + c\log_e t \quad . \tag{3.2.13}$$

This expression is a linear relationship, $y = k_0 + k_1 x$. Accordingly, a plot of the left-hand member of Eq. 3.2.13 versus $\log_e t$ should yield a straight line whose intercept and slope provide the values of b and c. Such a plot is shown in Fig. 3.2.3(b). A linear least squares computation gives $b = 34.89$ and $c = 8.46$.

Substitution of these numerical values into Eq. 3.2.1 yields the solid curve shown in Fig. 3.2.3.(a). From Eqs. 3.2.5 and 3.2.6 the inflection point is located at $t_i = 34.37$ (i.e., year 1952.37), $N_i = 0.59$. The maximum rate of substitution, $n_i = 9.0$ percent per year, occurred during that year.

Finally, Sharif and Islam define an "effective life span" (ELS). This quantity is defined as the time required for N/N_* to increase from a value of 0.10 to a value of 0.90.

We calculate the ELS by writing Eq. 3.2.1 in the form

$$t = b\left[\log_e\left(\frac{1}{1-N/N_*}\right)\right]^{1/c} \quad . \tag{3.2.14}$$

Substituting $N/N_* = 0.90$ in this equation to get t_{90} and $N/N_* = 0.10$ to obtain t_{10} and subtracting, yields the expression

$$\text{ELS} = b\left[(2.303)^{1/c} - (0.105)^{1/c}\right] \quad . \tag{3.2.15}$$

With $b = 34.89$ and $c = 8.46$ substituted into this equation, we determine $\text{ELS} = (38.50-26.74) = 11.76\,\text{years}$. This period extended over the years from

1944.74 to 1956.50. A final computation gives the result that at $t = 41.79$ (year: 1959.79), 99 percent of all locomotives in the United States were either diesel or electrically powered – the long era of the steam locomotive was over.

3.3 A Generalized Distribution

3.3.1 Cumulative and Density Distribution Functions

A somewhat generalized distribution is now introduced. As will be seen, this generalized case reduces to some special cases, including the logistic and normal probability distributions, by assigning specific values to one or more of the parameters in the distribution function. The analysis which follows stems from a model presented by Prentice (1976) in his examination of dose response relationships in biological assays.

The cumulative function for this generalized case is given by

$$U = \frac{1}{B(m_1, m_2)} \int_{-\infty}^{T} \frac{1}{(1 + e^{-z})^{m_1}} \frac{1}{(1 + e^{z})^{m_2}} dz \qquad (3.3.1)$$

in which $U = N/N_*$ and $T = at$. The quantity $B(m_1, m_2)$ is the beta function defined by the relationship

$$B(m_1, m_2) = \frac{\Gamma(m_1)\Gamma(m_2)}{\Gamma(m_1 + m_2)} \quad , \qquad (3.3.2)$$

where $\Gamma(z)$ is the gamma function. For positive integers of z, $\Gamma(z) = (z-1)!$ The parameters of the distribution function, m_1 and m_2, are positive though not necessarily integers.

The density distribution function corresponding to Eq. 3.3.1 is

$$u = \frac{1}{B(m_1, m_2)} \frac{1}{(1 + e^{-T})^{m_1}} \frac{1}{(1 + e^{T})^{m_2}} \quad . \qquad (3.3.3)$$

Setting the first derivative of this equation equal to zero shows that the maximum value of the density function, u_i, or equivalently, the inflection point of the cumulative function, is located at

$$T_i = \log_e(m_1/m_2) \quad . \qquad (3.3.4)$$

Clearly, T_i is zero, larger than zero, or less than zero depending on whether m_1/m_2 equals one, is greater than one, or is less than one. The value of u_i is

$$u_i = \frac{1}{B(m_1, m_2)} \frac{1}{(1 + m_2/m_1)^{m_1}} \frac{1}{(1 + m_1/m_2)^{m_2}} \quad . \qquad (3.3.5)$$

We note that time, T, is measured from the inflection point of the cumulative distribution.

3.3.2 A Generalized Symmetrical Function

If the values $m_1 = m_2 = m$ are substituted into Eq. 3.3.1, and we utilize the relationships $\cosh z = [\exp(z) + \exp(-z)]/2$ and $\cosh(z/2) = [(1 + \cosh z)/2]^{1/2}$, we obtain

$$u = \frac{1}{2^{2m}} \frac{\Gamma(2m)}{[\Gamma(m)]^2} \left(\operatorname{sech} \frac{T}{2} \right)^{2m} \tag{3.3.6}$$

and

$$U = \frac{1}{2^{2m}} \frac{\Gamma(2m)}{[\Gamma(m)]^2} \int_{-\infty}^{T} \left(\operatorname{sech} \frac{z}{2} \right)^{2m} dz \quad . \tag{3.3.7}$$

The first derivative of Eq. 3.3.6, when set equal to zero, indicates that the maximum value of the density function and the inflection point of the cumulative function, occur at $T_i = 0$. This result is obtained directly from Eq. 3.3.4. From Eq. 3.3.5 or Eq. 3.3.6, the maximum density is

$$u_i = \frac{1}{2^{2m}} \frac{\Gamma(2m)}{[\Gamma(m)]^2} \quad . \tag{3.3.8}$$

Setting the second derivative of Eq. 3.3.6 equal to zero shows that the inflection points of the density function are located at

$$T_c = \pm 2 \operatorname{arcsinh} \frac{1}{\sqrt{2m}} \quad . \tag{3.3.9}$$

The value of the density function at these points is

$$u_c = \frac{1}{2^{2m}} \frac{\Gamma(2m)}{[\Gamma(m)]^2} \left(\frac{2m}{1 + 2m} \right)^m \quad . \tag{3.3.10}$$

Case 1: $m = 1$

This first special case corresponds to the ordinary logistic distribution. Substituting $m = 1$ into Eq. 3.3.6, gives the density function

$$u = \frac{1}{4} \operatorname{sech}^2 \frac{T}{2} \tag{3.3.11}$$

as indicated by Eq. 2.2.21. The cumulative function, Eq. 3.3.7, reduces to

$$U = \frac{1}{2} \left(1 + \tanh \frac{T}{2} \right) \tag{3.3.12}$$

which is the same as Eq. 2.2.20. Clearly, $u_i = 1/4$ and $U_i = 1/2$. From Eq. 3.3.9, the inflection points of the density function are located at $T_c = \pm 2 \operatorname{arcsinh}(1/\sqrt{2})$; the corresponding value of the density, from Eq. 3.3.10 , is $u_c = 1/6$.

Case 2: $m = 2$ In a similar fashion, setting $m = 2$ in Eqs. 3.3.6 and 3.3.7 gives

$$u = \frac{3}{8}\operatorname{sech}^4\frac{T}{2} \tag{3.3.13}$$

and

$$U = \frac{1}{2}\left(1 + \frac{3}{2}\tanh\frac{T}{2} - \frac{1}{2}\tanh^3\frac{t}{2}\right) \ . \tag{3.3.14}$$

It is apparent that $u_i = 3/8$ and $U_i = 1/2$. We also establish that $T_c = \pm 2\operatorname{arcsinh}(1/2)$ and $u_c = 6/25$.

Case 3: $m \to \infty$

Although the algebra quickly becomes tedious, the analysis proceeds in the same way for the determination of the cumulative and density distribution functions for larger values of m. Thus, when $m = 3$, we obtain

$$u = \frac{15}{32}\operatorname{sech}^6\frac{T}{2} \tag{3.3.15}$$

and

$$U = \frac{1}{2}\left(1 + \frac{15}{18}\tanh\frac{T}{2} - \frac{5}{4}\tanh^3\frac{T}{2} + \frac{3}{8}\tanh^5\frac{T}{2}\right) \ . \tag{3.3.16}$$

The other quantities are: $u_i = 15/32$, $U_i = 1/2$, $T_c = \pm 2\operatorname{arcsinh}(1/\sqrt{6})$, and $u_c = 405/1372$.

As m takes on larger and larger values, the pattern of the distributions is clear enough. As Eq. 3.3.6 shows, the density function behaves like $\operatorname{sech}^{2m}(T/2)$. The cumulative function is a series of $\tanh(T/2)$ terms with coefficients that add to zero for $T = -\infty$ and to unity for $T = \infty$.

The maximum density, u_i, and inflection point density, u_c, increase according to Eqs. 3.3.8 and 3.3.10, respectively. The times corresponding to the inflection points of the cumulative function and of the density function, respectively, remain at $T_i = 0$ and decrease according to Eq. 3.3.9.

When m becomes very large, use can be made of the following expressions

$$\Gamma(z) = \sqrt{2\pi}\,e^{-z}z^{z-\frac{1}{2}} \ ; \tag{3.3.17}$$

$$\lim_{m\to\infty}\left(1 + \frac{z}{m}\right)^m = e^z \ . \tag{3.3.18}$$

$$\operatorname{arcsinh}\left(\frac{1}{z}\right) = \left(\frac{1}{z}\right) - \frac{1}{2\times 3}\left(\frac{1}{z}\right)^3 + \cdots \ . \tag{3.3.19}$$

With these approximations,

$$u_i = \frac{1}{\sqrt{2\pi}}\sqrt{\frac{m}{2}} \ ; \qquad T_i = 0 \ ; \tag{3.3.20}$$

$$u_c = \frac{1}{\sqrt{2\pi}} \sqrt{\frac{m}{2e}} \quad ; \qquad T_c = \pm \sqrt{\frac{2}{m}} \quad . \tag{3.3.21}$$

The density function, Eq. 3.3.6, becomes

$$u = \frac{1}{\sqrt{2\pi}} \sqrt{\frac{m}{2}} \operatorname{sech}^{2m} \frac{T}{2} \quad . \tag{3.3.22}$$

For comparison, a normal probability distribution is defined

$$u = \frac{1}{\alpha\sqrt{2\pi}} \exp\left(-\frac{1}{\beta}\frac{T^2}{2}\right) \quad , \tag{3.3.23}$$

where α and β are functions of m to be determined. Matching the expressions for the hyperbolic secant relationship, Eq. 3.3.22, and the normal probability relationship, Eq. 3.3.23, at the points, (T_i, u_i) and (T_c, u_c), we establish that $\alpha = (2/m)^{1/2}$ and $\beta = \alpha^2 = (2/m)$. Accordingly, Eq. 3.3.23 becomes

$$u = \frac{1}{\sqrt{2\pi}} \sqrt{\frac{m}{2}} \exp\left(-\frac{m}{2}\frac{T^2}{2}\right) \quad . \tag{3.3.24}$$

The standard form of the normal probability density function is

$$u = \frac{1}{\sigma\sqrt{2\pi}} \exp\left(-\frac{(T-\overline{T})^2}{2\sigma^2}\right) \quad . \tag{3.3.25}$$

Comparing this expression with Eq. 3.3.24, we see that the standard deviation is $\sigma = (2/m)^{1/2}$ and the variance, $\sigma^2 = 2/m$. The mean value, $\overline{T} = T_i = 0$.

Equating Eqs. 3.3.22 and 3.3.24 yields the approximation, $\operatorname{sech}(T/2) = \exp(-T^2/8)$. For comparison, these quantities are expanded in their respective series

$$\operatorname{sech}\frac{T}{2} = 1 - \frac{1}{8}T^2 + \frac{5}{384}T^4 - \frac{61}{46080}T^6 + \cdots \tag{3.3.26}$$

and

$$\exp(-T^2/8) = 1 - \frac{1}{8}T^2 - \frac{3}{384}T^4 - \frac{15}{46080}T^6 + \cdots \quad . \tag{3.3.27}$$

At $T = 1$, for example, the difference in the numerical values of these two equations, to the indicated four terms, is about 0.48 percent.

The density function given by Eq. 3.3.24 corresponds to the cumulative function

$$U = \frac{1}{\sqrt{2\pi}} \int_{-\infty}^{\theta} e^{-\frac{1}{2}z^2} dz \quad , \tag{3.3.28}$$

where $\theta = (m/2)^{1/2}T$. Accordingly, we have established that the generalized symmetrical distribution of Eq. 3.3.7 approximates the normal probability distribution for large values of m. In Fig. 3.3.1, density and cumulative distributions are shown for various values of m.

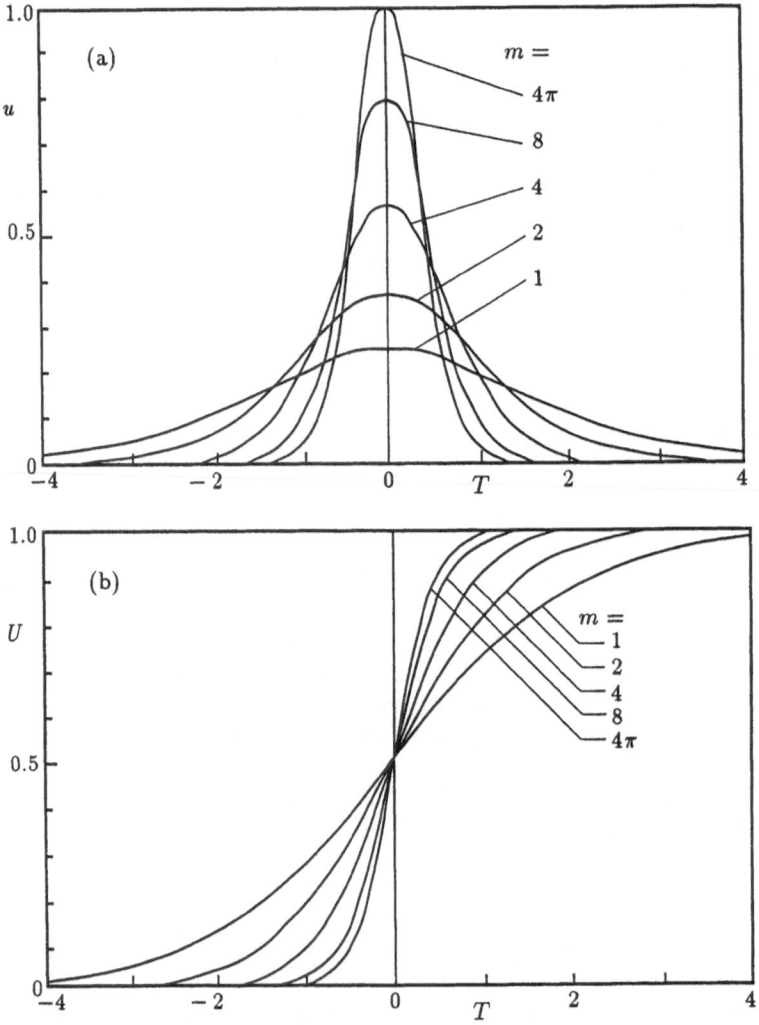

Fig. 3.3.1 A generalized symmetrical distribution, $m_1 = m_2 = m$, for various values of m. (a) Density distribution and (b) cumulative distribution

As θ becomes infinitely large, with positive values of T, the upper limit of Eq. 3.3.28 becomes positively infinite and hence $U = 1$. Alternatively, for negative values of T, the upper limit is negatively infinite and so $U = 0$. Thus, for infinite m, the cumulative distribution is simply a "step function" at $T = 0$. The density distribution reduces to a "pulse function" with infinite amplitude and zero width.

3.3.3 Extreme Maximum Value Distribution

Consider now the case in which m_1 has any positive value and $m_2 = 1$. Then Eq. 3.3.1 reduces to

$$U = m_1 \int_{-\infty}^{T} \frac{e^{-z}}{(1 + e^{-z})^{m_1 + 1}} dz \quad , \tag{3.3.29}$$

which simplifies to

$$U = \frac{1}{(1 + e^{-T})^{m_1}} \quad . \tag{3.3.30}$$

The corresponding density function is

$$u = \frac{m_1 e^{-T}}{(1 + e^{-T})^{m_1 + 1}} \quad . \tag{3.3.31}$$

From this expression we determine that the time of the cumulative inflection point is $T_i = \log_e(m_1)$. The corresponding values of u_i and U_i are

$$u_i = \frac{1}{(1 + 1/m_1)^{m_1 + 1}} \quad ; \qquad U_i = \frac{1}{(1 + 1/m_1)^{m_1}} \quad . \tag{3.3.32}$$

In Fig. 3.3.2, plots of Eqs. 3.3.30 and 3.3.31 are shown for several values of m_1. The ordinary logistic equation, $m_1 = 1$, represents a limiting case. It is observed that for $m_1 > m_2$, the density curves are positively skewed.

The time coordinate is now translated according to the relationship $T = T_i + T_*$, so that the origin of T_* is at the inflection point of the cumulative distribution. This relationship is substituted into Eqs. 3.3.30 and 3.3.31. Subsequently, letting $m_1 \rightarrow \infty$, the density distribution becomes

$$u = \exp[-T_* - \exp(-T_*)] \tag{3.3.33}$$

and the cumulative distribution

$$U = \exp[\exp(-T_*)] \tag{3.3.34}$$

The function described by Eqs. 3.3.33 and 3.3.34 is known as the *extreme maximum value* distribution. These equations are also shown in Fig. 3.3.2, although they have been translated from an infinite distance to the right for easier viewing. It is seen that both the maximum density, u_i, and the cumulative at the inflection point, U_i, have the value $1/e$ when m_1 is infinitely large

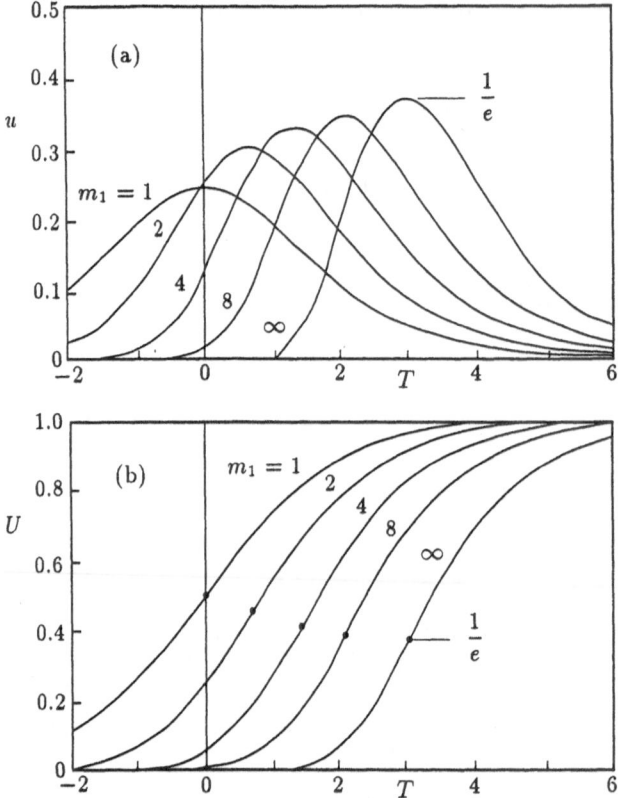

Fig. 3.3.2 A generalized distribution with $m_2 = 1$ for various values of m_1.(a) Density distribution and (b) cumulative distribution. The two curves for $m_1 = \infty$ are in fact an infinite distance to the right. This is the extreme maximum value distribution

Dropping the subscript on T, and recalling that $u = dU/dT$, it is easily established that the growth rate corresponding to the extreme maximum distribution is

$$\frac{dU}{dT} = U \log_e \frac{1}{U} \tag{3.3.35}$$

and the specific growth rate is

$$\frac{1}{U}\frac{dU}{dT} = \log_e \frac{1}{U} \quad . \tag{3.3.36}$$

These two equations, as well as Eqs. 3.3.33 and 3.3.34, are shown in Fig. 3.3.3. Finally, it is easily seen that

$$\frac{dU}{dT} = e^{-T} U \quad . \tag{3.3.37}$$

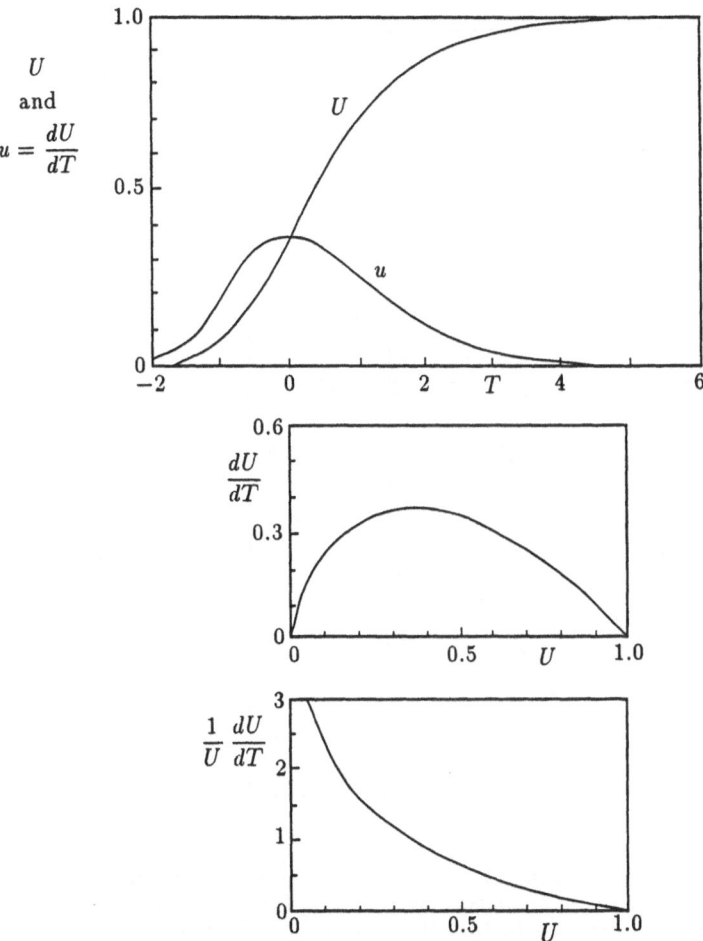

Fig. 3.3.3 Descriptions of the extreme maximum value distribution

From this equation, the similarity between the extreme maximum value distribution and the Gompertz distribution is noted. Symbolically, Eq. 3.3.37 is the same as Eq. 3.1.4, the basic differential equation of the Gompertz distribution.

3.3.4 Extreme Minimum Value Distribution

The case in which $m_1 = 1$ and m_2 has any positive value is analyzed in precisely the same way as in the preceding section. From Eq. 3.3.1, the cumulative function is

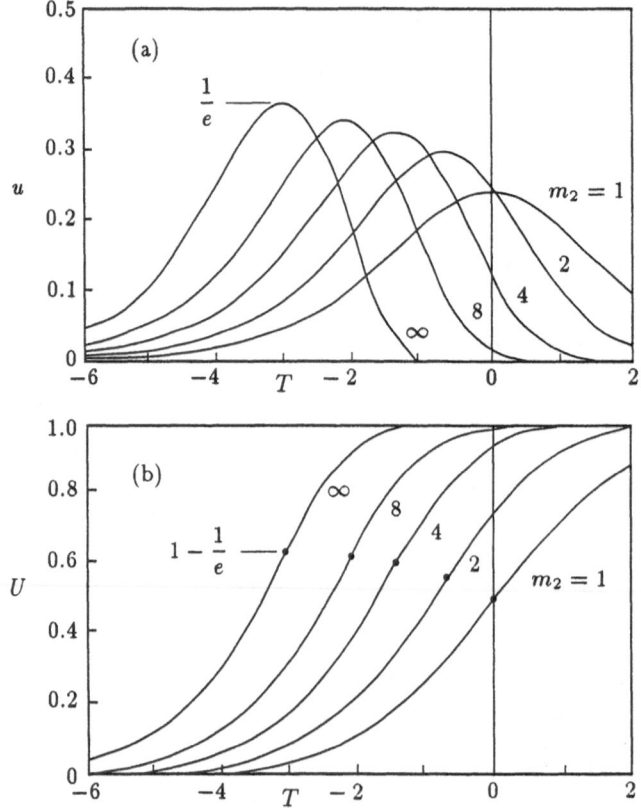

Fig. 3.3.4 A generalized distribution with $m_1 = 1$ for various values of m_2. (a) Density distribution and (b) cumulative distribution. The two curves for $m_2 = \infty$ are in fact an infinite distance to the left. This is the extreme minimum value distribution

$$U = m_2 \int_{-\infty}^{T} \frac{e^z}{(1+e^z)^{m_2+1}} dz \quad , \tag{3.3.38}$$

which can be written

$$U = 1 - \frac{1}{(1+e^T)^{m_2}} \quad . \tag{3.3.39}$$

The density function becomes

$$u = \frac{m_2 e^T}{(1+e^T)^{m_2+1}} \tag{3.3.40}$$

and the time corresponding to the inflection point of the cumulative function is $T_i = \log_e(1/m_2)$. The corresponding values of u_i and U_i are

$$u_i = \frac{1}{(1 + 1/m_2)^{m_2+1}} \quad ; \quad U_i = 1 - \frac{1}{(1 + 1/m_2)^{m_2}} \quad . \tag{3.3.41}$$

Plots of Eqs. 3.3.39 and 3.3.40 are presented in Fig. 3.3.4 for various values of m_2. Clearly, when $m_1 < m_2$, the density curves are negatively skewed.

As before, the time coordinate is shifted by the expression, $T = T_i + T_*$, so that the origin, $T_* = 0$, corresponds to the inflection point of the cumulative curve. Substituting this time translation into Eqs. 3.3.39 and 3.3.40 and letting $m_2 \to \infty$ gives the following asymptotic expressions

$$u = \exp(T_* - \exp T_*) \tag{3.3.42}$$

and

$$U = 1 - \exp(-\exp T_*) \quad . \tag{3.3.43}$$

These two equations describe the *extreme minimum value* distribution and are also shown in Fig. 3.3.4. In this case, it is noted that the maximum density is $u_i = 1/e$ and that at the inflection point the value of the cumulative function is $U_i = 1 - (1/e)$.

As in the previous section, dropping the subscript on T and using $u = dU/dT$, the growth rate for the extreme minimum value distribution is

$$\frac{dU}{dT} = (1 - U) \log_e \left(\frac{1}{1-U} \right) \tag{3.3.44}$$

and the specific growth rate is

$$\frac{1}{U} \frac{dU}{dT} = \frac{1-U}{U} \log_e \left(\frac{1}{1-U} \right) \quad . \tag{3.3.45}$$

These equations are shown in Fig. 3.3.5, along with plots of Eqs. 3.3.42 and 3.3.43.

To conclude, it is easy to establish that the basic differential equation for the extreme minimum value distribution is

$$\frac{dU}{dT} = e^T (1 - U) \quad . \tag{3.3.46}$$

The subject of extreme value distribution is covered at length by Gumbel (1958). Likewise, extensive treatment is given by Cox (1970) on the analysis of binary data, the topic of our next illustration.

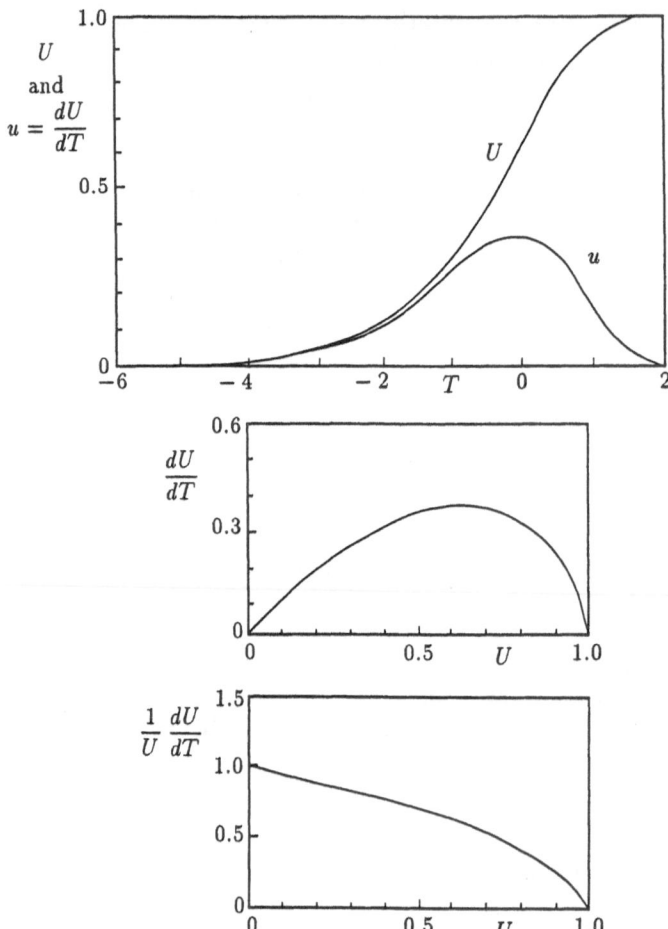

Fig. 3.3.5 Descriptions of the extreme minimum value distribution

3.3.5 An Illustration: Dose Response Analysis of Beetle Mortality Data

Quite a number of years ago, Bliss (1935) carried out experiments to measure the mortality of adult beetles following five hours exposure to gaseous carbon disulfide at various concentrations. The result of his quantal response ("alive or dead") experiments are presented in Table 3.3.1. These data have been studied at length by Prentice (1976) and by Dobson (1983) in conjunction with their generalized linear model analyses.

The data of Table 3.3.1 are shown in Fig. 3.3.6 in which the fraction killed, $U = N/N_*$, is plotted against the dose concentration, t.

Table 3.3.1 Experimental data on the mortality of adult beetles. From Bliss (1935), Prentice (1976) and Dobson (1983)

t (dosage $\log_{10} CS_2$ [mg/l])	N_* (number of insects)	N (number killed)	U N/N_*
1.6907	59	6	0.1017
1.7242	60	13	0.2167
1.7552	62	18	0.2903
1.7842	56	28	0.5000
1.8113	63	52	0.8254
1.8369	59	53	0.8983
1.8610	62	61	0.9839
1.8839	60	60	1.0000

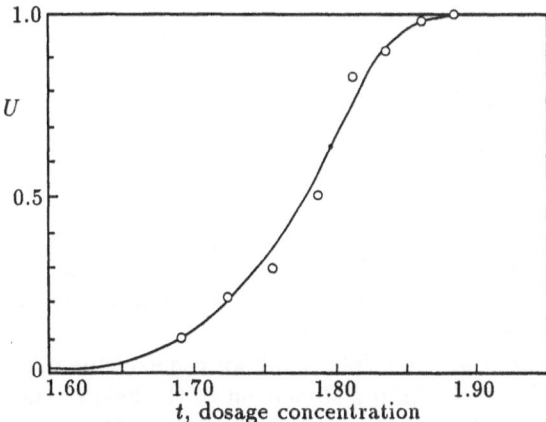

Fig. 3.3.6 Fractional mortality of adult beetles to varying concentrations of carbon disulfide. Solid curve is the extreme minimum value distribution. Data of Bliss (1935)

These data were analyzed on the basis of the following functions:
Logistic function

$$U = \frac{1}{1 + e^{-T}} \quad ; \tag{3.3.47}$$

Normal probability function

$$U = \frac{1}{\sqrt{2\pi}} \int_{-\infty}^{T} e^{-z^2/2} dz \quad ; \tag{3.3.48}$$

Extreme maximum value function

$$U = \exp[-\exp(-T)] \quad ; \tag{3.3.49}$$

Extreme minimum value function

$$U = 1 - \exp[-\exp T] \quad ; \tag{3.3.50}$$

in which

$$T = k + at \quad . \tag{3.3.51}$$

A least squares analysis was carried out in the framework of each of these four dose response models. The results are summarized in Table 3.3.2.

Table **3.3.2** Results of analysis of beetle mortality data. Values of correlation coefficient r and dosage coefficients, k and a, for selected functions. $T = k + at$

Function	r	k	a
Extreme minimum	0.994	−38.591	21.491
Normal	0.980	−34.996	19.795
Logistic	0.972	−62.212	35.221
Extreme maximum	0.940	−47.558	27.264

As indicated in Table 3.3.2, the correlation coefficient, r, has the largest value when the extreme minimum value distribution function is utilized. The curve of this function is shown in Fig. 3.3.6, using Eq. 3.3.50 and the values $k = -38.591$, $a = 21.491$.

Based on the numerical values shown in Table 3.3.2 for the extreme minimum value function, the lethal dose for killing 50 percent of the beetles (i.e., the LD$_{50}$ value) is $t_{50} = 1.7786$. The corresponding value of the 50 percent lethal concentration is $C_{50} = \text{antilog}_{10}(t_{50}) = 60.07 \, \text{mg/l} \, CS_2$. The values of C_{90} and C_{99} are, respectively, 68.31 and 73.58 mg/l CS$_2$. Some practical aspects of this topic of quantal response of insects to chemical dosage have been considered by Matthews (1984).

3.4 Hyperlogistic Distribution

3.4.1 The Differential Equation and Some Examples

A quite generalized form of the differential equation for growth and transfer phenomena has been presented by Peschel and Mende (1986). Labeling it the "hyperlogistic differential equation", they present the following definition

$$\frac{dU}{dT} = U^r (1 - U^s)^n \quad . \tag{3.4.1}$$

As before, $U = N/N_*$, $T = at$ and r, s, and n are non-negative constants. Assignment of particular values to these constants produces some of the differential equations we have examined in previous sections.

We can simplify Eq. 3.4.1 somewhat by making the substitution: $V = U^s$. This yields

$$\frac{dV}{dT} = sV^m (1 - V)^n \quad , \tag{3.4.2}$$

where $m = (r + s - 1)/s$.

Before proceeding to the task of obtaining the solution to this equation, we look at a few special cases. The simplest case is generated if $m = 1$ (and hence $r = 1$) and $n = 0$ is substituted in Eqs. 3.4.1 and 3.4.2. This yields the equation for exponential growth in both V and U.

An interesting result is obtained if $m = 0$ and $n = 1$ in Eq. 3.4.2. This time we get the confined exponential equation in V whose solution is

$$V = 1 - (1 - V_0)e^{-sT} \quad . \tag{3.4.3}$$

Accordingly

$$U = [1 - (1 - U_0^s)e^{-sT}]^{1/s} \quad . \tag{3.4.4}$$

In this case $r = 1 - s$, and so the differential equation in U reduces to

$$\frac{dU}{dT} = U^{1-s} - U \quad . \tag{3.4.5}$$

This is essentially the differential equation given by Bertalanffy (1968) in his mathematical model of animal growth.

If we set $m = 1$ and $n = 1$ in Eq. 3.4.2 then we have the differential equation for an ordinary logistic in V. Since $r = 1$, then Eq. 3.4.1 becomes a power law logistic in U. This particular equivalency between U and V has been pointed out by Nelder (1961) and by Turner et al. (1969).

An important category of the hyperlogistic differential equation, expressed by Eq. 3.4.1, has been presented and analyzed by Turner et al (1976). Utilizing our own symbols for the various quantities, Turner and his colleagues write

$$\frac{dU}{dT} = U^{1-s(n-1)} (1 - U^s)^n \quad . \tag{3.4.6}$$

If $n = 1$, this equation reduces to a power law logistic. If $n = 0$, it becomes essentially the power law exponential we examined in Section 2.1.

As before, letting $V = U^s$ in Eq. 3.4.6 yields

$$\frac{dV}{dT} = sV^{2-n} (1 - V)^n \tag{3.4.7}$$

In this case, if $n = 1$, we have the ordinary logistic function in V. If $n = 0$, we obtain the "coalition growth model" which we examine in Section 3.4.6.

The solution to Eq. 3.4.7, and hence Eq. 3.4.6, is straightforward. A simple integration yields

$$U = \frac{1}{\left\{1 + \left[\left(\frac{U_0^s}{1-U_0^s}\right)^{n-1} + s(n-1)T\right]^{-1/(n-1)}\right\}^{1/s}} \quad , \tag{3.4.8}$$

which agrees with the result of Turner et al. It is not difficult to show that when $n = 1$, Eq. 3.4.8 reduces to the ordinary logistic equation in V and the power law logistic in U.

When $s = 1$ is substituted into Eq. 3.4.8, we obtain

$$U = \frac{1}{1 + \left[\left(\frac{U_0}{1-U_0}\right)^+ (n-1)T\right]^{-1/(n-1)}} \quad . \tag{3.4.9}$$

This equation was utilized by Pruitt et al. (1974) in their study of complement mediated lysis of sensitized red cells. They concluded that Eq. 3.4.9 provides an accurate description of the quantity of cells lysed as a function of time.

3.4.2 Solution to the Differential Equation

We now seek the solutions to Eqs. 3.4.1 and 3.4.2. Taking the initial condition $V(0) = V_0 = U_0^s$, Eq. 3.4.2 can be written in the following form

$$sT = \int_{V_0}^{V} z^{-m}(1-z)^{-n}dz \quad . \tag{3.4.10}$$

This equation is essentially the incomplete beta function given by Abramowitz and Stegun (1965) and extensively tabulated by Pearson (1968):

$$I_x(p,q) = \frac{1}{B(p,q)} \int_0^x z^{p-1}(1-z)^{q-1}dz \quad , \tag{3.4.11}$$

in which $B(p,q)$ is the beta function defined as

$$B(p,q) = \int_0^1 z^{p-1}(1-z)^{q-1}dz = \frac{\Gamma(p)\Gamma(q)}{\Gamma(p+q)} \quad , \tag{3.4.12}$$

where $\Gamma(z)$ is the gamma function. Comparing Eqs. 3.4.10 and 3.4.11, we see that $p = 1 - m$ and $q = 1 - n$.

From the preceding, it is clear that the general solution of Eq. 3.4.2 and hence Eq. 3.4.1 is the following

$$sT = B(p,q)[I_V(p,q) - I_{V_0}(p,q)] \quad , \tag{3.4.13}$$

which says that $T = T_*(V) = T(U)$.

As indicated by Abramowitz and Stegun (1965) for values of $< x < 1$, the incomplete beta function can be expressed by the following infinite series:

$$I_x(p,q) = \frac{x^p(1-x)^q}{pB(p,q)} \left(1 + \sum_{j=0}^{\infty} \frac{B(p+1,j+1)}{B(p+q,j+1)} x^{j+1} \right) \quad . \tag{3.4.14}$$

From here on, we shall confine our interest to the case $n = 1$. Accordingly, the differential equations, Eqs. 3.4.1 and 3.4.2, become

$$\frac{dU}{dT} = U^r(1 - U^s) \tag{3.4.15}$$

and

$$\frac{dV}{dT} = sV^m(1 - V) \quad . \tag{3.4.16}$$

With $n = 1$ then $q = 0$ and Eq. 3.4.13 becomes

$$sT = B(p,0)[I_V(p,0) - I_{V_0}(p,0)] \quad . \tag{3.4.17}$$

The infinite series, Eq. 3.4.14, drastically simplifies to yield

$$sT = V^p \sum_{j=0}^{\infty} \frac{V^j}{p+j} - V_0^p \sum_{j=0}^{\infty} \frac{V_0^j}{p+j} \quad , \tag{3.4.18}$$

which we shall regard as our final solution.

For the following three cases, we have exact answers; these can be used to check the accuracy of Eq. 3.4.18:

Case 1. $m = 0$, $p = 1$

Confined exponential equation

$$sT = \log_e \left(\frac{1 - V_0}{1 - V} \right) \quad ; \tag{3.4.19}$$

Case 2. $m = 1/2$, $p = 1/2$

Exponential logistic equation

$$sT = \log_e \left(\frac{1 - \sqrt{V_0}}{1 + \sqrt{V_0}} \right) \left(\frac{1 + \sqrt{V}}{1 - \sqrt{V}} \right) \quad ; \tag{3.4.20}$$

Case 3. $m = 1$, $p = 0$

Ordinary logistic equation

$$sT = \log_e \left(\frac{1 - V_0}{V_0}\right) \left(\frac{V}{1 - V}\right) \quad .$$
(3.4.21)

Utilizing the relationship

$$\log_e z = (z - 1) - \frac{1}{2}(z - 1)^2 + \frac{1}{3}(z - 1)^3 - \ldots$$
(3.4.22)

we could show that the series solutions generated by Eqs. 3.4.19 through 3.4.21 are identical to those yielded by Eq. 3.4.18.

3.4.3 Some Properties of the Hyperlogistic Equation

Differentiating Eq. 3.4.15 with respect to T and setting the result equal to zero gives the ordinate of the inflection point

$$U_i = \left(\frac{r}{r + s}\right)^{1/s} \quad .$$
(3.4.23)

Substituting this result into Eq. 3.4.15 yields the slope at the inflection point

$$\left(\frac{dU}{dT}\right)_i = \left(\frac{r}{r + s}\right)^{r/s} \left(\frac{s}{r + s}\right) \quad .$$
(3.4.24)

For the three special cases given by Eqs. 3.4.19 to 3.4.21 (i.e., $m = 0$, $1/2$, 1), the inflection point time, T_i, can be computed directly. For other values of m, iterated solutions of Eq. 3.4.18 must be carried out to determine T_i .

We now examine the growth rate equation given by Eq. 3.4.15. We consider two general cases. Case I: s is constant, r is variable and Case II: r is constant, s is variable.

Case I: s constant, r variable. This display of Eq. 3.4.15 is presented in Fig. 3.4.1 with four plots, each with a constant value of s. For each plot, there are five curves corresponding to $r = 0$, $1/4$, $1/2$, $3/4$, and 1. In Fig. 3.4.1(a), the familiar parabola corresponding to the ordinary logistic equation is shown $(r = 1, s = 1)$. In the same plot, the linear relationship corresponding to the confined exponential equation is displayed $(r = 0, s = 1)$.

In each of the four plots, a dashed curve is shown which passes through the maximum points of the curve. The abscissas of these points, of course, define U_i. From Eq. 3.4.23 we note that $U_i = 1$ as $r \to \infty$.

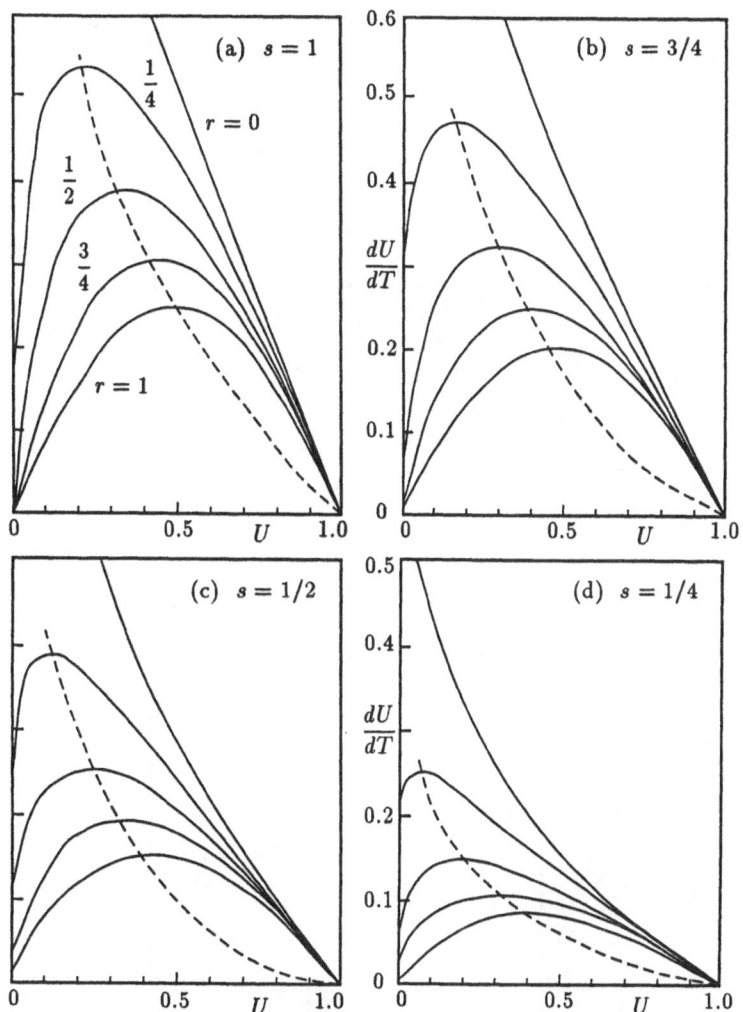

Fig. 3.4.1 Plots of dU/dT vs. U for the hyperlogistic equation. Case I: s constant, r variable

Case II: r constant, s variable. The presentation of Eq. 3.4.15 for this category is shown in Fig. 3.4.2. Again, four plots are displayed corresponding to four values of r. For each plot there are four curves corresponding to $s = 1$, 3/4, 1/2, and 1/4.

The dashed lines pass through the maximum points. From Eq. 3.4.23 we establish that $U_i = 1$ as $s \to \infty$. When $s = 0$, we obtain $U_i = \exp(-1/r)$.

From the preceding, we observe that the hyperlogistic growth equation provides a very wide range of inflection point ordinates.

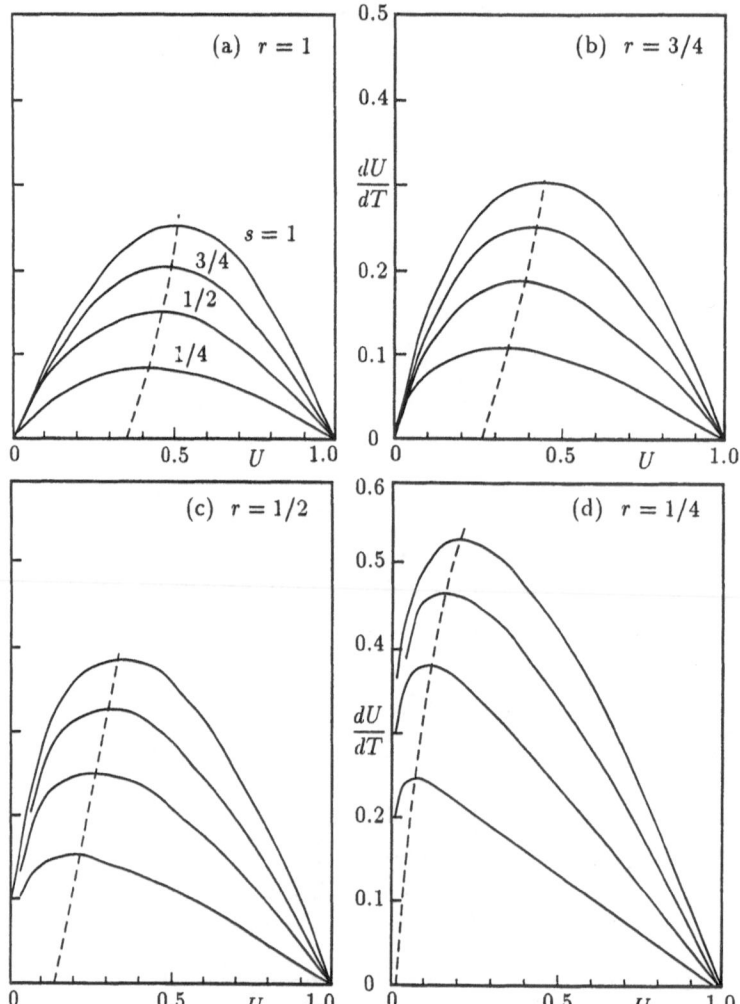

Fig. 3.4.2 Plots of dU/dT vs. U for the hyperlogistic equation. Case II: r constant, s variable

3.4.4 Numerical Examples of the Hyperlogistic Equation

Numerical example 1. We now utilize Eq. 3.4.18 for numerical computation. Selected values of r and s are shown in Table 3.4.1; the initial value $U_0 = 0.10$ is employed.

The solutions, $T = T_*(V) = T(U)$, are obtained from Eq. 3.4.18 and the relationship $U = V^{1/r}$. For the cases with $p = 1/2$, (i.e., curves b, d, and f), exact solutions are available from Eq. 3.4.20 to provide a computational check.

Table 3.4.1 Numerical example I of the hyperlogistic equation. Note: $m = (r+s-1)/s$; $p = 1 - m$; $U_0 = 0.10$

Curve	r	s	m	p	T_i	U_i	$(dU/dT)_i$
a	1/4	3	3/4	1/4	0.476	0.425	0.745
b	1/4	3/2	1/2	1/2	0.290	0.273	0.620
c	1/4	1	1/4	3/4	0.190	0.200	0.53
d	1/2	1	1/2	1/2	0.661	0.333	0.385
e	1/2	2/3	1/4	3/4	0.632	0.281	0.303
f	3/4	1/2	1/2	1/2	1.583	0.360	0.186
g	3/4	1/3	1/4	3/4	3.498	0.332	0.135

The results of the computations are shown in Fig 3.4.3. The solid dots identify the inflection points whose coordinates are listed in Table 3.4.1.

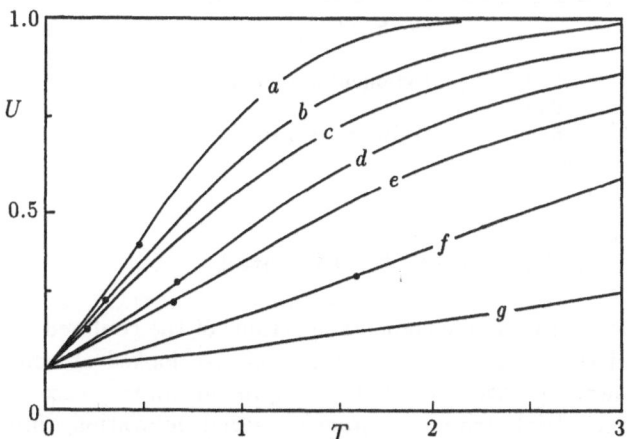

Fig. 3.4.3 Plots of the hyperlogistic equation for the values of (r, s) listed in Table 3.4.1

Numerical example II. As mentioned in Section 2.4.3, Wilson and Kirkby (1980) present the following differential equation in connection with some problems of mathematical modeling in geography

$$\frac{dN}{dt} = \lambda N^r (N_* - N) \quad , \tag{3.4.25}$$

which in dimensionless form becomes

$$\frac{dU}{dT} = U^r (1 - U) \quad . \tag{3.4.26}$$

This is a special case of Eq. 3.4.15 with $s = 1$. Accordingly, with $m = (r + s - 1)/s$, we have, $r = m = 1 - p$, and $V = U$.

For the purpose of numerical example and graphical display, the values of r listed in Table 3.4.2 were selected. For comparison with the numerical example of Section 2.4.6, values of $a = 0.20$, $N_0 = 10$ and $N_* = 100$ are utilized.

The results of calculations using Eq. 3.4.18 are shown in Fig. 3.4.4. Exact solutions provided by Eqs. 3.4.19 to 3.4.21 are utilized to check the series solutions for the curves corresponding to $r = 0$, $1/2$, and 1.

The solid dots in Fig. 3.4.4 identify the inflection points, computed from Eqs. 3.4.23; numerical values of t_i, N_i and $(dN/dt)_i$, are listed in Table 3.4.2.

Table 3.4.2 Numerical example II of the hyperlogistic equation. $a = 0.20$, $N_0 = 10$, $N_* = 100$

r	t_i	N_i	$(dN/dt)_i$	Remarks
0	–	–	–	Confined exponential
1/4	0.950	20.00	10.70	
1/2	3.310	33.33	7.70	Exponential-logistic
3/4	6.555	42.86	6.05	
1	10.990	50.00	5.00	Ordinary logistic

3.4.5 An Illustration: Adoption of a Tornado Warning Device Revisited

We recall that in Section 2.4 we made some comparisons of the confined exponential equation and the ordinary logistic equation. We looked at some examples where these two important distributions represent limiting cases in phenomena involving industrial technology transfer, social innovation diffusion, chemical reaction kinetics, and the psychology of learning processes.

Subsequently we linearly combined these two relationships to construct the following differential equation

$$\frac{dN}{dt} = a_*(N_* - N) + aN\left(1 - \frac{N}{N_*}\right) \quad , \tag{3.4.27}$$

which, in dimensionless form, becomes

$$\frac{dU}{dT} = \left(\frac{1 - w}{w}\right)(1 - U) + U(1 - U) \quad , \tag{3.4.28}$$

where $w = a/(a + a_*)$. If $w = 0$ (i.e., $a = 0$), this equation reduces to the confined exponential; if $w = 1$ (i.e., $a_* = 0$), it reduces to the ordinary logistic. Giving this matter an "innovation diffusion" interpretation, we say that

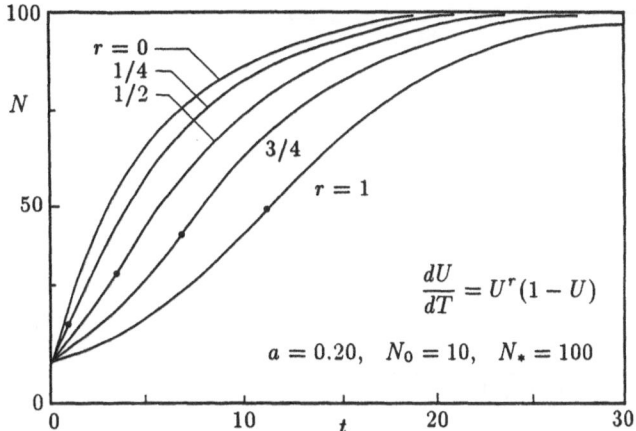

Fig. 3.4.4 Plots of the hyperlogistic equation for the case $s = 1$ for various values of r

values of w between zero and unity represent a blend of an externally influenced pure source model $(a = 0)$ and an internally influenced pure interaction model $(a_* = 0)$.

In the present section we have quite a different mathematical vehicle to handle this kind of consideration: the hyperlogistic equation. In this case the equation is

$$\frac{dU}{dT} = U^r(1 - U) \quad . \tag{3.4.29}$$

When $r = 0$, Eq. 3.4.29 reduces to the confined exponential (pure source model) and when $r = 1$ it becomes the ordinary logistic (pure interaction model). For $0 < r < 1$, we have a blend.

For a direct comparison of the "linearly combined logistic-confined exponential equation" and the "hyperlogistic equation", we use the same illustration as we did in Section 2.4.7: adoption of a tornado warning device. The data for the illustration are presented in Table 2.4.3 and are plotted in Fig. 3.4.5. As before, we assume that $N_* = 100$ percent. It is necessary to determine the values of a, r, and N_0.

We start by writing Eq. 3.4.29 in finite difference form

$$\frac{1}{1 - \overline{U}}\frac{\Delta U}{\Delta t} = a\overline{U}^r \tag{3.4.30}$$

and taking logarithms

$$\log_e\left(\frac{1}{1 - \overline{U}}\frac{\Delta U}{\Delta t}\right) = \log_e a + r\log_e \overline{U} \quad . \tag{3.4.31}$$

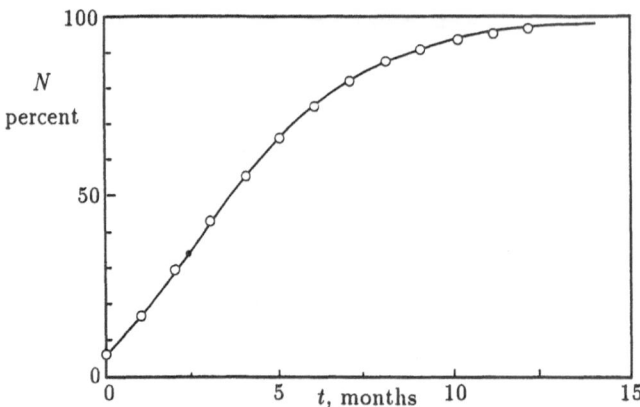

Fig. 3.4.5 Rate of adoption of a tornado warning device. Compare with Fig. 2.4.5

Utilizing the data of Table 2.4.3, we compute the variables indicated by Eq. 3.4.30. The results are displayed in the log-log plot of Fig. 3.4.6. A least-squares calculation based on Eq. 3.4.31 yields the values: $a = 0.377$ and $r = 0.517$ with a correlation coefficient of 0.939. The solid line shown in Fig. 3.4.6 is based on these values of a and r. Since $s = 1$, then $m = r = 0.517$ and $p = 1 - m = 0.483$.

The value of U_0 was determined by using Eq. 3.4.18 and the tabulated values of U for $t = 0$ to 3. This yielded $U_0 = 0.0508$ and so $N_0 = 5.08$.

With values of a, r, and N_0 known, the relationship $t = t(N)$ was computed from Eq. 3.4.18. The result is displayed as the solid curve in Fig. 3.4.5. The ordinate of the inflection point, shown as the solid dot in the figure, was calculated from Eq. 3.4.23; the abscissa was computed from Eq. 3.4.18. The results: $t_i = 2.40$ and $N_i = 34.08$. These values are in close agreement with those obtained from the combined logistic-confined exponential model of Section 2.4.7: $t_i = 2.35$, $N_i = 34.30$.

In the linearly combined model of Section 2.4, the parameter $w = a/(a + a_*)$ provided an index for judging the relative influences of "externally influenced pure source" effects and "internally influenced pure interaction" effects.

We arbitrarily set the criterion that when $w = 1/2$ (i.e., $a = a_*$), the two effects are equally important. In the tornado warning device illustration, $w = 0.306/(0.306 + 0.096) = 0.761$. Hence we concluded that the major factor affecting adoption of the device was an internally influenced communication mechanism.

In the hyperlogistic model, the parameter r serves as the index for scaling the relative influences of the two effects. When $r = 0$, a "pure source" mechanism controls the growth; when $r = 1$, a "pure interaction" mechanism determines the transfer. In our warning device illustration, $r = 0.517$ and so we conclude again that the interaction mechanism dominates. We note that r

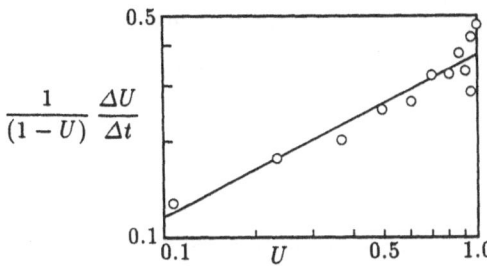

Fig. 3.4.6 Adoption of a tornado warning device. Plot to determine values of the parameters a and r

is not nearly as sensitive an index as is w in delineating the relative influences of the two mechanisms .

Finally, for the case $r = 1/2$, an exact solution is possible. Transposing Eq. 3.4.20 we obtain

$$U = \left(\frac{1 - \beta e^{-T}}{1 + \beta e^{-T}}\right)^2 \quad ; \quad \beta = \frac{1 - \sqrt{U_0}}{1 + \sqrt{U_0}} \quad . \tag{3.4.32}$$

If $\sqrt{U_0} \ll 1$, this equation becomes

$$U = \tanh^2\left(\frac{T}{2}\right) \quad . \tag{3.4.33}$$

We recall from Eq. 2.4.31 that a similar approximation, viz., $U = \tanh T$, was obtained for the case $w = 1/2$ in the combined logistic-confined exponential model.

3.4.6 Coalition and Modified Coalition Growth Models

In Section 2.1.6, in an example of power law exponential growth, we examined how the world's population has grown since the year 1650. We concluded that analysis with the observation that if the population of the world continues to increase at its present rate then the "doomsday" of an infinite population will occur in the year 2025 or thereabouts.

We recall that this troublesome forecast was based on the hyperbolic growth relationship of Eq. 2.1.21

$$\frac{N}{N_0} = \frac{1}{1 - at} \quad , \tag{3.4.34}$$

where $t = 0$ corresponds to the year 1650; a least squares computation gave $N_0 = 0.525$ billion and $a = 0.00267 \, \text{year}^{-1}$.

We now let $U = N/N_*$, $T = \alpha N_* t$, and $\alpha = a/N_0 = 0.005086$. Once again, the carrying capacity or equilibrium value is N_*.

With these definitions, Eq. 3.4.34 can be written in the form

$$T = \frac{1}{U_0} - \frac{1}{U} \quad . \tag{3.4.35}$$

It is easily established that Eq. 3.4.35 is the solution to the differential equation

$$\frac{dU}{dT} = U^2 \quad . \tag{3.4.36}$$

These two relationships comprise the so-called "coalition growth model" of von Foerster et al. (1960), considered briefly in Section 2.1.6. We will refer to this simple case as the "zero order analysis".

Substantially more realistic is the "modified coalition growth model" proposed by Austin and Brewer (1971). This point of view stipulates that an explosive hyperbolic growth can endure to a certain point; thereafter, a logistic-type quantity must be incorporated to assure a finite limit to growth. That is, to follow Austin and Brewer

$$\frac{dU}{dT} = \frac{1}{2c}\left(1 - e^{-2cU}\right)(1 - U)U \quad , \tag{3.4.37}$$

where $c = \alpha N_*/2A$ in which A is a parameter with the dimensions of a growth coefficient (i.e., 1/time); c is dimensionless.

In Eq. 3.4.37, for very large values of cU, the exponential term vanishes and the differential equation reduces to an ordinary logistic. On the other hand, for small values of cU, the first two terms of a series expansion of the exponential gives

$$\frac{dU}{dT} = U^2(1 - U) \quad . \tag{3.4.38}$$

Inclusion of the third term in the expansion yields

$$\frac{dU}{dT} = U^2(1 - cU)(1 - U) \quad . \tag{3.4.39}$$

It is not possible to obtain an exact solution of Eq. 3.4.37 and indeed we really do not need it. Instead, from here on, we will regard Eqs. 3.4.38 and 3.4.39 as growth models in their own right. Both equations contain the explosive U^2 term and both feature a stabilizing logistic component. Clearly Eq. 3.4.38 is a special case of Eq. 3.4.39 with $c = 0$. We will describe the two cases as the "first order analysis" and the "second order analysis", respectively.

First order analysis. The solution to Eq. 3.4.38 is

$$T = \left(\frac{1}{U_0} - \frac{1}{U}\right) + \log_e\left[\left(\frac{1-U_0}{U_0}\right)\left(\frac{U}{1-U}\right)\right] \quad . \tag{3.4.40}$$

The inflection point of this growth curve is obtained from Eq. 3.4.23 with $r = 2$ and $s = 1$ or more directly by setting the first derivative of Eq. 3.4.38 equal to zero. Either way, we obtain

$$T_i = \left(\frac{1}{U_0} - \frac{3}{2}\right) + \log_e\left[2\left(\frac{1}{U_0} - 1\right)\right] \quad ;$$

$$U_i = \frac{2}{3} \quad ; \qquad \left(\frac{dU}{dT}\right)_i = \frac{4}{27} \quad . \tag{3.4.41}$$

Second order analysis. Several plots of Eq. 3.4.39 for various values of c including $c = 0$ ("first order analysis"), are shown in Fig. 3.4.7. The range of c is restricted to $0 < c < 1$; the roots of Eq. 3.4.39 are $U = 0$, 1, and $1/c$. Maximum points and inflection points are identified by the solid dots in Fig. 3.4.7. At $U = 0$ the slopes of these rate-of-growth curves are all zero; at $U = 1$, the slopes are, in descending order in the figure, -1, $-2/3$, $-1/3$, and 0 .

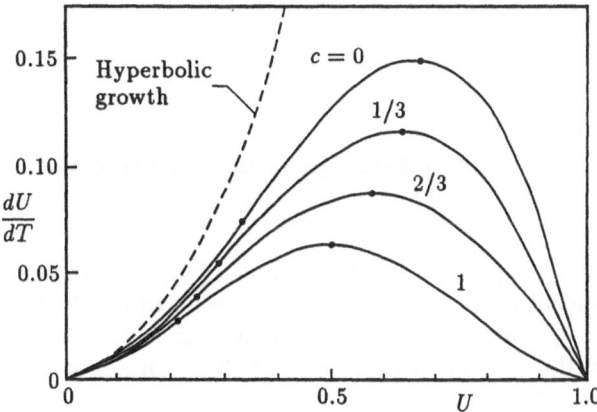

Fig. 3.4.7 Plot of rates of growth for modified coalition growth model

The solution to Eq. 3.4.39 is

$$T = \left(\frac{1}{U_0} - \frac{1}{U}\right) + \frac{c^2+1}{2(c-1)}\log_e\left[\left(\frac{1-cU_0}{1-U_0}\right)\left(\frac{1-U}{1-cU}\right)\right]$$

$$+ \left(\frac{c+1}{2}\right)\log_e\left[\frac{(1-cU_0)(1-U_0)}{U_0^2}\frac{U^2}{(1-cU)(1-U)}\right] \tag{3.4.42}$$

where, again, $T = \alpha N_* t$, $U = N/N_*$, and $c = \alpha N_*/2A$. If $c = 0$, Eq. 3.4.42 reduces to Eq. 3.4.40. If $c = 1$, we can establish that

$$T = \left(\frac{1}{U_0} - \frac{1}{U}\right) + \left(\frac{1}{1-U} - \frac{1}{1-U_0}\right) + 2\log_e\left[\left(\frac{1-U_0}{U_0}\right)\left(\frac{U}{1-U}\right)\right] .$$

$$(3.4.43)$$

Finally, setting the first derivative of Eq. 3.4.39 equal to zero provides the inflection point of the growth curve

$$U_i = \frac{3(c+1)}{8c}\left(1 - \sqrt{1 - \frac{32c}{9(c+1)^2}}\right) . \qquad (3.4.44)$$

This relationship is shown in Fig. 3.4.8. We determine that if $c = 0$ then $U_i = 2/3$.

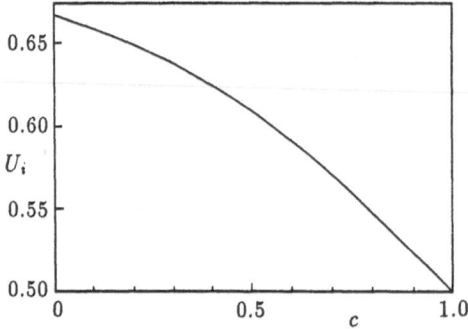

Fig. 3.4.8 Dimensionless ordinate of the inflection point as a function of the parameter c

Figure 3.4.9 displays several plots of Eq. 3.4.42 for different values of c. The inflection points are identified. The curve corresponding to $c = 1$, with an inflection point at $U_i = 1/2$, has the characteristics of an ordinary logistic, although its mathematical description, given by Eq. 3.4.43, is somewhat more complicated.

An important and desirable feature of the modified coalition growth model is that it permits an "explosive" initial growth phase and a "stabilized" final growth phase. The model also allows the inflection point of the growth curve to range from $U_i = 1/2$ to $U_i = 2/3$.

3.4.7 An Illustration: Population of the World

We are now equipped with the necessary mathematical tools to fend off the infinite population of the world forecasted for doomsday 2025. For numerical values we select the following

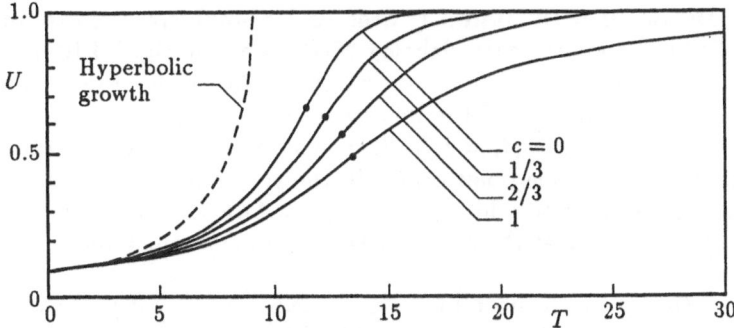

Fig. 3.4.9 Plots of growth curves for the modified coalition growth model

Year 1980 : $t = 330$; $N_0 = 4.415$; $N_* = 25.0$;

$$U_0 = 4.415/25.0 = 0.1766 \quad ; \quad A = 0.15 \quad ;$$

$$\alpha = 0.005086 \quad ; \quad c = \alpha N_*/2A = 0.424 \quad .$$

Hyperbolic or coalition growth model ("zero order analysis"). This computation, essentially a repetition of that of Section 2.1.6, utilizes Eq. 3.4.35. The resulting curve is shown as the dashed line in the semi-logarithmic plot of Fig. 3.4.10.

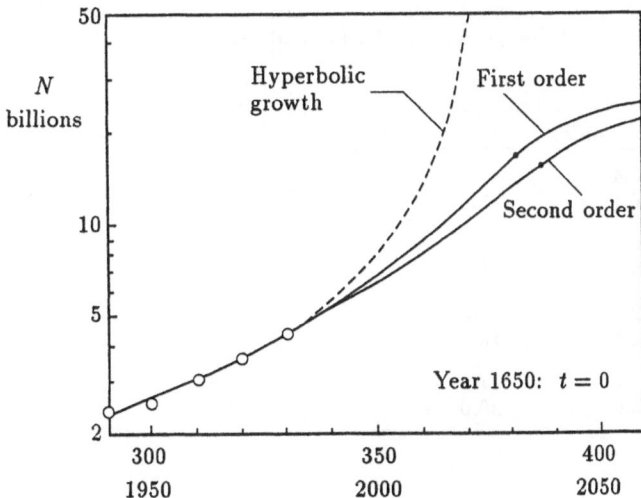

Fig. 3.4.10 Projections of the population of the world

Modified coalition growth model ("first order analysis"). Next we utilize Eq. 3.4.40 to obtain the curve labeled "first order" in Fig. 3.4.10. For this $c = 0$ case, the inflection point occurs at $U_i = 2/3$ which corresponds to $N_i = 16.67$ billion. The time of the inflection point, computed from Eq. 3.4.41, is $T_i = 6.40$. Accordingly, $t_i = T_i/\alpha N_* = 50.3$ years measured from the year 1980. Hence the inflection point occurs in the year 2030. At that time, from the rate of growth relationship given by Eq. 3.4.41, the population of the world will be increasing at the rate of 0.47 billion persons per year.

Modified coalition growth model ("second order analysis"). Finally, we employ Eq. 3.4.42 to produce the curve identified as "second order" in Fig. 3.4.10. In this instance, from Eq. 3.4.44, the ordinate of the inflection point is $U_i = 0.622$ and so $N_i = 15.55$ billion. The inflection time is $t_i = 56.1$ years measured from the year 1980. So, with this model, the inflection point of world population growth will occur in the year 2036.

Comments. It should be mentioned that the value of $N_* = 25$ billion was selected only to numerically illustrate the model. That number may be too large or it may be too small; it is probably reasonable. Austin and Brewer (1971) selected $N_* = 50$ billion and numerically integrated Eq. 3.4.37 to obtain $N = 38$ billion for the year 2026.

Table 3.4.3 summarizes the results we have acquired in our illustration.

Table **3.4.3** Population of the world based on modified coalition growth model (first order) and modified coalition growth model (second order)

Year	t [years]	N [billions]	
		first order	second order
1990	340	5.4	5.3
1995	345	6.1	5.8
2000	350	6.8	6.4
2010	360	8.9	8.1
2020	370	12.0	10.3
2030	380	16.5	13.2
2040	390	21.0	17.0
2050	400	23.0	20.0

3.4.8 An Illustration: Growth of the Public Debt of the United States

As a final illustration of hyperlogistic growth, we examine the matter of the public debt of the United States and the rate at which it has been growing since World War II.

The amount of the public debt is shown in Fig. 3.4.11(a) for each year commencing with 1950. For the moment, we disregard the solid curve shown in the figure. We let N represent the debt expressed in billions of U.S. dollars; time $t = 0$ corresponds to the year 1950. Table 3.4.4 indicates the precise amounts of the debt at five-year intervals.

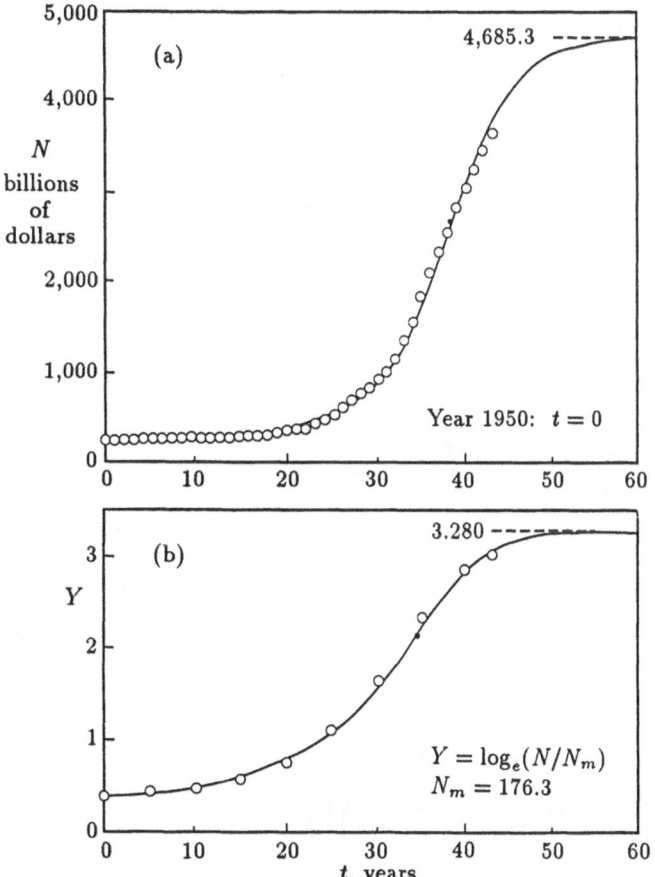

Fig. 3.4.11 Growth of the public debt of the United States. (a) Actual and projected growth curve and (b) transformed growth curve

Table **3.4.4** Public debt of the United States commencing with 1950. Amounts are in billions of U.S. dollars. Reference: U.S. Department of the Treasury, Bureau of the Public Debt, Washington, D.C., July 1990

Year	t	N	Year	t	N
1950	0	255.4	1975	25	534.2
1955	5	272.3	1980	30	908.7
1960	10	283.8	1985	35	1823.8
1965	15	314.1	1990	40	3051.3
1970	20	372.6			

Instead of selecting and fitting a cumulative distribution function for the $N(t)$ curve shown in Fig. 3.4.11(a), it is convenient to transform the ordinate, N, of this figure by the following equation

$$Y = \log_e \left(\frac{N}{N_m} \right) \tag{3.4.45}$$

where N_m is a quantity we consider in a moment. Since the ratio N/N_m is dimensionless, the particular unit of N (e.g., billions or trillions of dollars) is not important. Thus, we deal with the natural logarithm of N instead of N itself. With the ascertained value of N_m , the transformed five-year values of N are shown in Fig. 3.4.11(b).

We now assume that the growth of the quantity Y can be described by the modified coalition growth function expressed by Eq. 3.4.38. Taking $U = Y/Y_*$ and $T = (aY_*t)/Y_0$, Eq. 3.4.38 becomes

$$\frac{dY}{dt} = \frac{a}{Y_0} Y^2 \left(1 - \frac{Y}{Y_0} \right) \quad . \tag{3.4.46}$$

The solution to this equation, of course, is given by Eq. 3.4.40 in the following modified form

$$t = \frac{1}{a} \left[\left(1 - \frac{Y}{Y_0} \right) + \frac{Y_0}{Y_*} \log_e \left(\frac{Y}{Y_0} \right) \left(\frac{Y_* - Y_0}{Y_* - Y} \right) \right] \quad . \tag{3.4.47}$$

From Eqs. 3.4.40, the inflection point coordinates of $Y = Y(t)$ are

$$t_i = \frac{1}{a} \left\{ \left(1 - \frac{3Y_0}{2Y_*} \right) + \frac{Y_0}{Y_*} \log_e \left[2 \left(\frac{Y_*}{Y_0} - 1 \right) \right] \right\} \quad ;$$

$$Y_i = \frac{2}{3} Y_* \quad ; \qquad \left(\frac{dY}{dt} \right)_i = \frac{4}{27} \frac{aY_*^2}{Y_0} \quad . \tag{3.4.48}$$

We determine the numerical values of the various parameters as we have done in previous sections. In addition, using Eq. 3.4.47 for values of (t, Y)

near the origin, we establish the magnitude of the scaling constant N_m. The results are

$$a = 0.0329 \quad Y_0 = 0.368 \quad N_0 = 254.7$$
$$N_m = \quad 176.3 \quad Y_* = 3.280 \quad N_* = 4685.3$$

Using Eqs. 3.4.48 we get the $Y(t)$ inflection point coordinates: $t_i = 34.7$ (Year: 1984.7), $Y_i = 2.187$, and $(dY/dt)_i = 0.1425$.

The inflection point coordinates of the $N(t)$ curve are easily determined. We rewrite Eq. 3.4.45 in the form

$$N = N_m \exp(Y) \quad . \tag{3.4.49}$$

Setting the second derivative of this equation equal to zero and utilizing Eq. 3.4.46, we obtain the following expression for the value of Y at the N-inflection point

$$Y_{i(N)} = \frac{1}{2}(Y_* - 3)\left(\sqrt{1 + \frac{8Y_*}{(Y_* - 3)^2}} + 1 \right) \quad . \tag{3.4.50}$$

From this relationship, the time, ordinate, and slope corresponding to the $N(t)$ inflection point are calculated from the previous equations. We obtain the following results: $Y_i(N) = 2.705$, $t_i = 38.6$ (Year 1988.6), $N_i = 2636.2$, and $(dN/dt)_i = 302.4$. The inflection points for the $N(t)$ and $Y(t)$ curves are identified by the solid dots in Figs. 3.4.11(a) and (b).

Table **3.4.5** Public debt of the United States. Comparison of actual or estimated amounts with computed amounts. Computed projections. Amounts are in billions of U.S. dollars

Actual or estimated amounts		Computed amounts and projections	
Year	Amount	Year	Amount
1984	1573.0	1984.7	1570.3
1985	1823.8	1985.1	1672.4
1986	2111.0	1986.9	2147.4
1987	2336.0		
1988	2572.0		
1989	2819.1	1989.0	2757.3
1990	3051.3		
1991	3267.1	1991.8	3540.4
1992	3466.2		
1993	3651.5	1993.5	3912.8
		1996.5	4324.3
		2000.0	4546.0
		2003.8	4637.8
		2006.1	4661.0
		N_*	4685.3

Utilizing the equations and numerical values given above, we compute the two growth curves, $Y(t)$ and $N(t)$. These are shown as the solid lines in Figs. 3.4.11(a) and (b).

Results are summarized in Table 3.4.5. On the left-hand side of the table are the actual or estimated amounts of the public debt, N, commencing with the year 1984. On the right-hand side are the computed and projected amounts.

Decimal parts of years may be indicated in the table because Eq. 3.4.47 provides the implicit solution $t = t(N)$. If desired, iterated calculations can yield $N = N(t)$. On the basis of the preceding analysis we make the following observations: (a) In 1995 and 2000, the total public debt of the United States will be, respectively, 4146.4 and 4546.0 billion dollars and (b) the asymptotic or ultimate total public debt will be 4685.3 billion dollars.

3.5 Various Other Distributions

3.5.1 Comparison of Distribution Functions

There are many mathematical functions which can be utilized to provide frameworks to describe growth and transfer phenomena and processes. In previous sections we examined a number of these functions including the following three simple and useful relationships

- Logistic distribution
- Normal probability distribution
- Confined exponential distribution

Two other very important distributions have been considered. One of these, the exponential function, is a special case of the logistic distribution corresponding to an infinite carrying capacity. The other, the error function, discussed in Section 2.5.2, is simply an alternative form of the normal probability distribution. Thus if

$$U(T) = \frac{1}{\sqrt{2\pi}} \int_{-\infty}^{T} e^{-\frac{1}{2}z^2} dz \quad , \tag{3.5.1}$$

then

$$U(T) = \frac{1}{2}\left[1 + \operatorname{erf}\left(\frac{1}{\sqrt{2}}T\right)\right] \tag{3.5.2}$$

where

$$\operatorname{erf}(x) = \sqrt{\frac{2}{\pi}} \int_{0}^{x} e^{-z^2} dz \tag{3.5.3}$$

is the error function.

In these equations, and those to follow, the indicated dimensionless variables are: $U = N/N_*$, $u = dU/dT = n/aN_*$, $n = dN/dt$, and $T = at$.

With respect to the time variable, T, it is pointed out that the numerical values of the coefficient, a, are not necessarily the same when making a comparison of two different distribution functions. For precise comparison, a criterion must be selected such as (a) equating slopes of the cumulative curve at the cumulative distribution inflection point, (b) equating coordinates at the standard deviation points of the density distribution, or (c) equating coordinates at the density distribution inflection points. This topic was discussed in Section 2.5.4 where a comparison was made of the logistic and normal probability distributions.

3.5.2 Arctangent–Exponential Distribution

In a relatively little-known paper published years ago, Feller (1940) made comparisons of the logistic growth curve with the growth curves described by the error function and by an "arctangent–exponential" function.

In dimensionless form, with the time coordinate measured from the cumulative curve inflection point, the latter function has the definition

$$U = \frac{2}{\pi} \arctan\left[\exp\left(\frac{\pi T}{4}\right)\right] \quad . \tag{3.5.4}$$

The corresponding density distribution is

$$u = \frac{1}{4}\operatorname{sech}\frac{\pi T}{4} \tag{3.5.5}$$

From Eqs. 3.5.4 and 3.5.5, the basic differential equation is

$$\frac{dU}{dT} = \frac{1}{4}\sin \pi U \quad , \tag{3.5.6}$$

which is similar to the relationship for the logistic equation, $dU/dT = U(1 - U)$.

Equating the first derivative of Eq. 3.5.5 to zero gives the location of the inflection point of the cumulative distribution

$$T_i = 0 \quad ; \quad u_i = \frac{1}{4} \quad ; \quad U_i = \frac{1}{2} \quad . \tag{3.5.7}$$

Likewise, setting the second derivative of Eq 3.5.5 equal to zero identifies the inflection points of the density distribution

$$T_c = \pm\frac{4}{\pi}\operatorname{arcsinh}(1) \quad ; \quad u_c = \frac{1}{4\sqrt{2}} \quad ; \quad U_c = \frac{1}{4}, \frac{3}{4} \quad . \tag{3.5.8}$$

The arctangent–exponential distribution is almost indistinguishable from the logistic distribution.

3.5.3 Pearson Type VII Distribution

In their lucid presentation of the subject of probability distributions, Elderton and Johnson (1969) give thorough coverage to the so-called Pearson distributions. The subject is also examined by Jeffreys (1967) and by Peschel and Mende (1986).

In this regard we start with the following differential equation proposed originally by Pearson (1913).

$$\frac{1}{n}\frac{dn}{dt} = \frac{t+c}{b_0 + b_1 t + b_2 t^2} \qquad (3.5.9)$$

By assigning various values to the four coefficients it is possible to construct numerous types of density distributions. For example, setting $b_1 = b_2 = c = 0$ readily yields the normal probability distribution.

For the present purpose, taking $b_1 = c = 0$, reduces Eq. 3.5.9 to the form

$$\frac{1}{n}\frac{dn}{dt} = \frac{t}{b_0 + b_2 t^2} \qquad (3.5.10)$$

Integrating this equation and requiring the total area under the density distribution curve to be N_* gives

$$n = \frac{aN_*}{B(m - \frac{1}{2}, \frac{1}{2})} \left(1 + a^2 t^2\right)^{-m} \quad , \qquad (3.5.11)$$

in which $B(p,q)$ is the beta function, $m > 1/2$ is a parameter, and $a^2 = b_2/b_0$. In dimensionless form, Eq. 3.5.11 becomes

$$u = \frac{1}{B(m - \frac{1}{2}, \frac{1}{2})} \left(1 + T^2\right)^{-m} \qquad (3.5.12)$$

where $u = n/aN_*$ and $T = at$. Accordingly

$$U = \frac{1}{B(m - \frac{1}{2}, \frac{1}{2})} \int_{-\infty}^{T} (1 + z^2)^{-m} dz \quad . \qquad (3.5.13)$$

Utilizing a substitution of the form $\zeta = 1/(1 + z^2)$, we obtain

$$U = 1 - \frac{1}{2} I_\theta(m - \frac{1}{2}, \frac{1}{2}) \quad , \qquad (3.5.14)$$

where $I_\theta(p,q)$ is the incomplete beta function and $\theta = 1/(1 + T^2)$.

The expressions given by Eqs. 3.5.12 through 3.5.14 are identified as the Pearson Type VII distribution. The incomplete beta function is well-tabulated by Pearson (1968). A special case of the Type VII distribution is the Student's t-distribution.

Numerous years ago, Winsor (1932b) made comparisons of the logistic, the normal probability, and the Type VII distributions. For the comparison criterion he equated the standard deviations of the three distributions. This yielded $a = \pi/\sqrt{3}$ for the logistic, $a = 1$ for the normal probability, and $m = 5$, $a^2 = 1/7$ for the Type VII distribution. Winsor noted that the Type VII curve gave a much closer fit to the logistic curve than did the normal probability curve.

From Eq. 3.5.13 it is easily established that the coordinates corresponding to the inflection point of the cumulative curve are

$$T_i = 0 \quad ; \quad u_i = \frac{1}{B(m - \frac{1}{2}, \frac{1}{2})} \quad ; \quad U_i = \frac{1}{2} \quad . \tag{3.5.15}$$

At the inflection point of the density curve

$$T_c = \pm\frac{1}{\sqrt{2m+1}} \quad ; \quad u_c = \frac{1}{B(m - \frac{1}{2}, \frac{1}{2})}\left(\frac{2m+1}{2(m+1)}\right)^m \quad ;$$

$$U_c = 1 - \tfrac{1}{2}I_{\theta_c}(m - \tfrac{1}{2}, \tfrac{1}{2}) \quad ; \quad \theta_c = \frac{1}{1 + T_c^2} = \frac{2m+1}{2(m+1)} \quad . \tag{3.5.16}$$

3.5.4 Arctangent Distribution

We now consider a special case of the Pearson Type VII distribution: the arctangent distribution.

In Eq. 3.5.12 we let $m = 1$. Then since

$$B(p, q) = \frac{\Gamma(p)\Gamma(q)}{\Gamma(p+q)} \quad , \tag{3.5.17}$$

where $\Gamma(z)$ is the gamma function and since $\Gamma(1/2) = \sqrt{\pi}$, we obtain

$$u = \frac{1}{\pi}\frac{1}{1 + T^2} \tag{3.5.18}$$

This particular function has a special name: the Cauchy distribution. It also has an interesting geometrical interpretation: the historic Witch of Agnesi.

Integration of Eq. 3.5.18 yields the cumulative distribution function

$$U = \frac{1}{2}\left(1 + \frac{2}{\pi}\arctan T\right) \quad , \tag{3.5.19}$$

a result also provided by Eq. 3.5.13 with $m = 1$. From Eqs. 3.5.18 and 3.5.19 we obtain the corresponding differential equation

$$\frac{dU}{dT} = \frac{1}{\pi}\sin^2 \pi U \quad . \tag{3.5.20}$$

The inflection points of the cumulative and density distribution curves are

$$T_i = 0 \quad ; \quad u_i = \frac{1}{\pi} \quad ; \quad U_i = \frac{1}{2} \tag{3.5.21}$$

and

$$T_c = \pm\frac{1}{\sqrt{3}} \quad ; \quad u_c = \frac{3}{4\pi} \quad ; \quad U_c = \frac{1}{3}, \frac{2}{3} \quad . \tag{3.5.22}$$

By way of summary, values of the critical points (T_i, u_i, U_i) and (T_c, u_c, U_c) for the various distributions considered above are presented in Table 3.5.1.

Table **3.5.1** Comparison of coordinates corresponding to inflection points of (I) cumulative distribution curve and (II) density distribution curve

I Distribution	T_i	u_i	U_i
Logistic	0	$\frac{1}{4}$	$\frac{1}{2}$
Normal probability	0	$\frac{1}{\sqrt{2\pi}}$	$\frac{1}{2}$
Arctan–exponential	0	$\frac{1}{4}$	$\frac{1}{2}$
Pearson type VII	0	$\frac{\Gamma(m)}{\sqrt{\pi}\Gamma(m-\frac{1}{2})}$	$\frac{1}{2}$
Arctangent	0	$\frac{1}{\pi}$	$\frac{1}{2}$

II Distribution	T_c	u_c	U_c
Logistic	$\pm 2\operatorname{arcsinh}\frac{1}{\sqrt{2}}$	$\frac{1}{6}$	$\frac{1}{2}\left(1\pm\frac{1}{\sqrt{3}}\right)$
Normal probability	± 1	$\frac{1}{\sqrt{2\pi e}}$	$\frac{1}{\sqrt{2\pi}}\int_{-\infty}^{\pm 1}e^{-z^2/2}dz$
Arctan–exponential	$\pm\frac{4}{\pi}\operatorname{arcsinh}(1)$	$\frac{1}{4\sqrt{2}}$	$\frac{1}{2}\left(1\pm\frac{1}{2}\right)$
Pearson type VII	$\pm\frac{1}{\sqrt{2m+1}}$	(refer to Eq.3.5.16)	
Arctangent	$\pm\frac{1}{\sqrt{3}}$	$\frac{3}{4\pi}$	$\frac{1}{2}\left(1\pm\frac{1}{3}\right)$

We note that all five of these distributions possess cumulative curve inflection points at $U_i = 1/2$. With this constraint, the selection of which distribution to use as a framework for data analysis and display is frequently a matter of personal choice. Certainly the logistic and normal probability distributions are employed most frequently. Nevertheless, since the shapes of the various distribution curves are not exactly the same, we should keep in mind that the less well-known functions may be useful in some cases.

3.5.5 Gamma Distribution

Virtually all of the cumulative distribution functions we have considered thus far have featured a time variable that ranges from $t = -\infty$ to $t = +\infty$. This characteristic is precisely what we need in most growth and transfer phenomena.

Nevertheless, in earlier sections we have examined a couple of notable exceptions: the confined exponential distribution and the Weibull distribution. Both of these have, or can have, the feature that when $t = 0$ then $N(0) = N_0 = 0$. This characteristic is needed when considering growth or transfer processes in which we know that $N = 0$ at $t = 0$.

In this regard, we now examine another function which possesses this important feature: the gamma distribution. We start by setting $b_0 = b_2 = 0$ in the Pearson differential equation, Eq. 3.5.9. This yields

$$\frac{1}{n}\frac{dn}{dt} = \frac{t+c}{b_1 t} \quad . \tag{3.5.23}$$

Integrating this expression and using the condition that the total area under the density distribution curve must equal N_*, we obtain

$$n = \frac{aN_*}{\Gamma(\lambda)}(at)^{\lambda-1}e^{-at} \quad , \tag{3.5.24}$$

where $a = -1/b_1$ and $\lambda = 1 + c/b_1$. In dimensionless form this expression becomes

$$u = \frac{1}{\Gamma(\lambda)}T^{\lambda-1}e^{-T} \quad . \tag{3.5.25}$$

This is the Pearson Type III distribution or, essentially, the gamma distribution. The corresponding cumulative distribution termed the incomplete gamma function, has the definition

$$U = \frac{1}{\Gamma(\lambda)}\int_0^T z^{\lambda-1}e^{-z}\,dz \quad . \tag{3.5.26}$$

This function has been tabulated by Pearson (1934). In a slightly modified form, Eq. 3.5.26 becomes the chi-squared distribution.

We note from Eq. 3.5.26 that $U = 0$ when $T = 0$. The inflection points of the cumulative distribution and density distribution occur at

$$T_i = \lambda - 1 \quad ; \qquad T_c = (\lambda - 1)\left(1 \pm \frac{1}{\sqrt{\lambda - 1}}\right) \quad . \tag{3.5.27}$$

Plots of Eqs. 3.5.25 and 3.5.26 are shown in Fig. 3.5.1 for values of $\lambda = 1$, $2, 3$ and 5. We note that $\lambda = 1$ produces the confined exponential distribution. The cumulative distribution contains an inflection point which ranges from $U_i = 0$ for $\lambda = 1$ to $U_i = 1/2$ for $\lambda = \infty$. When $\lambda = 10$, for example, $U_i = 0.413$.

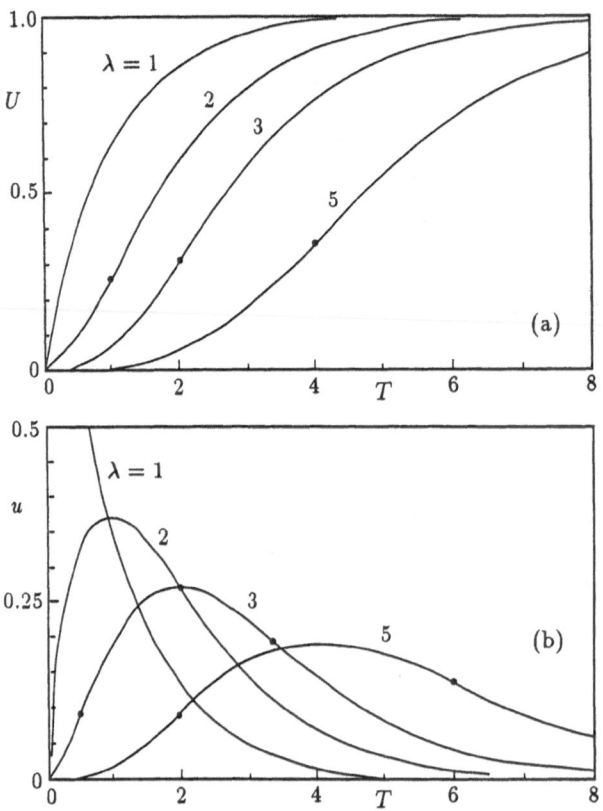

Fig. 3.5.1 Plots of the gamma distribution for $\lambda = 1, 2, 3$ and 5. (a) Cumulative distribution and (b) density distribution

3.5.6 Generalized Gamma Distribution

To conclude this section we examine briefly a generalized gamma distribution. This function has the definition

$$u = \frac{c}{\Gamma(\lambda/c)} T^{\lambda-1} e^{-T^c} \quad . \tag{3.5.28}$$

Case 1. If $\lambda = 1$, $c = 1$, we obtain the confined exponential distribution

$$u = e^{-T} \quad . \tag{3.5.29}$$

Case 2. If $c = 1$, Eq. 3.5.28 reduces to the gamma distribution we considered above

$$u = \frac{1}{\Gamma(\lambda)} T^{\lambda-1} e^{-T} \quad . \tag{3.5.30}$$

Case 3. If $\lambda = c$, the Weibull distribution, considered in Section 3.2, is produced

$$u = cT^{c-1} e^{-T^c} \quad . \tag{3.5.31}$$

A comprehensive presentation of numerous distribution functions is provided by Hastings and Peacock (1974). Additional information on this subject is given by Cooper and Weekes (1983) and Elderton and Johnson (1969).

3.5.7 An Illustration: Generation Times of Cells

We all know that cells – the microscopic structures of all plants and animals – possess mechanisms which enable them to divide to create new cells. A certain period of time is required for a "mother" cell to generate two "daughters" and, in turn, for these two new mothers to each generate two new daughters and so on. Now, in a colony of cells, if the time intervals between such births and divisions were always the same, we could easily determine the total number of cells in the colony at any particular time. We establish that

$$N/N_0 = 2^{t/t_g} \quad , \tag{3.5.32}$$

where N_0 is the number of cells at time $t = 0$ ($N_0 = 1$, say); t_g is the doubling time which, in this constant time interval case, is the same as the generation time.

The geometrical growth described by Eq. 3.5.32 is in the form contemplated by Malthus in his early discourses on population growth. We prefer to write this equation to express exponential, instead of geometrical, growth. That is

$$N/N_0 = e^{at} \tag{3.5.33}$$

where $a = (\log_e 2)/t_g$. Of course we know that the Malthusian or exponential growth of the cell colony indicated by Eq. 3.5.33, applies only if there are no restraints on unlimited growth. In reality, as we know, finite resources and crowding effects impose a certain carrying capacity. In this case, Eq. 3.5.33 gives way to the ordinary logistic equation.

Here comes the however. Microbiologists and others working in the field of cell kinetics established a long time ago, with experiments involving continually increasing cell populations, that cell colonies virtually never display equal values of individual generation times. For example, in his early classic, Gause (1934) examined this feature at length with his experiments involving the growth of *Paramecium caudatum* .

The phenomenon of cell growth and division, and the associated problem of distribution of generation times, has been studied by a great many researchers. Representative are the works of Marr et al. (1969), Powell and Errington (1963), Prescott (1959), and Takahashi (1966, 1968). Some of the mathematical analyses in these studies utilize the results of Kendall (1952) concerning variable generation times in stochastic birth processes. Rubinow (1968, 1980) analyzes the problem and provides numerous references.

For our illustration we use the experimental results of Prescott (1959) and the mathematical model presented by Rubinow (1968, 1980).

In his experiments, Prescott utilized a strain of *Tetrahymena geleii* as the growing cell population. Under carefully controlled and optimal growth conditions, he synchronized his experiments to assure that at time $t = 0$ all cells were new and of precisely the same age. He observed an initial period of dormancy; no cell divisions occurred for a period of time $t_0 = 85$ minutes.

Subsequently he determined a total of 766 separate generation times and prepared a histogram of their distribution; the time interval was $\Delta t = 2$ minutes.

The histogram devised by Prescott was modified slightly by Rubinow (1968) to yield the density distribution plot shown in Fig. 3.5.2; the ordinate, n, is a flux expressed in cells per minute. The time scale has an origin, $t = 0$, corresponding to the end of the dormant period, $t_0 = 85$ minutes.

The obvious skewing of the experimental data toward the larger generation times was noted by Prescott. This characteristic has been observed by numerous other researchers carrying out similar experiments utilizing other kinds of micro-organisms. Because of this skewness, almost invariably the Pearson III, i.e., gamma distribution, has been utilized as the mathematical framework for the analysis and display of experimental data on cell generation times.

For our illustration we examine the data of Fig. 3.5.2 with the gamma distribution as our framework. We start with the basic relationship of Eq. 3.5.24

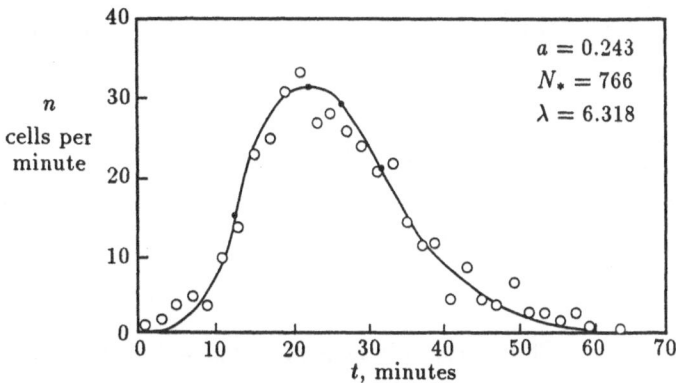

Fig. 3.5.2 Distribution of cell generation times for *Tetrahymen geleii*. Data of Prescott (1959). Gamma distribution is based on analysis of Rubinow (1968, 1980)

$$n = \frac{aN_*}{\Gamma(\lambda)}(at)^{\lambda-1}e^{-at} \quad . \tag{3.5.34}$$

Utilizing the method of least squares, Rubinow determined the value $\lambda = 6.318$. In addition we obtain: $a = 0.242$ and $N_* = 766$. Substituting these numbers into Eq. 3.5.34 yields the solid curve shown in Fig. 3.5.2. It is clear that the gamma distribution provides a good description of the experimental data.

We now calculate some of the important parameters which characterize density distribution functions.

Mean value or expectation. The definition of the mean value is given by Eq. 2.2.26

$$\bar{t} = \frac{1}{N_*} \int_0^\infty t\, n(t)\, dt \quad . \tag{3.5.35}$$

Substitution of Eq. 3.5.34 into this expression gives

$$\bar{t} = \lambda/a \tag{3.5.36}$$

as the mean value of the gamma distribution. In our example, with $a = 0.242$ and $\lambda = 6.318$, we obtain: $t = 26.0$ minutes, or 111.0 minutes after the start of the experiment. The corresponding value of n is 29.2 cells/minute.

Variance and standard deviation. We compute the variance from its definition in Eq. 2.2.28

$$\sigma^2 = \frac{1}{N_*} \int_0^\infty (t - \bar{t})^2 n(t) dt \quad . \tag{3.5.37}$$

The indicated integration yields

$$\sigma^2 = \lambda/a^2 \quad ; \qquad \sigma = \sqrt{\lambda}/a \quad . \tag{3.5.38}$$

We recall that the standard deviation, σ, is the square root of the variance. So, for our illustration, $\sigma^2 = 107.0$ and $\sigma = 10.34$ minutes. These values of $\bar{t}$ and σ agree with the values indicated by Prescott.

Maximum or modal value. If we set the first derivative of Eq. 3.5.34 equal to zero we obtain the time corresponding to the maximum value of n

$$t_m = (\lambda - 1)/a \quad . \tag{3.5.39}$$

For our illustration, $t_m = 21.88$ minutes. Consequently, $n_m = 31.7$ cells/min.

Inflection point values. Setting the second derivative of Eq. 3.5.34 equal to zero provides the inflection point times

$$t_i = \frac{\lambda - 1}{a} \left(1 \pm \frac{1}{\sqrt{\lambda - 1}} \right) \quad . \tag{3.5.40}$$

From this equation we compute $t_i = 12.39$ and 31.37 minutes. Accordingly, $n_i = 15.5$ and 21.5.

These various critical points are shown as solid dots in Fig. 3.5.2.

Coefficient of variation. By definition this parameter is equal to the standard deviation divided by the mean. That is

$$\text{C.V.} = \sigma/\overline{T} = 1/\sqrt{\lambda} \quad . \tag{3.5.41}$$

For our illustration, C.V. $= 0.398$.

Finally, we look at two other statistical parameters, important but less frequently utilized than those listed above.

Backing up briefly. If we compute the *first* moment of the density distribution about the origin, we obtain an equation for the mean value, $\bar{t}$. If we calculate the *second* moment about the mean, we get an expression for the variance, σ^2.

In the same fashion, computation of the *third* moment provides an equation which characterizes the *skewness* of the density distribution. A similar *fourth* moment calculation yields an index which describes the *peakedness* of the distribution.

Coefficient of skewness. From the third moment computation we obtain the following coefficient of skewness, γ_1, for the gamma distribution.

$$\gamma_1 = 2/\sqrt{\lambda} \quad . \tag{3.5.42}$$

If the maximum (or modal) time is *less* than the mean time, the coefficient of skewness is *positive* and the distribution is said to be *positively skewed* (the longer tail is in the positive direction). Clearly, the gamma function is always positively skewed. For our illustration, $\gamma_1 = +0.796$.

On the other hand, if the maximum (modal) time is *larger* than the mean time, the coefficient of skewness is *negative* and the distribution is *negatively skewed* (the longer tail is in the negative direction).

Coefficient of kurtosis. Finally, calculating the fourth moment of the density distribution produces the coefficient of kurtosis, γ_2. For the gamma distribution we have

$$\gamma_2 = 3 + 6/\lambda \ \ . \tag{3.5.43}$$

This parameter characterizes the "peakedness" or the "flatness" of a density distribution. The number 3 is the value of γ_2 for the normal probability distribution. So this parameter is a measure of how peaked or how flat a density distribution is when compared to the normal curve. For our illustration, $\gamma_2 = 3.95$. On this basis, the gamma function has a sharper peak than does the normal probability distribution.

A given distribution is said to be platykurtic, mesokurtic, or leptokurtic depending on whether γ_2 is less than, equal to, or greater than 3.

We continue this illustration concerning generation times of cells in Section 4.4.6.

3.5.8 An Illustration: Population of Great Britain

Throughout our analyses thus far we have examined and utilized the logistic distribution to considerable extent; this will continue in ensuing sections. However, at this point we take a diversion.

It is noted that there are close similarities between the basic differential equations for, respectively, the logistic and the arctangent– exponential distributions. That is,

$$\frac{dU}{dT} = U(1 - U) \quad \text{and} \quad \frac{dU}{dT} = \frac{1}{4}\sin \pi U \ \ . \tag{3.5.44}$$

The latter distribution was examined briefly in Section 3.5.2. From the above relationships we observe that in both distributions, the growth rate $dU/dT = 0$ when $U = 0$ and 1 and has a maximum value of $1/4$ when $U = 1/2$. Furthermore, from Table 3.5.1 we note that $U = 1/2$ at the inflection points of the cumulative curves of both.

So for a change, we utilize the arctangent–exponential distribution as the framework for our next illustration: the population of Great Britain.

Integration of Eq. 3.5.44 gives the following cumulative distribution function of the arctangent–exponential

$$N = \frac{2N_*}{\pi}\arctan\left(\beta e^{\pi at/4}\right) \tag{3.5.45}$$

where $\beta = \tan(\pi N_0 / 2N_*)$. The inflection point coordinates are

$$t_i = \frac{4}{\pi a} \log_e \left(\frac{1}{\beta}\right) \quad ; \quad N_i = \tfrac{1}{2}N_* \quad ; \quad \left(\frac{dN}{dt}\right)_i = \tfrac{1}{4}aN_* \qquad (3.5.46)$$

and the coordinates of maximum curvature are

$$t_c = \frac{4}{\pi a} \log_e \left[\frac{1}{\beta}(3 \pm 2\sqrt{2})\right] \quad ;$$

$$N_c = \tfrac{1}{2}N_*(1 \pm \tfrac{1}{2}) \quad ; \quad \left(\frac{dN}{dt}\right)_c = \frac{1}{4\sqrt{2}}aN_* \quad . \qquad (3.5.47)$$

The population of Great Britain is shown in Table 3.5.2 at decade intervals from 1801 to 1971. This is the same range of time considered by Leach (1981) in his analysis featuring the logistic distribution. The data of the table are displayed in Fig. 3.5.3.

Utilizing the arctangent–exponential framework, we determine the following values of the parameters: $a = 0.0191$, $N_0 = 10.66$ and $N_* = 65.75$. For comparison, Leach obtained: $a = 0.0196$, $N_0 = 10.00$ and $N_* = 63.65$. Substitution of our values into Eq. 3.5.45 yields the curve shown in Fig. 3.5.3.

Table 3.5.2 Population of Great Britain by decades, 1801 to 1971. Reference: Annual Abstracts of Statistics, 1975 edition, Central Statistical Office, Her Majesty's Stationery Office, London, 1976

Year	t years	N millions	Year	t years	N millions
1801	0	10.501	1891	90	33.029
	10	11.970		100	37.000
	20	14.092		110	40.831
1831	30	16.231	1921	120	42.769
	40	18.534		130	44.795
	50	20.817		140	46.877
1861	60	23.128	1951	150	48.854
	70	26.072		160	51.284
	80	29.710		170	53.979

From Eqs. 3.5.46 and 3.5.47 the coordinates of the critical points are

Inflection point	*Maximum curvature points*
$t_i = 89.80\,(1890.80)$	$t_c = 31.05\,(1832.05)$ and $148.55\,(1949.55)$
$N_i = 32.88$	$N_c = 16.44$ and 49.31
$(dN/dt)_i = 0.314$	$(dN/dt)_c = 0.222$

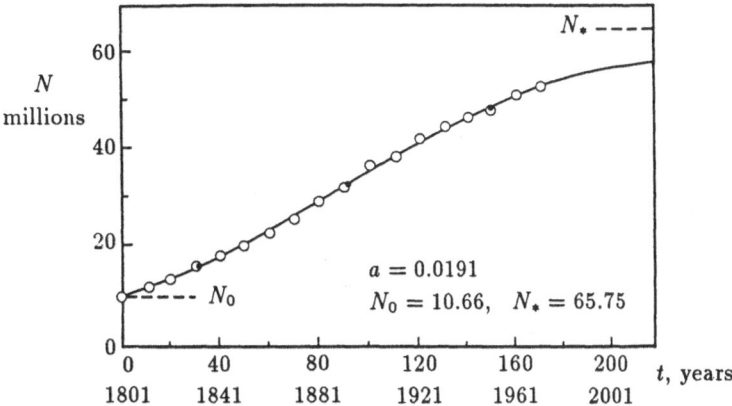

Fig. 3.5.3 Population of Great Britain

These three points are shown as solid dots in Fig. 3.5.3. On the basis of the above results we make the following obvservations:

– The population of Great Britain was increasing at its maximum rate ("maximum velocity") of 314 000 persons per year in 1891.
– In the vernacular of Section 2.2.4, the "take off" of Great Britain's population ("maximum acceleration") occurred in 1832; the "settling down" point ("maximum deceleration") took place in 1949.
– The projected population of Great Britain is 57.230 million in 1996, 57.831 million in 2001, and 58.913 million in 2011. On this point, Leach tabulates 12 values of the 2011 population made at various times during the period 1972 to 1981. Projections range from 53.54 to 64.28 million. The average value of 58.97 million is quite close to the above-indicated calculated value.
– The computed asymptotic population is $N_* = 65.75$ million.

Our main conclusion is that the arctangent–exponential distribution provides a satisfactory framework for the analysis and display of data on population growth.

4. Phenomena with Variable Growth Coefficients

In nearly all of the preceding sections, we have assumed that the growth or transfer coefficient, a, the crowding coefficient, b, and the carrying capacity, $N_* = a/b$, are constants. We now remove these restrictions and assume that, in general, these parameters are functions of time. Accordingly, the differential equation for the logistic is

$$\frac{dN}{dt} = a(t)N\left(1 - \frac{N}{N_*(t)}\right) \quad . \tag{4.0.1}$$

As before, the initial condition is $N(0) = N_0$.

This expression is a form of the Ricatti equation which is easily solved by letting $N = 1/C$. With this substitution we obtain

$$\frac{dC}{dt} + a(t)C = \frac{a(t)}{N_*(t)} \quad . \tag{4.0.2}$$

This equation is a first order, linear, complete (non-homogeneous) differential equation with variable coefficients. Using the same method we used in Section 2.3.3 with the the confined exponential equation, the following generalized answer is obtained

$$N(t) = \frac{N_0 \exp\left(\int_0^t a(\xi)d\xi\right)}{1 + N_0 \int_0^t \frac{a(\xi)}{N_*(\xi)} \exp\left(\int_0^\xi a(\eta)d\eta\right) d\xi} \quad , \tag{4.0.3}$$

where ξ and η are integration variables.

In principle, we can solve this equation for any specified functions, $a(t)$ and $N_*(t)$. Indeed, a number of studies have been carried out which commence with generalized forms of $a = a(t)$ and $N_* = N_*(t)$. For example, Nisbet and Gurney (1976) utilize sinusoidally variable forms for both $a(t)$ and $N_*(t)$ incorporating a phase angle, ϕ, between the two functions. They also include the effect of discrete time delay, a topic we consider in Chapter 6.

Investigations by Coleman (1979) and Cushing (1986) also feature oscillatory forms for both the growth coefficient and the carrying capacity. Arrigoni and Steiner (1985) utilize a sinusoidal-type function for $N_*(t)$ in a power law logistic.

We shall consider two main categories of problems: (1) variable growth coefficient with constant carrying capacity and (2) constant growth coefficient with variable carrying capacity. The first category of problems is presented in the headings of Chapter 4 and the second in those of Chapter 5.

We begin with the first category: the growth coefficient varies with time and the carrying capacity is constant. This kind of phenomenon has a broad range of application. Indeed, on a kind of micro-time scale basis it would encompass virtually all growth phenomena and transfer processes.

Most population growths in fact involve time-dependent growth coefficients. The rate of transfer of a particular technology could be, for example, time cyclical in nature. We have already seen in Section 3.1 that the growth coefficients of some kinds of tumors decrease exponentially with time. The actual growth of a tree during a year depends on time-varying ambient conditions of temperature and rainfall which change from month to month but are essentially repeated a year later. The primary productivity of a body of water, its biomass and oxygen generation by photosynthesis, are time-dependent due to diurnal variations of light intensity. The rate of growth of bacteria in soil is affected by daily and annual variations of temperature. And so on.

For this set of problems we utilize the generalized result of Eq. 4.0.3. We suppose that $a = a(t)$ and that $N_* = N_{*0}$. Accordingly, Eq. 4.0.3 reduces to

$$N = \frac{N_{*0}}{1 + \left(\frac{N_{*0}}{N_0} - 1\right) \exp\left(-\int_0^t a(\xi)d\xi\right)} \ . \tag{4.0.4}$$

This result is a rather general form for the logistic equation; clearly, there is a finite and constant carrying capacity, N_{*0}.

However, to keep our problem as simple as possible we will make another assumption: the carrying capacity, N_{*0}, is infinite. In this case, Eq. 4.0.4 becomes

$$N = N_0 \exp\left(\int_0^t a(\xi)d\xi\right) \tag{4.0.5}$$

which is a generalized exponential equation.

4.1 Linearly Variable Growth Coefficient

4.1.1 The Growth Curve and Its Properties

We consider the case in which the growth coefficient, $a(t)$, is either a constant or is linearly changing. If $a(t) = a_0 = $ constant, we obtain from Eq. 4.0.5

$$N = N_0 \exp(a_0 t) \ , \tag{4.1.1}$$

which is the familiar exponential growth equation. If $a(t)$ is linearly changing we have

$$a(t) = a_0(1 - ct) \quad . \tag{4.1.2}$$

A positive sign could be used in this equation . However, the outcome would be to simply produce a growth equation more explosive than the already explosive exponential equation; nothing interesting is gained.

Substitution of Eq. 4.1.2 into Eq. 4.0.5 gives

$$N = N_0 \exp\left[a_0(t - \tfrac{1}{2}ct^2)\right] \quad . \tag{4.1.3}$$

Now since $c > 0$, it is clear that for large values of t, N approaches zero regardless of the value of c. The relationship of Eq. 4.1.3 is shown in Fig. 4.1.1 for several values of c.

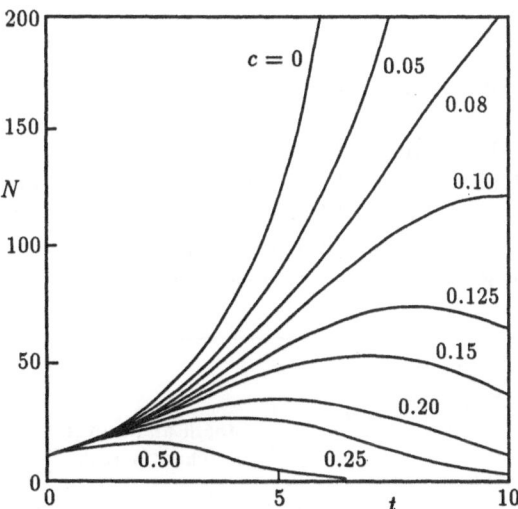

Fig. 4.1.1 Growth curves with linearly variable growth coefficients. $a_0 = 0.50$, $N_0 = 10$

The first derivative of Eq. 4.1.3 gives the slope of the growth curve

$$\frac{dN}{dt} = a_0 N_0 (1 - ct) \exp[a_0(t - \tfrac{1}{2}ct^2)] \tag{4.1.4}$$

from which we obtain the maximum point

$$t_m = \frac{1}{c} \quad ; \qquad N_m = N_0 \exp(a_0/2c) \quad . \tag{4.1.5}$$

It is observed from Eq. 4.1.2 that at this value, $t_m = 1/c$, the growth coefficient, $a(t)$, changes from a positive to a negative quantity.

When the second derivative of Eq. 4.1.3 is equated to zero, the inflection points are obtained

$$t_i = \frac{1}{c}\left(1 \pm \sqrt{\frac{c}{a_0}}\right) \quad ; \qquad N_i = N_0 \exp\left(\frac{a_0 - c}{2c}\right)$$

$$\left(\frac{dN}{dt}\right)_i = \sqrt{a_0 c}\, N_0 \exp\left(\frac{a_0 - c}{2c}\right) \quad . \tag{4.1.6}$$

Translating the origin from $t = 0$ to $\tau = 0$ by the equation $\tau = t - t_m$ yields

$$N = N_m \exp\left(-\frac{a_0 c}{2}\tau^2\right) \tag{4.1.7}$$

which is an interesting result. It indicates that in this linearly decreasing growth coefficient case, N is normally distributed about $\tau = 0$ with variance $\sigma^2 = 1/a_0 c$.

4.1.2 An Illustration: Growth and Decline of U. S. Sailing Vessels

The linearly variable growth coefficient model provides a suitable framework for the display and analysis of historical data concerning the growth and decline of sailing ships in America. In Table 4.1.1, the gross tonnages of U.S. merchant sailing vessels are listed, at five-year intervals, for the period from 1790 to 1965. These data are presented in graphical form in Fig. 4.1.2.

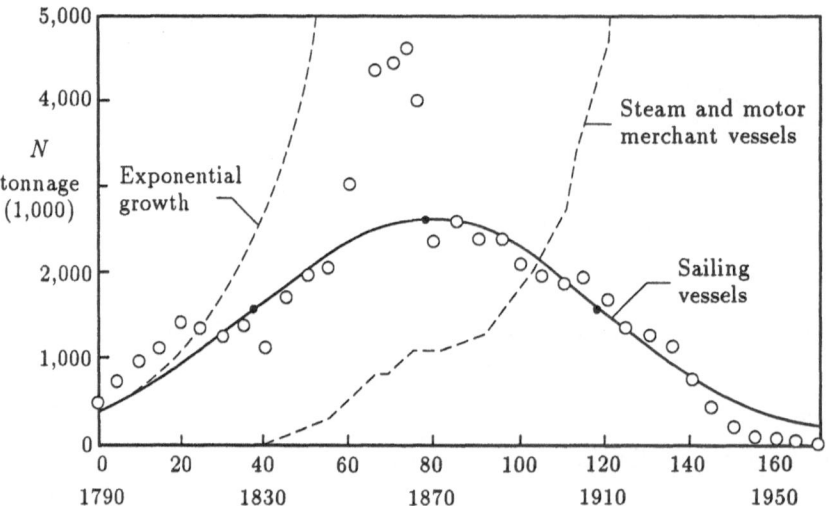

Fig. 4.1.2 Gross tonnage of U.S. sailing ships, 1790–1965. Example of exponential growth with a linearly variable growth coefficient

Table **4.1.1** Gross tonnage of U.S. sailing vessels, 1790–1965. Values of N are expressed in 1000 tons. Reference: Historical Statistics of the United States, Colonial Times to 1970. U.S. Department of Commerce, Government Printing Office, Washington, D.C., 1975

Year	t	N	Year	t	N
1790	0	478	1880	90	2366
1795	5	748	1885	95	2374
1800	10	972	1890	100	2109
1805	15	1140	1895	105	1965
1810	20	1424	1900	110	1885
1815	25	1365	1905	115	1962
1820	30	1258	1910	120	1655
1825	35	1400	1915	125	1384
1830	40	1127	1920	130	1272
1835	45	1702	1925	135	1125
1840	50	1978	1930	140	757
1845	55	2091	1935	145	441
1850	60	3010	1940	150	200
1855	65	4442	1945	155	115
1860	70	4486	1950	160	82
1865	75	4030	1955	165	40
1870	80	2363	1960	170	23
1875	85	2585	1965	175	8

It is observed in the table and the figure that in 1790, the first year for which data were available, there were about 500 thousand tons of sailing ships in America. This tonnage increased rapidly during the first half of the 19th century, especially during the years immediately preceding the Civil War. Tonnage reached a maximum of 4665 thousand in 1863. After the war, the amount decreased rapidly and drastically to around 2500 thousand tons. From then on, sailing vessel tonnage declined each year; wind power was being replaced by mechanical power. As the figure shows, the total tonnages of the two types of ships were approximately the same in 1893 when there were about 2150 thousand tons of each. By 1935, sail tonnage was about the same as it had been in 1790. By 1960, the sailing vessel had all but disappeared.

A number of growth models could be employed to analyze the data of this sailing ship case. For example, we could use the exponentially variable carrying capacity model of Section 5.1. Clearly, it is impossible to know enough about the social and economic factors involved in this kind of growth-decline phenomenon to specify the most appropriate analytical framework. So we select the linearly decreasing growth coefficient model because it is simple and because it satisfies our requirements for a growth phase, a maximum, and a declining phase for the cumulative distribution function.

In the analysis of the sailing vessel tonnages given in Table 4.1.1, we arbitrarily exclude the data points associated with the period just before and during the Civil War (1850, 1855, 1860, 1865).

Computations for the mean value, t_m, and the standard deviation, σ, of the data of Table 4.1.1 give the results: $t_m = 78.1$ and $\sigma = 40.2$. Since $c = 1/t_m$ and $\sigma = 1/a_0 c$, we obtain $c = 0.0128$ and $a_0 = 0.0483$.

The average value of N_0, computed from each data point using Eq. 4.1.3, yields $N_0 = 395$. The maximum tonnage is $N_m = 2606$ which occurs, of course, at $t_m = 78.1$ (i.e., year 1868.1). The inflection points are $t_{i1} = 37.9$, $t_{i2} = 118.3$; $N_i = 1581$. The slopes of the growth curve at the inflection points are $(dN/dt)_i = \pm 39.3$ thousand tons/year. These values are the maximum rates of growth and decline; they occurred in the years 1827.9 and 1908.3. The maximum and inflection points are shown in the figure.

This illustration involving the growth and decline of sailing ships in America is another example of technology substitution: steam ships replacing sailing vessels. The case is examined at greater length by Lanford (1972). A similar study involving the replacement of horses by tractors on Illinois farms is given by Mattingly (1987). We looked at the substitution of synthetic fibers for natural fibers in Section 2.2.3 and the substitution of diesel locomotives for steam locomotives in Section 3.2.3.

We examine another kind of replacement problem in Section 5.1.2: the "demographic substitution" of people living increasingly in the urban areas of America instead of on farms.

4.2 Hyperbolically Variable Growth Coefficient

4.2.1 The Growth Curve and Its Properties

An appropriate next problem is to examine the case in which the growth coefficient behaves hyperbolically. That is

$$a(t) = \frac{a_0}{1 + ct} \quad , \tag{4.2.1}$$

where c is a positive or negative constant. Substituting this relationship into Eq. 4.0.5, we obtain

$$N = N_0 (1 + ct)^{a_0/c} \quad . \tag{4.2.2}$$

The first and second derivatives of this equation are

$$\frac{dN}{dt} = a_0 N_0 (1 + ct)^{(a_0/c)-1} \tag{4.2.3}$$

and

$$\frac{d^2 N}{dt^2} = a_0 (a_0 - c) N_0 (1 + ct)^{(a_0/c)-2} \quad . \tag{4.2.4}$$

From these equations, we see that there are no maximum nor inflection points and that the slope is always positive. The curvature is positive (concave

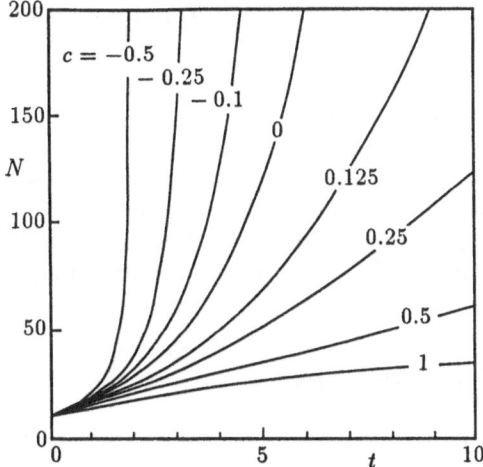

Fig. 4.2.1 Growth curves with hyperbolically variable growth coefficients. $a_0 = 0.5$, $N_0 = 10$

upward) for all negative values of c and is negative (concave downward) for positive values of c larger than a_0. Importantly, N becomes infinite, for finite values of t, for all negative values of c.

In Fig. 4.2.1, several curves are shown with hyperbolically variable growth coefficients for various values of c.

4.2.2 Relationship to Power Law Exponential Growth

The differential equation for this hyperbolic case is

$$\frac{dN}{dt} = \frac{a_0}{1 + ct} N \quad . \tag{4.2.5}$$

Letting $W = N/N_0$, $T = a_0 t$, and $r = 1 - (c/a_0)$, we obtain

$$\frac{dW}{dT} = \frac{1}{1 + (1 - r)T} W \tag{4.2.6}$$

whose solution is

$$W = [1 + (1 - r)T]^{1/(1-r)} \quad . \tag{4.2.7}$$

Consequently

$$\frac{dW}{dT} = W^r \quad . \tag{4.2.8}$$

These last two equations are exactly the same as Eqs. 2.1.13 and 2.1.14 which describe power law exponential growth. Thus, we note the identity of the frameworks of hyperbolically variable growth coefficient and power law exponential growth.

When $r = 0$ (i.e., $c = a_0$), we obtain the linear growth case, and when $r = 1$ (i.e., $c = 0$), the equations reduce to the case of simple exponential growth.

When $r = 2$ (i.e., $c = -a_0$), we acquire the equation for hyperbolic growth, $N = N_0/(1 - a_0 t)$. This is the relationship we examined in Section 2.1.6 in connection with the growth of the world's population. It is interesting to note that the most explosive of all growth equations, the hyperbolic equation, is also the simplest.

4.2.3 An Illustration: Population of the Great Plains States

In the wake of the Civil War, with the phenomenal westward expansion of America's railroads, the Great Plains states – Kansas, Nebraska, North Dakota, and South Dakota – began to increase rapidly in population. The early settlers of the region were people migrating from the eastern states of the nation along with a great many immigrants from the countries of northern Europe. These people settled on farms and in small towns. Agriculture was by far the main endeavor; wheat and corn were, and are, the most important crops of the region.

During the two decades, 1860 to 1880, the combined populations of the four states increased at an exponential rate: 141 000 in 1860 to 501 000 in 1870 to 1 583 000 in 1880. During that period, the average annual growth rate of population increase was approximately $a = 0.121$.

However, commencing around 1880 the annual growth rate began to decline. For example, during the decade 1880–1890, the annual growth rate dropped to about 0.065 and from 1900 to 1910, it fell to approximately 0.022. Indeed, during the depression–drought years of 1930 to 1940, the growth rate was negative, $a = -0.005$. This gradual decline of the growth rate of the Great Plains states continues at the present time; it appears that it will continue to decrease in the years to come. We use this geographic–demographic setting as the topic of our next illustration.

It is apparent from Eq. 4.2.1 that with a positive value of the parameter c, we have a very simple expression for quantitatively describing a continuously decreasing growth rate, a, with increasing time, t. A priori, there is no reason to believe that this equation should describe the Great Plains states phenomenon; however, it is worth a try.

The observed combined populations of the four states for the period 1880 to 1990 are shown in Table 4.2.1. We take the year 1880 to correspond to $t = 0$. Also shown in the table are the computed populations for the period 1880 to 2000; for the moment we disregard these. Plotted data are displayed in Fig. 4.2.2.

Table 4.2.1 Combined populations of the Great Plains states (Kansas, Nebraska, North Dakota, South Dakota), 1860 to 1990.Projections to year 2000. Populations in millions. Reference: The Universal Almanac 1990, Andrews and McMeel, Kansas City, Mo., 1989

Year	t	N_{obs}	N_{comp}	Year	t	N_{obs}	N_{comp}
1860		0.141		1940	60	4.402	4.569
1870		0.501		1950	70	4.504	4.746
1880	0	1.583	1.635	1960	80	4.903	4.905
1890	10	3.031	2.978	1970	90	5.018	5.050
1900	20	3.257	3.500	1980	100	5.278	5.183
1910	30	4.044	3.858	1990	110	5.448	5.307
1920	40	4.349	4.137	1995	115		5.366
1930	50	4.633	4.369	2000	120		5.423

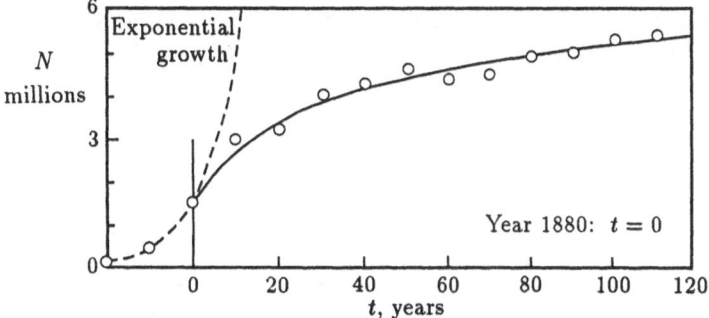

Fig. 4.2.2 Growth of population of the Great Plains States. Example of exponential growth with a hyperbolically variable growth coefficient

We need to determine the values of the parameters a_0, c, and N_0. The most direct way to do this is to write Eq. 4.2.5 in finite difference form

$$\frac{1}{\overline{N}} \frac{\Delta N}{\Delta t} = a = \frac{a_0}{1 + ct} \quad . \tag{4.2.9}$$

From the data of Table 4.2.1 we calculate the values of the left-hand member of this equation. We then rewrite Eq. 4.2.9 in the form

$$\frac{1}{a} = \frac{1}{a_0} + \frac{c}{a_0} t \quad , \tag{4.2.10}$$

which is a linear equation. A least squares computation provides the numerical values of a_0 and c. From Eq. 4.2.2, we calculate the average value of N_0 using the tabulated values of N.

Our numerical results are the following $a_0 = 0.25$, $c = 1.0$, $N_0 = 1.635$. Substituting these numbers into Eq. 4.2.2 gives the following equation

$$N = 1.635(1 + t)^{0.25} \quad . \tag{4.2.11}$$

This expression is shown as the solid curve in Fig. 4.2.2. The computed values of N, listed in Table 4.2.1, are obtained from this equation. There is fairly good agreement between observed and computed values of the population.

A comment or two. It is likely that the confined exponential distribution of Section 2.3.2 would be equally suitable as a framework for analyzing the 1880–1990 data of our illustration. In this case, it would be necessary to determine the asymptotic population, N_*. By the same token, had we included the 1860–1880 data in our analysis, we would have obtained an unsymmetrical S-shaped growth curve with an inflection point occurring around 1880 or 1885. This pattern is indicated in the figure. Then we might have employed the power law logistic of Section 2.6.2 or the modified logistic of Section 2.9.2 as our framework. Again, we would need to determine the carrying capacity, N_*.

All of which simply emphasizes a truism that, in general, there is more than one framework for data analysis and display. In our illustration of the population of the four states of the Great Plains, the remarkably simple hyperbolically variable growth coefficient model appears to provide a suitable framework.

4.3 Exponentially Variable Growth Coefficient

4.3.1 Extreme Maximum Value Distribution

Probably the most important category of growth equations with time-dependent growth coefficients is that in which the coefficient varies exponentially.

We start with the case in which

$$a(t) = a_0 e^{-kt} \quad , \tag{4.3.1}$$

where k is a positive constant. Accordingly the differential equation is

$$\frac{dN}{dt} = a_0 e^{-kt} N \tag{4.3.2}$$

and from the general solution provided by Eq. 4.0.5 we obtain

$$N = N_0 \exp\left[\frac{a_0}{k} \left(1 - e^{-kt}\right)\right] \quad . \tag{4.3.3}$$

As $t \to \infty$, this expression yields the asymptotic value $N_* = N_0 \exp(a_0/k)$, and so

$$N = N_* \exp\left(-\frac{a_0}{k} e^{-kt}\right) \quad . \tag{4.3.4}$$

This equation is the Gompertz distribution. In Section 3.1, numerous properties of the Gompertz relationship were obtained, including the coordinates of the inflection point

$$t_i = \frac{1}{k}\log_e\frac{a_0}{k} \quad ; \qquad N_i = \frac{1}{e}N_* \quad . \tag{4.3.5}$$

Translating the origin of Eq. 4.3.4 to the inflection point by the relationship $\tau = t - t_i$ and letting $U = N/N_*$ and $T = k\tau$, yields the expression

$$U = \exp[-\exp(-T)] \tag{4.3.6}$$

which we acquired from the generalized distribution considered in Section 3.3.3. Differentiating Eq. 4.3.6 gives

$$u = \exp[-T - \exp(-T)] \quad . \tag{4.3.7}$$

These last two equations define the cumulative function, $U(T)$, and the density function, $u(T)$, of the extreme maximum value distribution. They are displayed in Fig. 4.3.1 and some of their properties are listed in Table 4.3.1.

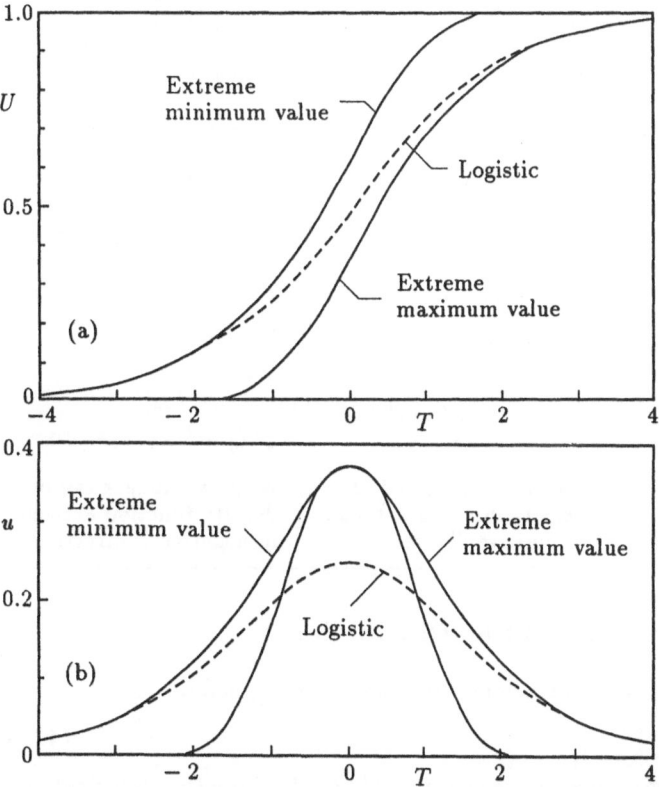

Fig. 4.3.1 Plots of the extreme value and logistic distributions. (a) Cumulative distribution and (b) density distribution

Table 4.3.1 Properties of the extreme value distributions

Property of distribution	Extreme maximum value distribution	Extreme minimum value distribution
Cumulative distribution	$U = \exp(-e^{-T})$	$U = 1 - \exp(-e^{T})$
Density distribution	$u = \exp(-T - e^{-T})$	$u = \exp(T - e^{T})$
$\dfrac{dU}{dT} = f(T, U)$	$e^{-T} U$	$e^{T}(1 - U)$
$\dfrac{dU}{dT} = g(U)$	$U \log_e \left(\dfrac{1}{U}\right)$	$(1 - U) \log_e \left(\dfrac{1}{1 - U}\right)$
Inflection point of cumulative distribution	$T_i = 0, \quad U_i = 1/e$ $u_i = \left(\dfrac{dU}{dT}\right)_i = 1/e$	$T_i = 0, \quad U_i = 1 - 1/e$ $u_i = \left(\dfrac{dU}{dT}\right)_i = 1/e$
Inflection point of density distribution; maximum curvature point of cumulative distribution	$T_c = \log_e m$ $U_c = \exp(-m)$ $u_c = \left(\dfrac{dU}{dT}\right)_i = m e^{-m}$ $m = \frac{1}{2}(3 \pm \sqrt{5})$	$T_c = \log_e m$ $U_c = 1 - \exp(-m)$ $u_c = \left(\dfrac{dU}{dT}\right)_i = m e^{-m}$ $m = \frac{1}{2}(3 \pm \sqrt{5})$
Inverse distribution	$-T = \log_e \log_e (1/U)$	$T = \log_e \log_e (1/1 - U)$
Median value	$-\log_e \log_e 2$	$\log_e \log_e 2$
Mean value	γ	$-\gamma$
	(γ: Euler constant = 0.57722)	
Variance	$\pi^2/6$	$\pi^2/6$
Skewness	Positive; longer tail of density function extends in positive T direction	Negative; longer tail of density function extends in negative T direction

4.3.2 Extreme Minimum Value Distribution

Next, we select a growth coefficient with the following definition

$$a = a_0 e^{kt} \tag{4.3.8}$$

in which k is again a positive constant. Thus, the growth coefficient is a positive instead of a negative exponential.

We saw in the previous section, specifically in Eq. 4.3.2, that the combination of an exponentially decreasing growth coefficient, $\exp(-kt)$, and an otherwise unconfined growth quantity, N, resulted in a finite asymptotic value, N_*.

At the other extreme, which we now consider, we combine an exponentially increasing growth coefficient, $\exp(kt)$, and a confined growth quantity, $(N_* - N)$. That is

$$\frac{dN}{dt} = a_0 e^{kt}(N_* - N) \quad . \tag{4.3.9}$$

The solution to this equation is

$$N = N_* \left[1 - \exp\left(-\frac{a_0}{k}e^{kt}\right)\right] \quad . \tag{4.3.10}$$

The initial and asymptotic values are related by

$$N_0 = N_* \left[1 - \exp\left(-\frac{a_0}{k}\right)\right] \quad . \tag{4.3.11}$$

Setting the second derivative of Eq. 4.3.10 equal to zero determines the inflection point of the cumulative distribution

$$t_i = \frac{1}{k}\log_e \frac{k}{a_0} \quad ; \qquad N_i = \left(1 - \frac{1}{e}\right) = N_* \quad . \tag{4.3.12}$$

As before, we shift the origin to the inflection point to obtain

$$U = 1 - \exp(-\exp T) \quad , \tag{4.3.13}$$

where $U = N/N_*$, $T = k\tau$, and $\tau = t - t_i$. The derivative of this equation yields

$$u = \exp(T - \exp T) \quad . \tag{4.3.14}$$

These two expressions, which we obtained in Section 3.3.4, define the cumulative function, $U(T)$, and the density function, $u(T)$, of the extreme minimum value distribution. Properties of these functions are listed in Table 4.3.1 and are plotted in Fig. 4.3.1.

The cumulative and density distributions of the ordinary logistic equation are, from Table 2.5.5

$$U = \frac{1}{1 + e^{-T}} \quad ; \qquad u = \frac{e^{-T}}{(1 + e^{-T})^2} \quad . \tag{4.3.15}$$

These expressions are shown as dashed lines in Fig. 4.3.1. Utilizing the relationship, valid for large T

$$\log_e(1 + e^{-T}) = e^{-T} \tag{4.3.16}$$

it is established that the extreme maximum (minimum) value distribution approaches the logistic for large positive (negative) T. From Table 2.5.5, we note that the variance of the logistic distribution is $\sigma^2 = \pi^2/3$. As indicated in Table 4.3.1, this is double the variance of the extreme value distribution, $\sigma^2 = \pi^2/6$.

4.3.3 An Illustration: Survival of Rats

To illustrate a phenomenon involving exponentially variable growth coefficients, we select some experimental data acquired by Miescher and reported by Batschelet (1979) concerning the survival of rats.

In his experiment, Miescher started with 144 rats all from the same breed and all seven months of age. He then simply recorded how many rats were alive each month thereafter. His data are listed in Table 4.3.2 and displayed in Fig. 4.3.2. For example, after 12 months the number of surviving rats was 108 and after 24 months, 19 remained alive.

Table **4.3.2** Survival of rats. Data of Miescher from Batschelet (1979)

t [months]	M	t [months]	M
1	143	16	70
2	143	17	55
3	141	18	50
4	139	19	48
5	138	20	36
6	137	21	30
7	136	22	29
8	131	23	22
9	122	24	19
10	118	25	11
11	114	26	8
12	108	27	5
13	97	28	3
14	90	29	3
15	80	30	2

In this illustration, we consider the so-called "survival function", $S(T)$, instead of the growth function, $U(T)$. They are simply related: $S(T) = 1 - U(T)$. As a matter of fact, as Impagliazzo (1985) describes, the early work by Gompertz was in connection with attempts to quantify survival rates and life expectancies of humans.

The framework for an examination of the rat survival data is provided by the extreme minimum value distribution of Eq. 4.3.13. The associated survival function is

$$S = \exp(-\exp T) \quad . \tag{4.3.17}$$

From this relationship we obtain the survival equation

$$M = M_0 \exp\left[-\frac{a_0}{k}\left(e^{kt} - 1\right)\right] \tag{4.3.18}$$

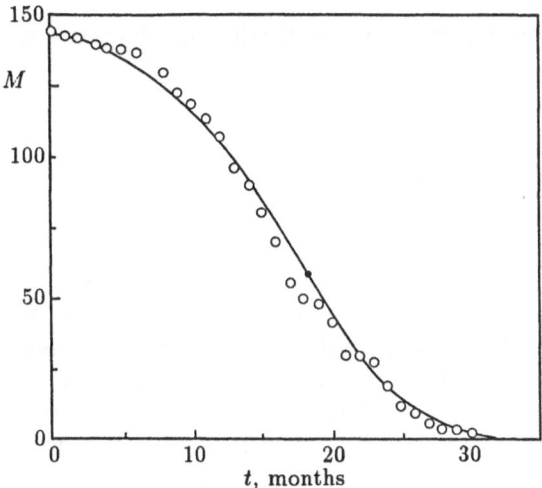

Fig. 4.3.2 Experimental data on survival of rats. Example of an extreme minimum value distribution. Data of Miescher from Batschelet (1979)

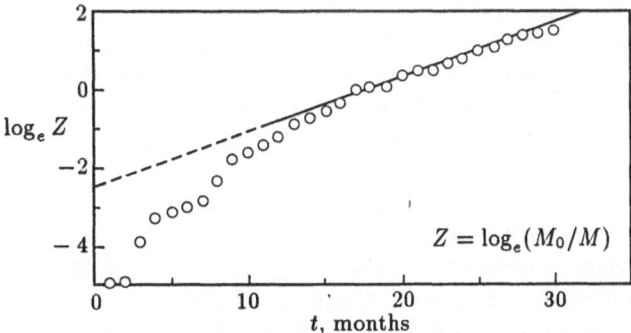

Fig. 4.3.3 Plot of rat survival data to determine values of a_0 and k

where M is the number of survivors at time t and $M_0 = M(0)$. This equation can be rewritten in the form

$$Z = \frac{a_0}{k} \left(e^{kt} - 1 \right) \tag{4.3.19}$$

where $Z = \log_e(M_0/M)$. For values of $\exp(kt) \gg 1$, this equation becomes

$$\log_e Z = \log_e \frac{a_0}{k} + kt \tag{4.3.20}$$

which is linear. The experimental data of Table 4.3.2 are shown in Fig. 4.3.3 with the coordinates of Eq. 4.3.20. A least squares computation for the data corresponding to $t = 15$ to $t = 30$ gives $a_0 = 0.0120$ and $k = 0.130$. The inflection point occurs at $t_i = 18.3$, $N_i = 58.09$. The survival curve, Eq. 4.3.18, is shown in Fig. 4.3.2.

It appears that the extreme minimum value distribution describes the data reasonably well.

4.4 Sinusoidally Variable Growth Coefficient

4.4.1 Some Examples of Oscillatory Phenomena

A great many growth phenomena are cyclic or oscillatory in behavior. For example, a small tree is planted, begins its growth, and its total mass – roots, trunk and limbs, and leaves – continually increases. For a certain period of time its growth tends to be exponential and then, as in the case of virtually all growth processes, the rate of increase of its overall mass slows down. Ultimately the tree attains its full size.

However, during each of the numerous years the tree is growing, its total mass during a year oscillates about an average value: leaves and possibly fruit appear in the spring, the tree is in full foliage during the summer, leaves are shed in the autumn, and the basic growth of the trunk and limbs is almost dormant during the winter. That is, a secondary growth pattern, oscillatory in nature, is superimposed on the primary average growth of the tree. Clearly, numerous environmental factors – temperature, sunlight, soil moisture, nutrients and the like – cause this annual cycle of variation in the growth of the tree.

Numerous phenomena of this type in plant life are examined by Sweeny (1969) and in animal life by Andrewartha and Birch (1964). Pielou (1977) considers cyclic phenomena in connection with ecology and Keyfitz (1968) examines this kind of problem in demography. Oscillatory behavior in economic trade and business cycles is reviewed by van der Ploeg (1984). A journal on cycle research is published by the Foundation for the Study of Cycles.

4.4.2 Simple Harmonic Growth Coefficient

We start with the simplest case: a simple harmonic growth coefficient. That is

$$a(t) = a_0 + a_m \sin(\omega t + \phi) \quad . \tag{4.4.1}$$

This equation says that the growth coefficient, $a(t)$, has an average value, a_0, and oscillates with amplitude, a_m, about this average at a frequency, $\omega = 2\pi/T$; T is the period of oscillation and ϕ is a phase angle.

Substituting Eq. 4.4.1 into Eq. 4.0.5, we obtain

$$N = N_0 \exp\left\{ a_0 t + \frac{a_m}{\omega} \left[\cos\phi - \cos(\omega t + \phi) \right] \right\} \tag{4.4.2}$$

which might be labeled a sinusoidal exponential growth equation.

Plots of Eqs. 4.4.1 and 4.4.2 are presented in Fig. 4.4.1 for various values of the parameters a_m and ϕ. In all four plots of the figure, $a_0 = 0.50$, $N_0 = 10$, $T = 5.0$ and $\omega = 1.257$. Two values of a_m are indicated: 0.25 and 0.50. The varying parameter among the four plots is the phase angle, ϕ, which has the respective values: 0, $\pi/2$, π, and $3\pi/2$.

In the four plots of the figure we observe that the magnitude of N after one cycle, i.e., $t = T$, is the same regardless of the value of the phase angle, ϕ. In Fig. 4.4.1(a), where $\phi = 0$ (the "rising tide" case), the value of N increases rapidly during the first half of the cycle and then grows more slowly during the second half. In (c), with $\phi = \pi$ ("falling tide"), the reverse is the case. As we would expect, in the intermediate cases shown in (b) $\phi = \pi/2$ ("high tide") and (d) $\phi = 3\pi/2$ ("low tide"), N grows at rates, during the period of oscillation, somewhat between the other two cases.

Setting the first derivative of Eq. 4.4.2 equal to zero gives the times for the maximum and minimum values of N for the case $\phi = 0$:

$$t_m = T \left(\frac{j}{2} + (-1)^{j+1} \frac{1}{2\pi} \arcsin \frac{a_0}{a_m} \right) \quad , \tag{4.4.3}$$

where odd values of integer j give the maxima and even values the minima.

In Fig. 4.4.2, several curves are shown in a semi-logarithmic display of Eq. 4.4.2, for values of a_m ranging from zero to 2.0, over two periods of oscillation. As we see in the figure, and as Eq. 4.4.3 confirms, maxima and minima exist only if a_m is greater than a_0. When $a_m = 0$, we obtain the familiar straight line of the exponential growth equation.

4.4.3 Exponentially Decreasing Growth Coefficient: Type I

Next we consider the case in which the growth or transfer coefficient has the definition

$$a(t) = [a_0 + a_m \sin(\omega t + \phi)] \exp(-kt) \quad . \tag{4.4.4}$$

Several plots of this equation are presented in Fig. 4.4.3 for various values of a_m. The curve $a_m = 0$ describes the Gompertz growth coefficient; we examined this case in Section 3.1.2.

Setting the phase angle $\phi = 0$, and substituting Eq. 4.4.4 into Eq. 4.0.5 gives

$$N = N_* \exp \left\{ -\left[\frac{a_0}{k} + \frac{a_m \omega}{k^2 + \omega^2} \left(\frac{k}{\omega} \sin \omega t + \cos \omega t \right) \right] e^{-kt} \right\} \tag{4.4.5}$$

in which, with $N(0) = N_0$

$$N_* = N_0 \exp \left(\frac{a_0}{k} + \frac{a_m \omega}{k^2 + \omega^2} \right) \quad , \tag{4.4.6}$$

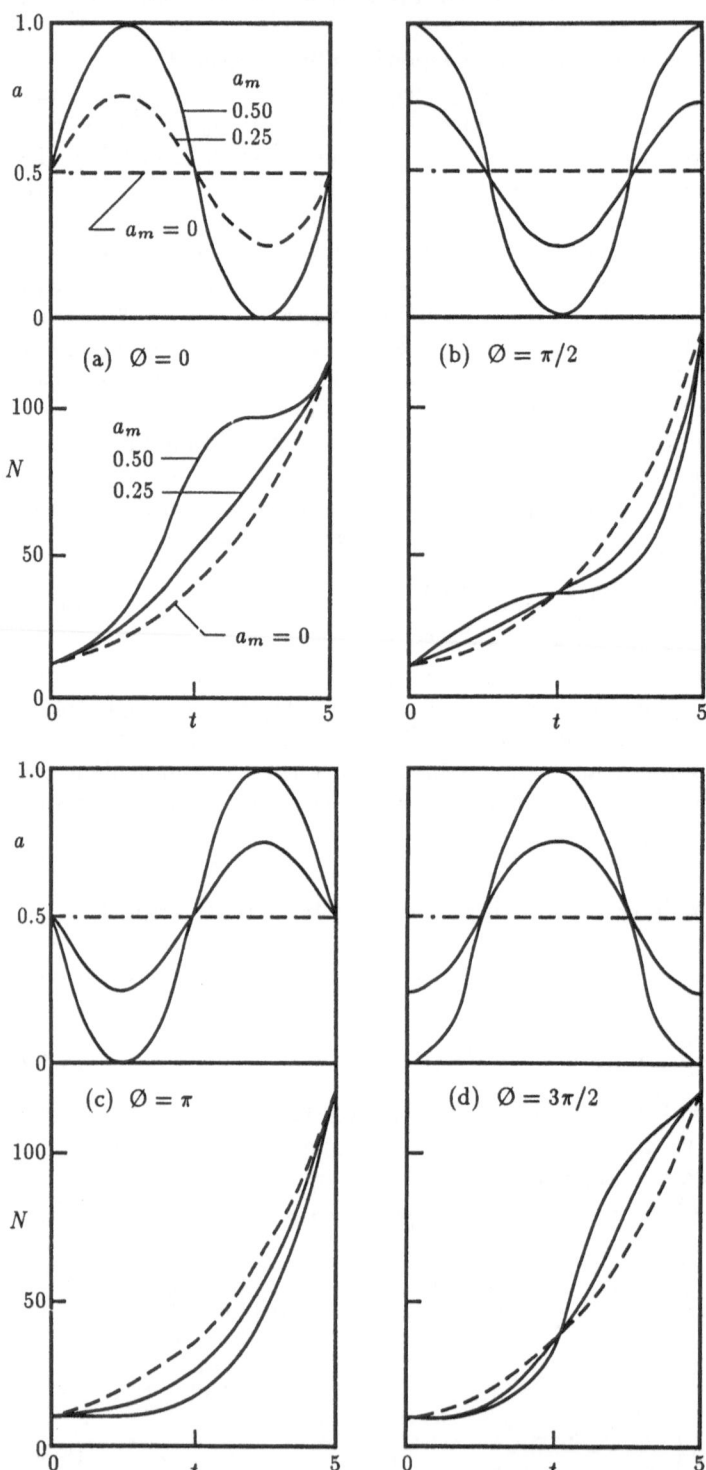

Fig. 4.4.1 Growth curves with sinusoidally variable growth coefficients for various values of phase angle ϕ

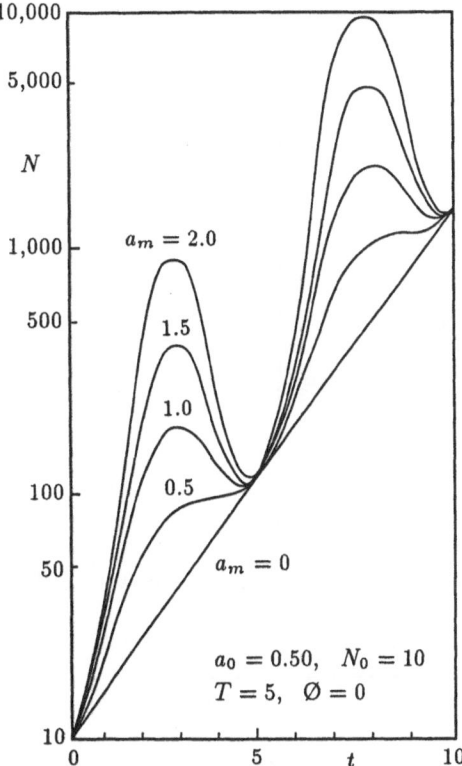

Fig. 4.4.2 Growth curves with sinusoidally variable growth coefficients

where N_* is the value of N as t becomes infinite. It would be logical to term Eq. 4.4.5 the sinusoidal Gompertz equation.

Equating the first derivative of Eq. 4.4.5 to zero gives the same equation we obtained in the previous section, viz., Eq. 4.4.3, to determine the times corresponding to maximum and minimum N. In Fig. 4.4.4 plots of Eq. 4.4.5 are shown for values of a_m ranging from zero to 1.0. The ordinary Gompertz equation is given by $a_m = 0$.

4.4.4 An Illustration: Growth of a Species of Land Snails

Interesting data are given by Bertalanffy (1968) concerning experiments which he and Muller carried out on seasonal variations of the metabolic rates and associated growth rates of a species of land snails (*Eulota fruticum*).

In these experiments, the weights of the land snails were determined at approximately regular intervals over a period of about 35 months. Observed average values of snail weights are listed in Table 4.4.1 and are presented graphically in Fig. 4.4.5.

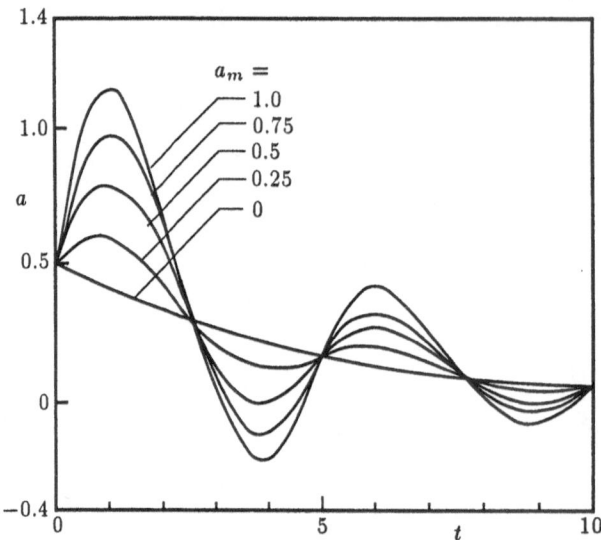

Fig. 4.4.3 Type I sinusoidally variable exponentially decreasing growth coefficient curves. $a_0 = 0.50$, $k = 0.217$, $T = 5.0$

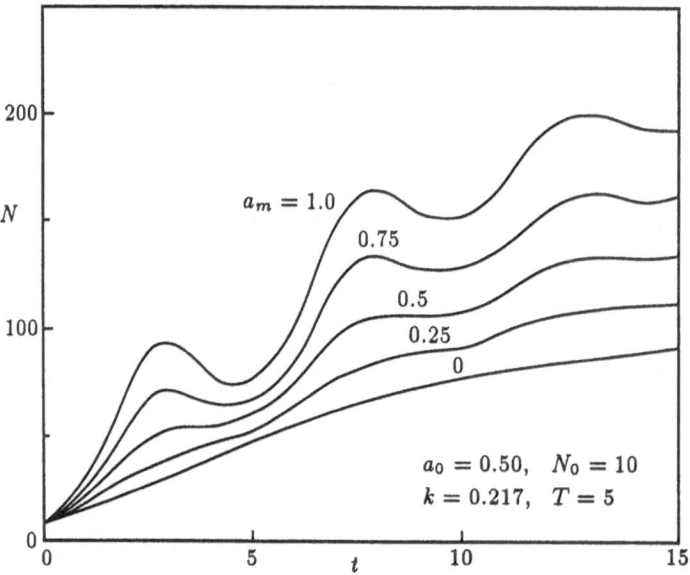

Fig. 4.4.4 Growth curves with Type I sinusoidally variable exponentially decreasing growth coefficients

Table 4.4.1 Growth of land snails. From Bertalanffy (1968)

Time t [months]	Weight N [mg]	Time t [months]	Weight N [mg]
1.5	40	16.1	770
3.1	90	17.7	900
4.1	100	19.8	1130
6.0	98	20.9	1250
7.8	140	22.9	1650
9.0	195	24.0	1880
10.0	505	27.1	1850
11.1	760	30.1	1650
12.5	1000	32.0	1750
13.5	1050	34.5	2410

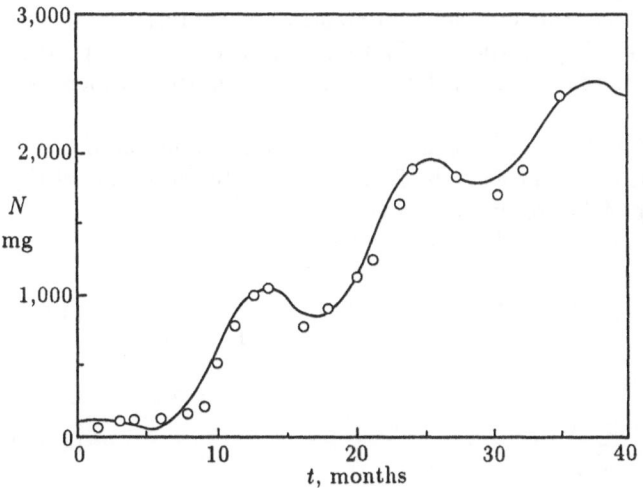

Fig. 4.4.5 Growth of a species of land snails. Comparison of computed Type I growth curve with experimental data. Data of Bertalanffy (1968)

It is assumed that the growth model of the previous section, Eq. 4.4.5, provides a suitable framework. From the plot of Fig. 4.4.5, we select the period of oscillation, $T = 12$ months. Values of the other coefficients (N_0, a_0, a_m, k) are determined by point-matching at maximum and minimum points. A phase angle $\phi = \pi$ is utilized. The following are the summarized results

$$N_0 = 27.6 \qquad N_* = 2976$$

$$a_0 = 0.277 \qquad a_m = 0.520$$

$$T = 12.0 \qquad k = 0.0747$$

$$\omega = 0.524 \qquad \phi = \pi \quad .$$

Substitution of these numbers into Eq. 4.4.5 (modified to include $\phi = \pi$) yields the curve shown in Fig. 4.4.5. The model seems to describe the data reasonably well.

4.4.5 Exponentially Decreasing Growth Coefficient: Type II

As another example, we consider the case in which the growth coefficient behaves as follows

$$a(t) = a_0 + a_m e^{-kt} \sin \omega t \quad . \tag{4.4.7}$$

We compare this Type II sinusoidally variable exponentially decreasing growth coefficient with the Type I considered in Section 4.4.3. In Type II we note that only the amplitude, a_m, is attenuated by the negative exponential. In contrast, in Type I the entire growth coefficient, a, is multiplied by $\exp(-kt)$. As we shall see, this modification creates a substantial difference in the behavior of $N(t)$. Type I allows a finite asymptotic value. Type II is essentially an exponential function which forces $N(t)$ to infinity as t increases without limit.

If $a_m = 0$, Eq. 4.4.7 reduces to the ordinary exponential equation; if $k = 0$, it reverts to the simple harmonic growth of Section 4.4.2. Several plots of Eq. 4.4.7 are shown in Fig. 4.4.6 for various values of a_m.

Substitution of Eq. 4.4.7 into Eq. 4.0.5 yields the solution

$$N = N_* \exp \left[a_0 t - \frac{a_m \omega}{k^2 + \omega^2} e^{-kt} \left(\frac{k}{\omega} \sin \omega t + \cos \omega t \right) \right] \quad , \tag{4.4.8}$$

where

$$N_* = N_0 \exp \left(\frac{a_m \omega}{k^2 + \omega^2} \right) \quad . \tag{4.4.9}$$

Displays of Eq. 4.4.8 are presented in the semi-logarithmic plot of Fig. 4.4.7 for several values of a_m.

From the derivative of Eq. 4.4.8 we obtain the following expression for the determination of the times, t_m, corresponding to maximum and minimum of N

$$e^{-kt_m} \sin \omega t_m = -\frac{a_0}{a_m} \quad . \tag{4.4.10}$$

The dashed line shown in Fig. 4.4.7 corresponds to the critical value, $a_m = 1.112$, for which the maximum and minimum points just converge. The coordinates of this critical point are $t_m = 3.614$, $N_m = 164.3$. Clearly, the slope of the growth curve is zero at this point.

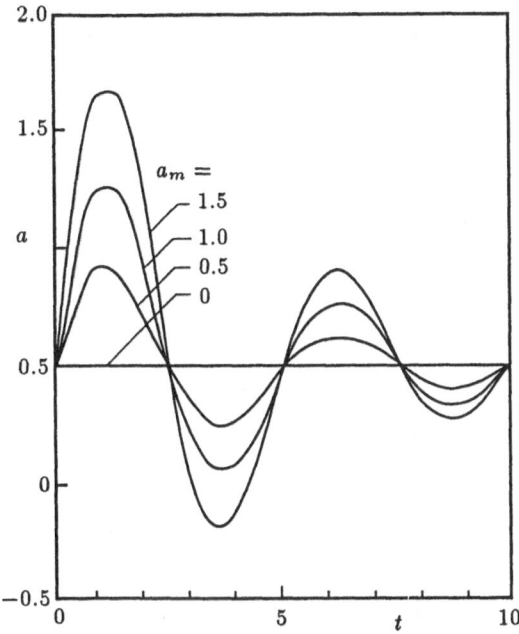

Fig. 4.4.6 Type II sinusoidally variable exponentially decreasing growth coefficient curves. $a_0 = 0.50$, $k = 0.217$, $T = 5.0$

4.4.6 An Illustration: Growth of Cell Populations

In Section 3.5.7, in our consideration of the gamma distribution, we utilized an illustration concerning the growth of cells. We saw that the generation times of individual cells in a growing cell population vary over a fairly wide range. Our main earlier result was that the distribution of generation times, in the particular case we examined, was well described by the gamma function.

We continue the topic of growth of cell populations as an illustration relating to Type II sinusoidally variable exponentially decreasing growth coefficients.

First we take a closer look at the so-called cell cycle. Our presentation is based on the model generally accepted by cell scientists and concisely described by Baserga (1981) and Knolle (1988).

At the moment of termination of the division of a mother cell to create two new daughters, the mother cell, of course, disappears and the daughter cells each begin a new cycle. This moment marks the beginning of the cycle; i.e., $t = 0$ on the generation time clock.

The cell cycle normally comprises four phases; in chronological order these are: (1) the first gap phase (G1) during which RNA and proteins are synthesized; the duration of the G1 phase is the most variable of the entire cycle; (2) the synthesis phase (S) during which DNA replication takes place in the

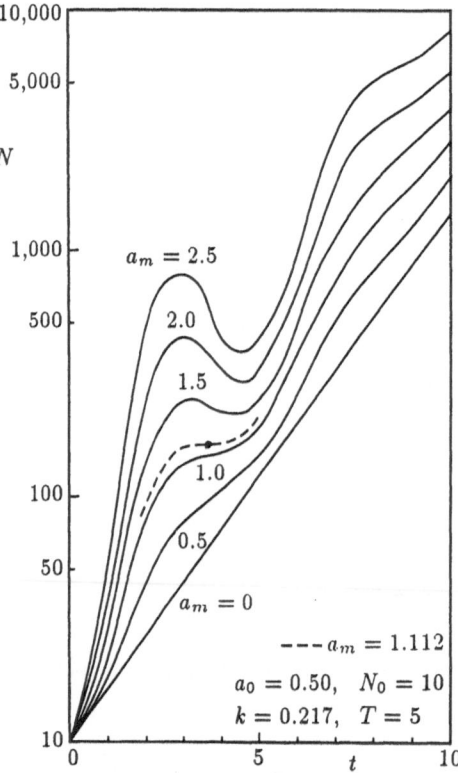

Fig. 4.4.7 Growth curves with Type II sinusoidally variable exponentially decreasing growth coefficients

growing cell; (3) the second gap phase (G2), and (4) the mitosis phase (M) which is the period when the mother divides into two daughters. The end of the mitosis phase marks the end of the cell cycle, i.e., $t = t_g$ on our generation time clock. During the period of the cycle, the cell has doubled its mass.

Some cells fail to divide. In these instances they have either died or they have entered a quiescent gap phase (G_0). In the latter case, sooner or later they re-enter the cell cycle.

Customarily, it is considered that there are two types of cells created in mitotic processes: (1) P cells, comprised of the proliferating (dividing) cells together with the G_0 quiescent gap cells and (2) Q cells, composed of the sterile (dead) cells. At any time, t, the total number of cells, $N(t) = P(t) + Q(t)$. The proliferating fraction is $F(t) = P(t)/N(t)$.

For our illustration we utilize the gamma function generation time distribution we examined in Section 3.5.7 to obtain an expression for the total number of cells, $N(t)$. We assume that all cells are proliferating, i.e., $Q(t) = 0$.

In our generation time illustration we used the data of Prescott (1959) involving a strain of *Tetrahymena geleii*. Prescott carried out a second series of experiments with the identical strain of *T. geleii* and precisely the same growth conditions. He devised his experiment to provide exactly $N_0 = 50$ perfectly synchronized cells at time $t = t_0$, the moment of growth initiation.

The results of Prescott are shown as the solid points in Fig. 4.4.8. The coordinates of an original plot were made dimensionless by Rubinow (1968); $t_0 = 85$ minutes is the duration of the synchronized first cycle.

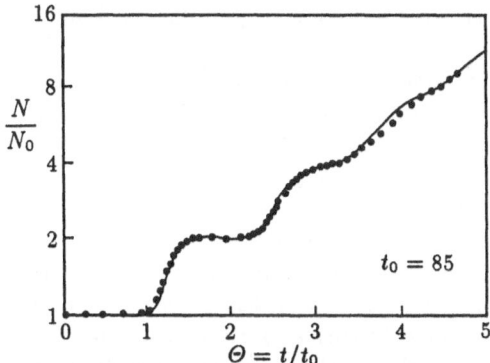

Fig. 4.4.8 Growth of cell population from $N_0 = 50$ cells at $t = 0$. Solid points are the data of Prescott (1959). Curve is the perfect memory model of Rubinow (1980)

One complicating feature in growth phenomena is that many such processes contain a so-called "age structure". That is, the density of the population depends not only on time, t, but also on the age, x, of each constituent of the population. This is the case in our illustration involving the growth of *Tetrahymena geleii*.

Indeed most growth phenomena in ecology and demography are age-structure processes. Fortunately, in many cases we can disregard this complication; criteria for such disregard relate to the ratio of a "delay time" to a characteristic "growth time". In Chapter 6 we examine some topics concerning time delays. The equivalence of an age structured growth process to a corresponding time delay process is pointed out by MacDonald (1978).

In our illustration the variable generation times, or equivalently, the ages of the growing population, makes our analysis rather difficult. Indeed it is necessary to start with the so-called McKendrick–von Foerster equation

$$\frac{\partial n}{\partial t} + \frac{\partial n}{\partial x} = -\mu n \quad , \tag{4.4.11}$$

where $n(x,t)$ is the age distribution density function, x is the age, and $\mu(x,t)$ is the mortality or death rate.

This equation, which is a kind of continuity or conservation relationship, is the equation for age-dependent population growth. As is true for all differential equations, including this partial differential equation, it is necessary to impose appropriate initial and boundary conditions in order to obtain the solution, $n = n(x, t)$. Incidentally, with certain simplifying assumptions, Hoppensteadt (1975) shows that Eq. 4.4.11 leads to expressions which describe ordinary exponential and logistic growth .

The subject of age-dependent growth has been examined by many investigators over a long period of time. Early studies were carried out by Sharpe and Lotka (1911) and by McKendrick (1926). Foundations of the closely related subject of renewal theory, mentioned in Section 3.2.1, were established by Feller (1941) and bases of matrix methodology in population dynamics were formulated by Leslie (1945).

It is not surprising that demographers have made substantial contributions to the subject of age-dependent population growth; classic references are the works of Keyfitz (1968), Coale (1972) and Pollard (1973). More recent are the contributions of Impagliazzo (1985), who gives a number of applications relating to the demography of Denmark, and Song and Yu (1988) who give numerous examples concerning demography in China.

Equally significant advances in age-dependent population dynamics have been made in the fields of biology and ecology. Noteworthy are texts by Pielou (1977) and Krebs (1978). Both Rubinow (1973) and Hoppensteadt (1975) give concise presentations on topics of cell growth and population dynamics. Numerous aspects of cell kinetics, with special focus on the subject of cancer chemotherapy, are examined by Knolle (1988). A contribution by Lebowitz and Rubinow (1974) elaborates the earlier analysis by Rubinow (1968) of the Prescott (1959) data of *T. geleii* growth.

Our illustration awaits us. The experimental data displayed in Fig 4.4.8 were analyzed by Rubinow (1980) on the basis of four mathematical models: (1) an age-time model, (2) a discrete compartment model, (3) a diffusion model and (4) a maturity-time perfect memory model.

The perfect memory model showed best agreement with the data; this result is displayed as the solid curve in Fig. 4.4.8. The salient assumption in this model is that each cell remembers perfectly and duplicates the generation time of its mother. In this case, the equation for $N(t)$, shown as the solid curve in the figure, is given by the expression

$$\frac{N}{N_0} = \sum_{j=1}^{\infty} 2^{j-1} \int_{t/j}^{t/(j-1)} u(z)dz \quad . \tag{4.4.12}$$

where

$$u(\theta) = \frac{\alpha^\lambda}{t_0 \Gamma(\lambda)} (\theta - 1)^{\lambda - 1} e^{-\alpha(\theta - 1)} \quad , \tag{4.4.13}$$

in which $\alpha = at_0$ and $\theta = t/t_0$. This equation is based on the generation time distribution given by Eq. 3.5.34. We note that Eq. 4.4.12 is an infinite series of incomplete gamma functions; this function is tabulated by Pearson (1934) and Abramowitz and Stegun (1965).

For the perfect memory model it is not difficult to obtain $N(t)$ with a numerical integration based on the density distribution of Eq. 4.4.13. In such an integration it is necessary to remember, for example, that fast-growing cells may be entering their $(j+1)$-th cycles as slow-growing cells pass through their j-th cycles. The results based on the perfect memory criterion, are shown as the solid points in Fig. 4.4.9.

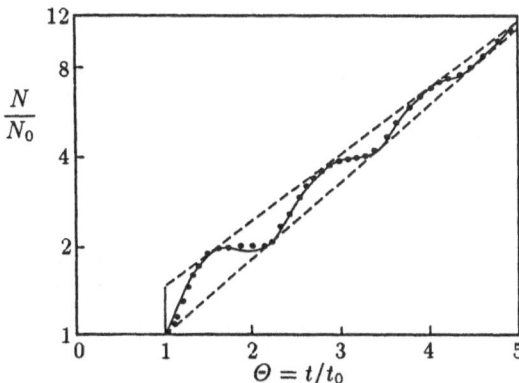

Fig. 4.4.9 Growth of cell population. Solid points computed by numerical integration of gamma distribution. Curves computed from a damped oscillation growth model

Several features are apparent in Figs. 4.4.8 and 4.4.9. First, we note that for large values of time the growth of the cell population is exponential; the initial synchronization of cells is ultimately lost. Second, the amplitudes of the maximum values of N above the exponential base line, decrease with time. Third, the time intervals between maxima seem to be reasonably constant. Similar patterns of "damped oscillation" bacterial growth have been reported and examined by various investigators , e.g., Powell (1956) and Marr et al. (1969).

Accordingly, with a measure of pragmatism, we fabricate a mathematical framework which easily accommodates the above indicated features. The simplest framework is the following

$$\log_e \frac{N}{N_0} = A(\theta - 1) + Be^{-k_*(\theta-1)}[1 - \cos\omega_*(\theta - 1)] \quad , \qquad (4.4.14)$$

where A, B, k_* and ω_* are dimensionless parameters. From the experimental data of Fig. 4.4.8 we determine the following values

$$A = 0.603 \quad ; \qquad B = 0.237 \quad ; \qquad k_* = 0.549 \quad ; \qquad \omega_* = 5.464 \quad .$$

Substitution of these quantities into Eq. 4.4.14 yields the solid curve shown in Fig. 4.4.9. The period of oscillation is $T_* = 2\pi/\omega_* = 1.150$. When $(\theta - 1) =$

$jT_*/2$ with even (odd) integer values of j, Eq. 4.4.14 yields the lower (upper) envelope shown as dashed lines in Fig. 4.4.9.

Recall that our illustration relates to the topic of sinusoidally variable exponentially decreasing growth coefficients. With the blunt presentation of Eq. 4.4.14 we have gone directly to an answer and, in doing so, have bypassed the definition of the growth coefficient, $a(t)$. Accordingly, as a matter of interest, we work the problem in reverse to establish that the functional form of $a(t)$ which yields Eq. 4.4.14 is

$$a(\tau) = a_0 + a_m e^{-k_* \tau} [\omega_* \sin \omega_* \tau - k_* (1 - \cos \omega_* \tau)] \quad , \tag{4.4.15}$$

where $\tau = (\theta - 1)$ and $\theta = t/t_0$. In addition, we obtain the relationships: $A = a_0 t_0$ and $B = a_m t_0$ where A and B are the dimensionless parameters appearing in Eq. 4.4.14. Plots of Eq. 4.4.15 are similar to those shown in Fig. 4.4.6.

The main result of our example is the observation that the relationship of Eq. 4.4.14 describes the experimental data of Fig. 4.4.8 reasonably well. Mathematically, this equation is easily manipulated; slopes and curvatures, maximum/minimum and inflection points are not difficult to determine.

Clearly, Eq. 4.4.14 is only a framework; there is no apparent phenomenological basis for the relationship. Of course, the selection of the gamma distribution as the framework for the analysis of generation times is itself somewhat arbitrary. As Powell and Errington (1963) indicate, various investigators have utilized the normal probability, the Pearson Type V, or other distributions as a "convenient description" of experimental data on bacterial generation times.

Indeed, throughout our entire examination of growth phenomena we have invariably selected convenient frameworks. For example, the growth rate of the logistic distribution is $dU/dT = U(1 - U)$. However, we noted in Section 3.5.2 that the arctangent–exponential distribution, $dU/dT = (1/4) \sin \pi U$, has very similar properties and hence could be an alternative framework. In a classic paper, Berkson (1951) explains why he prefers "logits" (logistic) to "probits" (normal probability). And so on .

Finally, in all mitotic processes, and without specification of a particular generation time distribution function, the problem of time delays is confronted. As we shall see in Chapter 6, the introduction of time delays in a process can lead to solutions involving summations of exponential functions with real and complex exponents; these exponents are the roots of a characteristic equation. This can give rise to oscillations with continuously decreasing amplitudes. The outcome is well-summarized by Knolle (1988):

"... an initially synchronized population approaches ultimately a stable age distribution, which admits exponential growth. This process of decay of synchrony exhibits damped oscillations of the phase indicies with a period near the mean cycle time and a damping factor that depends on the variance of the cycle time."

4.4.7 Sinusoidally Variable Growth Coefficient in a Power Law Exponential Equation

Our final topic combines the consideration of the previous section with the power law exponential equation we examined in Section 2.1.5. That is, we have the expression

$$\frac{dN}{dt} = a(t)N_0^{1-r}N^r \quad . \tag{4.4.16}$$

The quantity involving N_0 is introduced for dimensional reasons. For the time-dependent growth coefficient we use

$$a(t) = a_0 + a_m e^{-kt}\sin(\omega t + \phi) \quad . \tag{4.4.17}$$

For clarity, the solution is given in the following intermediate form

$$\frac{1}{1-r}\left[\left(\frac{N}{N_0}\right)^{1-r} - 1\right] = a_0 t - \frac{a_m\omega}{k^2+\omega^2} \times$$

$$\left[e^{-kt}\left(\frac{k}{\omega}\sin(\omega t + \phi) + \cos(\omega t + \phi)\right) - \left(\frac{k}{\omega}\sin\phi + \cos\phi\right)\right] \quad . \tag{4.4.18}$$

For the case $r = 1$, the left-hand side of this equation becomes $\log_e(N/N_0)$ as we require.

This solution is a rather generalized result. For example, suppose that $k = 0$, $\phi = 0$, and $r = 2$. Substituting these values into Eq. 4.4.18 yields

$$\frac{N}{N_0} = \frac{1}{1 - a_0 t - (a_m/\omega)(1 - \cos\omega t)} \quad , \tag{4.4.19}$$

which is a modest generalization of the hyperbolic growth equation we considered in Sections 2.1.5 and 4.2.2.

Clearly, numerous other special cases could be obtained from Eq. 4.4.18. Instead of pursuing these, we conclude the subject of sinusoidally variable growth coefficients with the following illustration.

4.4.8 An Illustration: Number of Patents Issued for Inventions

Commencing with the year 1790, the U.S. Patent and Trademark Office has kept records of the number of patents it has issued annually. Table 4.4.2 lists, at 10-year intervals beginning with 1800, the cumulative number of patents issued for inventions. The data given in the table are plotted in Fig. 4.4.10.

We observe in Fig. 4.4.10 that the growth of the cumulative number of patents, N, is unrestricted over the indicated period, i.e., there is no inflection point nor apparent "carrying capacity". However, we cannot assume that the growth is exponential ($r = 1$).

Table 4.4.2 Cumulative number of patents for inventions issued by the U.S. Patent and Trademark Office. Reference: The Universal Almanac 1990, Andrews and McMeel, Kansas City, Mo., 1989

Year	t	N	Year	t	N
1800	0	309	1900	100	673 954
1810	10	1 402	1910	110	989 147
1820	20	3 332	1920	120	1 372 264
1830	30	6 418	1930	130	1 795 353
1840	40	11 937	1940	140	2 235 216
1850	50	17 870	1950	150	2 543 652
1860	60	40 935	1960	160	2 973 772
1870	70	120 394	1970	170	3 558 887
1880	80	245 832	1980	180	4 246 687
1890	90	453 346	1988	188	4 797 987

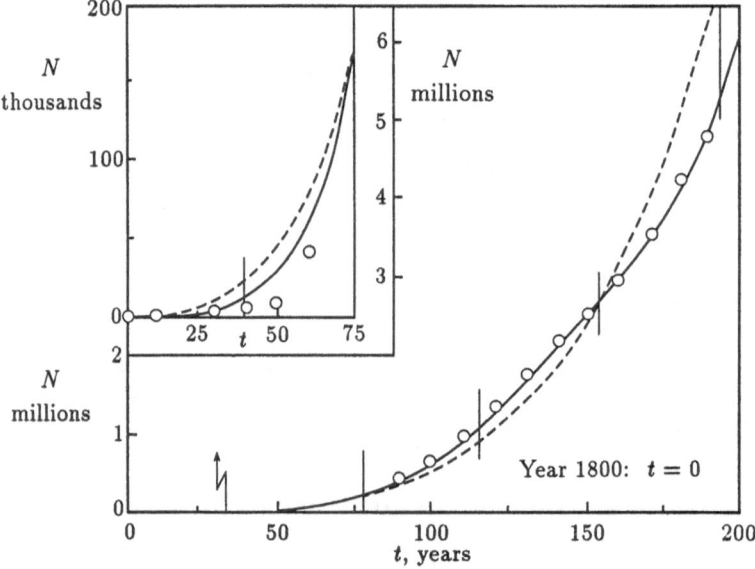

Fig. 4.4.10 Number of patents issued for inventions. Comparison of computed and observed values

At this stage in the analysis, we neglect sinusoidal variations (i.e., we set $a_m = 0$ in Eq. 4.4.17) and consider the "average" differential equation

$$\frac{dN}{dt} = a_0 N_0^{1-r} N^r \quad . \tag{4.4.20}$$

Expressing the derivative in finite difference form and taking logarithms gives

$$\log_e \frac{\Delta N}{\Delta t} = \log_e(a_0 N_0^{1-r}) + r \log_e \overline{N} \qquad (4.4.21)$$

which is linear. Computing the data of Table 4.4.2 in the form stipulated by Eq. 4.4.21, we obtain the plot shown in Fig. 4.4.11. A least squares computation provides: $r = 0.785$.

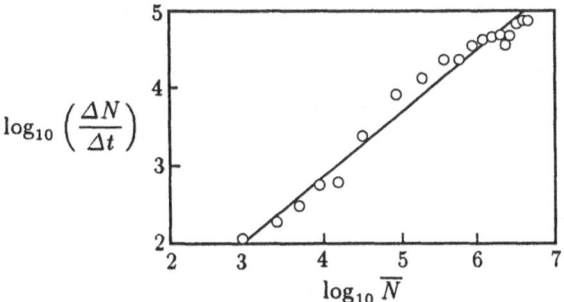

$$\log_{10}\left(\frac{\Delta N}{\Delta t}\right)$$

$$\log_{10} \overline{N}$$

Fig. 4.4.11 Number of patents issued for inventions. Plot to determine value of r

The solution to Eq. 4.4.20 is

$$N^{1-r} = N_0^{1-r} + a_0(1-r)N_0^{1-r}t \ , \qquad (4.4.22)$$

which again is of the form $y = k_0 + k_1 x$. Another least squares calculation gives $N_0 = 172.6$ and $a_0 = 0.211$.

Before proceeding with our illustration, it may be helpful to raise the following point. For an unrestricted growth phenomenon with no apparent inflection point it is advisable to follow the above procedure to establish the value of r at the outset. Such computations were carried out with the data of the following tables

a. Table 2.1.2 – Population of the United States. Data for the period 1790 through 1860. The result: $r = 0.998$. This value is extremely close to the assumed value, $r = 1$.

b. Table 2.1.3 – Population of the World. The result: $r = 1.953$. This value is very close to the assumed value, $r = 2$.

Returning to patents, we plot the final form of Eq. 4.4.22; this "average" curve is shown as the dashed line in Fig. 4.4.10. Comparing the position of this curve with the location of the data points gives the approximate value of a period of oscillation, $T = 154$.

Comparing the insert plot of Fig. 4.4.10 (for small t) with the growth patterns shown in Fig. 4.4.1 suggests that the value $\phi = 3\pi/2$ is reasonable. Point matching at $t = T/4$ and $t = 3T/4$ provides the numerical values of a_m and k. The magnitudes of the various parameters are listed below

$r = 0.785$ $N_0 = 172.6$

$a_0 = 0.211$ $a_m = 0.0531$

$\phi = 3\pi/2$ $k = -0.0040$

$T = 154$ $\omega = 0.0408$

The negative value of k indicates an exponentially increasing growth co-efficient. Substitution of these numerical values into Eq 4.4.18 produces the solid curve shown in Fig. 4.4.10. Again, the insert in the upper left corner of the figure shows the relationship for small values of time.

Except for the data points $t = 50$ and $t = 60$, it is observed that the computed curve is in good agreement with tabulated values.

The several short vertical lines in Fig. 4.4.10 define quarter-period inter-vals of $T/4 = 38.5$ years duration. We note that the actual growth curve falls above the average curve between $t = T/2 = 77$ and $t = T = 154$. This interval corresponds to the period between the years 1877 and 1954. Without reading too much into this observation, this period, as we all know, was an extremely productive era for the development of science and technology, discovery and invention in America.

5. Phenomena with Variable Carrying Capacities

Throughout the earlier analyses it has been assumed that the carrying capacity, N_*, has a constant value or indeed an infinite value. However, there are many situations involving growth phenomena and transfer processes in which the carrying capacity changes with time. A few examples in which N_* is time dependent are given below. We assume here and in the ensuing sections that the growth coefficient, a, is constant.

The biomass of an aquatic plant in a small lake begins to increase with the warm months of spring and during the hot summer season it begins to approach the carrying capacity őf the lake. However, during that same spring-summer period, the carrying capacity itself has increased, for various ecological reasons, from what it was during the cold months of winter. As the cool autumn season commences, the carrying capacity of the lake declines and so does the biomass.

A government authority begins the development of a large irrigation district which is to be further developed over a period of several years. After the project is initiated, individual farmers begin to occupy small areas of the district; the total area utilized for agriculture, N, increases with time. Concurrently, the government authority continually enlarges the overall area, N_*, of the irrigation district. Accordingly, both N and N_* are increasing functions of time.

A new service is introduced in a community in which the originally envisaged total number of customers for the service is N_*. Some time after the introduction of the service, the number of customers, N, has increased and consequently the number of non-customers, $N_* - N$, has decreased. However, suppose that for some reason (e.g., rising monthly charges for the service) the total number, N_*, begins to decrease. In this case, both N and N_* depend on time. Initially, the value of N increases but, because of the declining magnitude of N_*, the number of customers, N, reaches a maximum value and thereafter decreases.

These are examples of the kinds of problems we now examine. The function $N_* = N_*(t)$ in a particular case could be quite complicated; it is not necessarily even a continuous function. However, we shall consider only relatively simple continuous functions for N_*.

In the following sections, we assume that the carrying capacity is (a) exponentially variable with time, (b) logistically variable, (c) linearly variable, (d) hyperbolically variable, and (e) sinusoidally variable. Two additional cases are examined: a sinusoidally variable exponentially changing carrying capacity and a power law logistic with a power law logistically variable carrying capacity.

For these analyses, we utilize the general result of Eq. 4.0.3. Letting $a(t) = a_0 = $ constant in that equation, our basic expression for time dependent carrying capacity problems is

$$N = \frac{N_0 \exp(a_0 t)}{1 + a_0 N_0 \int_0^t \frac{\exp(a_0 \xi)}{N_*(\xi)} d\xi} \quad . \tag{5.0.1}$$

5.1 Exponentially Variable Carrying Capacity

5.1.1 The Growth Curve and Its Properties

We consider the case in which the carrying capacity is increasing or decreasing exponentially with time. In this instance

$$N_*(t) = N_{*0} e^{ct} \quad , \tag{5.1.1}$$

where the parameter c is positive or negative. The quantity $N_{*0} = $ constant is the value of the carrying capacity at $t = 0$. It is assumed that the growth coefficient, $a = a_0$, is a positive constant.

Substitution of Eq. 5.1.1 into Eq. 5.0.1 yields the solution

$$N = N_{*0} \left\{ \left(\frac{a_0}{a_0 - c} \right) e^{-ct} + \left[\frac{N_{*0}}{N_0} - \left(\frac{a_0}{a_0 - c} \right) \right] e^{-a_0 t} \right\}^{-1} \quad . \tag{5.1.2}$$

This equation behaves in several ways, depending on the relative values of the growth coefficient, a_0, and the carrying capacity coefficient, c.

First, if $c = a_0$, then Eq. 5.1.2 reduces to the exponential growth equation

$$N = N_0 e^{a_0 t} \quad . \tag{5.1.3}$$

Dividing Eq. 5.1.3 by Eq. 5.1.1 gives $N = (N_0/N_{*0})N_*$. In this case, the growth curve and the carrying capacity curve, increasing at the same rate, are separated by the constant amount N_0/N_{*0}. Clearly, the carrying capacity curve is increasing at a rate just sufficient to prevent any restraint on the growth curve. That is, both functions are growing exponentially at the same rate.

Next, suppose that $c = a_0(1 - N_0/N_{*0})$. This is precisely the value of c needed to force the second term in the denominator of Eq. 5.1.2 to zero. Accordingly, Eq. 5.1.2 becomes

$$N = N_0 e^{ct} \quad .$$

(5.1.4)

This result indicates that the growth curve is again exponential but is increasing at a rate less than in the case where $c = a_0$.

As we would expect, when $c = 0$, Eq. 5.1.2 reduces to the ordinary logistic equation

$$N = \frac{N_{*0}}{1 + \left(\frac{N_{*0}}{N_0} - 1\right) e^{-a_0 t}} \quad .$$

(5.1.5)

When $c < 0$, the growth equation, $N = N(t)$, is confined to the region between zero and N_{*0}. In this case, as we establish below, there is a maximum value of N.

It is recalled that a_0 is a positive constant; c may be positive or negative. For large values of time, t, the second term in the denominator of Eq. 5.1.2 vanishes. Accordingly, for large t we have

$$N = N_{*0} \left(\frac{a_0 - c}{a_0}\right) e^{ct} \quad .$$

(5.1.6)

If $c = 0$, then $N = N_{*0}$ as expected. If $c > 0$ (but less than a_0), then $N \to \infty$ as $t \to \infty$. Alternatively, if $c < 0$, then $N \to 0$ as $t \to \infty$.

Dividing Eq. 5.1.6 by Eq. 5.1.1 gives $N/N_* = (a_0 - c)/a_0$. From this result, it is observed that when $c > 0$, the value of N, at large values of time, is always smaller than N_* by a constant amount; alternatively, when $c < 0$, the value of N is always larger than N_* by a constant fraction.

We now examine some of the properties of the growth curve. Differentiating Eq. 5.1.2 and equating the result to zero, gives

$$t_m = \frac{1}{a_0 - c} \log_e \left(\frac{a_0 m}{-c}\right)$$

(5.1.7)

where

$$m = \left(\frac{a_0 - c}{a_0}\right) \frac{N_{*0}}{N_0} - 1 \quad .$$

(5.1.8)

The quantity t_m is the time corresponding to the maximum value, N_m; clearly, a maximum exists only for $c < 0$. Substituting Eq. 5.1.7 into Eq. 5.1.2 gives

$$N_m = N_{*0} \left(\frac{a_0 m}{-c}\right)^{c/(a_0 - c)} \quad .$$

(5.1.9)

This same result is obtained when the value of t_m, given by Eq. 5.1.7, is substituted into Eq 5.1.1. Thus, the two curves, $N = N(t)$ and $N_* = N_*(t)$, intersect at the point (t_m, N_m). We get the same result from Eq. 4.0.1 which stipulates that dN/dt is zero when $N = N_*$.

Setting the second derivative of Eq. 5.1.2 equal to zero determines the inflection points. The answer is

$$t_i = \frac{2}{a_0 - c} \log_e \left[\pm \frac{\sqrt{m}}{2} \left(\frac{a_0 - c}{c} \right) \left(1 \pm \sqrt{1 - \frac{4a_0 c}{(a_0 - c)^2}} \right) \right] \quad . \qquad (5.1.10)$$

An interesting case arises when $c = -a_0$. This corresponds to the situation in which the carrying capacity curve is decreasing exponentially at the same rate that the growth curve is initially increasing exponentially. The substitution $c = -a_0$ in Eq 5.1.2 gives

$$N = \frac{N_{*0}}{\frac{1}{2} e^{a_0 t} + \left(\frac{N_{*0}}{N_0} - \frac{1}{2} \right) e^{-a_0 t}} \quad . \qquad (5.1.11)$$

Likewise, with $c = -a_0$, we obtain

$$t_m = \frac{1}{2a_0} \log_e m \quad ; \qquad m = 2\frac{N_{*0}}{N_0} - 1 \quad ; \qquad (5.1.12)$$

$$t_i = \frac{1}{a_0} \log_e [\pm \sqrt{m}(1 \pm \sqrt{2})] \quad ; \qquad N_m = \frac{N_{*0}}{\sqrt{m}} \quad . \qquad (5.1.13)$$

From this last equation, it is established that the magnitudes of N at the two inflection points are both $N_i = N_{*0}/\sqrt{2m}$.

We rewrite Eq. 5.1.11 in the form

$$N = \frac{N_{*0}}{\frac{1}{2}\left(e^{a_0 t} + me^{-a_0 t}\right)} \quad . \qquad (5.1.14)$$

In this expression we make the substitution $\tau = t - t_m$. Omitting some details, the following answer is obtained

$$N = \frac{N_{*0}}{\sqrt{m}} \operatorname{sech}(a_0 \tau) \quad . \qquad (5.1.15)$$

From this result we note, in this $c = -a_0$ case, that the cumulative distribution is a "bell-shaped" curve symmetrical about τ.

Several plots are shown in Fig. 5.1.1 to illustrate the pattern of the growth curves. The lower-most curve, $c = -0.25$, corresponds to the $a_0 = -c$ case which led to the hyperbolic secant relationship of Eq. 5.1.15.

The dashed lines identify the carrying capacity curves and, for comparison, the exponential growth curve is shown.

For the various plots displayed in the figure, Table 5.1.1 lists the corresponding values of values of t_m (time of maximum N), N_m (magnitude of maximum N), and t_i (times of inflection points). These critical points are shown in the figure.

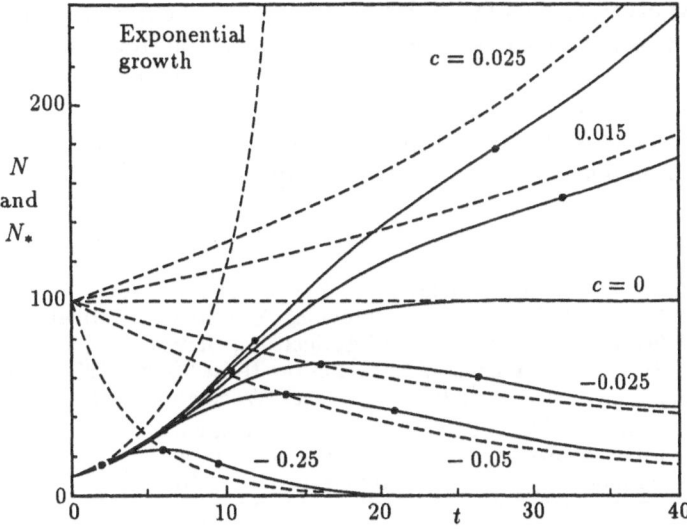

Fig. 5.1.1 Growth curves with exponentially variable carrying capacity for various values of c. $a_0 = 0.25$, $N_0 = 10$, $N_{*0} = 100$. Solid lines are the growth curves, N. Dashed lines are the carrying capacity curves, N_*.

Table 5.1.1 Values of growth curve parameters for various values of the carrying capacity coefficient. See Fig. 5.1.1

c	m	t_m	N_m	t_{i1}	t_{i2}
+0.025	8.00	–	–	11.55	27.39
+0.015	8.40	–	–	10.23	31.83
0.000	9.00	–	–	8.79	–
−0.025	10.00	16.75	65.79	7.14	26.35
−0.050	11.00	13.36	51.28	6.00	20.72
−0.250	19.00	5.89	22.94	2.36	9.41

5.1.2 An Illustration: Farm Population of the United States

In the early decades of the United States, most of the people lived on farms, earning their livelihoods by growing crops and raising livestock. However, even from the start, say in 1790, as the population of America grew at an annual rate of about 3.0 percent, the percentage of the farm population to the total population began to decline from the just-over 90 percent figure determined from the first census in 1790.

By 1850, when the total population of the United States was around 23 million, the percentage of farm to total population was almost exactly 50 percent. In 1910, the farm population reached a peak of slightly over 32 million but the percentage, farm to total, had dropped to about 35 percent. By 1930, the farm population was around 25 percent of the total, by the mid-1950s it was 10 percent, and by 1970 it was less than 5 percent.

Magnitudes of the total population, farm population, and percentage of farm to total population for the decade years, 1790 to 1980, are presented in Table 5.1.2. The farm populations for these years are shown in Fig. 5.1.2.

Table 5.1.2 Total population, farm population, and ratio of farm to total population, United States, 1790–1980. References: A Chronology of American Agriculture, Economic Research Service, U.S. Department of Agriculture, Government Printing Office, Washington, D.C.,1971; Statistical Abstract of the United States 1985. U.S. Department of Commerce, Government Printing Office, Washington, D.C.

Year	t [years]	total population N_t [millions]	farm population N [millions]	ratio N/N_t [%]
1790	0	3.929	3.600	91.6
1800	10	5.308	4.671	88.0
1810	20	7.240	5.828	80.5
1820	30	9.638	6.939	72.0
1830	40	12.861	7.720	60.0
1840	50	17.064	9.012	52.8
1850	60	23.192	11.680	50.4
1860	70	31.443	15.141	48.2
1870	80	38.558	18.373	47.7
1880	90	50.189	22.981	45.8
1890	100	62.980	26.379	41.9
1900	110	76.212	29.414	38.6
1910	120	92.228	32.077	34.8
1920	130	106.021	31.974	30.2
1930	140	123.203	30.529	24.8
1940	150	132.165	30.547	23.1
1950	160	151.326	23.048	15.2
1960	170	179.323	15.669	8.74
1970	180	203.302	9.712	4.78
1980	190	226.546	7.241	3.20

It is quite clear that there were compelling social and economic pressures which caused this continual migration from the farms to the towns and cities. Regardless of what those pressures may have been, the fact is that the socio-economically imposed carrying capacity was continually decreased.

For the purpose of our illustration, it is assumed that the carrying capacity can be described by an exponentially decreasing function of the form given by Eq. 5.1.1. In this case, the solution, $N = N(t)$, is provided by Eq. 5.1.2 with $c < 0$.

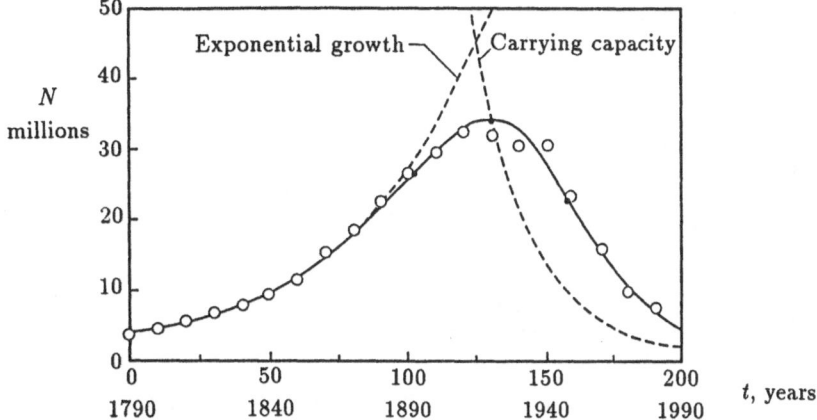

Fig. 5.1.2 U.S. farm population, 1790–1980. Example of logistic growth with an exponentially decreasing carrying capacity

In order to fit Eq. 5.1.2 to the observed data, we need to determine the numerical values of the parameters a_0, c, N_0, and N_{*0}.

We start by obtaining the values of a_0 and N_0. To do this, we assume small values of time, t, and that $N_{*0}/N_0 \gg a_0/(a_0 - c)$. As a consequence, Eq. 5.1.2 reduces to the exponential growth equation, $N = N_0 \exp(a_0 t)$. This expression is written in the form

$$\log_e N = \log_e N_0 + a_0 t \tag{5.1.16}$$

which is a linear relationship. In Fig. 5.1.3(a) the farm populations, N, given in Table 5.1.2 are plotted in terms of Eq. 5.1.16 for the decade years, 1790 through 1940. It is seen that a linear relationship extends over the approximate range $t = 0$ to 100. In other words, America's farm population grew at an exponential rate during the 100-year period from 1790 to 1890. A least squares computation corresponding to the linear range gives $a_0 = 0.0198 \, \text{year}^{-1}$ and $N_0 = 3.697 \, \text{million}$.

There remains the task of determining the values of c and N_{*0}. This can be done by writing Eq. 5.1.2 in the following form

$$\left(\frac{e^{a_0 t}}{N} - \frac{1}{N_0} \right) = \frac{1}{N_{*0}} \left(\frac{a_0}{a_0 - c} \right) \left(e^{(a_0 - c)t} - 1 \right) \quad . \tag{5.1.17}$$

Recalling that c is negative, we assume that for large values of time, the first term in the brackets of the right-hand member, $\exp(a_0 - c)t \gg 1$. Accordingly, Eq. 5.1.17 can be written

$$\log_e \left(\frac{e^{a_0 t}}{N} - \frac{1}{N_0} \right) = \log_e \left[\frac{1}{N_{*0}} \left(\frac{a_0}{a_0 - c} \right) \right] + (a_0 - c)t \tag{5.1.18}$$

which again is linear. The data of Table 5.1.2 for the period $t = 110$ through 190, computed in terms of Eq. 5.1.18, are shown in Fig. 5.1.3(b). A least

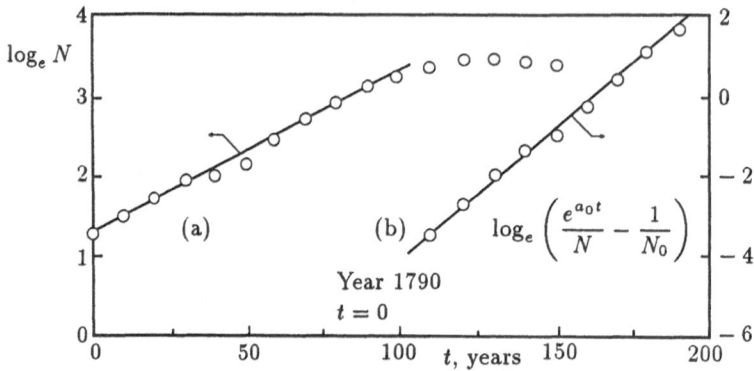

Fig. 5.1.3 U.S. farm population, 1790 to 1980. Plots to determine values of parameters: (a) small values of time: a_0 and N_0; (b) large values of time: c and N_{*0}

squares calculation involving these large time data yields $c = -0.0462\,\text{year}^{-1}$ and $N_{*0} = 14\,010$ million. The magnitude of N_{*0} is, of course, ridiculous but fictitious.

To summarize, we have: $a_0 = 0.0198$, $c = -0.0462$, $N_0 = 3.697$, and $N_{*0} = 14\,010$. Substituting these numerical values into Eq. 5.1.2 produces the solid curve shown in Fig. 5.1.2. The dashed lines show the carrying capacity curve and, for comparison, the exponential growth curve.

The time corresponding to the maximum value, computed from Eq. 5.1.7, is $t_m = 130.25$ (i.e., around 1920). From Eq. 5.1.9, we obtain $N_m = 34.122$. These computed values agree fairly well with observed values.

The inflection points, calculated from Eq. 5.1.10, appear at $t_{i1} = 101.61$ ($N = 25.959$) and $t_{i2} = 158.87$ ($N = 22.406$). These results indicate that the farm population was increasing at its maximum rate in 1892 and decreasing at its maximum in 1949. These critical points are identified in Fig. 5.1.2 by the small dots.

An extrapolation of Eq. 5.1.2, using the above-indicated parameter values, predicts a farm population of about 4.30 million in 1990 and around 2.82 million by the year 2000. The growth model considered in Section 5.6.3 indicates that the total population of the United States in the year 2000 will be approximately 270 million. On this basis, the percentage of American farm population to total population will be about one percent by the turn of the century.

5.2 Logistically Variable Carrying Capacity

5.2.1 Some Previous Studies

We have just considered the case in which the carrying capacity increases or decreases exponentially with time. We now examine the more general case in which the carrying capacity changes with time according to a logistic function.

This problem has been studied by several investigators. As described in Section 5.6, Turner et al (1969) obtained an answer to the problem of power law logistic growth, $dN/dt = aN[1 - (N/N_*)^s]$, with a power law logistically changing carrying capacity. They illustrated their result with an analysis of U.S. census data for the period 1790 to 1960. Beck (1982) also examined the growth of U.S. population, assuming a logistically increasing carrying capacity, in her study of the population genetics of cystic fibrosis.

Sharif and Ramanathan (1981, 1984) developed a number of innovation diffusion models. They considered three cases: (a) confined exponential growth, (b) logistic growth, and (c) generalized growth involving a combination of the two. For each of these models, they obtained solutions, $N = N(t)$, using various functions for the time dependent carrying capacity, $N_* = N_*(t)$, including the cases of exponentially variable and logistically variable carrying capacities.

In the light of these analyses, Sharif and Ramanathan examined several problems of innovation diffusion concerning the adoption and use of (a) fluoridated water in the U.S., (b) credit card banking the the U.S., (c) oral contraceptives in Thailand and (d) farm machinery in Thailand. For each of these cases they identified the diffusion model and the carrying capacity time function which yielded the most satisfactory results. They conclude that the assumption of a constant carrying capacity may lead to incorrect forecasts. They indicate that an exponentially variable carrying capacity is useful for short and medium term forecasts but that the logistically increasing carrying capacity is more realistic for long term forecasts.

5.2.2 The Growth Equation and Its Properties

If the carrying capacity is logistically changing, we have

$$\frac{dN_*}{dt} = cN_* \left(1 - \frac{N_*}{N_{**}}\right) \quad , \tag{5.2.1}$$

in which c is a non-negative constant. The quantity N_{**} is the carrying capacity of the carrying capacity. This above equation is simply the differential equation for ordinary logistic growth; its solution is

$$N_* = \frac{N_{**}}{1 + \left(\frac{N_{**}}{N_{*0}} - 1\right)e^{-ct}} \; . \tag{5.2.2}$$

Substituting this result into Eq. 5.0.1 gives

$$N = \frac{N_{**}}{1 + \left(\frac{N_{**}}{N_{*0}} - 1\right)\left(\frac{a_0}{a_0 - c}\right)e^{-ct} + \left[\left(\frac{N_{**}}{N_0} - 1\right) - \left(\frac{N_{**}}{N_{*0}} - 1\right)\left(\frac{a_0}{a_0 - c}\right)\right]e^{-a_0 t}} \; . \tag{5.2.3}$$

This result agrees with the solutions of Turner et al. and Sharif and Ramanathan. The results of a numerical example are presented in Fig. 5.2.1 for several values of c and N_{**}. The growth curves are shown as solid lines; the carrying capacity curves are dashed lines. If we let

$$m_1 = \left(\frac{N_{**}}{N_0} - 1\right) - \left(\frac{N_{**}}{N_{*0}} - 1\right)\left(\frac{a_0}{a_0 - c}\right) \tag{5.2.4}$$

and

$$m_2 = \left(\frac{N_{**}}{N_{*0}} - 1\right)\left(\frac{a_0}{a_0 - c}\right) \; , \tag{5.2.5}$$

then Eq 5.2.3 becomes

$$N = \frac{N_{**}}{1 + m_2 e^{-ct} + m_1 e^{-a_0 t}} \; . \tag{5.2.6}$$

The first derivative of Eq. 5.2.6 with respect to time is

$$\frac{dN}{dt} = \frac{N_{**}(m_2 c e^{-ct} + m_1 a_0 e^{-a_0 t})}{(1 + m_2 e^{-ct} + m_1 e^{-a_0 t})^2} \; . \tag{5.2.7}$$

From this equation it is established that N has a maximum or a minimum value when

$$t_{\mathrm{m}} = \frac{1}{a_0 - c} \log_e \left(-\frac{m_1 a_0}{m_2 c}\right) \; . \tag{5.2.8}$$

Setting the derivative of Eq. 5.2.7 equal to zero determines the inflection points. It is not possible to obtain an explicit solution for the inflection point time, t_i. However, this quantity can be determined by an iterated solution of the following equation

$$a_0^2 \phi^2 - a_0^2 \phi - (a_0^2 - 4a_0 c + c^2)\phi\psi - c^2\psi - c^2\psi^2 = 0 \; , \tag{5.2.9}$$

in which

$$\phi = m_1 e^{-a_0 t_i} \; ; \qquad \psi = m_2 e^{-ct_i} \; . \tag{5.2.10}$$

Table 5.2.1 lists the numerical values of the maximum points and inflection points of the curves of Fig. 5.2.1. These critical points are shown in the figure.

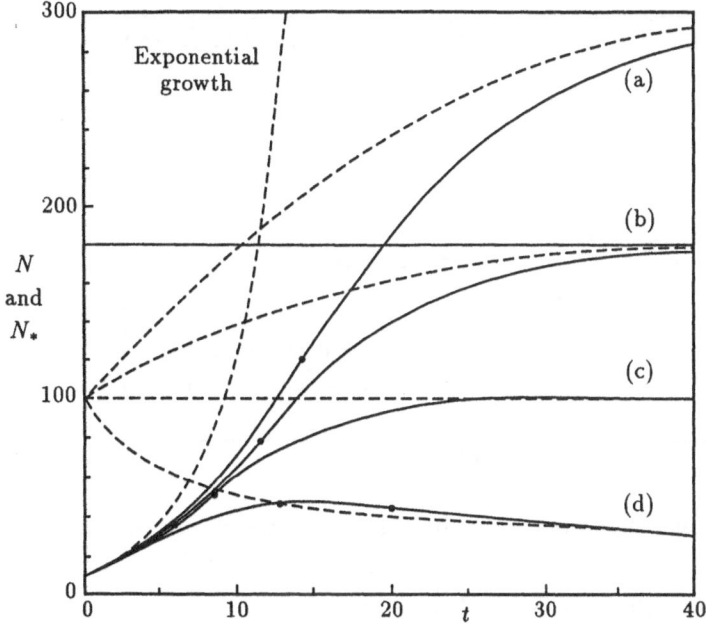

Fig. 5.2.1 Growth curves with logistically variable carrying capacity for various values of c and N_{**}. $a_0 = 0.25$, $N_0 = 10$, $N_{*0} = 100$. Values of parameters and critical points are listed in Table 5.2.1

Table 5.2.1 Computed values of the maximum points and inflection points of the curves of Fig. 5.2.1. $a_0 = 0.25$, $N_0 = 10$, $N_{*0} = 100$

Curve	c	N_{**}	Maximum point t_m	N_m	Inflection point t_i	N_i
a	0.100	300	–	–	14.45	121.0
b	0.100	180	–	–	11.61	79.0
c	0.000	100	–	–	8.79	50.0
d	0.025	20	13.58	46.47	5.33	27.8
					20.04	42.2

Another display of Eq. 5.2.3 is given in Fig. 5.2.2 in which the parameter of the various curves is N_{*0}. When $N_{*0} = 100$ the curve becomes the ordinary logistic.

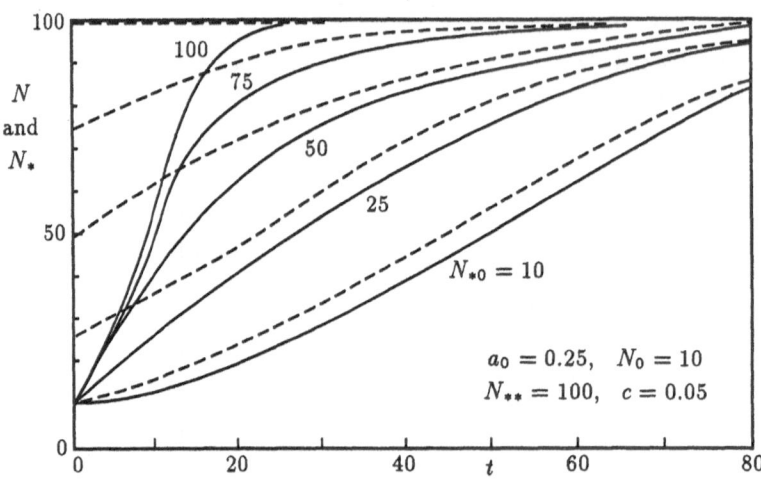

Fig. 5.2.2 Examples of growth curves with logistically variable carrying capacities

5.2.3 Relative Values of Growth Parameters

In addition to the two growth coefficients, a_0 and c, three other parameters establish the shape of the growth equation for this logistically variable case: N_0, N_{*0}, and N_{**}. There are six possible combinations of the relative magnitude of these three parameters.

The following numerical example illustrates this point. Suppose that $a_0 = 1.0$ and $c = 0.20$. Selected values of N_0, N_{*0}, and N_{**} are listed in Table 5.2.2.

Table 5.2.2 Example using various values of the gowth parameters. Values of the maximum and minimum points are listed. $a_0 = 1.0$, $c = 0.20$. See Fig. 5.2.3

Case	N_0	N_{*0}	N_{**}	t_m	N_m
a	10	100	200	–	–
b	200	100	10	–	–
c	10	100	50	4.514	62.71
d	50	100	10	0.460	55.83
e	150	100	200	1.624	116.10
f	200	100	150	2.432	114.73

The curves corresponding to the six cases indicated in the table are shown in Fig. 5.2.3. The solid lines refer to the growth curves and the dashed lines to the carrying capacity.

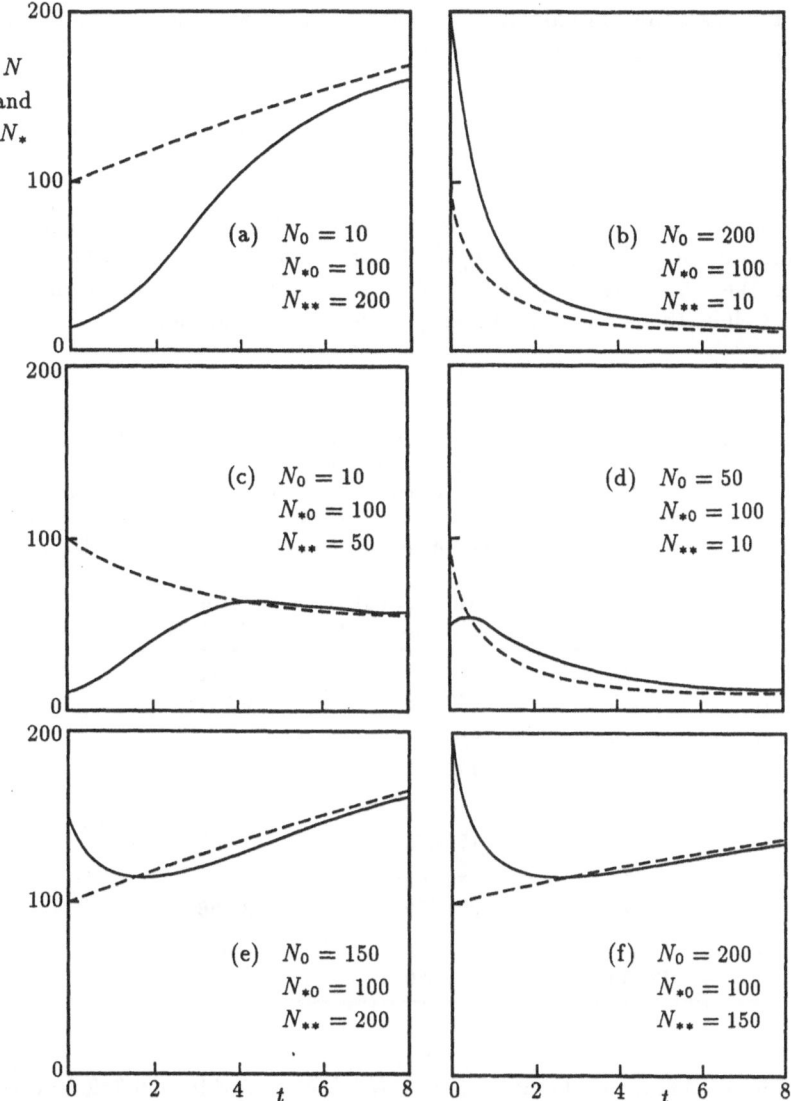

Fig. 5.2.3 Growth curves with logistically variable carrying capacity for various values of N_0, N_{*0}, and N_{**}. In all cases $a_0 = 1.0$ and $c = 0.20$. See Table 5.2.2

Of course, case (a) is the one we generally have in mind when we think of a variable carrying capacity although case (c) might be almost as relevant. It is observed that in cases (a) and (b) there are neither maximum nor minimum N and the two curves, $N = N(t)$ and $N_* = N_*(t)$, do not intersect. In cases (c) and (d), N has a maximum value and the curves intersect at the point of maximum N. In cases (e) and (f), N has a minimum value and the curves intersect at the point of minimum N. The coordinate of these maximum and minimum points are listed in Table 5.2.2.

5.2.4 An Illustration: Enrollments in Universities in the United States

In Table 5.2.3, the number of students enrolled for degree credit in American universities is shown at ten year intervals for the period 1900 to 1990. Also shown in the table is the population of the United States for corresponding years as well as the ratio of enrollment, N, to total population, N_t.

Table **5.2.3** Degree credit enrollment in U.S. universities, total population, and ratio of enrollment to total population, 1900 to 1990.
References: Historical Statistics of the United States, Colonial Times to 1970. U.S. Department of Commerce, Government Printing Office, 1975. Statistical Abstract of the United States, 1990, U.S. Department of Commerce, Washington, D.C.

Year	t years	Enrollment N, thousands	Total population N_t, thousands	Ratio N/N_t %
1900	0	238	76 212	0.312
1910	10	355	92 228	0.385
1920	20	598	106 021	0.564
1930	30	1 101	123 203	0.894
1940	40	1 494	132 165	1.130
1950	50	2 659	151 326	1.757
1960	60	3 583	179 323	1.998
1970	70	7 920	203 302	3.896
1980	80	11 387	226 546	5.026
1990	90	12 130	250 550	4.841

It is observed in the table that the ratio N/N_t increased from 0.312 percent in 1900 to 4.841 percent in 1990. We note that if the initial ($t = 0$) percentage had remained constant throughout the years, then the enrollment in 1990 would have been $N = (0.00312)(250\,550) = 782$ thousand instead of 12 130 thousand.

Accordingly we use Eq. 5.2.6 to describe the phenomena. The data of Table 5.2.3 are shown in Fig. 5.2.4. From these data, we obtain the following values of various parameters

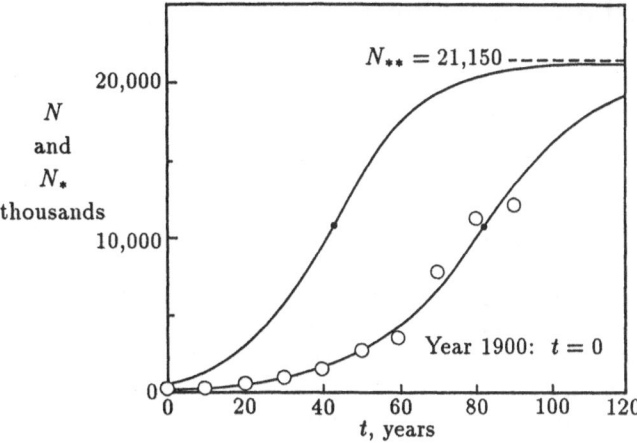

Fig. 5.2.4 Degree credit enrollment in U.S. universities, 1900–1990. Example of logistic growth with a logistically increasing carrying capacity

$a_0 = 0.0633$ $c = 0.0802$

$N_0 = 218$ $N_{*0} = 691$ $N_{**} = 21\,150$

$m_1 = 210.7$ $m_2 = -114.6$.

Substitution of these values into Eq 5.2.6 yields the solid line curve shown in Fig. 5.2.4. Utilizing Eqs. 5.2.9 and 5.2.10, the inflection point occurs at $t_i = 82.7$, $N_i = 10\,730$.

The carrying capacity, also shown in Fig. 5.2.4, is obtained from Eq. 5.2.2. The inflection point is located at $t_i = 42.3$, $N_i = 10\,575$.

We note that the inflection point of the growth curve occurred at $t_i = 82.7$ (year 1982.7). At that time, as computed from Eq. 5.2.7, the maximum rate of enrollment increase was 320.7 thousand per year.

In Section 5.6.3, we determine that the projected total population of the United States in the year 2000 will be $N_t = 270\,000$ thousand. From Eq. 5.2.6, university enrollment will be $15\,800$ thousand in 2000. So the ratio in that year will be $N/N_t = 5.85$ percent.

From the shape of the growth curve, $N(t)$, in Fig. 5.2.4 it is apparent that we could have analyzed these enrollment data with an ordinary logistic equation with a constant carrying capacity. However, in phenomena in which the two growth relationships , $N(t)$ and $N_*(t)$, are both involved, as in the present illustration, it is more meaningful and insightful to utilize the time dependent carrying capacity model.

5.3 Linearly Variable Carrying Capacity

5.3.1 The Growth Equation and Its Properties

We now examine the problem in which the carrying capacity is described by the linear function

$$N_*(t) = N_{*0}(1 + ct) \quad , \tag{5.3.1}$$

where the coefficient c may be positive or negative. Surprisingly, this very simple description of the time dependent carrying capacity produces an answer which is somewhat more complicated than the answers for the exponentially and logistically variable cases.

Substituting Eq 5.3.1 into the general solution of Eq. 5.0.1 yields the following answer for the case in which c is positive

$$N = \frac{N_{*0}}{\frac{1}{1+ct}\left(re^{-r}\mathrm{Ei}(r)\right) + \left(\frac{N_{*0}}{N_0} - r_0 e^{-r_0}\mathrm{Ei}(r_0)\right)e^{-a_0 t}} \tag{5.3.2}$$

where $r_0 = a_0/c$ and $r = r_0(1 + ct)$. The quantity, $\mathrm{Ei}(z)$, is the exponential integral function defined as

$$\mathrm{Ei}(z) = \int_{-\infty}^{z} \frac{e^{\zeta}}{\zeta} d\zeta \quad . \tag{5.3.3}$$

For the more interesting case in which c is negative, the solution is expressed in the alternative form

$$N = \frac{N_{*0}}{\frac{1}{1+ct}\left(\rho e^{\rho}\mathrm{E}_1(\rho)\right) + \left(\frac{N_{*0}}{N_0} - \rho_0 e^{\rho_0}\mathrm{E}_1(\rho_0)\right)e^{-a_0 t}} \quad , \tag{5.3.4}$$

where $\rho_0 = -r_0 = -a_0/c$ and $\rho = \rho_0(1+ct)$. The quantity $\mathrm{E}_1(z)$ is a modified exponential integral with the definition

$$\mathrm{E}_1(z) = \int_{z}^{\infty} \frac{e^{-\zeta}}{\zeta} d\zeta \quad . \tag{5.3.5}$$

It is not difficult to establish that $\mathrm{E}_1(z) = -\mathrm{Ei}(-z)$. The exponential integral functions are well tabulated; Abramowitz and Stegun (1965), for example, present tables giving numerical values of $\mathrm{Ei}(z)$ and $\mathrm{E}_1(z)$.

For large z, the asymptotic expansion of $\mathrm{Ei}(z)$ is: $z \exp(-z)\mathrm{Ei}(z) = 1$. Using this relationship, Eq. 5.3.2 reduces to the ordinary logistic equation when $c = 0$. Also, for large values of time, $N = N_* = N_{*0}(1 + ct)$. Thus the growth curve approaches the carrying capacity curve for large t.

As in the case of the exponentially changing carrying capacity, when c is negative there is a maximum value of N. Unlike the exponential case, expressions for the time and magnitude of maximum N are not very simple. Taking the derivative of Eq. 5.3.4 and setting the result equal to zero, gives the following expression for the time, t_m, corresponding to maximum N

$$\frac{e^{-\rho_m}}{\rho_m}\left(1 - \rho_m e^{\rho_m}E_1(\rho_m)\right) = \frac{e^{-\rho_0}}{\rho_0}\left(\frac{N_{*0}}{N_0} - \rho_0 e^{\rho_0}E_1(\rho_0)\right) \quad , \qquad (5.3.6)$$

where $\rho_m = \rho_0(1 + ct_m)$.

Results of computations involving growth or transfer with linearly variable carrying capacities are shown in Fig. 5.3.1 for various values of c.

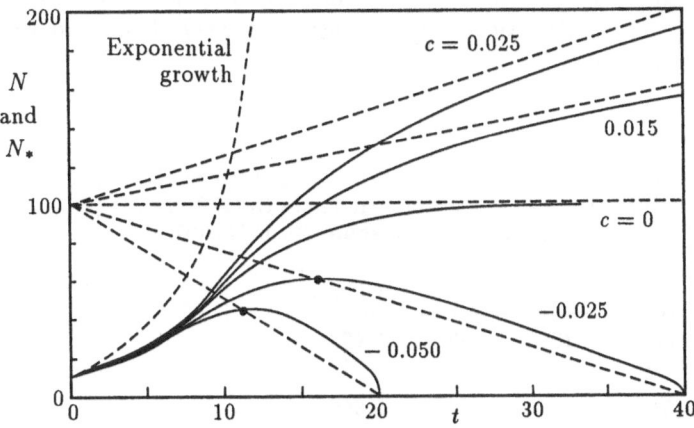

Fig. 5.3.1 Growth curves with linearly variable carrying capacity for various values of c. $a_0 = 0.25$, $N_0 = 10$, $N_{*0} = 100$

For positive c, with increasing values of time, the growth curves approach the carrying capacity curves and together head off to infinity. For negative c, N has a maximum value; the growth curve and carrying capacity curve intersect at the point of maximum N. When $c = -0.025$, utilizing Eq. 5.3.6, $t_m = 15.22$ and $N_m = 61.95$. When $c = -0.050$, then $t_m = 11.06$ and $N_m = 44.70$.

In the case of the exponentially variable carrying capacity, with negative c, the growth function asymptotically approaches the limiting value, $N = 0$. However, with a linearly decreasing carrying capacity, the behavior is quite different. When c is negative, the carrying capacity is definitely zero at time $1/c$. In the neighborhood of $t_e = -1/c$, the value of N decreases rapidly to zero. In fact, as indicated in Fig. 5.3.1, the growth curve approaches the point $(1/c, 0)$, in a direction perpendicular to the time axis.

5.3.2 An Illustration: Horses and Mules on U.S. Farms

From the earliest years of rural America and until the development and ever-increasing use of the tractor and other machinery, the prime source of motive power for the American farmer was the horse or mule. By 1865 there were already approximately 7.0 million horses and mules on American farms and by the turn of the century the number had increased to 20 million.

In 1915, the number of these work animals on America's farms reached its maximum: nearly 26.5 million. In the years that followed, the number declined at an increasing rate; by 1960, with only 3.1 million and still decreasing, the Department of Agriculture discontinued its periodic census of horses and mules. These statistical data are listed in Table 5.3.1 and are shown graphically in Fig. 5.3.2.

Table 5.3.1 Number of horses and mules on U.S. farms, 1865 to 1960. Reference: Historical Statistics of the United States, Colonial Times to 1970, U.S. Department of Commerce, Government Printing Office, Washington, D.C., 1975

Year	t	$N \ (10^6)$	Year	t	$N \ (10^6)$
1865	0	7.000	1915	50	26.493
1870	5	8.270	1920	55	25.742
1875	10	10.881	1925	60	22.569
1880	15	12.170	1930	65	19.124
1885	20	14.802	1935	70	16.683
1890	25	17.518	1940	75	14.478
1895	30	20.557	1945	80	11.950
1900	35	20.004	1950	85	7.781
1905	40	22.077	1955	90	4.309
1910	45	24.211	1960	95	3.089

This phenomenon of the "rise and fall" of horses and mules on U.S. farms is an illustration of a process in which initially there is ordinary logistic growth to a maximum magnitude followed by an accelerated descent to extinction. These characteristics are precisely stipulated by the linearly decreasing carrying capacity model.

With reference to Eq. 5.3.4, we note that there are four parameters involved: a_0, c, N_0, and N_{*0}. The approximate values of these parameters can be obtained as follows.

Initial value, N_0: A linear least squares computation involving the first several entries of the table gives $N_0 = 7.08$.

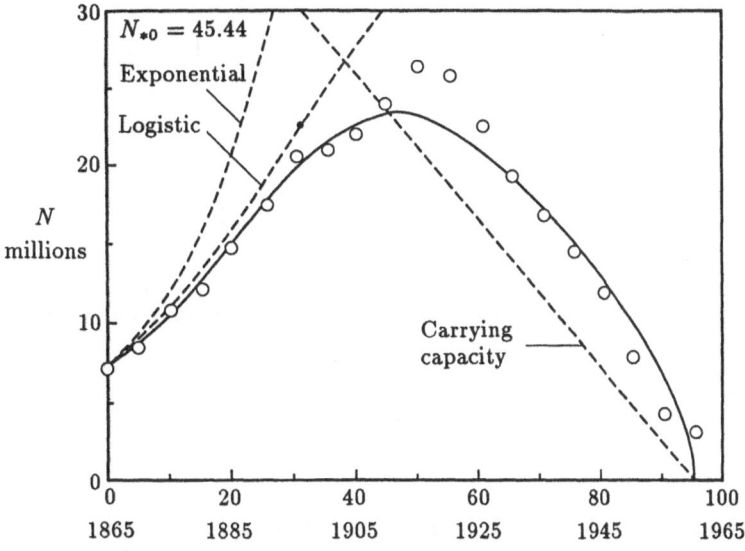

Fig. 5.3.2 Number of horses and mules on U.S. farms. Example of logistic growth with a linearly decreasing carrying capacity

Carrying capacity coefficient, c: An approximate value of c is obtained by taking $N_* = 0$ at $t_e = 95.0$. This yields $c = -0.0105$.

Growth coefficient, a_0, and carrying capacity, N_{*0}: The basic differential equation for this case is

$$\frac{dN}{dt} = a_0 N \left(1 - \frac{N}{N_{*0}(1 + ct)}\right) \quad , \tag{5.3.7}$$

which can be written in the form

$$\frac{1}{\overline{N}} \frac{\Delta N}{\Delta t} = a_0 - \frac{a_0}{N_{*0}} \frac{\overline{N}}{1 + ct} \quad . \tag{5.3.8}$$

This is a linear relationship. Casting the data of Table 5.3.1 into the form indicated by Eq. 5.3.8 yields the plot shown in Fig. 5.3.3. From this we obtain $a_0 = 0.055$ and $N_{*0} = 45.44$.

Summarizing the above, we have the following values: $a_0 = 0.055$, $c = -0.0105$, $N_0 = 7.08$, $N_* = 45.44$, and $r_0 = a_0/c = -5.238$. Since r_0 is negative, we utilize Eq. 5.3.4 for computation. This gives the solid line shown in Fig. 5.3.2. The maximum point is located at $t_m = 46.0$ and $N_m = 23.5$. The carrying capacity relationship is shown by the dashed line. For comparison, the exponential and logistic growth curves are indicated. The inflection point of the logistic occurs at $t_i = 30.72$, $N_i = 22.72$.

In concluding this illustration, we can say that the logistic equation with a linearly decreasing carrying capacity describes the growth and decline of the horse and mule population reasonably well except in the region of maximum

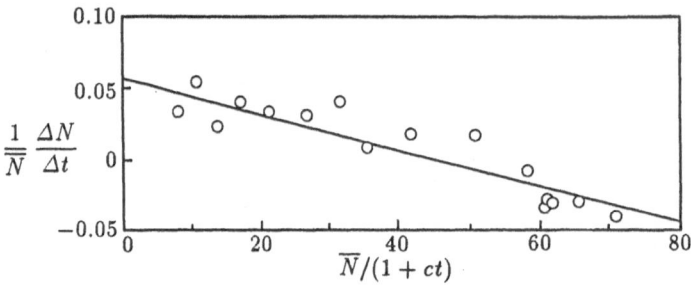

$$\frac{1}{\overline{N}} \frac{\Delta N}{\Delta t}$$

$$\overline{N}/(1 + ct)$$

Fig. 5.3.3 Number of horses and mules on U.S. farms. Plot to determine values of a_0 and N_{*0}

N. This region corresponds approximately to the years 1915 to 1920. This period, we recall, were the World War I years for America.

5.3.3 An Illustration: Steam Locomotives on U.S. Railroads

A phenomenon very similar to the "rise and fall" of horses and mules on U.S. farms was the "growth and decline" of steam locomotives on U.S. railroads. The two phenomena occurred at about the same time and the reasons for the adoption and subsequent rejection of the respective technologies were virtually identical: technology substitution. Tractors replaced horses and mules; diesel locomotives replaced steam locomotives.

Statistical data concerning the growth and decline of steam locomotives on U.S. railroads are shown in Table 5.3.2 and Fig 5.3.4.

Table 5.3.2 Number of steam locomotives on U.S. railroads, 1875 to 1960. Reference: Historical Statistics of the United States, Colonial Times to 1970, U.S. Department of Commerce, Government Printing Office, Washington, D.C., 1975

Year	t	N (10^6)	Year	t	N (10^6)
1875	0	15.50	1920	45	68.55
1880	5	17.90	1925	50	67.71
1885	10	25.70	1930	55	59.41
1890	15	30.14	1935	60	48.48
1895	20	35.70	1940	65	42.41
1900	25	37.66	1945	70	41.02
1905	30	48.36	1950	75	26.68
1910	35	60.02	1955	80	6.27
1915	40	66.23	1960	85	0.37

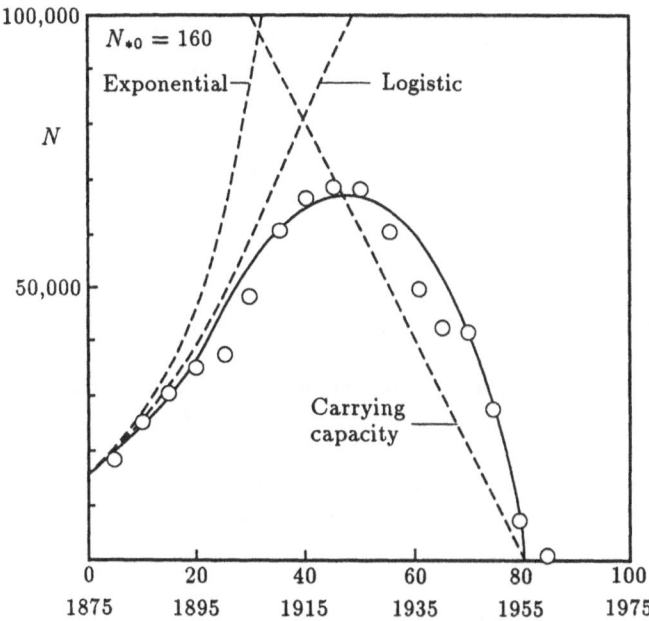

Fig. 5.3.4 Number of steam locomotives on U.S. railroads. Example of logistic growth with a linearly decreasing carrying capacity

Following the Civil War in America, there was enormous expansion of the nation's railway system. With that came a rapid increase in the number of steam powered locomotives needed to haul the country's ever-growing passenger and freight traffic. The number of locomotives increased from about 15 000 in 1875 to nearly 40 000 by the turn of the century; by the mid-1920s, a maximum of almost 70 000 was reached.

During the depression years, from about 1930 to 1940, there was a large decrease in the number of steam locomotives. In the late 1930s, diesel powered locomotives and, to a lesser extent, electrically powered locomotives, began to appear on the nation's railroads. The extremely heavy traffic caused by World War II temporarily halted further demise of the steam engine. However, the ensuing 15 years, 1945 to 1960, witnessed the virtually complete extinction of the steam locomotive. As we saw in Section 3.3.2, technology was transferred to the diesel.

We observed above that the quite simple linearly changing carrying capacity model produces analytical expressions that are somewhat awkward to handle. For example, the expression for the time, t_m, corresponding to maximum N, viz., Eq. 5.3.6, is not very convenient to use. However, to compensate for this, we note that approximate values of the four parameters are relatively easy to determine.

Utilizing the same methodology as in the previous illustration, we obtain the following values: $a_0 = 0.0570$, $c = -0.0125$, $N_0 = 15.1$, and $N_{*0} =$

160.0. Substitution of these quantities into Eq. 5.3.4 produces the solid curve shown in Fig. 5.3.4. The coordinates of the maximum point are $t_m = 47.0$ and $N_m = 66.7$. The carrying capacity and the exponential and ordinary logistic growth curves are shown in the figure. The inflection point of the logistic occurs at $t_i = 39.7$, $N_i = 80.0$. On this basis, if ordinary logistic growth had continued there would have been, eventually, about 160 000 steam locomotives on America's railroads!

As in our previous illustration involving technology transfer from horses and mules to tractors, we conclude that the linearly decreasing carrying capacity model adequately describes the technology transfer from steam locomotives to diesel locomotives.

Perhaps we can generalize to some extent. Whenever we have a growth or transfer phenomenon sequentially characterized by periods of initial increase, retarded growth, maximum magnitude, and, most importantly, accelerated descent to abrupt extinction, then evidently the linearly decreasing carrying capacity model provides the mathematical framework we need.

5.4 Hyperbolically Variable Carrying Capacity

5.4.1 Linearly Changing Crowding Coefficient

Suppose that the time dependent carrying capacity is described by the relationship

$$N_*(t) = \frac{N_{*0}}{1 + ct} \; , \tag{5.4.1}$$

where c may be positive or negative. We shall call this a hyperbolically variable carrying capacity.

Recall that the carrying capacity, by definition, is the ratio of the growth coefficient to the crowding coefficient; that is, $N_* = a/b$. So, with $a = a_0$, we have $b(t) = a_0/N_*(t) = (a_0/N_{*0})(1 + ct)$. In other words, a hyperbolically changing carrying capacity is equivalent to a linearly changing crowding coefficient.

If c is negative then at a certain time, $t_e = 1/c$, the crowding coefficient becomes zero, the carrying capacity becomes infinite and consequently the growth is exponential. On the other hand, if c is positive, the crowding coefficient increases linearly with time and eventually becomes infinite. In this case, the carrying capacity decreases hyperbolically and for large values of time both N_* and N asymptotically approach zero.

These remarks are made in order to not lose sight entirely of the crowding coefficient, b. We tend to always regard the growth coefficient, a, and the carrying capacity, N_*, as the parameters of the problem.

5.4.2 The Growth Curve and Its Properties

If Eq. 5.4.1 is substituted into Eq. 5.0.1 the following solution is obtained

$$N = \frac{N_{*0}}{\left(1 + ct - \dfrac{c}{a_0}\right) + \left(\dfrac{N_{*0}}{N_0} - 1 + \dfrac{c}{a_0}\right) e^{-a_0 t}} \ . \tag{5.4.2}$$

When $c = 0$, this answer reduces to the logistic equation. Differentiating Eq. 5.4.2 gives the time at which N has its maximum value

$$t_m = \frac{1}{a_0} \log_e \left(\frac{a_0 m}{c}\right) \tag{5.4.3}$$

where

$$m = \frac{N_{*0}}{N_0} - 1 + \frac{c}{a_0} \ . \tag{5.4.4}$$

The corresponding value of N is

$$N_m = \frac{N_{*0}}{1 + \frac{c}{a_0} \log_e \left(\frac{a_0 m}{c}\right)} \ . \tag{5.4.5}$$

We note that a maximum occurs only when c is greater than zero. The results of some computations are shown in Fig. 5.4.1. The maximum points for the two cases with positive c are: (1) $c = 0.05$, $t_m = 15.31$, $N_m = 56.63$ and (2) $c = 0.25$, $t_m = 9.21$, $N_m = 30.28$.

5.4.3 An Illustration: Growth Rates of Wheat Plant Components

As an illustration of a growth phenomenon involving a hyperbolically changing carrying capacity, we consider the growth of a wheat plant and its several components.

Obviously this hyperbolic relationship has no intrinsic basis in the actual growth mechanisms of the plant. It is selected simply because it appears to have all the features we need for a framework for analysis and display of data.

A plot prepared by Fischer (1983) is presented in Fig. 5.4.2 which shows the dry weights of the components of a spring wheat plant as functions of time. The abscissa of the plot is time, t, in days after sowing; the ordinate, total dry matter, has the units of grams per square meter of planted area.

The plot indicates that after 120 days, for example, the total dry weight of the wheat plants is about $1\,500\,\text{g/m}^2$. The uppermost curve, the "grain" curve, gives the cumulative weight of all components of the plant (i.e., grain, spike, stem, leaf, and root). By the same token, when $t = 120$ days, the "spike" curve gives $N = 950\,\text{g/m}^2$, which is the cumulative weight of the remaining components (i.e., spike, stem, leaf, and root). And so on.

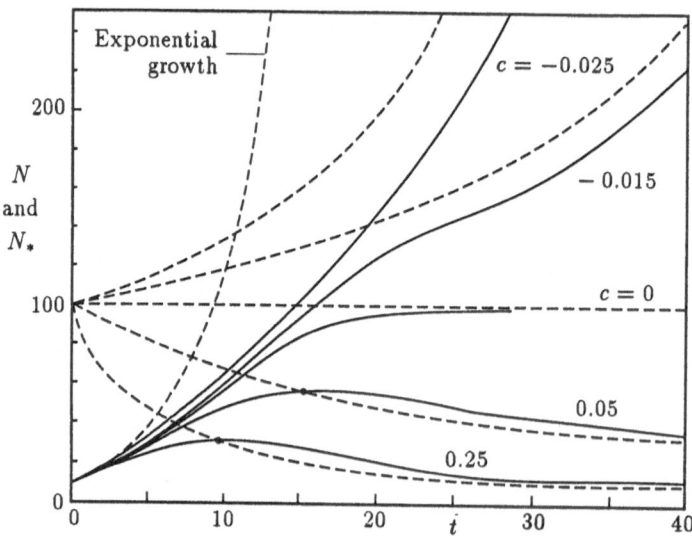

Fig. 5.4.1 Growth curves with hyperbolically variable carrying capacity for various values of c. $a_0 = 0.25$, $N_0 = 10$, $N_{*0} = 100$

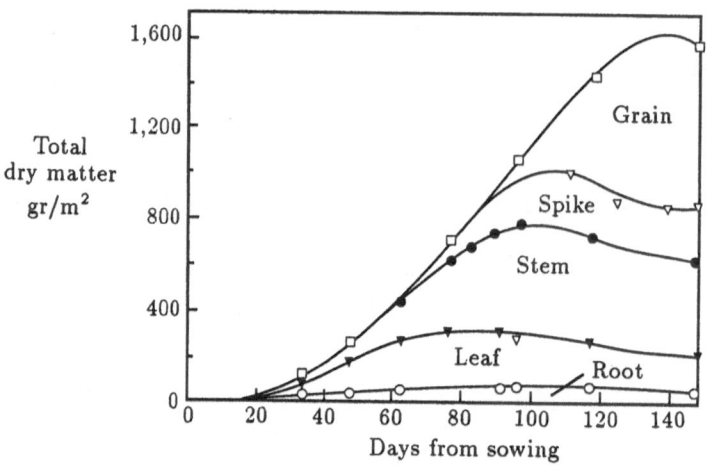

Fig. 5.4.2 Growth curves of a wheat plant and its components. From Fischer (1983)

To begin our analysis it is assumed that the grain curve is an ordinary logistic even though Fig. 5.4.2 shows a maximum near $t = 140$. From Eq. 5.4.2 with $c = 0$, we have for the grain curve

$$N = \frac{N_{*0}}{1 + \left(\frac{N_{*0}}{N_0} - 1\right) e^{-a_0 t}} \qquad . \tag{5.4.6}$$

As we have done numerous times before, values of a_0 and N_{*0} are determined from the observed data by the finite difference method; N_0 is then established from Eq. 5.4.6. The results: $a_0 = 0.0505$, $N_0 = 25.67$ and $N_{*0} = 1656$. Substitution of these numbers into Eq. 5.4.6 yields the grain curve shown in Fig. 5.4.3. The inflection point of this curve is located at $t_i = 82.2$ days, $N_i = 828\,\mathrm{g/m}^2$. The maximum rate of growth is $(dN/dt)_i = 20.9\,\mathrm{g/m}^2\mathrm{day}$. These figures indicate that 80 days after sowing, wheat is growing at its maximum rate of around 200 kg/hectare per day and has attained half of its final weight of 16 metric tons per hectare.

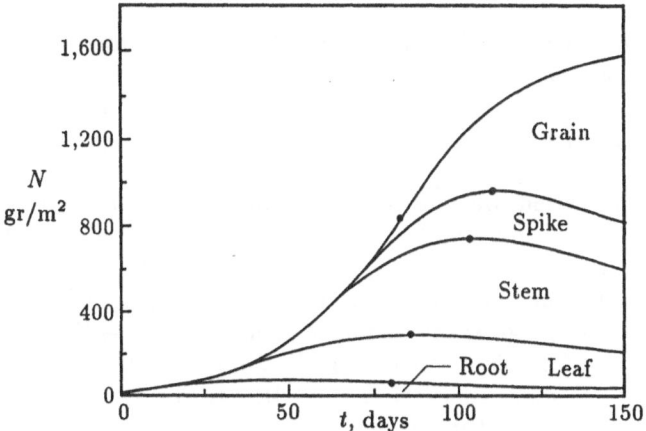

Fig. 5.4.3 Computed growth curves of a wheat plant and its components. Example of logistic growth with hyperbolically changing carrying capacity

The experimental results shown in Fig. 5.4.2 indicate that the curves for the spike, stem, leaf, and root data all have maximum values. With regard to Fig. 5.4.1, we note that when c is positive, the curve $N = N(t)$ has a maximum. So we decide to use this hyperbolic model as the framework for analysis of each of the components of the plant. We assume that a_0 and N_0 are the same for all five of the component curves but that N_{*0} and c have different values for each.

Utilizing the data of Fig. 5.4.2, the numerical values of the parameters of each component are determined. Results are summarized in Table 5.4.1. The table lists the maximum point coordinates of each component.

Substituting the respective numerical values of the parameters into Eq. 5.4.2 yields the curves shown in Fig. 5.4.3. The close similarity of Figs. 5.4.2 and 5.4.3 is noted.

We continue our illustration a bit further to put the results into a practical form. Agriculturalists use the term "economic yield" to express that portion of a crop which provides the primary economic product; in the example, the weight of the grains of wheat per unit planted area is the economic yield.

Table 5.4.1 Values of growth parameters of wheat plant components. $a_0 = 0.0505\,\mathrm{day}^{-1}$; $N_0 = 25.67\,\mathrm{g/m^2}$

Component	N_{*0}	c	m	t_m	N_m
Grain	1656	0	63.51	–	1655.7
Spike	4922	0.03778	191.49	109.8	956.0
Stem	5366	0.05987	209.22	102.4	752.3
Leaf	820	0.02046	31.35	86.1	296.9
Root	89	0.00243	2.52	78.3	74.8

The term "biological yield" is the total dry weight of the entire plant producing the economic yield. The difference between the biological yield and the economic yield might be termed the "redundant yield". The ratio of the economic yield to the biological yield, expressed as a percentage, is called the "harvest index".

To complete our illustration, we substitute the values of the parameters given in Table 5.4.1 corresponding to the grain component into Eq. 5.4.2 (with $c = 0$) to obtain the biological yield, N_b. We then substitute the values corresponding to the spike component to determine the redundant yield, N_r. The difference between the two gives the economic yield, $N_e = N_b - N_r$. The harvest index is computed from its definition, H.I. $= (N_e/N_b) \times 100$. In Fig. 5.4.4, two curves show the magnitudes of N_e and H.I. as functions of the number of days after sowing the wheat seeds.

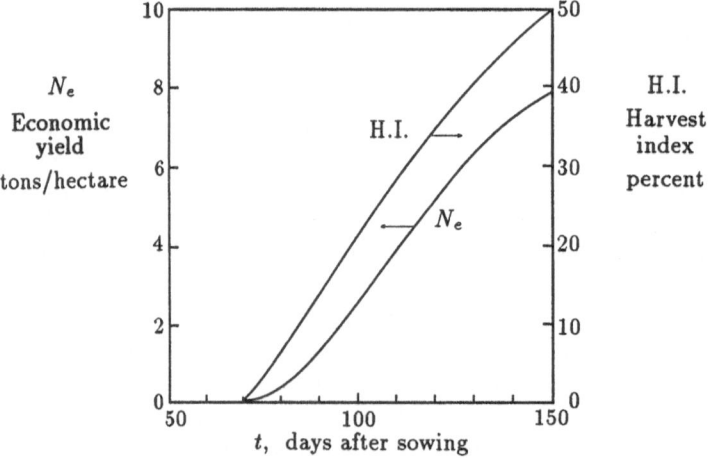

Fig. 5.4.4 A wheat plant and its components. Economic yield and harvest index of a wheat crop

5.5 Sinusoidally Variable Carrying Capacity

5.5.1 Cyclic Variations in Growth and Transfer Phenomena

Next we consider the case which, in some ways, is the most interesting: the sinusoidally variable carrying capacity.

Our interest stems from the fact that a great many growth and transfer processes are cyclic in nature. For example, periodic variations of diurnal or annual ambient temperature may produce associated periodic variations in biological activity. Cyclic changes of socio-economic parameters might cause corresponding cyclic changes in technology or innovation acceptance or rejection. Oscillatory variations of the agricultural or vegetative yield of a geographical area could produce variations in the human or animal population the area is able to sustain.

In the cases of the immediately preceding sections, we saw that the various problems were essentially those which involved only "transient" regions. That is, after a certain period of time the quantity N either asymptotically approached zero, or a constant value, or went off to infinity. In contrast, in this sinusoidally variable case, we again have a transient region, but now this transient region is followed by a "quasi-steady state" regime. Indeed, as we shall see, this steady state regime is more interesting and relevant than the strictly transient zone.

The mathematical solution to this problem, transient and steady state, is presented below. An exact answer is obtained for the case in which the amplitude of the sinusoidal variation is small. The steady state solution has been presented by May (1981) and Frauenthal (1980) who give special attention to the extreme cases in which the product of the growth coefficient and the cyclic period is either very small or very large.

5.5.2 The Growth Curve and Its Properties

For this case, the carrying capacity is described by the following equation

$$N_*(t) = N_{*0} + N_a \sin \omega t \quad , \tag{5.5.1}$$

in which N_{*0} and N_a are, respectively, the average value and the amplitude of N_*. The frequency of the oscillation is $\omega = 2\pi/T$ where T is the period.

As before, the general solution to the problem of time dependent carrying capacities is given by Eq. 5.0.1. Substitution of Eq. 5.5.1 into that equation produces an integral of the form

$$I = \int_0^t \frac{e^{a_0 \xi}}{N_{*0} + N_a \sin \omega \xi} d\xi \quad . \tag{5.5.2}$$

It is not possible to integrate this expression exactly. So we write it in the form

$$I = \frac{1}{N_{*0}} \int_0^t e^{a_0 \xi} \left(1 + \frac{N_a}{N_{*0}} \sin \omega \xi \right)^{-1} d\xi \quad . \tag{5.5.3}$$

At this point it is assumed that the ratio N_a/N_{*0} is small compared to unity. With reference to the binomial series

$$(1 + z)^{-1} = 1 - z + z^2 - \dots \tag{5.5.4}$$

and neglecting the quadratic and higher order terms, Eq. 5.5.3 becomes

$$I = \frac{1}{N_{*0}} \int_0^t e^{a_0 \xi} \left(1 - \frac{N_a}{N_{*0}} \sin \omega \xi \right) d\xi \quad . \tag{5.5.5}$$

This integral is easily evaluated. After some algebra Eq. 5.0.1 yields the answer

$$N = \frac{N_{*0}}{1 - \left(\dfrac{\theta}{\theta^2 + 1} \right) \dfrac{N_a}{N_{*0}} (\theta \sin \omega t - \cos \omega t) + \left[\dfrac{N_{*0}}{N_0} - 1 - \left(\dfrac{\theta}{\theta^2 + 1} \right) \dfrac{N_a}{N_{*0}} \right] e^{-a_0 t}} \tag{5.5.6}$$

where $\theta = a_0/\omega = a_0 T/2\pi$. In this expression, as $t \to \infty$, the quantity in the final brackets of the denominator vanishes. This quantity describes the transient region of the growth; the remaining terms in the denominator characterize the quasi-steady state regime. Accordingly, we write

$$N_s = \frac{N_{*0}}{1 - \left(\dfrac{\theta}{\theta^2 + 1} \right) \dfrac{N_a}{N_{*0}} (\theta \sin \omega t - \cos \omega t)} \tag{5.5.7}$$

in which N_s is the steady state value of $N(t)$. Noting that N_a/N_{*0} is small compared to unity and, regardless of the value of $\theta = a_0/\omega$, Eq. 5.5.7 is inverted to obtain the solution

$$N_s = N_{*0} + \left(\frac{\theta}{\theta^2 + 1} \right) N_a(\theta \sin \omega t - \cos \omega t) \quad . \tag{5.5.8}$$

We now ask the question: after how many cycles, measured from $t = 0$, does the steady state region effectively begin? To answer this, we return to the complete solution given by Eq. 5.5.6 and arbitrarily set the transient term of the denominator to, say, 0.01. Solving for the corresponding time gives

$$t_s = \frac{1}{a_0} \left\{ 4.605 + \log_e \left[\left(\frac{N_{*0}}{N_0} - 1 \right) - \left(\frac{\theta}{\theta^2 + 1} \right) \frac{N_a}{N_{*0}} \right] \right\} \quad . \tag{5.5.9}$$

The maximum value of the quantity $\theta/(\theta^2 + 1) = 1/2$ when $\theta = a_0/\omega = 1.0$. We have already assumed that N_a/N_{*0} is small. Accordingly, as long as N_{*0}/N_0 is reasonably large, Eq. 5.5.9 can be written as simply

$$t_s = \frac{1}{a_0} \left[4.605 + \log_e \left(\frac{N_{*0}}{N_0} - 1 \right) \right] \quad . \tag{5.5.10}$$

Dividing this equation by the period, T, yields

$$\frac{t_s}{T} = \frac{1}{\theta}\frac{1}{2\pi}\left[4.605 + \log_e\left(\frac{N_{*0}}{N_0} - 1\right)\right] \quad . \tag{5.5.11}$$

For example, if $\theta = a_0/\omega = 1$, $N_0 = 10$ and $N_{*0} = 100$, then this equation gives $t_s/T = 1.083$. This indicates that the transient has essentially disappeared after one cycle; thereafter the magnitude of N is described by the steady state relationship of Eq 5.5.8.

From the derivative of Eq. 5.5.8 we determine the times corresponding to the maximum and minimum values of N in the steady state region. The result is

$$\frac{t_m}{T} = \frac{1}{2}\left(j - \frac{1}{\pi}\arctan\theta\right) \quad , \tag{5.5.12}$$

where odd values of j refer to maximum N and even values to minimum. The corresponding magnitudes of N_m are

$$N_m = N_{*0} \pm \frac{\theta}{\sqrt{\theta^2 - 1}}N_a \quad . \tag{5.5.13}$$

The carrying capacity curve intersects the growth curve at the maximum and minimum points of $N(t)$.

It is interesting to examine the behavior of $N(t)$, in the steady state region, for the cases in which takes on some special values.

Case 1. $\theta = a_0/\omega = a_0T/2\pi \ll 1$. From Eq. 5.5.8, we obtain

$$N = N_{*0} - \theta N_a \cos\omega t \tag{5.5.14}$$

which can be written

$$N = N_{*0} + \theta N_a \sin(\omega t - \pi/2) \quad . \tag{5.5.15}$$

This result indicates that when the product of the growth coefficient and the period of oscillation is small, the growth curve, $N(t)$, is out of phase with the carrying capacity curve, $N_*(t)$, by $\pi/2$ radians. Since θ is a small quantity, we note that in the variation of $N(t)$ during an oscillation is small.

Case 2. $\theta = a_0/\omega = a_0T/2\pi = 1$. Again from Eq. 5.5.8 we get

$$N = N_{*0} + \tfrac{1}{2}N_a(\sin\omega t - \cos\omega t) \tag{5.5.16}$$

which can be expressed in the form

$$N = N_{*0} + \frac{1}{\sqrt{2}}N_a(\sin\omega t - \frac{\pi}{4}) \quad . \tag{5.5.17}$$

In this case, $N(t)$ lags $N_*(t)$ by $\pi/4$ radians. Also, the amplitude of $N(t)$ is less by an amount $1/\sqrt{2}$ than the amplitude of $N_*(t)$.

Case 3. $\theta = a_0/\omega = a_0 T/2\pi \gg 1$. Finally we obtain from Eq. 5.5.8,

$$N = N_{*0} + N_a \sin \omega t \quad . \tag{5.5.18}$$

This answer shows that when the product of the growth coefficient and the period of oscillation is large, the growth curve coincides with the carrying capacity curve. In this case, to follow the description of May (1981), we say that N "tracks" N_*.

To illustrate the preceding analysis, we consider a numerical example in which $a_0 = 0.25$, $N_0 = 10$, $N_{*0} = 100$, and $N_a = 10$. Values of $T = 5$, 25, and 100 are selected.

The resulting plots are shown in Figs. 5.5.1, 5.5.2, and 5.5.3. These plots show both the transient and steady state regimes. The steady state curves corresponding to the three values of T are shown in Fig. 5.5.4. A summary of the numerical values of the various parameters is presented in Table 5.5.1.

Table 5.5.1 Values of the parameters for the steady state region. $a_0 = 0.25$, $N_0 = 10$, $N_{*0} = 100$, $N_a = 10$

Case	T	ω	$\theta = \frac{a_0}{\omega}$	$\frac{\theta}{\theta^2+1}$	$\frac{\theta}{\sqrt{\theta^2+1}}$
a	5	1.257	0.2	0.192	0.196
b	25	0.251	1.0	0.500	0.707
c	100	0.063	4.0	0.235	0.970

Case	t_s/T	$(t/T)_{max}$	N_{max}	$(t/T)_{min}$	N_{min}
a	5.42	0.469	101.95	0.969	98.05
b	1.08	0.375	107.07	0.875	92.93
c	0.27	0.289	109.70	0.789	90.30

5.5.3 Phase Plane Display

Our consideration of the sinusoidally variable carrying capacity, in the steady state region, provides an opportunity to introduce the subject of phase planes. A number of references on the subject are given in Section 7.6.1.

In the preceding analysis we examined the growth equation and the carrying capacity equation as functions of time, i.e., $N(t)$ and $N_*(t)$. We now eliminate the time variable to obtain an equation of the form $F(N, N_*) = 0$. The quantities N and N_* define the coordinates of the so-called phase plane.

In the present problem the function $F(N, N_*)$ is easily obtained. From Eqs. 5.5.1 and 5.5.8 we get

$$\left(1 + \frac{1}{\theta^2}\right)\xi^2 - 2\xi\eta + \eta^2 = \frac{N_a^2}{\theta^2 + 1} \tag{5.5.19}$$

in which $\xi = N - N_{*0}$ and $\eta = N_* - N_{*0}$.

In this instance, the phase plane relationship was not difficult to establish, because of the simplicity of the two equations. In most cases the analysis is considerably more complicated.

Suppose we regard the (N, N_*) relationship as a kind of "prey–predator" phenomenon, where N represents the prey and N_* the predator. In general, in this kind of "two-species" problem, an interacting relationship exists in which the growth rate and magnitude of one species determines the growth rate and magnitude of the other.

This kind of prey-predator phenomenon is described by the so-called Lotka–Volterra equations; over the years, it has been studied by many investigators including, of course, Lotka (1924) and Volterra (1926). More recently Keyfitz (1968), Maynard Smith (1968,1974) and van der Ploeg (1984) have presented analyses from viewpoints of demography, biology, and economics respectively.

In their usual form, the Lotka–Volterra equations are, for a two-species system

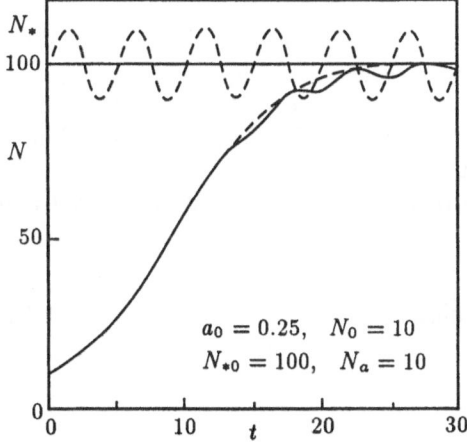

Fig. 5.5.1 Growth curves with a sinusoidally variable carrying capacity. $T = 5$, $\theta = 0.2$

$$\frac{dN}{dt} = a_1 N - b_1 N^2 - c_1 N N_* \tag{5.5.20}$$

and

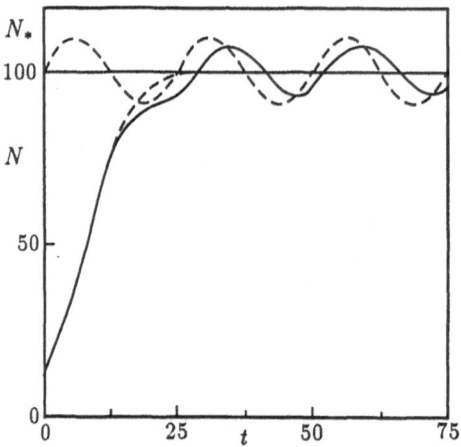

Fig. 5.5.2 Growth curves with a sinusoidally variable carrying capacity. $T = 25$, $\theta = 1.0$

$$\frac{dN_*}{dt} = -a_2 N_* + b_2 N_*^2 + c_2 N_* N \qquad (5.5.21)$$

where the constants are all positive. The quadratic terms (N^2, N_*^2, NN_*) represent the effects of intra- and inter-species crowding.

In our present problem, things are not so complicated. The Lotka–Volterra equations are presented here simply to illustrate the form in which an interacting problem is generally stated.

To display our problem in the same style, we differentiate Eqs. 5.5.1 and 5.5.8 to obtain

$$\frac{dN}{dt} = \left(\frac{\theta}{\theta^2 + 1}\right) N_a \omega (\theta \cos \omega t + \sin \omega t) \qquad (5.5.22)$$

and

$$\frac{dN_*}{dt} = N_a \omega \cos \omega t \quad . \qquad (5.5.23)$$

Eliminating the trigonometric terms, we reduce the equations to the forms: $dN/dt = f_1(N, N_*)$ and $dN_*/dt = f_2(N, N_*)$. Specifically we have, for the prey (N) and the predator (N_*),

$$\frac{dN}{dt} = a_0 N_* - a_0 N \qquad (5.5.24)$$

and

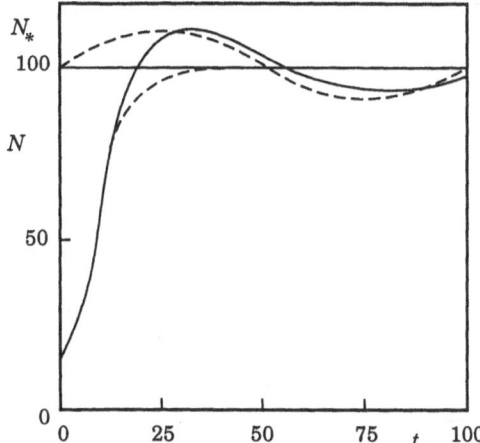

Fig. 5.5.3 Growth curves with a sinusoidally variable carrying capacity. $T = 100$, $\theta = 4.0$

$$\frac{dN_*}{dt} = a_0 N_* - \left(a_0 + \frac{\omega^2}{a_0}\right) N + \frac{\omega^2}{a_0} N_{*0} \quad . \tag{5.5.25}$$

Now suppose that, for some reason, this is where we started the problem and that we did not know the final answers, Eqs. 5.1 and 5.8. By differentiation and appropriate substitutions, Eqs. 5.5.24 and 5.5.25 can be written as second order differential equations

$$\frac{d^2 N}{dt^2} + \omega^2 N = \omega^2 N_{*0} \tag{5.5.26}$$

and

$$\frac{d^2 N_*}{dt^2} + \omega^2 N_* = \omega^2 N_{*0} \quad . \tag{5.5.27}$$

These equations have solutions of the form $A \sin t + B \cos t + N_{*0}$. Utilizing the initial conditions, we readily get the solutions, Eq. 5.5.1 and 5.5.8.

To return to the phase plane point of view, we divide Eq. 5.5.25 by 5.5.24 to obtain

$$\frac{dN_*}{dN} = \frac{N_* - (1 - 1/\theta^2)N + (1/\theta^2)N_{*0}}{N_* - N} \tag{5.5.28}$$

where $\theta = a_0/\omega$. This is a first order differential equation of the form

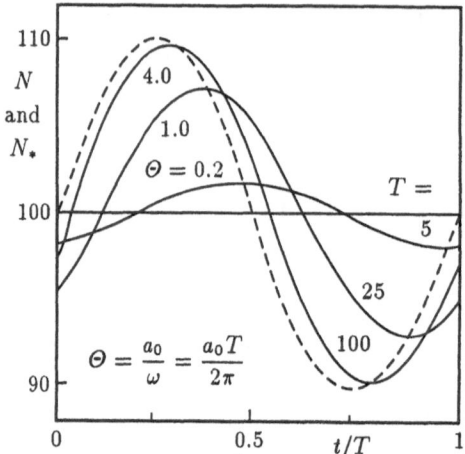

Fig. 5.5.4 Growth curves in the steady state region for various values of θ. See Table 5.5.1

$$\frac{dy}{dx} = \frac{Ax + By + E}{Cx + Dy + F} \quad . \tag{5.5.29}$$

Following the method given by Davis (1962), we obtain the answer given by Eq 5.5.19. That is

$$\left(1 + \frac{1}{\theta^2}\right)\xi^2 - 2\xi\eta + \eta^2 = \frac{N_a^2}{\theta^2 + 1} \tag{5.5.30}$$

where, again, $\xi = N - N_{*0}$ and $\eta = N_* - N_{*0}$. Clearly, this lengthy process of obtaining the answer which we had at the outset was carried out only to describe the method that would be necessary for more complicated "prey-predator" systems.

As shown in Fig. 5.5.5, the relationship of Eq. 5.5.30 is an ellipse whose major axis intersects the ξ-axis at an angle σ. We rotate the coordinate system from the (ξ, η) axes to the principal axes (x, y) by the transformation

$$\xi = x\cos\sigma - y\sin\sigma \tag{5.5.31}$$

and

$$\eta = x\sin\sigma + y\cos\sigma \quad . \tag{5.5.32}$$

Now we substitute these equations into Eq. 5.5.30 and collect the x^2, xy, and y^2 terms. To put the equations of the ellipse into its standard form

$$\frac{x^2}{a^2} + \frac{y^2}{b^2} = 1 \tag{5.5.33}$$

we equate the collected coefficients of the xy terms to zero. After some algebra we get

$$\sigma = \arctan\left(\frac{1}{2\theta^2}(1 + \sqrt{4\theta^2 + 1})\right) \quad . \tag{5.5.34}$$

When $\theta = 0$, we obtain $\sigma = \pi/2$ and when $\theta = \infty$, we get $\sigma = \pi/4$.

The numerical examples of Fig. 5.5.4 are displayed in the phase plane diagram of Fig. 5.5.5. The three ellipses refer to $T = 5$, 25, and 100 corresponding, respectively, to $\theta = 0.2$, 1.0, and 4.0. Time positions corresponding to various values of t/T are indicated at points along the $T = 25$ ellipse.

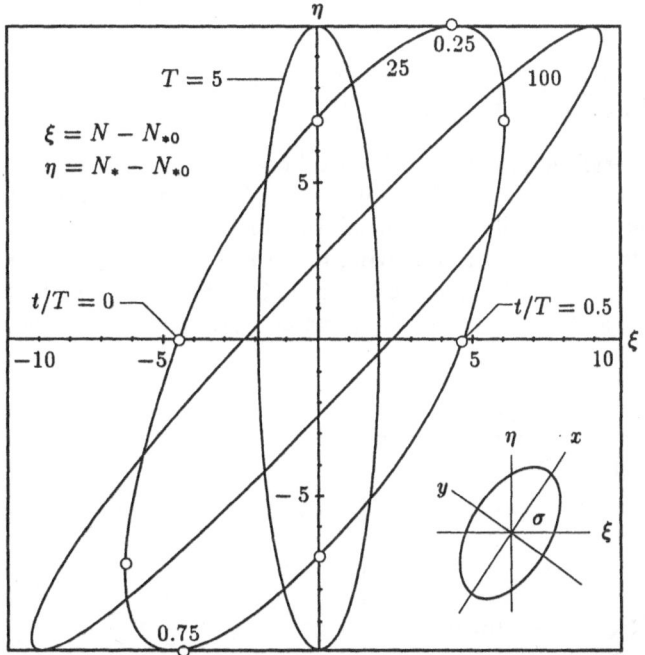

Fig. 5.5.5 Phase plane display of $N(t)$ and $N_*(t)$ for oscillations with periods $T = 5$, 25 and 100

5.5.4 Exponentially Changing Carrying Capacity

Finally we consider the case in which the carrying capacity varies simultaneously as a sinusoidal and an exponential function of time. Although this case probably does not find a great many practical applications, it is mathematically interesting and its solution is quite easy to obtain.

In this case, the carrying capacity has the following definition

$$N_*(t) = (N_{*0} + N_\text{a} \sin \omega t) e^{ct} \quad , \tag{5.5.35}$$

in which c is a positive or negative constant. The equation for the growth curve, $N(t)$, is obtained the same way as before; we again assume that N_a/N_{*0} is small compared to unity. The answer is

$$N = \frac{N_{*0}}{F_1(t) + F_2(t)} \tag{5.5.36}$$

where, with $\theta_* = (a_0 - c)/\omega$,

$$F_1(t) = \left[\left(\frac{a_0}{a_0 - c} \right) - \left(\frac{\theta_* + c/\omega}{\theta_*^2 + 1} \right) \frac{N_\text{a}}{N_{*0}} (\theta_* \sin \omega t - \cos \omega t) \right] e^{-ct} \tag{5.5.37}$$

and

$$F_2(t) = \left[\frac{N_{*0}}{N_0} - \left(\frac{a_0}{a_0 - c} \right) - \left(\frac{\theta_* + c/\omega}{\theta_*^2 + 1} \right) \frac{N_\text{a}}{N_{*0}} \right] e^{-a_0 t} \quad . \tag{5.5.38}$$

This is quite a generalized result. On the one hand, if $N_\text{a} = 0$ it reduces to the exponentially variable answer given by Eq. 5.1.2 On the other hand, if $c = 0$ it becomes the sinusoidally variable solution of Eq. 5.5.6.

By way of example, we select the following numerical values: $a_0 = 0.25$, $N_0 = 10$, $N_{*0} = 100$, $N_\text{a} = 10$, $T = 25$, and $c = -0.025$. Substituting these values into Eqs. 5.5.35 and 5.5.36 yields the plots of $N_*(t)$ and $N(t)$ shown in Fig 5.5.6.

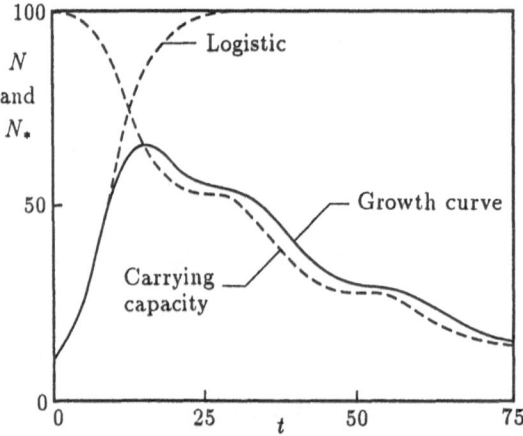

Fig. 5.5.6 Growth curve with a sinusoidally variable exponentially decreasing carrying capacity

5.5.5 An Illustration: Railway Mileage in the United States

In 1830 there were precisely 22 miles of railroads in the United States. However, this new technology was so quickly adopted that only a decade later, in 1840, there were more than 2800 miles of railways and by 1850 just over 9000 miles had been constructed. During the ensuing several decades, railroads were built at a very rapid rate and by the turn of the century, more than 200 000 miles of railway lines were in operation.

At about that time, 1900, the rate of increase began to decline. Indeed in 1915 the total length of railway lines in the U.S. reached its peak: over 258 000 miles. From that time on railway mileage decreased.

The miles of railroads in the United States are shown in Table 5.5.2 for the period 1840 to 1985. This information is displayed graphically in Fig. 5.5.7.

Table **5.5.2** Railway mileage in the United States, 1840-1985.
References: "Statistics of Railways in the U.S., 1953", Interstate Commerce Commission, Washington D.C., 1956; "Yearbook of Railroad Facts", 1981 Edition, Association of American Railroads, Washington, D.C., 1981; "California Almanac, 1986-1987 Edition", Pacific Data Resources, Novato, California, 1985

Year	t	$N\,(10^3)$	Year	t	$N\,(10^3)$
1840	0	2.82	1930	90	249.05
1850	10	9.02	1935	95	241.98
1860	20	22.18	1940	100	233.67
1870	30	45.38	1945	105	226.89
1880	40	87.63	1950	110	223.78
1890	50	148.20	1955	115	220.81
1900	60	205.91	1960	120	218.05
1905	65	230.12	1965	125	211.86
1910	70	255.34	1970	130	206.27
1915	75	258.73	1975	135	196.66
1920	80	252.85	1980	140	183.19
1925	85	249.65	1985	145	159.91

The plot of railway mileage versus time, $N(t)$, shown in the figure looks quite similar to the curve displayed in Fig. 5.5.6. Particularly interesting in Fig. 5.5.7 is an apparent cyclic variation, from the average line of decrease, of approximately one full cycle extending from about $t = 85$ (1925) to $t = 145$ (1985). In this cycle, with period $T = 60$ years, minimum and maximum variations from the mean appear to occur at $t = 100$ (1940) and $t = 130$

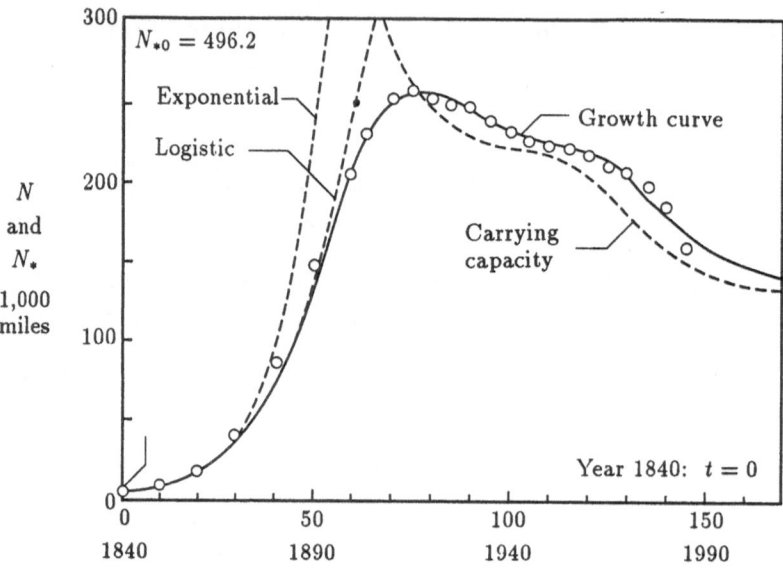

Fig. 5.5.7 Railway mileage in the United States, 1840–1985

(1970). The maximum rates of decrease, occurred around $t = 95$ (1935) and $t = 145$ (1985).

Obviously, there were a great many socio-economic-geographic factors relating to the growth and decline of America's railway network. With respect to its decline, we note a sharp decrease during the years of the depression and World War II (1930 to 1945, say). During the ensuing 15 years, 1945 to 1960, the rate of decline was less.

However, the 25-year period from 1960 to 1985 saw a very large reduction – almost 30 percent – of the nation's railway mileage. This period, of course, was the era of major expansion of America's inter-state highway system, enormous increase in the production of automobiles, substantial enlargement of the trucking industry, and incredible growth of air transportation. All of these developments had major effects on reducing the extent of the country's railway network.

Apart from these numerous social factors and explanations, it is interesting to place these railway mileage data into the framework of the sinusoidal-exponentially variable carrying capacity model.

As we did in Section 4.4.2, it is necessary to modify the carrying capacity equation to allow a phase angle, ϕ. That is

$$N_*(t) = [N_{*0} + N_a \sin(\omega t + \phi)]e^{ct} \quad . \tag{5.5.39}$$

Utilizing the various curve fitting techniques employed in previous sections, the following numerical values are determined

$$N_0 = 4.0 \qquad N_{*0} = 496.2 \qquad N_a = 34.3$$

$$a_0 = 0.080 \qquad c = -0.0080 \qquad \phi = -4.394$$

$$T = 60 \qquad \omega = 0.105 \qquad \theta_* = 0.838$$

Substitution of these numbers into Eqs 5.5.36 through 5.5.39 yields the curves shown in Fig. 5.5.7. For comparison, plots corresponding to exponential growth and ordinary logistic growth are shown.

On the basis of these results, the total railway mileage in the United States will be about 152 000 in 1995 and approximately 145 000 by the turn of the century.

5.6 Power Law Logistic with a Power Law Logistically Variable Carrying Capacity

5.6.1 The Power Law Logistic

In Sections 5.1 through 5.5 we analyzed the ordinary logistic equation for several cases in which the carrying capacity is a function of time. For these analyses we utilized the general solution of the logistic given by Eq. 5.0.1.

We now examine, in a rather generalized way, the power law logistic equation which we considered in Section 2.6. In this case the differential equation is

$$\frac{dN}{dt} = aN\left[1 - \left(\frac{N}{N_*}\right)^s\right] \quad . \tag{5.6.1}$$

If the carrying capacity, N_*, is a constant, we obtain the solution

$$N = \frac{N_*}{\left\{1 + \left[\left(\frac{N_*}{N_0}\right)^s - 1\right]e^{-ast}\right\}^{1/s}} \quad . \tag{5.6.2}$$

Now suppose that the carrying capacity of the power law logistic is a function of time. Specifically, suppppose that the carrying capacity is itself described by a power law logistic. We examine this problem in the following section.

5.6.2 The Differential Equation and Its Solution

Consider the case in which the carrying capacity, $N_*(t)$, is described by the equation

$$\frac{dN_*}{dt} = cN_*\left[1 - \left(\frac{N_*}{N_{**}}\right)^s\right] \tag{5.6.3}$$

where N_{**} is a constant and the initial condition is $N_*(0) = N_{*0}$. Accordingly, we have

$$N_* = \frac{N_{**}}{(1 + m_* e^{-cst})^{1/s}} \tag{5.6.4}$$

in which

$$m_* = \left(\frac{N_{**}}{N_{*0}}\right)^s - 1 \quad . \tag{5.6.5}$$

The growth equation, $N(t)$, is given by the expression

$$\frac{dN}{dt} = aN\left[1 - \left(\frac{N}{N_*}\right)^s\right] \quad . \tag{5.6.6}$$

It would be preferable to assign different exponents, say s_1 and s_2, to the right-hand members of Eqs 5.6.3 and 5.6.6. However, doing so would yield a differential equation impossible to solve in closed form and produce yet another constant to be evaluated. So necessarily we use s as the exponent in both equations.

Substitution of Eq. 5.6.4 into Eq. 5.6.6 yields a Bernoulli equation of the form

$$\frac{dy}{dx} = f(x)y + g(x)y^p \quad . \tag{5.6.7}$$

Using the same method as in Section 2.6.2, the following solution is obtained

$$N = \frac{N_{**}}{\left[1 + \frac{am_*}{a-c}e^{-cst} + \left(n_* - \frac{am_*}{a-c}\right)e^{-ast}\right]^{1/s}} \tag{5.6.8}$$

where

$$n_* = \left(\frac{N_{**}}{N_0}\right)^s - 1 \quad . \tag{5.6.9}$$

We note that Eq. 5.6.8 is a quite generalized answer. For example, if $N_{**} \to \infty$, the carrying capacity function, Eq. 5.6.4, reduces to the simple exponential function

$$N = N_{*0}e^{ct} \quad . \tag{5.6.10}$$

Accordingly, Eq. 5.6.8 becomes

$$N = \frac{N_{*0}}{\left\{\frac{a}{a-c}e^{-cst} + \left[\left(\frac{N_{*0}}{N_0}\right)^s - \frac{a}{a-c}\right]e^{-ast}\right\}^{1/s}} \quad . \tag{5.6.11}$$

When $s = 1$ this expression reduces to Eq. 5.1.2 corresponding to an exponentially variable carrying capacity in an ordinary logistic.

For a numerical example of the above we use the values: $a = 0.25$, $c = 0.10$, $N_0 = 10$, $N_{*0} = 100$, and $N_{**} = 300$. Three values of the parameter s are selected: $s = 2.0, 1.0$, and 0.5. Substituting these numbers into Eqs. 5.6.4 amd 5.6. 8 yields the carrying capacity and growth curves shown in Fig. 5.6.1.

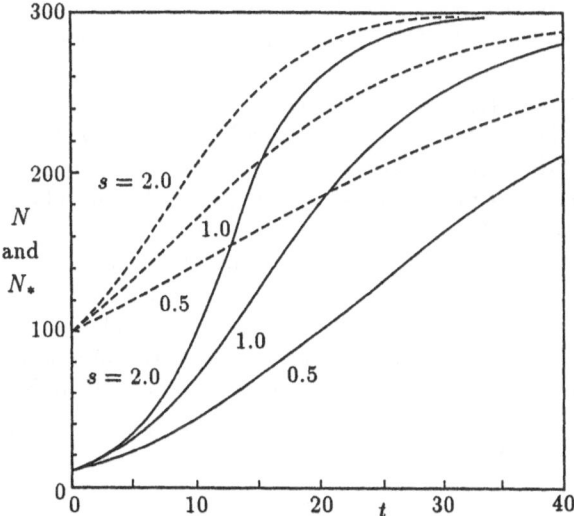

Fig. 5.6.1 Growth curves and carrying capacity curves for power law logistic growth with power law logistic carrying capacity. $a = 0.25$, $c = 0.10$, $N_0 = 10$, $N_{*0} = 100$, $N_{**} = 300$

5.6.3 An Illustration: Population of the United States

In our consideration of the exponential function in Section 2.1.3, we examined the growth of the population of the United States. It was seen that the population increased at an exponential rate during the period from 1790 to about 1870. However, in subsequent years the rate of increase was less than that predicted by the exponential.

It was mentioned in the earlier section that crowding effects began to retard the rate of increase after 1870 and that the logistic equation took these effects into account. Although we did not return to the topic of U.S. population growth in Section 2 .2, where we considered the logistic function, we do so now.

A number of years ago, Pearl and Reed (1920) published their results of fitting U.S. population data, for the census counts from 1790 through 1910, to the ordinary logistic equation

$$N = \frac{N_*}{1 + \left(\frac{N_*}{N_0} - 1\right) e^{-at}} \; . \tag{5.6.12}$$

To determine the three constants, (a, N_0, N_*) they "point-matched" at the years 1790, 1850, and 1910; this yielded the following expression

$$N = \frac{197.273}{1 + 49.209\,e^{-0.03134t}} \quad , \tag{5.6.13}$$

in which $t = 0$ corresponds to the year 1790. The inflection point occurs at $t_i = 124.32$ (year 1914.32) and $N_i = 98.637$. Clearly, the ultimate population or carrying capacity, is 197.273 million.

In his early classic, Lotka (1924) examined the matter of U.S. population growth at some length, including the Pearl–Reed results. Lotka concluded that "... if the population of the United States continues to follow this growth curve in future years, it will reach a maximum of some 197 million souls, about double its present population, by the year 2060 or so."

The population of the United States is shown in Table 5.6.1 by decades for the years 1790 through 1990. Included in the table are the populations computed from the Pearl–Reed relationship. This equation, developed before 1920, predicts the populations with surprising accuracy through 1950. However, as we see from the table, for subsequent years it substantially underestimates. Indeed, the Pearl–Reed prediction of an asymptotic population of 197 million was already exceeded before 1970.

Table 5.6.1 Population of the United States. Observed and computed values

Year	t	N Observed	N Pearl–Reed	N Computed
1790	0	3.929	3.929	3.956
1800	10	5.308	5.336	5.313
1810	20	7.240	7.228	7.137
1820	30	9.638	9.757	9.585
1830	40	12.861	13.109	12.874
1840	50	17.064	17.506	17.292
1850	60	23.192	23.192	23.224
1860	70	31.443	30.412	31.110
1870	80	38.558	39.372	41.333
1880	90	50.189	50.177	53.978
1890	100	62.980	62.769	68.054
1900	110	76.212	76.870	82.041
1910	120	92.228	91.972	95.572
1920	130	106.021	107.394	109.396
1930	140	123.203	122.397	124.467
1940	150	132.165	136.318	141.081
1950	160	151.326	148.678	159.604
1960	170	179.323	159.230	180.318
1970	180	203.302	167.945	202.951
1980	190	226.546	174.941	226.779
1990	200	250.550	180.438	250.205

An examination was made by Turner et al. (1969) of the problem of a power law logistic with a power law carrying capacity. They obtained the same solution for $N(t)$ as that given by Eq. 5.6.8. These investigators used the growth of U.S. population as a numerical example; they determined the value of the parameter, $s = 6.467$.

From Section 2.1.3 we established that $a = 0.0295$ and $N_0 = 3.956$. The magnitudes of all the constants are listed below.

$$a = 0.0295 \qquad c = 0.0125 \qquad s = 6.467$$

$$N_0 = 3.956 \qquad N_{*0} = 23.615 \qquad N_{**} = 302.293$$

$$m_* = 1.452\,(10^7) \qquad n_* = 1.508\,(10^{12})$$

Substitution of these numbers into Eq. 5.6.8 produces the population figures indicated in the last column of Table 5.6.1.

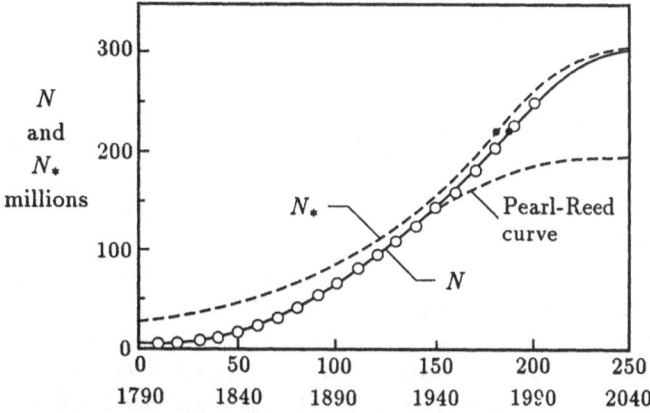

Fig. 5.6.2 Comparison of the observed population of the United States with computed values. Arithmetic plot

The results of our computations are shown as the solid curve in Fig. 5.6.2. We note fairly good agreement between the observed data and the calculated curve. In the figure the Pearl–Reed logistic and carrying capacity curve are indicated.

The inflection points of the carrying capacity and growth curves are shown. For the carrying capacity, the coordinates of the inflection point are

$$t_{i*} = \frac{1}{cs} \log_e \left(\frac{m_*}{s} \right) \quad ; \quad N_{i*} = \frac{N_{**}}{(1+s)^{1/s}} \ . \tag{5.6.14}$$

In our population problem, $t_{i*} = 180.91$ (year 1970.9), $N_{i*} = 221.54$.

The inflection point coordinates of the growth curve are given approximately by

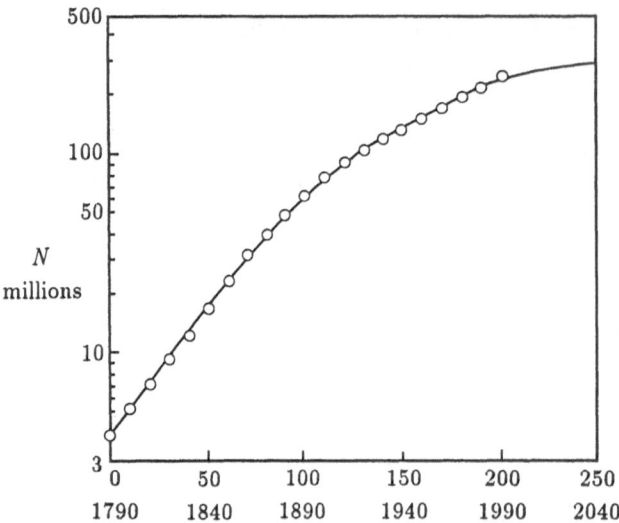

Fig. 5.6.3 Comparison of the observed population of the United States with computed values. Semi-logarithmic plot

$$t_i = \frac{1}{cs} \log_e \left(\frac{am_*}{s(a-c)} \right) \quad ; \quad N_i = \frac{N_{**}}{(1+s)^{1/s}} \quad . \tag{5.6.15}$$

Accordingly, $t_i = 187.73$ (year 1977.7), $N_i = 221.54$. We note that the ordinates of the two inflection points are the same. The inflection point of the carrying capacity precedes that of the growth curve by the time interval $\Delta t = (1/cs) \log_e[a/(a-c)]$. In our illustration, $t = 6.82$ years.

A semi-logarithmic display of the observed and computed populations is presented in Fig. 5.6.3. We note two rather distinct regimes of exponential growth: (a) from $t = 0$ to 80 (years 1790 to 1870) with a growth coefficient $a = 0.0295$ and (b) $t = 120$ to 190 (years 1910 to 1980) with an average growth coefficient $c = 0.0125$.

In the transition region, $t = 80$ to 120 (years 1870 to 1910), we can establish that the growth coefficient decreases almost linearly. Finally, for $t > 190$ (year 1980), the growth coefficient apparently descends exponentially. Though not a great deal would be gained, we could splice these four regions of different time dependent growth coefficients, as we did in Section 2.4.9 with the population of California, to obtain a growth curve virtually the same as that shown in Figs. 5.6.2 and 5.6.3.

Now to projections of future populations. Using the indicated numerical values of the parameters in Eq. 5.6.8, we project a U.S. population of 260.6 million for 1995 and 270.1 million for 2000. Beyond that, though long-term projections are risky, our framework predicts the following: 284.5 million (year 2010), 296.1 (2025), and 301.4 (2050).

To conclude, we look briefly at our illustration of population growth from the viewpoint of the demographer. This approach is based on "age structure" concepts in which a population depends not only on time, t, but also on the age, x, of each component of the population. We discussed this topic in Section 4.4.6 in connection with the growth of cell populations.

A term which arises in demographic analysis is "total fertility rate". This is an index which expresses the average number of births per woman in a specified range of ages, say 15 to 49, in a particular population. Westhoff (1986) reports that the fertility rate in America, which was 2.5 in 1970, fell to "well below replacement to a new historic low of 1.7 by 1976".

In making various projections of U.S. population, Westhoff assumes an annual net immigration of 450 000. In addition, with a total fertility rate, $r = 1.6$, the population of the nation will increase to a maximum of about 275 million in 2025 and thereafter decline. If it is assumed that $r = 2.3$, then the 2025 population will be approximately 330 million and increase to 450 million by 2080.

In between these two extremes of assumed fertility rates, if $r = 1.9$, then the population will be around 300 million in 2025 and stabilize at about 310 million by 2050. The projections of this $r = 1.9$ case are very close to the results of our power law logistic analysis.

In addition to the references given in Section 4.4.6 on age dependent population growth, texts dealing with demographic analysis are provided by Barclay (1958), Cox (1976) and Spiegelman (1968).

6. Phenomena with Time Delays

6.0.1 Introduction

So far, in our analyses, the assumption has always been made or implied that the rate at which an item is growing at time t depends on the magnitude of the item at that same time.

For example, for exponential growth we write

$$\frac{dN(t)}{dt} = aN(t) \tag{6.0.1}$$

and, with an initial condition specified, we easily establish that the growth is exponential.

Now what happens if we know or suspect that the present growth rate depends, not on the present magnitude but on the magnitude at an earlier time? For example, the present growth rate of a colony of flies depends not on the number of flies right now but rather on the number of flies laying a certain number of eggs a week or so ago. In this case we would write

$$\frac{dN(t)}{dt} = aN(t - \tau) \quad , \tag{6.0.2}$$

where τ, the average incubation period of the eggs, is a time delay or a time lag. As we shall see, this almost trivial change in the differential equation, greatly complicates the analysis and can produce drastic changes in the final answer.

Another example, somewhat more complicated: Suppose some particular item is growing and we know that, sooner or later, it will reach an equilibrium magnitude or carrying capacity. So we write the differential equation for the logistic

$$\frac{dN(t)}{dt} = aN(t) \left(1 - \frac{N(t)}{N_*}\right) \tag{6.0.3}$$

and, with an initial condition, get the well-known answer. This equation states that the growth rate is proportional to the fraction of the carrying capacity still available for growth at time t. However, suppose that this fraction available depends not on the present time but on a specified range of previous times. For example, consider a herbivore–plant system – animals grazing on vegetation. The areal density, N, depends on the amount of vegetation available. In turn the vegetation, once grazed, takes certain periods of time to recover. The outcome of this scenario is that the growth rate is determined not by Eq. 6.0.3 but instead by the relationship

$$\frac{dN(t)}{dt} = aN(t)\left(1 - \frac{1}{N_*}\int_{-\infty}^{t} K(t - \tau)N(\tau)d\tau\right) \tag{6.0.4}$$

where the dependent variable, $N(t)$, now appears also in an integral term. The quantity $K(z)$ is called the delay function. Again, the solution is $N = N(t)$. As is also true for Eq. 6.0.2, in order to solve a problem involving time delays we must specify not only an initial *point* condition, $N(0) = N_0$, but also an initial *interval* condition, $N(t) = g(t)$ for $\tau < t < 0$.

6.0.2 Types and Features of Delay Equations

The examples we just looked at illustrate the two types of delay equations we want to examine: (a) discrete time delay and (b) distributed or continuous time delay.

a. Discrete time delay equations. A somewhat generalized relationship for the first type is the linear equation

$$\frac{dN(t)}{dt} = a_1 N(t) + a_2 N(t - \tau) \quad , \tag{6.0.5}$$

where a_1 and a_2 are constants. This is an example of a differential–difference equation; more specifically, it is a so-called retarded delay equation.

Another important category of discrete delay equations, of course, is the logistic

$$\frac{dN(t)}{dt} = aN(t)\left(1 - \frac{N(t - \tau)}{N_*}\right) \quad . \tag{6.0.6}$$

b. Distributed time delay equations. For the second type of delay equation we restrict our attention to the modified logistic equation given by Eq. 6.0.4 and to a slightly different form

$$\frac{dN(t)}{dt} = aN(t)\left(1 - \frac{N(t)}{N_*} - \frac{c}{a}\int_{-\infty}^{t} K(t - \tau)N(\tau)d\tau\right) \tag{6.0.7}$$

in which we have added a so-called toxicity or pollution term.

The expressions given by Eqs. 6.0.4 and 6.0.7 are examples of integro-differential equations. They are not as formidable as they appear, especially if one can use a simple delay function, $K(z)$. In fact, and almost paradoxically as May (1974) points out, solutions for distributed delay equations are generally easier to obtain than solutions for discrete delay equations.

Furthermore, in many instances, distributed delay equations more accurately describe actual real world phenomena. Earlier we mentioned the colony of flies laying eggs. Now, perhaps it is possible to adequately depict the growth of the colony by using a discrete time delay based on the average incubation time of the eggs. However, we really do know that not all the eggs were laid at the same instant nor hatched at the same moment. Consequently, a distributed delay relationship, with the quite versatile delay function, $K(z)$, undoubtedly provides a more accurate answer.

As we shall see, the most important parameter in delay phenomena is the dimensionless quantity $\tau_* = a\tau$. We can define the reciprocal of the growth coefficient, $t_g = 1/a$, to be a kind of characteristic growth time. Indeed, for exponential growth we introduced, in Section 2.1.2, the notion of a doubling time, $t_2 = \log_e 2/a$. Accordingly, we can write $\tau_* = \tau/t_g$. If the delay time is small in comparison with the characteristic growth time then τ_* is small (with respect to unity, say). In this case, nothing much happens to the system. If the growth is exponential in character, the magnitude of N still proceeds to infinity or zero, depending on whether a is positive or negative, even though a discrete delay is providing a "feedback" signal.

On the other hand, if τ is large compared to t_g, then τ_* is large and the feedback signal, or multiplier, is excessive. If the growth is logistic in character and otherwise heading for its stable equilibrium magnitude, the long time delay drastically alters the growth pattern. In general, excessively large time delays create instabilities in a growth system resulting in convergent or divergent oscillations. They tend to lengthen the periods of the cycles and may produce highly unsymmetrical oscillation patterns.

The subject of time delays, with most of its mathematical base in differential–difference equations and integro-differential equations, is not an easy one. Yet it is apparent that the foundation for much of the framework for time-delayed growth and transfer processes is provided by these areas of mathematics. A number of references are cited in ensuing sections; the contributions of Murray (1989) and MacDonald (1989) are especially noteworthy.

6.1 Discrete Time Delay in the Exponential Equation

6.1.1 The Delay Differential Equation and Its Solution

We begin our analysis of time delays in growth phenomena and transfer processes by considering discrete time delays.

Commencing with the simplest case, we write the equation for exponential growth in the form

$$\frac{dN(t)}{dt} = aN(t - \tau) \tag{6.1.1}$$

in which the growth coefficient, a, and delay time, τ, are positive real constants. In previous sections it was implied that the dependent variable, N, was always $N(t)$. With the introduction of a time delay, it is necessary to specify the argument, i.e., $N(t - \tau)$, to avoid any confusion.

This modified differential equation for exponential growth states that the rate of increase of N depends not on the magnitude of N at time t but on the magnitude of N at a precisely specified previous time, $t - \tau$. For example, the rate of growth of an animal population depends not on the present population but on the population a gestation time earlier.

The relationship of Eq.6.1.1 is a simple illustration of a differential–difference equation or a so-called delay differential equation. These kinds of equations have been studied by Bellman and Cooke (1963), Cunningham (1958), Driver (1977), Pinney (1958), Saaty (1981) and numerous others.

If we set $\tau = 0$ and specify $N(0) = N_0$ as the initial condition, the solution to Eq. 6.1.1 is $N(t) = N_0 \exp(at)$. If the delay time is very small then we would still expect an exponential growth relationship. So, using a Taylor series expansion

$$N(t - \tau) = N(t) - \tau \frac{dN(t)}{dt} + \frac{1}{2}\tau^2 \frac{d^2 N(t)}{dt^2} - \cdots \quad , \tag{6.1.2}$$

and retaining only the linear term in τ, we obtain

$$N(t) = N_0 \exp\left(\frac{a}{1 + a\tau}t\right) \quad . \tag{6.1.3}$$

Thus, we again have an exponential relationship, though now with a slightly smaller effective growth coefficient.

It is well known that utilization of Taylor series to obtain solutions to delay differential equations can lead to erroneous answers; this caution has been made by Cunningham (1958), Goel et al. (1971), Saaty (1981) and others. This brief introduction involving Taylor series is employed only to signal that the appearance of discrete time delays can substantially alter and complicate solutions to equations, even those as simple as Eq. 6.1.1. We now take a different approach.

A straightforward way to solve linear differential equations is the method of Laplace transformations. Since there are many books available which deal with this subject, we will not go into detail; suggested references are Boyce and DiPrima (1986), Van Iwaarden (1985) and Spiegel (1965).

In order to obtain the solution to Eq. 6.1.1 it is necessary to specify not only an initial condition but also an interval condition. Thus, the complete problem is defined as follows

$$\frac{dN(t)}{dt} = aN(t - \tau) \quad ; \qquad t \geq 0 \tag{6.1.4}$$

$$N(0) = N_0 \quad ; \qquad N(t) = g(t) \quad ; \qquad -\tau \le t \le 0 \quad , \tag{6.1.5}$$

in which $g(t)$ describes the behavior of $N(t)$ in the initial interval.

The Laplace transform of $N(t)$ has the definition

$$\mathcal{L}\{N(t)\} = \overline{N}(s) = \int_0^\infty e^{-st} N(t)\, dt \quad . \tag{6.1.6}$$

Utilizing this equation, it can be established that Eqs. 6.1.4 and 6.1.5 are transformed into the relationship

$$\overline{N}(s) = \frac{N(0) + ae^{-\tau s} \int_{-\tau}^0 g(\zeta) e^{-s\zeta}\, d\zeta}{s - ae^{-\tau s}} \quad . \tag{6.1.7}$$

In principle, substitution of this equation into the complex inversion formula

$$N(t) = \frac{1}{2\pi i} \int_{c-i\infty}^{c+i\infty} e^{st} \overline{N}(s)\, ds \tag{6.1.8}$$

provides the desired solution, $N(t)$.

At this point, it is useful to specify two typical and simple initial conditions required for Eq. 6.1.5.

Case 1. $g(t) = 0$, $N(0) = N_0$. In this simplest case, Eq. 6.1.7 becomes

$$\overline{N}(s) = \frac{N_0}{s - ae^{-\tau s}} \quad . \tag{6.1.9}$$

Case 2. $g(t) = N_0$, $N(0) = N_0$. In this event, Eq. 6.1.7 reduces to

$$\overline{N}(s) = \frac{N_0(s + a - e^{-\tau s})}{s(s - ae^{-\tau s})} \quad . \tag{6.1.10}$$

Substitution of either Eq. 6.1.9 or Eq. 6.1.10 into Eq. 6.1.8 then provides the solution we seek.

The denominator of either of these equations is termed the characteristic function. The only singularities of the integrand of Eq. 6.1.8 are the poles corresponding to the characteristic functions of Eqs. 6.1.9 or 6.1.10. In the latter equation, a pole occurs at $s = 0$; in both equations, poles occur at the roots of the expression

$$s - ae^{-\tau s} = 0 \quad . \tag{6.1.11}$$

Once the roots of this equation, s_n, are determined, the complex inversion formula can be employed to obtain $N(t)$. It can be established that the integral of Eq. 6.1.8 is equal to the sum of the residues of $\exp(st)N(s)$ at the poles. In turn, Bellman and Cooke (1963) have shown that the sum of the residues is such that the solution is given by

$$N(t) = N_0 \sum_{n=1}^{\infty} c_n e^{s_n t} \qquad (6.1.12)$$

where the coefficients, c_n, are determined by the initial conditions. We will regard Eq 6.1.12 as the final solution.

It is quite amazing to observe that the introduction of a simple discrete time delay can change the almost trivial Eq. 6.1.1, with $\tau = 0$, into a very complicated problem.

6.1.2 Roots of the Characteristic Equation

With an eye on an illustration we consider below, let us change Eq. 6.1.1 from its exponential growth form to one of exponential decay. That is

$$\frac{dN(t)}{dt} = -a_* N(t - \tau) \quad , \qquad (6.1.13)$$

where a_*, a positive real constant, is the decay coefficient. Let us also put this equation and a typical initial condition into the following dimensionless form

$$\frac{dW(T)}{dT} = -W(T - \tau_*) \quad ; \qquad T \geq 0 \quad ; \qquad (6.1.14)$$

$$W(0) = 1 \quad ; \qquad W(T) = 0 \quad ; \qquad -\tau_* \leq T \leq 0 \quad ; \qquad (6.1.15)$$

in which $W(T) = N(t)/N_0$, $T = a_* t$, and $\tau_* = a_* \tau$. The solution, Eq. 6.1.12, becomes

$$W(T) = \sum_{n=1}^{\infty} c_n e^{s_n T} \quad , \qquad (6.1.16)$$

where s_n are the roots of the characteristic function or equation

$$s + e^{-\tau_* s} = 0 \quad . \qquad (6.1.17)$$

Since these roots are, in general, complex quantities we let $s = \alpha + i\beta$. Substituting this relationship into Eq. 6.1.17 and equating both real terms and imaginary terms to zero yields

$$\alpha = -e^{-\tau_* \alpha} \cos \tau_* \beta \quad ; \qquad \beta = e^{-\tau_* \alpha} \sin \tau_* \beta \quad . \qquad (6.1.18)$$

One solution of the second of these two equations is $\beta = 0$. Consequently, real roots of Eq. 6.1.17 are provided by

$$\alpha + e^{-\tau_* \alpha} = 0 \quad ; \qquad \beta = 0 \qquad (6.1.19)$$

as long as $\tau_* < 1/e$.

Dividing the first of Eqs. 6.1.18 by the second and letting $x = \tau_*\alpha$ and $y = \tau_*\beta$ gives

$$x = -y\cot y \quad ; \qquad \frac{1}{\tau_*} = e^{y\cot y}\frac{\sin y}{y} \quad . \tag{6.1.20}$$

From the second of these two equations, values of y can be determined for various values of τ_*; hence β can be computed. The first of the equations then yields x and, in turn, α.

A list of roots corresponding to the range $\tau_* = 0$ to $\pi/2$ has been prepared by Gumowski (1981); this list is presented in Table 6.1.1.

Table 6.1.1 Roots of the equation $s + \exp(-\tau_* s) = 0$ and ratio of period of oscillation to delay time. From Gumowski (1981)

τ_*	$-\alpha$	β	T_{osc}/τ_*	τ_*	$-\alpha$	β	T_{osc}/τ_*
0	1	0		1.0	0.31813	1.33724	4.699
0.1	1.11833	0		1.1	0.22866	1.26549	4.514
0.2	1.29586	0		1.2	0.15872	1.19935	4.366
0.3	1.63134	0		1.3	0.10314	1.13883	4.244
0.36	2.23912	0		1.4	0.05836	1.08356	4.142
0.366	2.46495	0		1.5	0.02186	1.03310	4.055
1/e	e	0		1.52	0.01538	1.02354	4.039
0.368	2.71680	0.06956	245.456	1.54	0.00914	1.01414	4.023
0.374	2.64437	0.48528	34.619	1.56	0.00315	1.00491	4.009
0.4	2.36022	1.01817	15.428	1.568	0.00081	1.00127	4.002
0.5	1.58805	1.54022	8.159	1.57	0.00023	1.00036	4.001
0.6	1.11715	1.60413	6.528	$\pi/2$	0	1	4
0.7	0.80696	1.56323	5.742	1.572	-0.00035	0.99946	3.999
0.8	0.59120	1.49187	5.265	1.574	-0.00092	0.99855	3.998
0.9	0.43492	1.41371	4.938	1.576	-0.00149	0.99765	3.996

As Maynard Smith (1974) and Pinney (1958) point out, if we seek the solution to the delayed exponential *growth* equation, i.e., Eq. 6.1.1, we need the roots of the characteristic equation for values of y in the ranges $(\pi,2\pi)$, $(3\pi,4\pi)$, $(5\pi,6\pi)$, and so on. On the other hand, for the delayed exponential *decay* equation, Eq. 6.1.13, we require the roots for y in the ranges $(0,\pi)$, $(2\pi,3\pi)$, $(4\pi,5\pi)$, etc. A tabulation given by Pinney (1958) provides roots of the characteristic equation, Eq. 6.1.17, for the range $(0,7\pi)$.

For values of $\tau_* > 1/e$, oscillatory solutions are obtained. Table 6.1.1 lists the ratios of the period of oscillation T_{osc} to the dimensionless delay time.

6.1.3 Behavior of the Solutions

On the basis of the preceding, we summarize the form and behavior of the solutions to the delay differential equation, Eq. 6.1.14, in the following ranges of τ_*:

1. $0 < \tau_* < 1/e$; $\beta = 0$;

$$W(T) = e^{\alpha T} ;   \alpha + e^{-\tau_* \alpha} = 0 .$$ (6.1.21)

In this case, since $\alpha < 0$, W decreases monotonically to zero as $T \to \infty$.

2. $1/e < \tau_* < \pi/2$;

$$W(T) = e^{\alpha T}(A \cos \beta T + B \sin \beta T) .$$ (6.1.22)

In this range, $\alpha < 0$. It is clear that W oscillates and approaches zero as $T \to \infty$.

3. $\tau_* = \pi/2$; $\alpha = 0$, $\beta = 1$

$$W(T) = A \cos T + B \sin T .$$ (6.1.23)

In this case, T oscillates with neither amplification nor attenuation. The period of oscillation is $T_{\mathrm{osc}} = 4\tau_*$.

4. $\tau_* > \pi/2$

$$W(T) = e^{\alpha T}(A \cos \beta T + B \sin \beta T) .$$ (6.1.24)

In this range, $\alpha > 0$. We note that W oscillates and goes to infinity as $T \to \infty$.

Useful relationships for direct computation of α and β for values of τ_* near $\pi/2$ have been given by Pinney (1958). He shows that

$$\alpha = \frac{2\pi}{\pi^2 + 4}\left(1 - \frac{\pi}{2\tau_*}\right) ;   \beta = \frac{1}{\pi^2 + 4}\left(4 + \frac{\pi^3}{2\tau_*}\right) .$$ (6.1.25)

In the neighborhood of $\tau_* = \pi/2$, the values of α and β computed from these equations agree closely with those listed in Table 6.1.1.

Finally, we return to the Taylor series given by Eq. 6.1.2. We observed that when only the linear term in the expansion is retained, we acquired Eq. 6.1.3. Qualitatively, this is the same as Eq. 6.1.21. Had we retained the quadratic term as well, we would have produced a second order differential equation with constant coefficients. The solution to that equation is qualitatively similar to those given by Eqs. 6.1.22 to 6.1.24.

It is interesting to observe that the introduction of a discrete time delay into an extremely simple first order differential equation, such as Eq. 6.1.13, yields results similar to those of a second order linear differential equation without time delay.

6.1.4 An Illustration: Tinbergen's Shipbuilding Cycle

For an illustration of a discrete delay phenomenon we go back quite a few years to select a problem considered by Tinbergen (1931). He was interested in the cyclic behavior between the rate of construction of merchant marine vessels and the tonnage of such vessels at a previous time.

In his study of these shipbuilding cycles, Tinbergen noted "the relationship between the increase in total tonnage and the volume of the total tonnage of about two years earlier." He presented the information displayed in Figs. 6.1.1, 6.1.2, and 6.1.3.

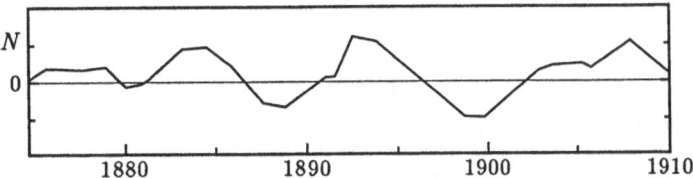

Fig. 6.1.1 Shipbuilding cycle of Tinbergen (1931). Total tonnage (deviation from mean) of merchant marine fleets of Great Britain, U.S.A., and Germany

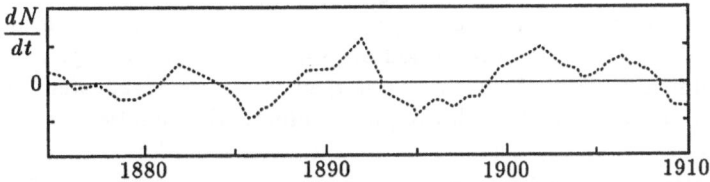

Fig. 6.1.2 Annual increases (deviation from mean) of new construction of merchant fleets of the three countries

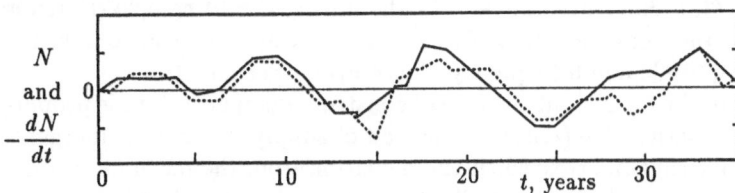

Fig. 6.1.3 Comparison of negative of annual increase of new construction with existing tonnage two years earlier

Fig. 6.1.1 shows the total tonnage of the merchant marine fleets of Great Britain, United States and Germany for the period 1875 to 1910. The ordinate of the plot is expressed in percentage of deviation from a mean. Tinbergen did not indicate the precise percentages.

Fig. 6.1.2 gives the annual increases in merchant marine tonnage of the three countries also scaled in terms of a deviation from a mean.

Tinbergen postulated that the change in total tonnage is linearly related to the magnitude of the tonnage approximately two years earlier. Accordingly, he assumed

$$\frac{dN(t)}{dt} = -a_* N(t - \tau) \quad , \tag{6.1.26}$$

which is our Eq. 6.1.13.

Fig. 6.1.3 displays the justification of Tinbergen's assumption. The solid line in the plot is essentially $N(t-\tau)$ transposed, with $\tau = 2.0$, from Fig. 6.1.1. The dotted line is the negative of the ordinate of Fig. 6.1.2, i.e., $-dN(t)/dt$. Tinbergen noted the close similarity of these two plots.

Much of the analysis and most of the results presented in our examination of discrete time delay were provided by Tinbergen. He indicated that the dimensionless delay time had a value $\tau_* = 1$ to 2. Accordingly, as shown in Table 6.1.1, the period of oscillation, $t_{osc} = 4\tau$. With an observed value $\tau = 2$ years, he concluded that an endogenous, or self-generating, cycle of about eight years existed during the period corresponding to his study.

The shipbuilding cycle analyzed by Tinbergen was later elaborated by Pinney (1958) who examined the following delay differential equation

$$\frac{dN(t)}{dt} + a_* N(t - \tau) = \epsilon N^3(t - \tau) \quad . \tag{6.1.27}$$

He showed that the solution to this equation behaves in the same way as the solution to Eq. 6.1.26 in the range $0 < \tau_* < \pi/2$. However, when τ_* is larger than $\pi/2$, a difference appears. In the linear ($\epsilon = 0$) case, $N(t)$ tends to infinity. In the non-linear ($\epsilon \neq 0$) case, $N(t)$ stabilizes in an oscillation with finite amplitude and a frequency of $1/4$ cycles per unit time.

We see from Eq. 6.1.27 that the rate of construction of new ships is proportional to the quantity $N - (\epsilon/a_*)N^3$, instead of simply N, as in Tinbergen's analysis. Pinney suggests that when the deviation from the mean of existing tonnage becomes quite large, shipbuilders will react in a relatively more conservative manner because of construction costs. Accordingly, the non-linear model of Eq. 6.1.27, given by Pinney, may be more realistic than the linear model of Eq. 6.1.26.

6.2 Discrete Time Delay in the Logistic Equation

6.2.1 Introduction to the Delay Differential Equation

A significant fraction of our overall consideration so far has been devoted to the logistic equation and its many ramifications. Hence an appropriate next topic to examine is the effect of time delays on the logistic. We start by writing

$$\frac{dN(t)}{dt} = aN(t - \tau_1)\left(1 - \frac{N(t - \tau_2)}{N_*}\right) \quad , \tag{6.2.1}$$

where, in the vernacular of biological growth phenomena, τ_1 is termed the reproduction time delay and τ_2 the reaction time delay. The former represents a gestation time and may be important during the early phases of a growth process. The latter, the reaction time delay, is a precisely specified time interval between a change in the environment and the consequent change in the rate of growth.

Virtually all of the many studies that have been carried out on the logistic equation with discrete time delay have discarded the reproduction delay time, τ_1, and have considered only the reaction time delay, τ_2. We shall do the same. Accordingly, Eq. 6.2.1 becomes, with $\tau_2 = \tau$,

$$\frac{dN(t)}{dt} = aN(t)\left(1 - \frac{N(t - \tau)}{N_*}\right) \quad . \tag{6.2.2}$$

As pointed out in the previous section, the subject of delay differential equations has received much attention over a period of many years. Some of the earliest work was carried out by researchers in the field of economics. Noteworthy are the studies of Tinbergen (1931), whose shipbuilding cycle we have already examined, and Kalecki (1935) who investigated the effect of time delay of investments on business cycles. An historic early contribution to the study of delay differential equations was made by Frisch and Holme (1935) in their examination of problems in dynamic economics.

With few exceptions, the early investigators of delay equations were concerned with the linear relationships we considered in Section 6.1. This is also true in the case of later studies by Jones (1961, 1962), Kakutari and Markus (1958) and Wright (1946, 1955). Although, these researchers were concerned primarily with the non-linear delayed logistic equation, they examined the linearized relationships in close detail.

In the fields of biology and ecology, the first appearance of the discrete delayed logistic, Eq. 6.2.2, was in a publication by Hutchinson (1948). In that epic paper, Hutchinson expressed the view that the observed oscillations in some kinds of biological populations could be explained by a discrete time delay in the crowding or resource term.

In the wake of these earlier studies, numerous investigators examined various aspects of delayed logistic phenomena. Extensive coverage and numerous references on the subject are given by Cushing (1977), Gopalsamy (1985, 1986), MacDonald (1978, 1989), May (1974) and May et al. (1974). Noteworthy are studies by Brauer and Sanchez (1975) who examined the problem of time delay in the logistic-with-harvesting equation, Caughley (1976) who considered the influence of time delay in the growth of ungulate populations, and Nisbet and Gurney (1976) who carried out an extensive analysis of discrete time delays in periodically varying environments.

6.2.2 Solution to the Discrete Delay Logistic Equation

If we make the following substitutions in Eq. 6.2.2

$$y(\theta) = \frac{N(t)}{N_*} - 1 \quad ; \qquad \theta = \frac{t}{\tau} \quad ; \qquad \tau_* = a\tau \tag{6.2.3}$$

and impose an initial condition we obtain

$$\frac{dy(\theta)}{d\theta} = -\tau_* y(\theta - 1)[1 + y(\theta)] \quad , \tag{6.2.4}$$

$$y(\theta) = g(\theta) \quad ; \qquad -1 \le \theta \le 0 \quad . \tag{6.2.5}$$

Expanding Eq. 6.2.4 and retaining only the linear term yields

$$\frac{dy(\theta)}{d\theta} = -\tau_* y(\theta - 1) \quad , \tag{6.2.6}$$

which is the same as Eq 6.1.14 with a change of variables. This equation has been studied by Wright (1955) and numerous others with the expectation that its solution possesses properties similar to those of the much more complicated Eq. 6.2.4.

Utilizing a Taylor series expansion, Cunningham (1954) converted the discrete delay logistic equation, Eq. 6.2.2, into the second order non-linear differential equation expressed by Eq 2.9.15. Again, we heed caution concerning the use of Taylor series in delay differential equations; Mazanov and Tognetti (1974) consider this matter at some length. By means of an analog computer, Cunningham also obtained an "exact" solution to the logistic equation with discrete delay. His results are shown in Fig. 6.2.1.

Comprehensive studies of Eqs. 6.2.4 and 6.2.5 have been carried out by a number of investigators over the years. Saaty (1981) has summarized the following important features of the solutions of these equations. For values of time, θ, approaching infinity: (a) $y(\theta)$ converges to zero monotonically for $0 < \tau_* < 1/e$, (b) $y(\theta)$ converges to zero in an oscillating pattern for $1/e < \tau_* < \pi/2$, and (c) $y(\theta)$ oscillates in a stable limit cycle pattern for $\tau_* > \pi/2$.

An extensive analysis of the delayed logistic equations was made by Jones (1961, 1962). An important part of his study was the numerical solution of these equations for values of τ_* ranging from 1.58 to 3.0. In the case $\tau_* = 2.0$,

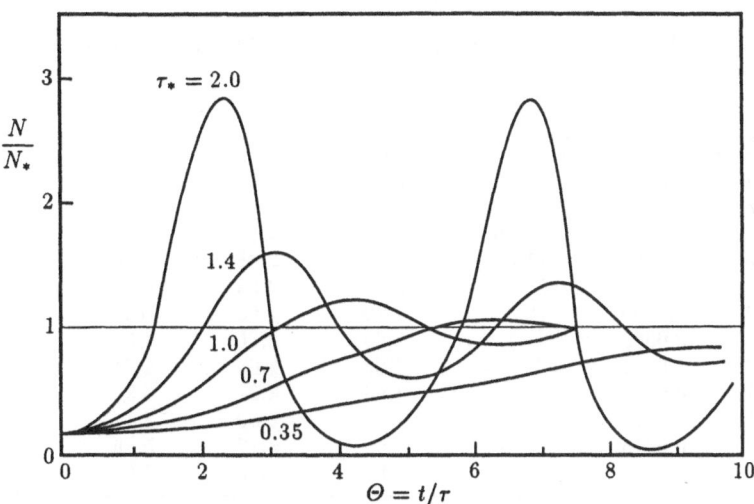

Fig. 6.2.1 Computer solution of the logistic equation with discrete time delay. Compare with linear plot of Fig. 2.9.4. From Cunningham (1954)

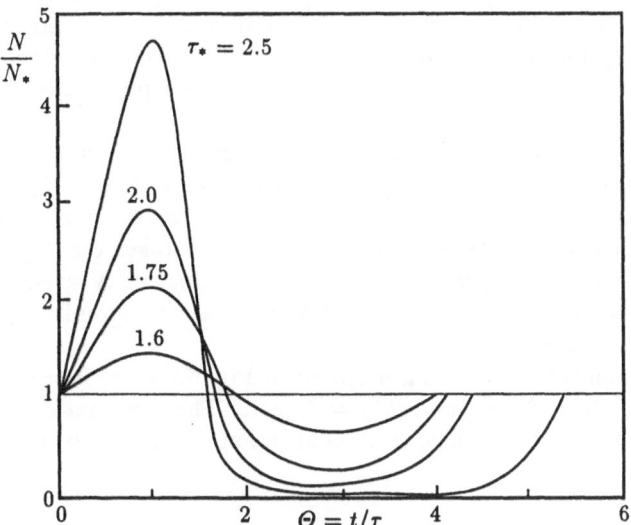

Fig. 6.2.2 Numerical solution of the logistic equation with discrete time delay. Post-transient regime. From Jones (1961)

Jones utilized several initial interval functions, $g(\theta)$, as stipulated by Eq. 6.2.5. He showed that regardless of the initial condition, $y(\theta)$ converges to the same stable limit cycle pattern, with amplitude and frequency determined by the value of τ_*. Plots of post-transient periodic solutions are shown in Fig. 6.2.2 for various values of τ_*.

Table 6.2.1 Global dimensions of limit cycle solutions of the logistic equation with discrete delay. Adapted from May (1974)

τ_*	$U_{\max}$	$U_{\min}$	$U_{\max}/U_{\min}$	$\theta_{osc}(+)$	$\theta_{osc}(-)$	θ_{osc}
$\pi/2$	1.00	1.00	1.00	2.00	2.00	4.00
1.6	1.45	0.566	2.56	1.91	2.12	4.03
1.7	1.91	0.332	5.76	1.85	2.24	4.09
1.8	2.30	0.198	11.6	1.80	2.38	4.18
1.9	2.63	0.118	22.2	1.76	2.53	4.29
2.0	2.92	0.0690	42.3	1.72	2.68	4.40
2.1	3.22	0.0383	84.1	1.69	2.85	4.54
2.2	3.57	0.0201	178	1.66	3.05	4.71
2.3	3.91	0.0096	408	1.63	3.27	4.90
2.4	4.28	0.0041	1040	1.61	3.50	5.11
2.5	4.69	0.0016	2930	1.59	3.77	5.36
3.0	7.62			1.52	5.53	7.05

In his classic publication, Hutchinson (1948) suggested that the presence of a substantial time delay in the growth of a single species population could produce "... a succession of high peaks alternate with very wide troughs, producing a biologically unstable situation in which short periods of excessive over-crowding alternate with long periods in which the population is dangerously rarefied."

The plots of Jones displayed in Fig. 6.2.2 bear this out. In these curves of N/N_* vs. $\theta = t/\tau$ we observe that as τ_* increases the maximum value, $N_{\max}$, increases markedly, the minimum value, $N_{\min}$, gets very close to zero, and the period of the limit cycle lengthens. Furthermore, as τ_* takes on larger values, the portion of the limit cycle corresponding to values of N greater than the carrying capacity, N_*, is reduced and the portion corresponding to values less than N_* is enlarged.

The global dimensions of the cycles illustrated in Fig. 6.2.2 are listed in Table 6.2.1. As before, $U = N/N_*$, and $\theta_{osc} = T_{osc}/\tau_*$. The information in the table, based on the results of Jones, is an adaptation of a table given by May (1974).

6.2.3 Numerical Example

For a numerical example we consider one presented by May (1974) and extend it somewhat. This example involves a single species population growing logistically with a reaction time delay. Accordingly, the growth relationship is given by Eq. 6.2.2. The initial condition is $N = N_0$ for $t < 0$.

Following May, we select the following numerical values: $N_0 = 1000$, $N_* = 2000$, $a = 1.0$, and $\tau = 3\pi/5$. His solution is shown in Fig. 6.2.3.

We start with the easy part. Suppose there were no delay, i.e., $\tau = 0$. In this case we obtain the ordinary logistic growth curve shown as the dotted

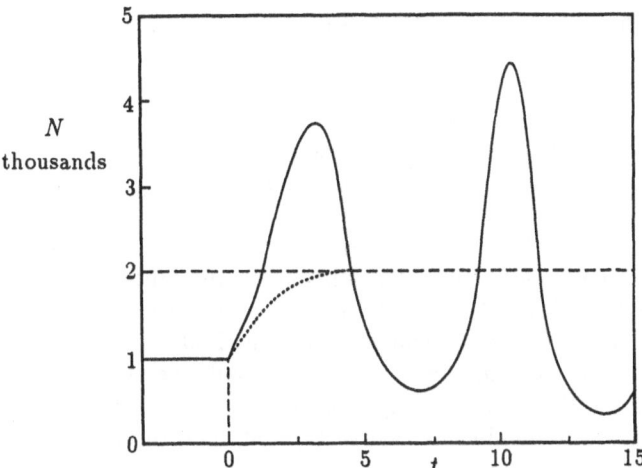

Fig. 6.2.3 Numerical example of logistic growth with discrete time delay, $\tau = 3\pi/5$, sufficiently large to produce limit cycles. From May (1974)

line in Fig. 6.2.3. The inflection point of this curve is located at $t = 0$, $N = N_0 = 1000$; N essentially reaches the carrying capacity, $N_* = 2000$, when $t = 5.0$.

Back to the hard part. Since $\tau_* = a\tau = 3\pi/5$ is greater than $\pi/2$ we anticipate a growth which, after a few oscillations, achieves a stable limit cycle pattern. This is shown in Fig. 6.2.3 although the "steady state" has not yet been reached.

From the information given in Table 6.2.1 we determine that after another cycle or two, the growth pattern will have the following global dimensions

$$N_{\max} = 5250 \qquad N_{\min} = 260 \qquad N_{\max}/N_{\min} = 20.2$$

$$t_{\text{osc}} = 8.06 \qquad t_{\text{osc}}(+) = 3.33 \qquad t_{\text{osc}}(-) = 4.73$$

We note that because of the time delay, an otherwise stable carrying capacity, $N_* = 2000$, gives way to a maximum population, $N_{\max} = 5250$ and minimum population, $N_{\min} = 260$. To again quote Hutchinson (1979): "Under such circumstances the population is likely to become extinct if an unfavorable, random, external circumstance intervenes during a depression."

If the discrete time delay had been even larger than $3\pi/5$ the amplitude and period of oscillation would, of course, be still greater. If the time delay had been somewhat smaller, then the growth patterns would be either (a) oscillatory convergent or (b) monotonically convergent to N_*, depending on the value of τ_*. These various growth patterns are illustrated in Fig. 6.2.1.

6.2.4 An Illustration: Nicholson's Blowflies

A series of experiments carried out by Nicholson (1954, 1957) during the mid-1950s involving the Australian sheep blowfly, *Lucilia cuprina*, has become a classic illustration of single species growth dynamics. As we shall see, these experiments exhibit surprisingly clear patterns of sustained large amplitude oscillations of nearly constant frequency.

A brief description of Nicholson's experiments. Cultures of *L. cuprina* were placed in large cages under constant conditions in the laboratory. The temperature was maintained at 25° C; ample supplies of water and dry sugar were maintained. Food for larvae, to which adult flies had no access, was provided in excess supply. Each day, 50 grams of ground liver were placed in each cage as food for the adults. Daily counts were made of the total adult population and of the number of eggs generated. The duration of an experiment was of the order of a year.

Typical experimental results are shown as the broken solid lines in Fig. 6.2.4. The average length of time between cycles is 38.7 days; the average number of adult flies is 2520.

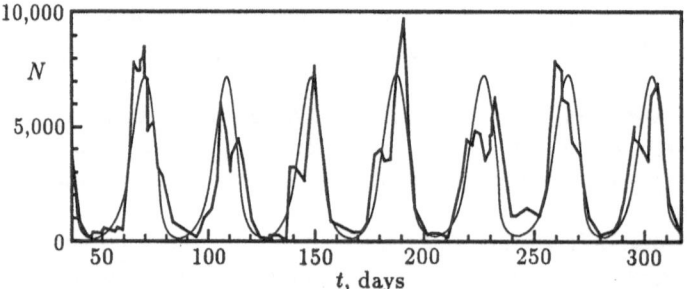

Fig. 6.2.4 Nicholson's experiments on growth and decline of Australian blowflies. Broken lines are results of Nicholson (1954); smooth curve is limit cycle solution given by May (1974)

In some experiments the amount of ground liver food supply was changed to lower levels. When it was reduced from 50 to 25 grams per day, the average number of adults was decreased by approximately one-half. When it was further reduced to 10 grams per day, the average number fell to 527. These results suggest that the equilibrium carrying capacity is directly proportional to the amount of daily food supply; this seems to be reasonable.

More importantly, for our illustration, is the pattern of the growth and decline cycles shown in the figure. In his analysis of Nicholson's experimental results, May (1974) determined the approximate ratio of maximum to minimum population, $N_{\max}/N_{\min}$.

Using this ratio he obtained, from Table 6.2.1, the value $\tau_* = 2.1$ and, accordingly, $\theta_{\text{osc}} = 4.54$. From the measured duration of the cycle, $t_{\text{osc}} = 38.7$

days, May computed the egg-to-adult development time, $\tau = 8.52$ days. This result agrees fairly well with directly measured values of 11 to 14 days.

Utilizing the above numerical results and additional information obtained from Fig. 6.2.4 and Table 6.2.1, we determine the following

$$N_{\max} = 7500 \qquad N_{\min} = 89 \qquad N_* = 2330$$

$$a = 0.246 \qquad t_{\mathrm{osc}}(+) = 14.4 \qquad t_{\mathrm{osc}}(-) = 24.3$$

If the growth of the adult fly population had been purely exponential, $N = N_0 \exp(at)$, with $N_0 = 89$, the population would have reached $N = 7500$ at $t = 18.0$ days. This is about one-half of the observed period of oscillation; $t_{\mathrm{osc}} = 38.7$ days.

Had the growth been simply logistic, the population would have reached the inflection point value, $N_{\mathrm{i}} = N_*/2 = 1165$, at $t = 13.1$ days. After $t = 38.7$ days, the population would have been 2325 which is essentially the equilibrium value, $N_* = 2330$.

We return to the main theme of our illustration. Since $\tau_* = 2.1$ is greater than $\pi/2$, we can expect a sustained limit cycle growth pattern if our discrete time delay model is a correct one. It appears that it is. The smooth solid curves in Fig. 6.2.4 show the computed oscillations.

Nicholson presents the following explanation of the growth and decline process. He reported that significant egg production occurred only during the period of low values of the adult population and that approximately 14 days were needed for an egg to become an adult. During the ensuing "recruitment period" of new adults, the population increased sharply. Intraspecies competition for the limited food supply became so intense that decreasingly few adults received sufficient food to develop eggs. Normal mortality inevitably caused the population to decrease. With reduced competition for food, increasing numbers of surviving adults were able to produce eggs. A period followed during which the adult population was at or near the minimum value. During this period the rate of egg production increased each day until it reached a maximum and then quickly fell to zero. A new cycle of growth and decline then began. Slobodkin (1961) presents an extensive discussion of Nicholson's experiments and their results.

Other investigators, for example, Gurney et al. (1980), have devised more complicated models in efforts to resolve the discrepancy between computed and observed values of the time delay in the experiments. Also, the invariable "notches" in the experimental plot shown in Fig. 6.2.4 – which were observed also in the egg generation data – suggests the existence of another time delay in the growth process. This notch phenomenon escapes prediction in the present model.

Even so, the above-described method of analysis devised by May has the obvious merit of providing an easy approximate solution to a difficult problem.

Ecologists have reported numerous instances of growth and decline cycles of animal populations in their natural habitats. Apparently, four year cycles are not infrequent, especially in high latitude regions.

In a very gross way, we can say that in such regions a time delay of approximately one year is imposed by the seasonal environment. A particular animal species whose inherent growth coefficient, a, is sufficiently large to make $\tau_* = a\tau$ greater than $\pi/2$, is evidently vulnerable to a cyclic growth pattern oscillating between large and small values at a period about four times larger than the time delay, i.e., four years.

It has also been reported that the populations of numerous species of mammals of the colder temperate zones oscillate in approximately 10 year cycles. The classic case of the snowshoe hare–lynx oscillation pattern is evidently one of these.

A number of references are given by May (1981) concerning wildlife cycles and Pollard (1973) discusses cyclic environments in demography. An extensive catalog of observed cycles in economics, agriculture, and manufacturing has been compiled by Wilson (1964).

6.3 Distributed Time Delay: Delay Integral in the Crowding Term

6.3.1 The Integro-differential Equation

In the immediately preceding sections we examined in some detail the subject of *discrete* time delays. We considered phenomena in which the growth rate of a particular quantity depends on the magnitude of the quantity at a specified previous moment.

In many instances, a more realistic point of view is to assume that the effects of a time delay are not concentrated at a precise earlier instant but are, instead, *distributed* over a specified range of time.

This problem of distributed time delays has been studied by many investigators commencing with the early work of Volterra (1926). Both Frauenthal (1980) and May (1974) provide lucid introductions to the subject; Cushing (1977, 1979) and MacDonald (1978, 1989) examine the topic at greater length. Brauer (1976) incorporates the effect of constant rate harvesting in distributed delay phenomena; he includes as an example the sandhill crane problem we considered in Section 2.7.6. Lal et al. (1988) examine a quite generalized case of innovation diffusion: distributed time delay, with adopter rejection, in a population affected by both internally and externally influenced communication mechanisms.

We begin by writing the differential equation for logistic growth in the following form

$$\frac{1}{N(t)}\frac{dN(t)}{dt} = a - b\int_{-\infty}^{t} K(t-z)N(z)dz \quad .$$ (6.3.1)

We observe that the quantity we want to determine, $N(t)$, appears in the integral. This expression is a Volterra-type integro-differential equation. The quantity $K(t)$ has various names: difference, displacement or delay kernel, memory function, heredity function, weighing function, and so on. We shall call it the delay function. This function specifies the influence which previous values of N have on the present growth rate. References dealing with Volterra equations are Linz (1985), Miller (1971), and Rabotnov (1980).

We recall that the differential equation for the ordinary logistic is

$$\frac{1}{N(t)}\frac{dN(t)}{dt} = a - bN(t) \quad .$$ (6.3.2)

In this case, the growth coefficient linearly decreases as the present value of N increases. In contrast, we note from Eq. 6.3.1 the growth coefficient decreases by an amount proportional to the integrated value of N over the entire indicated time range.

In order to solve Eq. 6.3.1, the precise form of the delay function, $K(t)$, must be specified. In some cases, it may be that previous values of N exert the same influence, on the current growth rate, over the entire time interval specified by the integral limits. In other cases, perhaps the current growth rate is affected mostly by the recent values of N with diminishing influence due to earlier values. In many cases, we might say that the present growth rate is only weakly influenced by the magnitude of N in the recent past and distant past but that it is strongly affected by values of N corresponding to a certain intermediate time interval. One thing is clear: mathematical difficulties encourage us to keep $K(t)$ as simple as possible.

Three examples of delay functions are shown in Fig. 6.3.1. The Dirac delta function is shown in Fig. 6.3.1(a). Substituting this function, $K(t) = \delta(t-\tau)$, into Eq. 6.3.1 and utilizing the relationship

$$\int_{-\infty}^{\infty} \delta(z-\zeta)N(z)dz = N(\zeta)$$ (6.3.3)

we obtain the result

$$\frac{1}{N(t)}\frac{dN(t)}{dt} = a - bN(t-\tau) \quad .$$ (6.3.4)

This expression, of course, is the logistic differential equation with discrete time delay. If $\tau = 0$, we regain the ordinary logistic.

The delay function shown in Fig. 6.3.1(b) is the negative exponential function

$$K(t) = re^{-st} \quad .$$ (6.3.5)

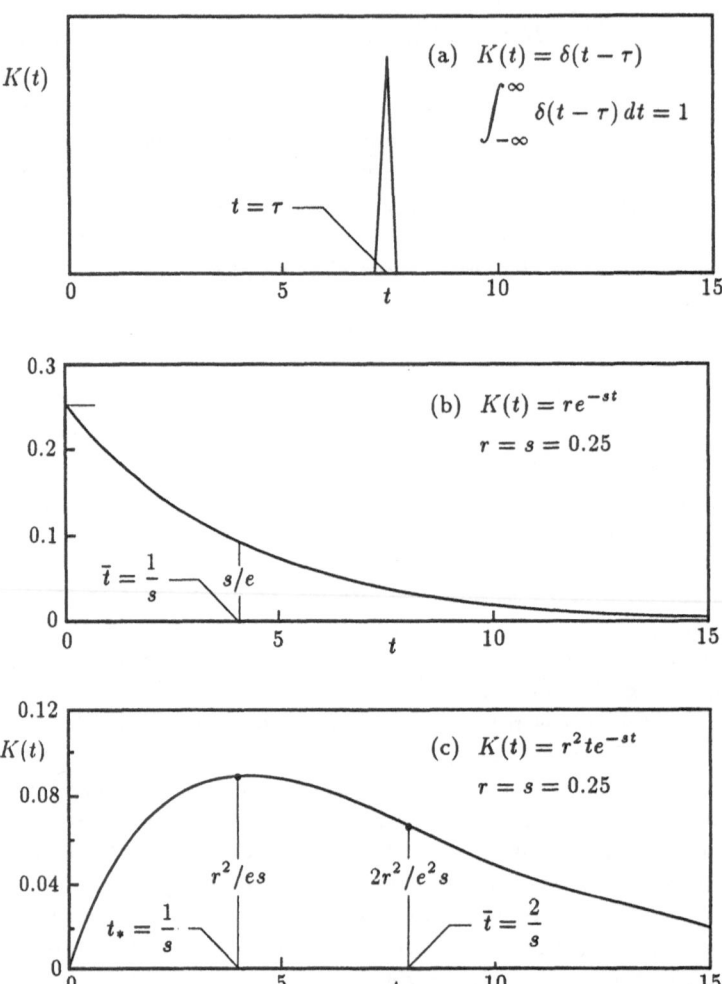

Fig. 6.3.1 Plots of three delay functions. (a) Dirac delta function; (b) negative exponential function; (c) gamma function

In this case, an effective or equivalent discrete delay time is given by the mean value, $\bar{t} = 1/s$. In Fig. 6.3.1(c), the delay function is a gamma function

$$K(t) = r^2 t e^{-st} \quad . \tag{6.3.6}$$

In this instance, the mean delay time is $\bar{t} = 2/s$; the modal delay time is $t_* = 1/s$.

6.3.2 Solution to the Integro-differential Equation

We begin our analysis by writing Eq. 6.3.1 in the following form

$$\frac{dN}{dt} = N\left(a - b\int_0^t K(t-z)N(z)dz\right) \quad , \tag{6.3.7}$$

in which the lower limit has been changed to $z = 0$ from the $z = -\infty$ value appearing in Eq. 6.3.1. This change implies that there is no "memory" of events prior to time $t = 0$. Rabatnov (1980) discusses the topic of "fading memory" delay functions.

Next, the negative exponential function, given by Eq. 6.3.5, is selected as the delay function. Borsellino and Torre (1974) employed this function in their analysis of the interaction effects between population growth and pollution. Accordingly, we obtain

$$\frac{dN}{dt} = N\left(a - br\int_0^t e^{-s(t-z)}N(z)dz\right) \quad . \tag{6.3.8}$$

Making the substitution

$$P = \int_0^t e^{-s(t-z)}N(z)dz \tag{6.3.9}$$

and differentiating this expression yields

$$\frac{dP}{dt} = N - sP \quad . \tag{6.3.10}$$

As a result, instead of the integro-differential equation given by Eq. 6.3.8, we have the following two first order differential equations

$$\frac{dN}{dt} = N(a - brP) \quad , \tag{6.3.11}$$

$$\frac{dP}{dt} = N - sP \quad . \tag{6.3.12}$$

It is convenient to express these equations in terms of the following dimensionless variables and parameters.

$$U = \frac{N}{N_*} \quad ; \qquad N_* = \frac{a}{b} \quad ; \qquad T = at \quad ;$$

$$M = bP \quad ; \qquad \sigma = \frac{s}{a} \quad ; \qquad \Gamma = \frac{r}{a} \quad . \tag{6.3.13}$$

Substitution of these quantities into Eqs. 6.3.11 and 6.3.12 gives

$$\frac{dU}{dT} = U(1 - \Gamma M) \quad , \tag{6.3.14}$$

$$\frac{dM}{dT} = U - \sigma M \quad . \tag{6.3.15}$$

Eliminating M from these equations yields the following second order non-linear differential equation

$$U\frac{d^2U}{dT^2} = \left(\frac{dU}{dT}\right)^2 - \sigma U\frac{dU}{dT} + \sigma U^2 - \Gamma U^3 \quad . \tag{6.3.16}$$

Before an attempt is made to solve this horrendous equation, we obtain the following special case solution.

Case I: $\sigma = 0$. In this case the negative exponential delay function, Eq. 6.3.5, takes on the constant value $K(t) = r$. Accordingly, Eq 6.3.16 reduces to

$$U\frac{d^2U}{dT^2} = \left(\frac{dU}{dT}\right)^2 - \Gamma U^3 \quad . \tag{6.3.17}$$

A first integration of this equation gives

$$\frac{dU}{dT} = U\sqrt{c - 2\Gamma U} \quad . \tag{6.3.18}$$

With the initial condition $U(0) = U_0$, and (since $M(0) = 0$) $(dU/dT)_0 = U_0$, we determine the integration constant to obtain

$$\frac{dU}{dT} = U\sqrt{(1 + 2\Gamma U_0) - 2\Gamma U} \quad . \tag{6.3.19}$$

This expression is integrated to yield the final answer

$$U = \frac{\lambda^2}{2\Gamma}\text{sech}^2(\tfrac{1}{2}\lambda T - \phi) \quad , \tag{6.3.20}$$

where

$$\lambda = \sqrt{1 + 2\Gamma U_0} \quad ; \qquad \phi = \text{arctanh}\frac{1}{\lambda} \quad . \tag{6.3.21}$$

Setting the first derivative of Eq. 6.3.20 equal to zero gives the maximum point

$$T_m = \frac{2\phi}{\lambda} \quad ; \qquad U_m = \frac{\lambda^2}{2\Gamma} \quad . \tag{6.3.22}$$

Equating the second derivative to zero provides the inflection points

$$T_i = \frac{2}{\lambda}\left(\phi \pm \text{arctanh}\frac{1}{\sqrt{3}}\right) \quad ;$$

$$U_i = \frac{\lambda^2}{3\Gamma} = \frac{2}{3}U_m \quad ; \qquad \left(\frac{dU}{dT}\right)_i = \pm\frac{\lambda^3}{3\sqrt{3}\Gamma} \quad . \tag{6.3.23}$$

Now at $T = 0$, we have $U = U_0$ and $(dU/dT) = U_0$. In addition, we establish that the curvature at $T = 0$ is $(d^2U/dT^2)_0 = U_0(1 - U_0)$. So when $\Gamma U_0 < 1$, the curvature is positive (concave upward) at the origin and when $\Gamma U_0 > 1$, the curvature is negative (concave downward).

It is not difficult to determine the other dependent variable in the problem, $M(T)$. Combining Eqs. 6.3.14 and 6.3.15, with $\sigma = 0$, we obtain

$$\frac{dM}{dT} = M(1 - \tfrac{1}{2}\Gamma M) + U_0 \quad . \tag{6.3.24}$$

As we saw in Section 2.7.4, this is the differential equation for logistic growth with immigration. Integrating either Eq. 6.3.24 or Eq. 6.3.15 (with $\sigma = 0$) yields

$$M = \frac{1}{\Gamma}[1 + \lambda\tanh(\tfrac{1}{2}\lambda T - \phi)] \quad . \tag{6.3.25}$$

The $\sigma = 0$ solution expressed by Eq. 6.3.20 has been given by Volterra (1934), MacDonald (1978), Reid (1952) and Small (1987). In their analyses, the delay function was assumed constant at the outset, i.e., the negative exponential delay function was not introduced. The acquisition of the final answer is not so complicated with that approach.

Case II: $\sigma \neq 0$. We return to Eq. 6.3.16. This equation is vastly simplified if we make the assumption, affecting the final term on the right-hand side, that $U^3 = U_r U^2$ where U_r is a constant reference value. We select U_0 as this value and give this selection a measure of justification later on. This step is not greatly different from the temperature approximation, $T^4 - T_0^4 = T_0^3(T - T_0)$, frequently made in problems involving radiation heat transfer.

With this assumption Eq 6.3.16 becomes

$$U\frac{d^2U}{dT^2} = \left(\frac{dU}{dT}\right)^2 - \sigma U\frac{dU}{dT} + (\sigma - \Gamma U_0)U^2 \quad . \tag{6.3.26}$$

Making the substitution $(dU/dT) = Uw$, this expression reduces to the simple first order linear differential equation

$$\frac{dw}{dT} + \sigma w = \sigma - \Gamma U_0 \quad . \tag{6.3.27}$$

Utilizing the same initial conditions as before gives $w(0) = 1$. This provides the solution for Eq. 6.3.27 which, in turn, gives

$$\frac{1}{U}\frac{dU}{dT} = 1 - \frac{\Gamma U_0}{\sigma}\left(1 - e^{-\sigma T}\right) \quad . \tag{6.3.28}$$

We pause at this point to obtain the solution for the other dependent variable, $M(T)$. Utilizing Eq. 6.3.14 to provide an expression for M and then substituting Eq. 6.3.28 into that expression yields

$$M = \frac{U_0}{\sigma} \left(1 - e^{-\sigma T}\right) \tag{6.3.29}$$

which is a confined exponential-type equation. Returning to Eq 6.3.28, a simple integration provides the final answer

$$U = U_0 \exp\left[(1 - \rho)T + \frac{\rho}{\sigma}\left(1 - e^{-\sigma T}\right)\right] \quad, \tag{6.3.30}$$

in which

$$\rho = \frac{\Gamma U_0}{\sigma} = \frac{br N_0}{as} \quad. \tag{6.3.31}$$

This is an interesting result. If $\rho = 0$, the growth is exponential. If $\rho = 1$, the growth follows the Gompertz equation which we examined in Section 3.1. If $\rho > 1$, the quantity U attains a maximum value and then goes to zero as $T \to \infty$.

The first derivative of Eq. 6.3.30 gives the coordinates of the maximum point

$$T_m = \frac{1}{\sigma}\log_e\left(\frac{\rho}{\rho - 1}\right) \quad; \quad U_m = U_0 e^{(1/\sigma)}\left(\frac{\rho - 1}{\rho}\right)^{\left(\frac{\rho-1}{\sigma}\right)} \quad. \tag{6.3.32}$$

The second derivative establishes the positions of the inflection points. It also allows us to compute the curvature at $T = 0$. The result is $(d^2U/dT^2)_0 = U_0(1 - \Gamma U_0)$ which is the same answer we obtained in the Case I: $\sigma = 0$ analysis. This means that selection of U_0 as the reference value, U_r, in the earlier assumption yields the same curvature at the origin in both the $\sigma = 0$ and $\sigma \neq 0$ analyses. Already we had imposed equal values of the ordinate and equal values of the slope at $T = 0$ in the two cases.

Several plots of $U(T)$ are displayed in Fig. 6.3.2 for selected values of ρ and σ. The curves shown in Fig. 6.3.2(a), corresponding to $\rho = 1$, are plots of the Gompertz equation.

6.3.3 An Illustration: Growth and Decline of the Populations of Northeast and East North Central American Cities

A. Central city populations

During the several decades following the Civil War, the industrial cities of the Northeast and East North Central regions of the United States increased rapidly in population. This period of fast growth of these cities coincided with the period of substantial increases in all aspects of industrialization, sharp rises in immigration from Europe, a fast-growing farm population and the associated rise in agricultural production, and phenomenal expansion of railways in the region.

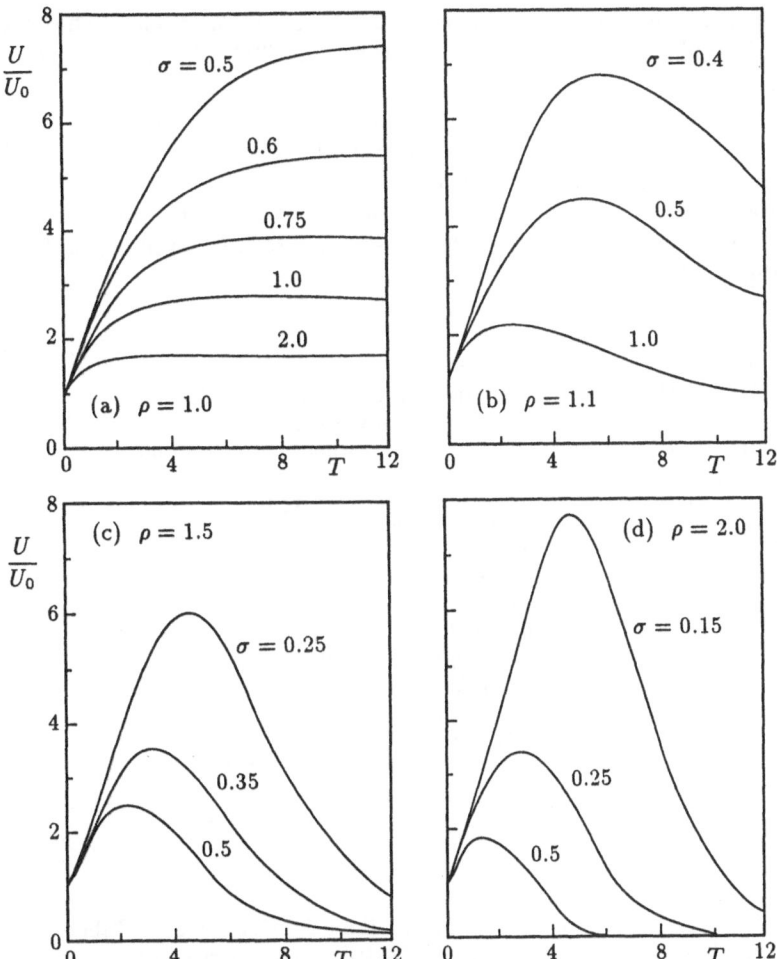

Fig. 6.3.2 Graphical displays of logistic growth with distributed time delay; negative exponential delay function

The rapid increases in populations of these cities continued until after World War I when the rate of increase began to decline. Towns and villages close to all of these cities began to appear with increasing frequency and to grow with increasing rates. Consequently, although the associated "metropolitan area" population of a particular city continued rapid growth, the population of the "central city" finally reached a maximum and then began to decrease. Of the 16 central cities we examine in our illustration, 14 attained maximum populations around 1950; the other two reached their peaks about a decade later. The central city populations of all 16 have continued to decline in subsequent years.

The 16 cities contained in our analysis are listed in Table 6.3.1. They are tabulated in rank order in population for the year 1900. At that time, these were the 16 largest cities in America, with two exceptions, and each had a population of 200 000 or more. The two exceptions are San Francisco (1900 population: 343 000) and New Orleans (287 000). These are excluded from our illustration for geographical reasons.

The total central city populations of all 16 are listed in Table 6.3.2 at decade intervals commencing with 1860. These data are displayed in graphical form in Fig. 6.3.3.

Table **6.3.1** Central city populations in rank order for year 1900. Populations are in thousands. References: Statistical Abstract of the United States, various editions, U.S. Department of Commerce, Government Printing Office, Washington D.C.

City	Population	City	Population
New York	3437	Buffalo	353
Chicago	1699	Cincinnati	326
Philadelphia	1294	Detroit	286
St. Louis	575	Milwaukee	285
Boston	561	Washington	279
Baltimore	509	Newark	246
Pittsburgh	452	Jersey City	206
Cleveland	382	Louisville	205

Table **6.3.2** Total central city population of the 16 cities listed in Table 6.3.1. By year, 1860-1986. References: Statistical Abstract of the United States, various editions, U.S. Department of Commerce, Government Printing Office, Washington, D.C.

Year	t years	Total population N, millions	Year	t years	Total population N, millions
1860	0	3.084	1930	70	20.960
1870	10	4.395	1940	80	21.733
1880	20	5.905	1950	90	23.265
1890	30	8.292	1960	100	22.484
1900	40	11.095	1970	110	21.564
1910	50	14.310	1980	120	18.638
1920	60	17.531	1986	126	18.425

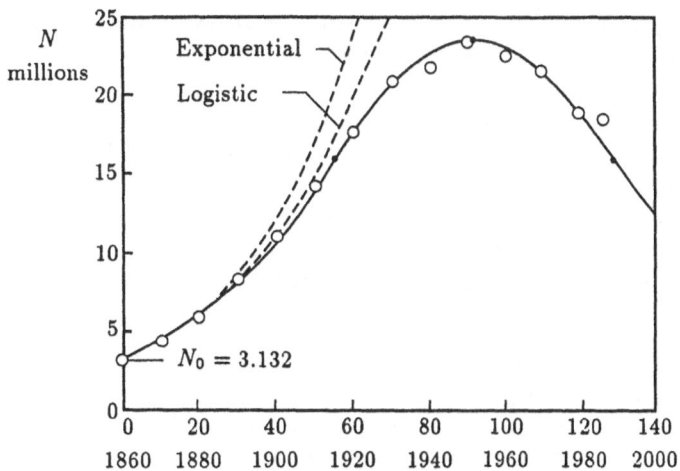

Fig. 6.3.3 Growth and decline of central city populations of 16 cities in the Northeast and East North Central regions of the United States

Frameworks for fitting the data of Table 6.3.2 were provided by (a) Case I: $\sigma = 0$, Eq. 6.3.20 and by (b) Case II: $\sigma \neq 0$, Eq. 6.3.30. Interestingly, the former, i.e., the "sech-squared" solution, gave the better fit. The following numerical values of the various parameters were determined:

$$a = 0.0337 \qquad N_0 = 3.132 \qquad s = 0$$

$$br = 0.0000279 \qquad \lambda = 1.0741 \qquad \phi = 1.6657$$

As indicated in Eq. 6.3.8, the two parameters b and r appear as a product say, $c = br$, and cannot be resolved without further information. Even so, it is desirable to retain them as separate constants; by doing so, we can better remember what they represent.

Substitution of the above-indicated numerical values into the dimensional form of Eq. 6.3.20 yields the solid line shown in Fig. 6.3.3. The observed values of N are in good agreement with the computed curve, at least for values of $t < 120$ (year: 1980). The critical points, calculated from Eqs. 6.3.22 and 6.3.23, are shown as solid dots in the figure. The numerical values are the following

$$t_m = 92.0 \quad (1952.0) \qquad N_m = 23.5$$

$$t_i = 55.65 \quad (1915.65) \quad \text{and} \quad 128.42 \quad (1988.42)$$

$$N_i = 15.67 \qquad (dN/dT)_i = 0.327$$

On the basis of these calculations we make the following observations

- The combined population of the central cities of these 16 major cities of the Northeast and East North Central States reached a maximum of approximately 23.5 million in 1952.
- In mid-1915, the combined population was around 15.7 million and was increasing at the maximum rate of about 327 000 per year. This corresponds to a specific growth rate of 0.0209.
- In mid-1988, the combined population was again approximately 15.7 million and was decreasing at the maximum rate of 327 000 annually. These are computed values. Quite possibly, observed values will reveal a smaller rate of decrease.

In Fig. 6.3.3, one of the dashed lines identifies the exponential growth curve, based on $a = 0.0337$ and $N_0 = 3.132$.

The other dashed line designates a logistic growth curve. This curve was obtained by employing the ordinary logistic equation

$$N = \frac{N_*}{1 + \left(\frac{N_*}{N_0} - 1\right) e^{-at}} \tag{6.3.33}$$

to compute the value of N_*. The data points corresponding to $t = 0$ through $t = 50$, i.e., prior to the inflection point shown in the figure, yield $N_* = 99.33$.

B. Metropolitan area populations

However well defined and socio-economically important a "central city" may be, it is the "metropolitan area" population that provides the most meaningful quantitative scale for the size of a city.

Accordingly, an analysis was made of the same 16 cities utilizing the Standard Metropolitan Statistical Area (SMSA) populations of these cities. (Newark and Jersey City were included under the New York-Northern New Jersey-Long Island SMSA commencing with 1970.)

These metropolitan area populations are listed in Table 6.3.3 at 10-year intervals beginning with 1940. They are shown as the solid points in Fig. 6.3.4; the open points are the central city data of Table 6.3.2.

Table **6.3.3** Total metropolitan area population of the 16 cities listed in Table 6.3.1. By year; 1940-1986. References: Statistical Abstract of the United States, various editions, U.S.Department of Commerce, Government Printing Office, Washington, D.C.

Year	t years	Total population N, millions	Year	t years	Total population N, millions
1940	80	34.038	1970	110	59.006
1950	90	37.741	1980	120	58.397
1960	100	45.296	1986	126	59.324

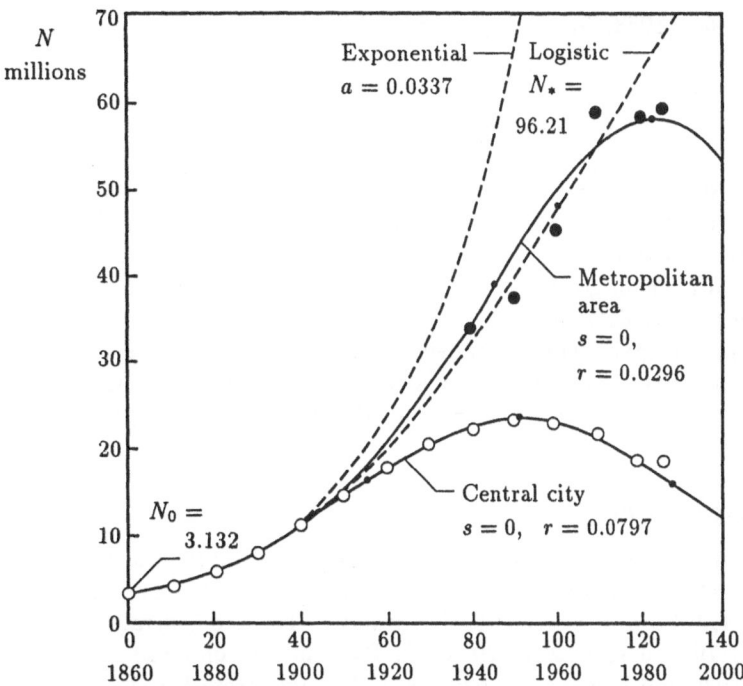

Fig. 6.3.4 Growth and decline of central city and metropolitan area populations of 16 cities in the Northeast and East North Central regions

The metropolitan area population data displayed in Fig. 6.3.4 show an interesting extension of the central city data corresponding to $t = 50$ (1910) and prior years. This is not surprising; the central cities were, in fact, the metropolitan areas until 1910 or so.

The central city data for $t \leq 50$, along with the tabulated metropolitan area data, were analyzed with the frameworks of (a) the sech-squared distribution and (b) the ordinary logistic distribution.

(a) Sech-squared distribution. The following numerical values of the parameters contained in Eq. 6.3.20 were calculated:

$$a = 0.0337 \qquad N_0 = 3.132 \qquad s = 0$$

$$br = 0.00001035 \qquad \lambda = 1.0281 \qquad \phi = 2.1389$$

The computed curve is shown as the solid line in the upper part of Fig. 6.3.4. The critical points, computed from Eqs. 6.3.22 and 6.3.23, have the following values:

$t_m = 123.45$ (1983.45) $N_m = 58.0$

$t_i = 85.46$ (1945.46)   and   161.48   (2021.48)

$N_i = 38.67$ $(dN/dT)_i = 0.773$

(b) Logistic distribution. As before, the logistic equation, Eq. 6.3.33, was employed to determine N_*. This time, all of the metropolitan area data were utilized as well as the central city data for $t < 50$. The resulting average value, $N_* = 96.21$, is in good agreement with the value $N_* = 99.33$ obtained from the central city data alone.

The computed logistic curve is shown as the dashed line in Fig. 6.3.4. The inflection point of the curve is located at $t_i = 100.65$ (1960.65), $N_i = 48.11$. Logistic growth projections are: $t = 135$ (1995), $N = 70.13$ and $t = 140$ (2000), $N = 76.03$.

C. Concluding remarks

We are now in a position to compute one more thing. From the calculated values, $N_* = a/b = 96.21$ and $a = 0.0337$, the value of the crowding coefficient is $b = 0.000350$. So now we can resolve b and r. For the central city data, $br = 0.0000279$ and so $r = 0.0797\,yr^{-1}$. For the metropolitan area data, $br = 0.00001035$ and hence $r = 0.0296\,yr^{-1}$. We note from Eq. 6.3.8 that in the limiting case, $r = 0$, growth is exponential.

The sech-squared distribution is symmetrical about its maximum point. Accordingly, this framework does confirm the observed decline in central city population. However, it predicts a 16-central city total population of 11.95 million by the year 2000, which is doubtful, and 5.82 million by 2025, which is unlikely.

It is premature to forecast whether the metropolitan area population will increase logistically or decline according to the sech-squared equation. Something in between seems probable. In the long run, it may be that both sets of data, central city and metropolitan area, can be better described by the non-symmetrical Case II: $\sigma \neq 0$ model expressed by Eq 6.3.30.

It might be interesting to extend this illustration to encompass a demographic stage three: growth and perhaps decline of the "megalopolis belts" originated by the 16 cities.

A concluding word. As we shall see in more detail in the next section, the integral term of the integro-differential equation, Eq. 6.3.1, represents an accumulating quantity which retards and eventually decreases the magnitude of N. In his early studies on mathematical biology, Kostitzin (1939) describes this accumulation effect as a toxicity created by catabolic production. Apparently there is an analogous effect in demography.

6.4 Distributed Time Delay: Delay Integral in a Pollution Term

6.4.1 The Integro-differential Equation and Its Solution

We extend our examination of distributed time delays by considering the following integro-differential equation

$$\frac{1}{N}\frac{dN}{dt} = a - bN - c\int_0^t K(t-z)N(z)dz \quad . \tag{6.4.1}$$

Comparing this expression with Eq 6.3.1, it is seen that the delay integral now appears, not in the crowding term, but in an arbitrarily included third term which we will call a toxicity or pollution term. We note that it also represents a kind of harvesting or emigration.

The relationship of Eq. 6.4.1 was the starting point in early studies by Volterra (1934) and Kostitzin (1939) concerning applications of integro-differential equations in biological phenomena. Over the years, considerable attention has been given to this equation including analyses by Davis (1962), who allowed c to be either positive or negative, Borsellino and Torre (1974), MacDonald (1978), Cohen et al. (1979), Small (1987) and numerous others.

To make the problem as simple as possible, we select the delay function $K(t) = 1$. Then Eq. 6.4.1 becomes

$$\frac{dN}{dt} = N\left(a - bN - c\int_0^t N(z)dz\right) \quad . \tag{6.4.2}$$

As before,

$$P = \int_0^t N(z)dz \tag{6.4.3}$$

and differentiating gives

$$\frac{dN}{dt} = N(a - bN - cP) \quad ; \quad \frac{dP}{dt} = N \quad . \tag{6.4.4}$$

Dividing the first of these equations by the second provides the following first order linear equation

$$\frac{dN}{dP} + bN = a - cP \tag{6.4.5}$$

With the initial conditions $N(0) = N_0$ and $P(0) = 0$, the solution is

$$N = N_{*c} - \frac{c}{b}P - (N_{*c} - N_0)e^{-bP} \quad , \tag{6.4.6}$$

where

$$N_{*c} = N_* \left(1 + \frac{c}{ab}\right) \quad ; \quad N_* = \frac{a}{b} \quad . \tag{6.4.7}$$

At this point we examine two special cases. Setting $c = 0$ in Eqs. 6.4.6 and 6.4.7 and integrating we obtain

$$P = \frac{1}{b} \log_e \left[1 + \frac{N_0}{N_*} (e^{at} - 1)\right] \quad . \tag{6.4.8}$$

Differentiating this expression yields, as we would expect, the ordinary logistic equation

$$N = \frac{N_*}{1 + \left(\frac{N_*}{N_0} - 1\right) e^{-at}} \quad . \tag{6.4.9}$$

Alternatively, we set $b = 0$ in Eqs. 6.4.6 and 6.4.7. Expanding the exponential term in a Taylor series we get

$$\frac{dP}{dt} = P(a - \tfrac{1}{2}cP) + N_0 \quad . \tag{6.4.10}$$

This result is the same as the dimensional form of Eq. 6.3.24, with br of that equation replaced by c. Consequently, as we would anticipate, Eq. 6.4.10 leads to the sech-squared solution given by Eq. 6.3.20.

Writing Eq. 6.4.6 in dimensionless form with the following variables and parameters

$$U = \frac{N}{N_*} \quad ; \quad M = bP \quad ; \quad T = at \quad ; \quad \Gamma = \frac{c}{ab} \quad , \tag{6.4.11}$$

we obtain

$$U = \frac{dM}{dT} = (1 - \Gamma) - \Gamma M - (1 + \Gamma - U_0)e^{-M} \quad , \tag{6.4.12}$$

or alternatively

$$T = \int_0^M \frac{dM}{(1 - \Gamma) - \Gamma M - (1 + \Gamma - U_0)\exp(-M)} \quad . \tag{6.4.13}$$

It is not possible to acquire a closed form solution to this equation. However, numerical integration is straightforward and provides $M = M(T)$ without any difficulty.

We calculate the maximum value, U_m, by utilizing the derivative of Eq. 6.4.12 to obtain

$$M_m = \log_e \left(1 + \frac{1 - U_0}{\Gamma}\right) \quad , \tag{6.4.14}$$

where M_m corresponds to

$$U_m = 1 - \Gamma \log_e \left(1 + \frac{1 - U_0}{\Gamma}\right) \quad , \tag{6.4.15}$$

which is the same as the expression given by Volterra and Kostitzin (1939).

It is worthwhile to introduce a carrying capacity into the analysis. In Eq. 6.4.2, we absorb the pollution term into the growth term to create a time dependent growth coefficient $a(t)$. Dividing this by the crowding coefficient, b, we obtain a time dependent carrying capacity $N_*(t)$. Defining $U_*(T) = N_*(t)/N_{*0}$, we establish the relationship

$$U_*(T) = 1 - \Gamma M \quad , \tag{6.4.16}$$

where U_* is a dimemsionless carrying capacity.

If $\Gamma = 0$ (i.e., $c = 0$), we have the ordinary logistic equation for which $U_* = 1$. When $U_* = 0$, i.e., when the carrying capacity has been depleted, M takes on the limiting value

$$M_e = \frac{1}{\Gamma} = \int_0^{T_e} U(T)dT \tag{6.4.17}$$

and from Eq. 6.4.12, the corresponding value of U is

$$U_e = \Gamma - (1 + \Gamma - U_0)e^{-1/\Gamma} \quad . \tag{6.4.18}$$

The coordinates $U_* = 0$, $T = T_e$, define the depletion or exhaustion point of the carrying capacity.

Differentiating Eq. 6.4.16 yields

$$\frac{dU_*}{dT} = -\Gamma \frac{dM}{dT} = -\Gamma U \quad . \tag{6.4.19}$$

Thus the slope of the carrying capacity curve is always negative and is proportional to the ordinate of the growth curve. We also establish that the ordinate of the inflection point of U_* is equal to U_m.

6.4.2 An Approximate Sech-Squared Solution

Although we cannot obtain an exact answer to Eq. 6.4.13, it is possible to acquire an approximate solution. Expanding the exponential term in the denominator in a Taylor series and rearranging we get

$$T = \int_0^M \frac{dM}{U_0 + (1 - U_0)M - \frac{1}{2}(1 + \Gamma - U_0)M^2} \tag{6.4.20}$$

which is essentially the same as Eq. 6.3.24. The solution to Eq. 6.4.20 is

$$M = \frac{\beta}{2\gamma}[1 + \lambda_* \tanh(\tfrac{1}{2}\beta\lambda_* T - \phi)] \tag{6.4.21}$$

where

$$\alpha = U_0 \quad ; \qquad \beta = 1 - U_0 \quad ; \qquad \gamma = \tfrac{1}{2}(1 + \Gamma - U_0) \quad ;$$

$$\Gamma = \frac{c}{ab} \quad ; \qquad \lambda_* = \sqrt{1 + \frac{4\alpha\gamma}{\beta^2}} \quad ; \qquad \phi = \operatorname{arctanh}\frac{1}{\lambda_*} \quad . \tag{6.4.22}$$

Differentiating this expression gives

$$U = \frac{\beta^2 \lambda_*^2}{4\gamma} \operatorname{sech}^2(\tfrac{1}{2}\beta\lambda_* T - \phi) \quad . \tag{6.4.23}$$

The maximum point of the growth curve is located at

$$T_{\mathrm{m}} = \frac{2\phi}{\beta\lambda_*} \quad ; \qquad U_{\mathrm{m}} = \frac{\beta^2 \lambda_*^2}{4\gamma} \quad . \tag{6.4.24}$$

6.4.3 Numerical Examples

For some numerical examples we select the following values: $a = 0.25$, $b = 0.0025$, $N_0 = 10$, $N_{*0} = 100$. Three values of the parameter c are employed, as tabulated below. Magnitudes of λ_* and ϕ are listed; these are utilized in the approximate solution given by Eq. 6.4.23.

Example	c	Γ	λ_*	ϕ
(i)	0.0001	0.16	1.1233	1.4232
(ii)	0.001	1.60	1.2717	1.0618
(iii)	0.01	16.0	2.2743	0.4718

The results of computations are shown in Fig. 6.4.1. The solid lines correspond to the answers provided by numerically integrating Eq. 6.4.13. The dashed lines refer to the approximate solution given by Eq. 6.4.23. It is noted that the two solutions come into closer agreement with increasing values of Γ. For comparison, the ordinary logistic is shown.

The solid lines shown in Fig. 6.4.2 are the same as those displayed in Fig. 6.4.1. The dashed lines are the carrying capacity curves, U_*, given by Eq 6.4.16. Tabulated below are the values of T, U, and M at the maximum points of U and exhaustion points of U_* for the three examples.

Γ	T_{m}	U_{m}	M_{m}	T_{e}	U_{e}	M_{e}
0.16	4.380	0.697	1.890	15.505	0.158	6.25
1.60	2.138	0.286	0.446	2.769	0.262	0.625
16.0	0.535	0.125	0.0547	0.536	0.124	0.0625

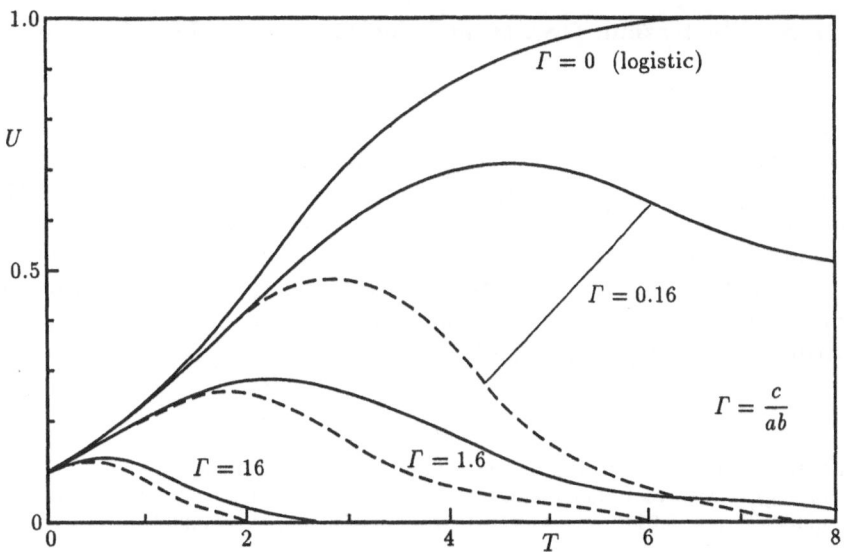

Fig. 6.4.1 Logistic growth with distributed time delay. Exact (solid) and approximate (dashed) solutions

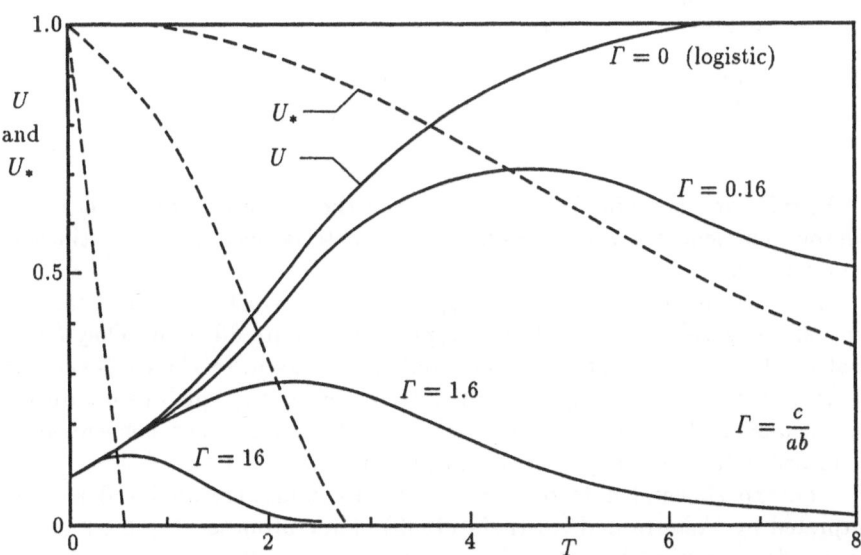

Fig. 6.4.2 Growth curves, U, and carrying capacity curves, U_*, for various values of Γ

6.4.4 An Illustration: Growth and Self-Contamination of Bacteria

Quite a number of years ago, Régnier and Lambin (1938) carried out experiments involving the growth of *Bacillus coli* in confined saline media containing various initial concentrations of a peptone nutrient.

In their experiments, the nutrient concentration ranged from $q = 0$ to 20.0 grams per liter. The temperature was maintained constant at 37°C; the initial bacteria concentration in the 12 series of experiments was approximately $N_0 = 400\,000$ per cubic centimeter. Concentrations, N, were determined from $t = 0$ to as long as $t = 21$ days. Measurements obtained in a typical experiment are listed in Table 6.4.1.

Table 6.4.1 Number of *Bacillus coli* vs. time. Experiment 8. Peptone concentration, $q = 1.0$ g/l. Data of Régnier and Lambin (1938). Note: $a = 1.13$, $b = 0.005$, $c = 0.000067$, $\Gamma = 0.0119$, $N_{*0} = a/b = 226$, $N_0 = 0.485$, $U_0 = 0.00215$

Time, t hours	N, million per cubic cm	Time, t hours	N, million per cubic cm
0	0.485	30	206.0
2	0.800	48	162.0
4	28.9	72	88.0
6	67.5	120	45.0
8	126.6	192	21.2
24	214.0		

Régnier and Lambin observed that when the initial concentration of the peptone nutrient was zero, the number of bacteria decreased to extinction very rapidly.

However, in the main series of experiments in which the initial concentration was varied up to 20 g/l, the response was quite different. They found that the bacteria concentration immediately increased, reached a maximum value within one to four days, and then decreased to very small values during the ensuing 14 to 21 days. They observed that the maximum concentration increased with increasing initial amounts of nutrient.

The experimental data of Regnier and Lambin were analyzed and interpreted by Volterra and Kostitzin (1939). They proposed Eq. 6.4.2 as the appropriate framework to describe the phenomenon. They reasoned that an otherwise logistic growth of the bacteria would be retarded and, sooner or later, decrease because of the exhaustion of the nutrient source and the accumulation of toxic substances generated by the bacteria. This effect of self-contamination, reflected in the third term of Eq. 6.4.2, provides a logical explanation of the observed growth and subsequent decline of the *B. coli* concentration.

From the experimental data of Régnier and Lambin, including the initial and maximum values of bacteria concentration, N_0 and N_m, Volterra and Kostitzin determined the magnitudes of the growth coefficient, a, the crowding coefficient, b, and the pollution coefficient, c. They obtained Eq. 6.4.15, in its dimensional form, and demonstrated the very close agreement between calculated and observed values of maximum concentration, N_m. They also noted that "the curve of these maxima as a function of the peptone concentration appears to be a logistic curve." Indeed, the plot shown in Fig. 6.4.3 does indicate this.

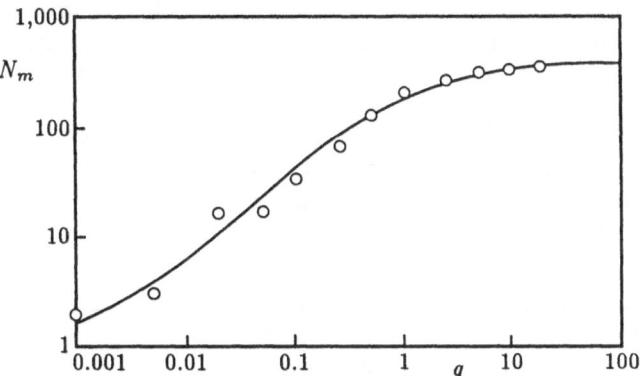

Fig. 6.4.3 Correlation between maximum concentration of *Bacillus coli* and initial amount of nutrient. Data of Régnier and Lambin (1938)

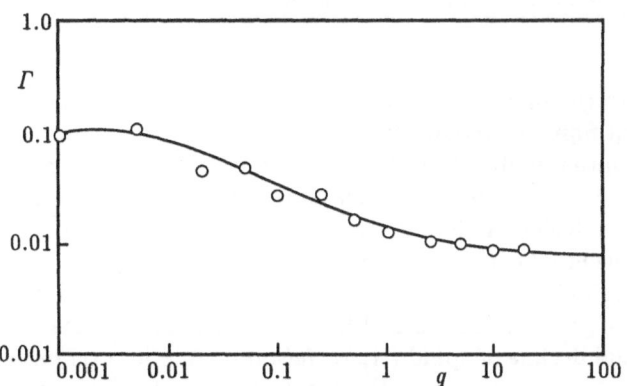

Fig. 6.4.4 Correlation between the parameter $\Gamma = c/ab$ and initial amount of nutrient. Data of Régnier and Lambin (1938)

As Fig. 6.4.4 displays, there is also a good correlation between the initial nutrient concentration, q, and the dimensionless parameter, $\Gamma = c/ab$.

The mathematical framework provided by Volterra and Kostitzin was employed to analyze the data of Régnier and Lambin. Experiment 8, with $q = 1.0$ g/l and listed in Table 6.4.1, serves as an example. Employing the calculated values of a, b, and c, and the dimensionless parameters of Eq. 6.4.11, the experimental data were computed as displayed in Fig. 6.4.5.

The solid line in this figure was obtained by a numerical integration of Eq. 6.4.13. It is noted that there is good agreement between the experimental data and the theoretical relationship. Although the calculated time is somewhat in error, the computed maximum value, U_m, agrees closely with the observed value. This is equally true for the other 11 experiments. Fig. 6.4.6 shows the magnitudes of the observed maxima; the solid curve is Eq. 6.4.15 with the assumption that the initial bacteria concentration, U_0, is negligibly small.

Finally, the dashed line in Fig. 6.4.5 shows the carrying capacity curve, $U_*(T)$, given by Eq. 6.4.16. When $U_* = 0$ we have, from Eq. 6.4.17, the limiting value, $M_e = 1/\Gamma$; this quantity represents the integrated value of U from $T = 0$ to $T = T_e$, the exhaustion time of the carrying capacity.

For Experiment 8, we determine the following values for the maximum point of the growth curve and exhaustion point of the carrying capacity curve.

Maximum point	Exhaustion point
$T_m = 12.43$	$T_e = 389.52$
$t_m = 11.0$ hours	$t_e = 344.7$ hours
$U_m = 0.947$	$U_e = 0.0119$
$N_m = 214.02$ million bacteria	$N_e = 2.689$ million bacteria
$M_m = 4.44$	$M_e = 84.03$
$P_m = 888$ million bacteria hours	$P_e = 18\,806$ million bacteria hours

A brief conclusion to the illustration. We have seen that when there is a source of nutrient, albeit finite in amount, the bacteria concentration initially increases and does so quite rapidly. Then, because of the combined effects of creation of toxic products and depletion of the nutrient, the concentration reaches a maximum and begins a slow decrease toward exhaustion of the carrying capacity and eventual extinction of the bacteria.

A study somewhat similar to our bacteria illustration was carried out by Kindlmann (1985). Using a mathematical framework similar to ours, he examined the growth and decline of aphid (*Aphis fabae*) populations on bean plants. After reaching a maximum value, the total number of aphids in a temperature-controlled cage declined sharply due to deterioration of their environment caused by honeydew excretions of the aphids. The dew formed a cover on the leaf surfaces which restricted or prevented feeding and movement by the aphids; starvation followed.

There are other analogous studies in ecology. For example, Klein (1968) has considered the phenomenon of the introduction, increase, and crash of

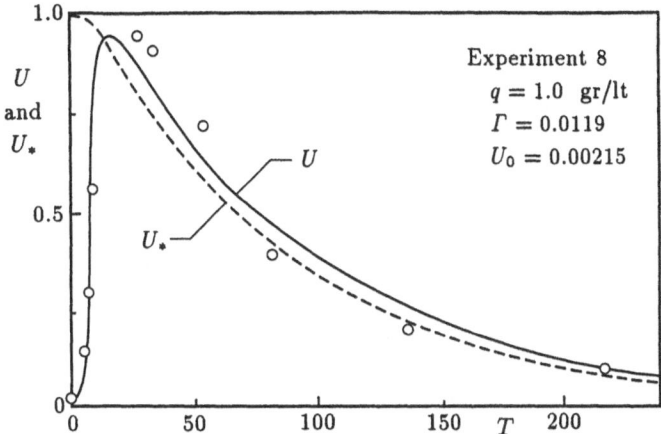

Fig. 6.4.5 Comparison of computed growth curve, U (solid), and experimental results; carrying capacity curve U_* (dashed). Experiment 8. Data of Régnier and Lambin (1938)

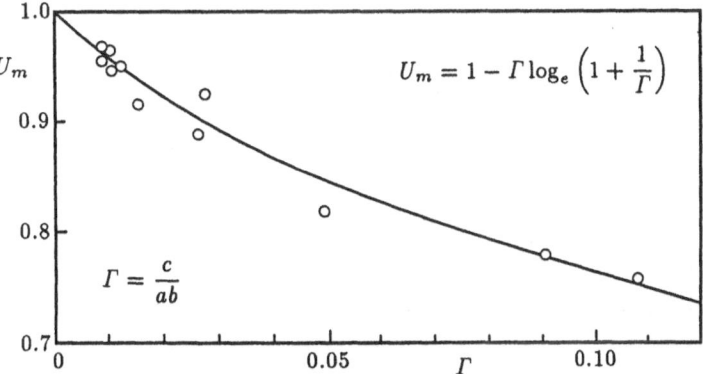

Fig. 6.4.6 Comparison of computed and experimental results for maximum concentration of *Bacillus coli* as a function of the parameter Γ. Data of Régnier and Lambin (1938)

reindeer on St. Matthew Island in the Bering Sea. Caughley (1970) has examined the growth and decline of wild goat populations in New Zealand. There are many other examples. Although these ecological analogs cannot always be described nor interpreted in precisely the way we have done with the *Bacillus coli* experiments, there are obviously similarities.

Perhaps the closest analogy of all is mankind's own environment: our depletion of non-renewable resources with simultaneous creation of self-contaminating non-degradable products.

7. Phenomena with Spatial Diffusion

7.0.1 Introduction

In all of the previous sections we have considered only those phenomena in which the growth or transfer process has been a function of a single independent variable: time. This point of view resulted in our examination of a number of problems of mathematical interest and practical importance.

Now we make our task considerably more complicated by introducing another independent variable: space. So instead of seeking solutions of the form $N = N(t)$, we now look for answers in the category $N = N(x, y, z, t)$, where the additional independent variables define, in this case, a rectangular coordinate system. Clearly, it is necessary to incorporate these spatial variables if we are to obtain answers to many kinds of problems. For example, an epidemic not only sweeps through a given locale on a day-to-day or week-to-week basis but also, and in general concurrently, spreads spatially through an entire community or across a wide region. The temperature of the soil some distance below the ground surface depends not only on the time of the day or season of the year but also on the depth of the particular point in question.

The rate of adoption of a new kind of corn seed developed in a certain part of Iowa is one matter; the temporal-spatial diffusion of this new technology to adjacent states and beyond is another.

During recent years, much progress has been made by many investigators on problems of spatial diffusion in growth and transfer processes. It is safe to say that a great deal of continued effort will be needed in the future.

7.0.2 The Diffusion Equation

We begin our analysis with the simplest model; even this one becomes very complicated soon enough. We introduce Fick's law of diffusion

$$q = -D\frac{\partial N}{\partial x} \quad . \tag{7.0.1}$$

This equation says: the amount, q, of a diffusing substance, N, crossing unit area per unit time is proportional to the gradient of the substance in the x-direction. The parameter D is called the diffusion coefficient or simply the diffusivity; the negative sign is utilized because diffusion occurs in the direction of decreasing gradient.

A similar relationship occurs in heat conduction based on Fourier's law

$$f = -K\frac{\partial \theta}{\partial x} \quad , \tag{7.0.2}$$

where f is the flux of heat in the x-direction, θ is the temperature, and K is the thermal conductivity.

Expressions identical to Eqs. 7.0.1 and 7.0.2 describe momentum transfer in a viscous fluid (Newton's law), current flow in an electric field (Ohm's law), and fluid flow in a porous medium (Darcy's law).

Consider an elemental volume of unit cross-sectional area and length, dx, in the direction of diffusion. A conservation statement or mass balance gives

$$\frac{\partial N}{\partial t} = \frac{\partial}{\partial x}\left(D\frac{\partial N}{\partial x}\right) \quad , \tag{7.0.3}$$

where, in general, $D = D(x, t; N)$. If D is a constant we obtain

$$\frac{\partial N}{\partial t} = D\frac{\partial^2 N}{\partial x^2} \quad . \tag{7.0.4}$$

In its simplest form, this is the so-called diffusion equation or equation of heat conduction. Its solution, $N = N(x,t)$, can be obtained if we know the initial and boundary conditions. This equation is easily extended to the three dimensions of a rectangular coordinate system

$$\frac{\partial N}{\partial t} = D\left(\frac{\partial^2 N}{\partial x^2} + \frac{\partial^2 N}{\partial y^2} + \frac{\partial^2 N}{\partial z^2}\right) \quad . \tag{7.0.5}$$

A mass balance on an elemental cylindrical volume gives the following form of the diffusion equation in an axially-symmetric cylindrical system

$$\frac{\partial N}{\partial t} = D\left(\frac{\partial^2 N}{\partial r^2} + \frac{1}{r}\frac{\partial N}{\partial r}\right) \quad , \tag{7.0.6}$$

where $r = (x^2 + y^2)^{1/2}$ is measured from the origin, $r = 0$.

Less frequent are the occasions to consider diffusion phenomena in a spherical system. Nevertheless, for completeness, the following is the diffusion equation in spherical coordinates

$$\frac{\partial N}{\partial t} = D\left(\frac{\partial^2 N}{\partial r^2} + \frac{2}{r}\frac{\partial N}{\partial r}\right) \quad , \tag{7.0.7}$$

where $r = (x^2 + y^2 + z^2)^{1/2}$.

In addition to the diffusive mechanism that spreads out a substance in the x- or r-direction, suppose there is a bulk movement of the medium in which the diffusion process is occurring, e.g., a diffusing fluid is in motion. This bulk movement is termed convection, or advection, or drift. Then the one-dimensional equation becomes

$$\frac{\partial N}{\partial t} + u_0 \frac{\partial N}{\partial x} = D \frac{\partial^2 N}{\partial x^2} \quad , \tag{7.0.8}$$

where u_0 is the velocity of the medium.

Finally, suppose that the substance is being added to or is being removed from the medium at a rate described by $F(N)$. Then the diffusion equation becomes

$$\frac{\partial N}{\partial t} + u_0 \frac{\partial N}{\partial x} = D \frac{\partial^2 N}{\partial x^2} + F(N) \quad . \tag{7.0.9}$$

It may be helpful to present the following generalized equation

$$\frac{\partial N}{\partial t} + u_0 \left(\frac{r_0}{r} \right)^m \frac{\partial N}{\partial r} = D \left(\frac{\partial^2 N}{\partial r^2} + \frac{m}{r} \frac{\partial N}{\partial r} \right) + F(N) \quad , \tag{7.0.10}$$

where, respectively, $m = 0$ corresponds to the rectilinear case (with $x = r$), $m = 1$ to the cylindrical case, and $m = 2$ to the spherical case. In this equation, the quantities r_0 and u_0 refer to a reference distance and a reference velocity. The four terms of this equation can be designated, respectively, as the (a) storage, (b) convection, (c) diffusion, and (d) reaction terms.

In the following sections we shall obtain various special case solutions to Eq. 7.0.10 and examine some examples and illustrations. A great many problems and solutions to Eq. 7.0.10 are given by Carslaw and Jaeger (1959) in connection with heat conduction and by Crank (1975) and Jost (1952) with regard to diffusion in solids, liquids and gases. A good deal of information is provided by Cannon (1984) on numerous aspects of the diffusion equation.

Many of the analytical models and mathematical frameworks we shall be considering have their rational bases in the laws of Fick and Fourier. It should be mentioned, however, that when we refer to diffusion in ensuing sections, we should have in mind a broader sense of the word. It is not only molecular diffusion or its mathematical equivalent that concerns us; we are equally interested in phenomena we call dispersion, dispersal, spreading, dissemination, and the like.

7.1 Diffusion from Instantaneous Sources

7.1.1 Plane Source, Line Source and Point Source

We start with the simplest problem. Suppose that, at time $t = 0$, a total amount of substance, M, is released instantaneously from a plane source of unit area located at the origin, $x = 0$, in the $y - z$ plane and allowed to diffuse from the plane in both the positive and negative x-directions. Then the solution to Eq. 7.0.4 is

$$N = \frac{M}{2\sqrt{\pi Dt}} \exp\left(-\frac{x^2}{4Dt}\right) \quad . \tag{7.1.1}$$

This is the density function of the normal probability distribution. We observe that the substance spreads out in such a way that the variance of the density distribution is $\sigma^2 = 2Dt$.

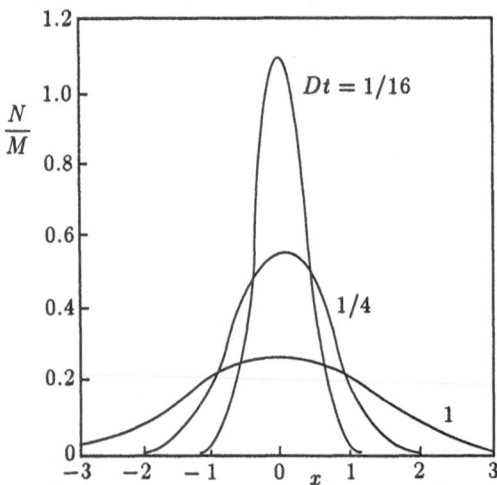

Fig. 7.1.1 Diffusion from a plane source for various values of Dt

Plots of concentration distributions for diffusion from a plane source are shown in Fig. 7.1.1 for various values of Dt. When $t = 0$, the concentration distribution is essentially the Dirac delta function we considered in Section 6.3.1. As shown in the figure, the distribution curves flatten out with increasing values of Dt because of diffusion. Inflection points occur at $x_c = \pm\sqrt{2Dt}$, $(N/M)_c = \frac{1}{2}\sqrt{\pi e Dt}$. A word of caution: neither Dt nor N/M are dimensionless quantities: be careful about units.

If diffusion in the negative x-direction is prevented by an insulated boundary, (i.e., $\partial N/\partial x = 0$ at $x = 0$), then the amount diffusing in the positive direction is doubled

$$N = \frac{M}{\sqrt{\pi Dt}} \exp\left(-\frac{x^2}{4Dt}\right) \quad . \tag{7.1.2}$$

Next, we consider the instantaneous release of an amount of substance, M, from a line source at $r = 0$ of unit length along the vertical z-axis. The substance diffuses in the radial direction. In this cylindrical geometry, the solution to Eq. 7.0.6 is

$$N = \frac{M}{4\pi Dt} \exp\left(-\frac{r^2}{4Dt}\right) \quad . \tag{7.1.3}$$

Finally, suppose that the amount of substance, M, is released instantaneously from a point source located at $r = 0$. In this spherical system the solution to Eq. 7.0.7 is

$$N = \frac{M}{8(\pi Dt)^{3/2}} \exp\left(-\frac{r^2}{4Dt}\right) \quad .$$

(7.1.4)

We establish from Eqs 7.1.3 and 7.1.4 that the concentration distributions for diffusion from a line source and a point source flatten out very quickly with increasing distance, as shown in Fig. 7.1.2. This is logical; the diffusing substance has correspondingly more space to occupy.

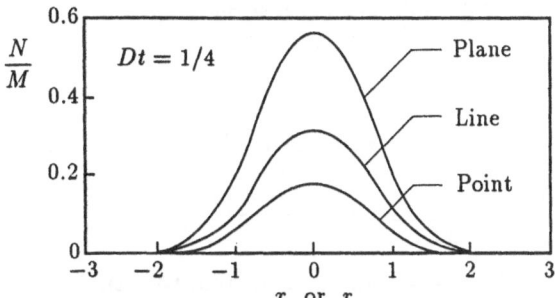

Fig. 7.1.2 Comparison of diffusion from a plane source, a line source, and a point source

7.1.2 Rectilinear Diffusion with Convection

We now examine a phenomenon which has many practical applications including problems involving the convection and dispersion of pollutants in the atmosphere, in rivers, and in underground aquifers. It has numerous applications in various types of chemical engineering processes and in the transport of petroleum products in pipelines.

We select Eq. 7.0.8, or the more generalized Eq. 7.0.10 with $m = 0$ and $F(N) = 0$, as the framework for the analysis of our problem. Suppose that a plane source located at $x = 0$, instantaneously emits an amount of substance, M, into a fluid moving in the positive x-direction at a velocity u_0. The solution to Eq. 7.0.8 can be obtained in two ways.

a. Change of dependent variable. In this case we let

$$N(x,t) = \Gamma(x,t) \exp\left(\frac{u_0 x}{2D} - \frac{u_0^2 t}{4D}\right) \quad .$$

(7.1.5)

Substituting the derivatives of this expression into Eq. 7.0.8 gives

$$\frac{\partial \Gamma}{\partial t} = D\frac{\partial^2 \Gamma}{\partial x^2} \quad ,$$

(7.1.6)

which is identical in form to Eq. 7.0.4. Accordingly, with the Dirac delta initial condition, we obtain the solution

$$\Gamma = \frac{M}{2\sqrt{\pi Dt}} \exp\left(-\frac{x^2}{4Dt}\right) \quad . \tag{7.1.7}$$

Returning to the original dependent variable, $N(x,t)$, we get the final answer

$$N = \frac{M}{2\sqrt{\pi DSt}} \exp\left(-\frac{(x - u_0 t)^2}{4Dt}\right) \quad . \tag{7.1.8}$$

b. **Change of independent variable.** For this method, we let $z = x - u_0 t$. This substitution transforms Eq. 7.0.8 into the expression

$$\frac{\partial N}{\partial t} = D\frac{\partial^2 N}{\partial z^2} \quad , \tag{7.1.9}$$

which once again is the same as Eq. 7.0.4. So we immediately have the solution

$$N = \frac{M}{2\sqrt{\pi Dt}} \exp\left(-\frac{z^2}{4Dt}\right) = \frac{m}{2\sqrt{\pi Dt}} \exp\left(-\frac{(x - u_0 t)^2}{4Dt}\right) \quad . \tag{7.1.10}$$

In problems involving convection it is sometimes helpful to visualize the diffusion process as one taking place with respect to a coordinate system moving, in this case, in the positive x-direction at a velocity, u_0. The transformation $z = x - u_0 t$ describes the motion of a wave or front moving at velocity u_0. The moving coordinate reference frame, attached to this front, converts the equation for convection-with-diffusion, given by Eq. 7.0.8, to the equation for diffusion only, expressed by Eq. 7.1.9.

7.1.3 An Illustration: Dispersion in Pipelines

In the long distance transport of petroleum and petroleum products by pipeline, it is frequently necessary to pump a certain type of product (e.g., crude oil) for a specified period of time and then follow it with another type (e.g., kerosene). Generally these petroleum products have different densities and viscosities; invariably the two liquids are miscible. Consequently, there will be a zone, on each side of the interface or front originally delineating the two liquids, in which there is a mixture. It is of practical interest to determine the length of the mixed zone. As we shall see, this phenomenon is a kind of diffusion process. However, it is not on the scale of molecular diffusion. Rather, the mixing is due to so-called turbulent diffusion and to variation of velocity across the diameter of the pipe.

This problem of longitudinal dispersion in circular pipes is analyzed at length by Taylor (1953, 1954) for both laminar flow (small Reynolds numbers) and turbulent flow (large Reynolds numbers). The corresponding problem for natural streams and rivers has been examined by Fischer (1968), Sooky

(1969), and numerous others. Ogata (1970), Fried and Combarnous (1971) and Greenkorn (1983) have extended the analysis to flow in porous media.

For our illustration we select a study carried out by Hull and Kent (1952) concerning dispersion in pipelines. In particular, we examine data they obtained from tests on a 10-inch diameter pipeline extending for about 180 miles from Rangely, Wyoming to Salt Lake City, Utah.

The two successive petroleum products in the pipe were crude oil followed by crude oil plus gasoline. An instantaneous plane source injection of radioactive barium was made at the interface between the two liquids at the Rangely pumping station (milepost: 0). Geiger counter measurements were made of the radioactive tracer front as it passed the eight monitoring stations terminating at Salt Lake City (milepost: 182.5). The mean velocity and Reynolds number were $u_0 = 2.68$ ft /sec, $Re = 24\,000$.

For our illustration, we use the data measured at Green River (milepost: 43.1). These data are listed in Table 7.1.1 and are displayed in Fig. 7.1.3.

Table 7.1.1 Dispersion of radioactive tracer wave, Green River, Wyoming ($L = 43.1$ miles). From Hull and Kent (1952). Notes: (a) $t_* = t - (x/u_0)$; $x = L = 227\,568$ ft; $u_0 = 2.68$ ft/s. (b) The units of N were reported as "counts per second". To illustrate the computation for determination of M, these units are arbitrarily changed to microcuries per cubic foot

t_* [s]	N	t_* [s]	N	t_* [s]	N
−400	2	−70	23	110	20
−270	8	−20	27	170	15
−190	14	0	27	225	10
−140	19	45	25	320	5
−105	22	80	23	490	2

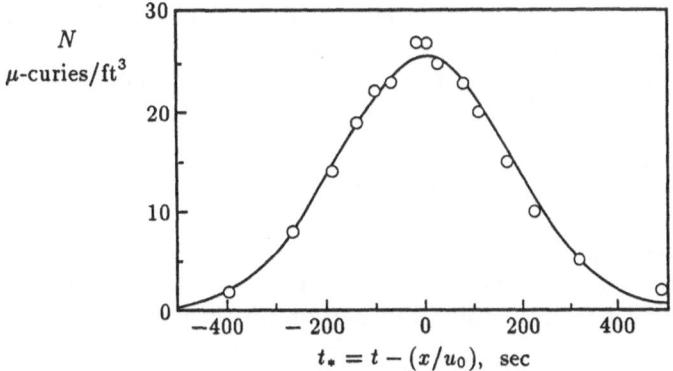

Fig. 7.1.3 Longitudinal dispersion in a pipeline. Radioactive tracer concentrations, Green River monitoring station. Data of Hull and Kent (1952)

We use Eq. 7.1.10 as the framework for analyzing the data. The two unknown parameters are (a) the longitudinal dispersion coefficient, D, and (b) the amount of radioactive barium, M.

Hull and Kent determined the value of the dispersion coefficient by measuring the time interval corresponding to a change in tracer concentration to one-half of the maximum value, N_m. Utilizing Eq. 7.1.10 this approach leads to the formula

$$D = \frac{u_0 L_*^2}{(16 \log_e 2)L} \; , \tag{7.1.11}$$

where L_* is the length of the tracer profile corresponding to $N_m/2$. Substituting into this relationship the measured value of $L_* = 1077$ ft and the values of L and u_0 given in Table 7.1.1 yields $D = 1.23 \, \text{ft}^2/\text{sec}$.

An alternative method of computation, which utilizes all the data measured at the monitoring station and also provides the value of M, is to write Eq. 7.1.10 in the following form

$$\log_e(N\sqrt{T}) = \log_e\left(\frac{M}{2\sqrt{\pi D}}\right) - \frac{1}{4D}\frac{(x - u_0 t)^2}{t} \; . \tag{7.1.12}$$

This is a linear correlation. Converting the data of Table 7.1.1 into the parameters stipulated by Eq. 7.1.12, produces the plot shown in Fig. 7.1.4. A least squares calculation yields the values: $D = 1.32 \, \text{ft}^2/\text{sec}$ and $M = 30.14$ millicuries/ft^2. With a pipe diameter, $d = 10$ inches, we determine that a total amount of radioactive barium, $M_T = M(\pi d^2/4) = 16.43$ millicuries, passed through the Green River monitoring station during the experiment. This computation for M_T at each of the monitoring stations would allow one to determine tracer losses due to decay or conceivably to pipe boundary adsorption.

There were eight monitoring stations along the Rangely pipeline. At each station, the half-value lengths, L_*, and the maximum values, N_m, were measured. Table 7.1.2 compares these observed values with those computed from the relationships

$$L_* = 4\sqrt{(\log_e 2)DL/u_0} \; ; \qquad N_m = \frac{M}{2}\sqrt{\frac{u_0}{\pi DL}} \tag{7.1.13}$$

obtained, respectively, from Eqs. 7.1.11 and 7.1.10.

From this illustration we conclude that for the relatively simple geometry of a circular pipe, our model for rectilinear dispersion with convection appears to be adequate.

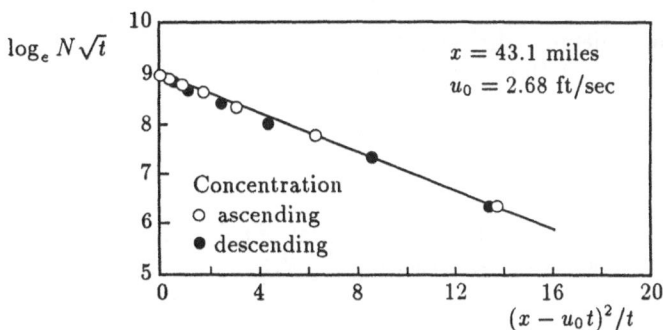

Fig. 7.1.4 Longitudinal dispersion in a pipeline. Plot to determine dispersion coefficient, D, and amount of radioactive tracer, M

Table 7.1.2 Comparison of observed and computed half-value lengths and maximum values of tracer concentration. Rangely Pipe Line. From Hull and Kent (1952)

Station	L [miles]	Half-value lengths L_* [ft]		Maximum values N_m [curies]	
		obs.	comp.	obs.	comp.
Rangely	0.062	48	42		670
White River	1.9	180	234	121	121
Bonanza	13.8	565	630	54	45
Green River	43.1	1077	1114	27	25
Hanna	108.5	1804	1769	16	16
Summit	125.0	2000	1898	15	15
Provo River	130.4	2064	1939	13	14
Salt Lake	182.5	2398	2294		12

7.1.4 Radial Diffusion with Exponential Growth

We now consider the problem of an instantaneous line source with radial diffusion and simultaneous exponential growth; there is no convection. In this cylindrical coordinate system, we set $u_0 = 0$, $m = 1$, and $F(N) = aN$ in the generalized expression of Eq. 7.0.10. These substitutions yield the differential equation

$$\frac{\partial N}{\partial t} = D \left(\frac{\partial^2 N}{\partial r^2} + \frac{1}{r} \frac{\partial N}{\partial r} \right) + aN \quad . \tag{7.1.14}$$

The bracketed term in the right-hand member of this equation describes the radial diffusion of the substance; the final term on the right expresses exponential growth. The solution to this equation is

$$N = \frac{M}{4\pi Dt} \exp \left(at - \frac{r^2}{4Dt} \right) \quad . \tag{7.1.15}$$

For this problem we define a diffusion front or wave in the following manner. Let us integrate Eq. 7.1.15 over the radial area indicated by the expression

$$P_* = \int_R^\infty N(r,t) 2\pi r \, dr \quad .$$ (7.1.16)

This relationship states that P_* is the total amount of substance outside the boundary $r = R(t)$ at any time t. Substituting Eq. 7.1.15 into Eq. 7.1.16 and integrating gives

$$P_* = M \exp\left(at - \frac{R^2}{4Dt}\right) \quad .$$ (7.1.17)

We define the front by arbitrarily setting the exponent of the exponential equal to zero. Accordingly, $R^2 = 4aDt^2$. It follows that $P_* = M$. This causes no difficulty; no matter how large M, and hence P_* may be, they are nevertheless small quantities compared to the amount of substance generated by the exponential growth, viz., $M \exp(at)$. Virtually all of the substance is within the circle of radius, R.

Consequently, we have the following relationships to describe the location, R, and velocity, u_*, of the diffusion wave

$$R = 2\sqrt{aD}\,t \quad ; \quad u_* = \frac{dR}{dt} = 2\sqrt{aD} \quad .$$ (7.1.18)

This problem and more complicated versions of it have been studied by numerous investigators, especially in connection with the dispersal of biological populations. Among these are the studies of Kendall (1948), Skellam (1951), Gurtin and MacCamy (1977), and Shigesada (1980). A comprehensive treatment of the subject has been carried out by Okubo (1980). In the following section we look at some examples of biological dispersal based on the simple model outlined above.

7.1.5 An Illustration: Biological Dispersion

A. Muskrats. One of the earliest and most noteworthy studies of the quantitative aspects of animal dispersal was the classic work of Skellam (1951). Among the numerous theoretical and applied topics he presented in that publication was an interesting example involving the spread of muskrats (*Ondatra zibethica*) in Central Europe.

The muskrat, an aquatic rodent, thrives in many parts of the temperate regions of Europe and North America. An adult is about 60 centimeters in length, including a 20 centimeter tail, and weighs about two kilograms. On the average, a female muskrat has three litters each year with from 4 to 8 in each litter. Gestation time is about 30 days; average life span is around five years.

According to a study published in 1930 by Ulbrich, referenced by Skellam (1951), in 1905 several muskrats found their way to freedom in a wilderness near the Moldau River about 50 kilometers southwest of Prague. Radial dispersion and exponential growth followed. Although the muskrat has many natural predators, most notably the mink, in this instance there was evidently no nature-imposed carrying capacity during the ensuing years. The muskrat population grew rapidly and dispersed widely.

Over the years, Ulbrich kept records of the spread of the muskrat. His map of the locations of the dispersion front is shown in Fig. 7.1.5 for the period 1905 to 1927. Cumulative areas defined by the contours of the map are listed in Table 7.1.3.

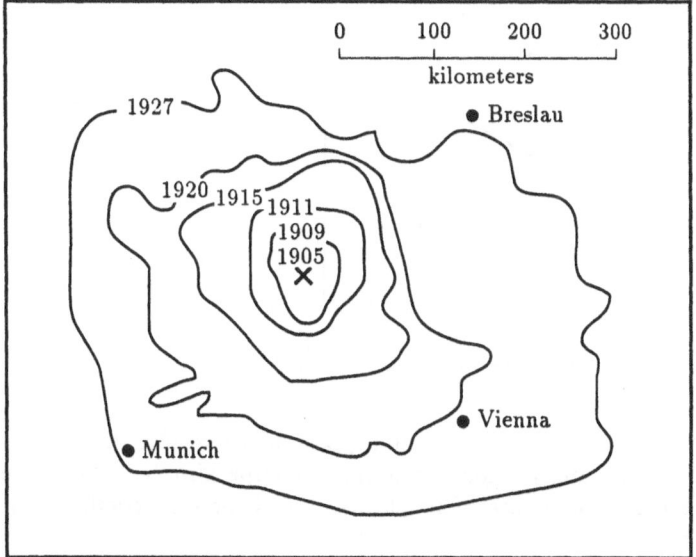

Fig. 7.1.5 Spread of muskrats (*Ondatra zibethica*) in Central Europe, 1905 – 1927. From Skellam (1951)

Table 7.1.3 Spread of muskrats in Central Europe

Year	A [km²]	Year	A [km²]
1905	0	1915	37 700
1909	5 400	1920	79 300
1911	14 000	1927	201 600

In the above model of radial dispersion with exponential growth we obtained the result: $R^2 = 4aDt^2$ in which R, the radius of the expanding dispersion front, defines a circular area which contains virtually all of the spreading substance. With the muskrats being the spreading substance and supposing that the map contours are almost circles we obtain

$$A = \pi R^2 = 4\pi aDt^2 \quad ; \qquad \sqrt{A} = 2\sqrt{\pi aD}\, t \quad . \tag{7.1.19}$$

This result indicates that the square root of the area is directly proportional to time t. This relationship is confirmed in Fig. 7.1.6. The computed slope of the line is $2\sqrt{\pi aD} = 20.16$ which yields $aD = 32.3\,(\mathrm{km/yr})^2$. From Eq. 7.1.18, the velocity of the front is $u_* = 2\sqrt{aD} = 11.4\,\mathrm{km/yr}$. From the map of Fig. 7.1.5 this velocity seems reasonable.

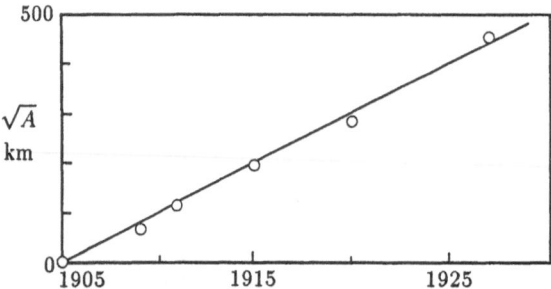

Fig. 7.1.6 Total area covered by the muskrats, 1905–1927

As Okubo (1980) points out, it is not possible to resolve between a and D without further information. Selecting an approximate value, $a = 2.65\,\mathrm{yr}^{-1}$, as the net growth rate of muskrats, the dispersion coefficient is $D = 12.2\,\mathrm{km^2/yr}$.

B. Gypsy Moths. Our second illustration is not greatly different from the first, except that it involves an insect instead of a mammal. It is concerned with the spread of the gypsy moth (*Lymantria dispar*) throughout the New England states during the early period of the twentieth century.

The female gypsy moth has a wing spread of about 1.5 to 2 inches; the male is somewhat smaller. The female lays about 400 eggs once a year in tree trunks or similar places. The eggs overwinter for about 270 days, followed by 65 days as larva, 15 as pupa and about 15 as adult. The moths, especially the female, are very weak fliers. Dispersal of the species is accomplished almost entirely by the highly buoyant wind-blown larva. The insects are voracious plant feeders; the larva are especially fond of willow, birch and oak trees. That is what caused all the trouble.

Gypsy moths were brought to Massachusetts from Europe around 1870 in connection with silkworm development research. Needless to say, some of the moths escaped from breeding cages but somehow large growths and widespread dispersal were kept under control for a number of years.

However, around 1900 there was a drastic increase of gypsy moth population in the Boston area which quickly spread to adjacent regions. By 1925 or so, when dispersal was finally halted, gypsy moths covered all of New England and parts of New York state and Canada. There was severe damage to forests throughout the region.

A map of the northeastern part of the United States is presented in Fig. 7.1.7. The map shows the locations of the dispersion front of the gypsy moth for the period 1900 to 1925. According to Elton (1958), there was no significant expansion of the front after 1925. Cumulative areas corresponding to the indicated fronts are shown in Table 7.1.4.

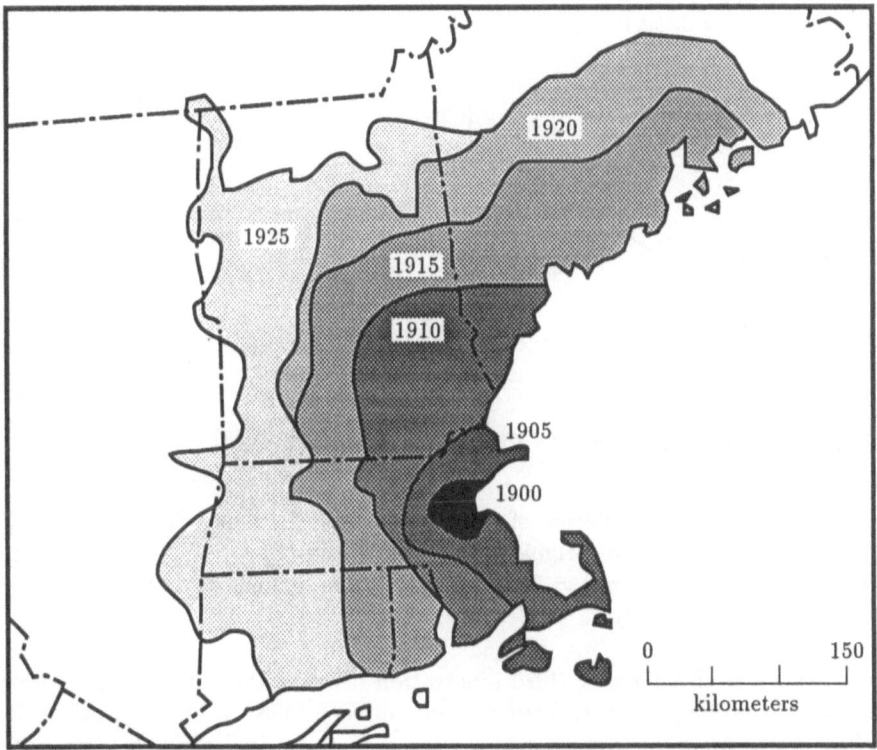

Fig. 7.1.7 Spread of gypsy moths (*Lymantria dispar*) in New England, 1900–1925. From Elton (1958)

Table 7.1.4 Spread of gypsy moths in New England

Year	A [km^2]	Year	A [km^2]
1900	1 290	1915	58 840
1905	9 080	1920	79 770
1910	26 960	1925	113 320

We follow the same procedure we did in the muskrat dispersion example. However, stretching our imagination, this time we assume the areas to be semicircles. Accordingly,

$$A = \frac{\pi R^2}{2} = 2\pi a D t^2 \quad ; \qquad \sqrt{A} = \sqrt{2\pi a D}\, t \quad . \tag{7.1.20}$$

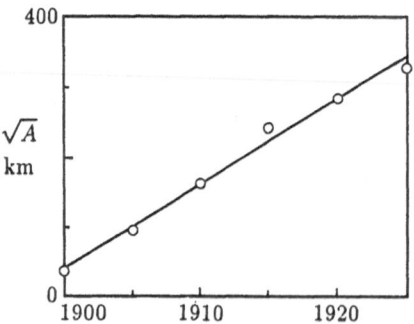

Fig. 7.1.8 Total area covered by the gypsy moths, 1900–1925

A least squares calculation of the data of Table 7.1.4 and displayed in Fig. 7.1.8 yields (a) a virtual origin at $t = 1896.8$ and (b) $\sqrt{2\pi a D} = 12.25$. The latter gives the result $aD = 23.9 \, (\text{km/yr})^2$. The dispersion wave velocity is $u_* = 2\sqrt{aD} = 9.8 \, \text{km/yr}$.

C. Japanese Beetles. A third illustration involves the spread of Japanese beetles (*Popillia japonica*) from a plant nursery in New Jersey starting in 1916. However, since this episode is similar to those involving the muskrats and gypsy moths, we shall not go into detail; instead we simply summarize some observations reported by Elton (1958).

Following their first appearance in mid-1916, the Japanese beetles grew quickly in number and spread rapidly. By 1920, the dispersion front covered 250 km^2, by 1930 nearly 15 000 km^2, and by 1941 over 53 000 km^2. At that time, the beetles had spread throughout all of New Jersey and Delaware as well as large sections of eastern Pennsylvania, northern Maryland and southern New York. In the years following 1941, the area of spread extended much further.

Analysis of the area–time data given by Elton, and assuming semicircular dispersion geometry, gave the following results: (a) growth-dispersion parameter $aD = 14.6 \, (\text{km/yr})^2$ and (b) dispersion front velocity $u_* = 2\sqrt{aD} = 7.6 \, \text{km/yr}$.

After looking at these three examples of biological dispersion we conclude that our very simple analytical framework describes some main features of a complex phenomenon rather well.

It is interesting that the values of the growth-dispersion parameter, aD, and the dispersion front velocity, u_*, are roughly the same for the muskrats, gypsy moths, and Japanese beetles. That we assumed circular or semicircular areas in our computations is not greatly significant. More importantly, we have established that the diffusion velocities, u_*, for the three quite different species are all of the order of 10 km/yr and the growth-dispersion parameters, aD, are of the order of $20 \, (\text{km/yr})^2$.

7.2 Diffusion from Continuous Sources

7.2.1 Rectilinear Diffusion with Constant Boundary Condition

As we shall see, mathematical solutions to problems of diffusion can become quite difficult and the answers rather complicated. The Laplace transformation is a very useful method for solving many of these problems. The methodology is well presented by Carslaw and Jaeger (1959) and in the references cited in Section 6.1.1. A lucid presentation of the diffusion equation and techniques for its solution are given by Farlow (1982).

We will start with a relatively simple problem: rectilinear diffusion in a semi-infinite region of constant diffusivity, D, with no convection nor reaction. In this case, Eq. 7.0.4 is the appropriate differential equation

$$\frac{\partial N}{\partial t} = D \frac{\partial^2 N}{\partial x^2} \ . \tag{7.2.1}$$

The solution to this equation is $N = N(x,t)$. We impose the initial condition $N(x,0) = 0$, and require that the magnitude of N remains finite as x goes to infinity. We select two types of boundary conditions at $x = 0$: (1) the magnitude of N is a specified function of time, including a constant value and (2) the flux of N is a specified function of time, including a constant value.

Problem 1. Suppose that $N(0,t) = N_*$. The answer to this problem is given by Carslaw and Jaeger (1959) and by Crank (1975)

$$N = N_* \text{erfc} \frac{x}{2\sqrt{Dt}} \ , \tag{7.2.2}$$

in which erfc(z) is the complementary error function; it has the following definition

$$\text{erfc } z = 1 - \text{erf } z = 1 - \frac{2}{\sqrt{\pi}} \int_0^z e^{-\zeta^2} d\zeta \quad , \tag{7.2.3}$$

where erf(z), the error function, is uniquely related to the normal probability function as we discussed in Section 2.5.2.

A plot of Eq. 7.2.2 is shown in Fig. 7.2.1. Utilizing Fick's law of diffusion, Eq. 7.0.1, the flux of N at $x = 0$ is

$$q(0,t) = N_* \sqrt{\frac{D}{\pi t}} \quad . \tag{7.2.4}$$

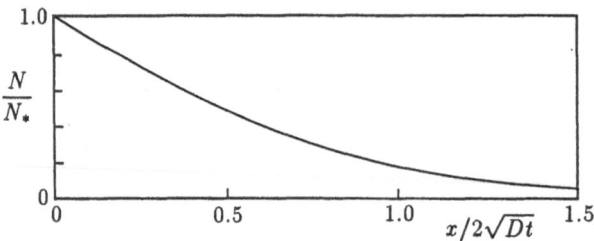

Fig. 7.2.1 Plot of the complementary error function

Problem 2. Suppose that $q = -DN_x(0,t) = $ constant where $N_x = \partial N / \partial x$. The solution to this problem is

$$N = 2q\sqrt{\frac{t}{D}} \left(\frac{1}{\sqrt{\pi}} \exp\left(-\frac{x^2}{4Dt}\right) - \frac{x}{2\sqrt{Dt}} \text{erfc} \frac{x}{2\sqrt{Dt}} \right) \quad . \tag{7.2.5}$$

In this case the magnitude of N at $x = 0$ is

$$N(0,t) = 2q\sqrt{\frac{t}{\pi D}} \quad . \tag{7.2.6}$$

Problems 1 and 2 have numerous applications in problems of heat conduction and chemical diffusion. Ghez (1988) utilizes Eq. 7.2.2 in connection with various crystal growth phenomena. Beckman (1970) discusses the utilization of the one-dimensional diffusion equation, Eq. 7.2.1, in problems involving innovation and expenditure diffusion, worker migration, commodity flows, and price waves. Maynard Smith (1968) and Segel (1980) present some applications of the diffusion equation in biology.

In the field of fluid mechanics there are problems analogous to the two we examined above. As mentioned in Section 7.0.2, such analogy is based on the diffusion of momentum instead of the diffusion of mass or heat. For example, an infinite plate in the $y - z$ plane and passing through the origin suddenly begins to move in the vertical z-direction at constant velocity, v_*.

As Batchelor (1967) shows, the velocity of the viscous fluid bounded by this "impulsive plate" is given by the expression

$$v = v_* \mathrm{erfc} \frac{x}{2\sqrt{\nu t}} \quad , \tag{7.2.7}$$

in which ν is the kinematic viscosity of the fluid. The shear stress, τ, exerted by the fluid on one side of the plate is

$$\tau = -\mu \left(\frac{\partial v}{\partial x} \right)_{x=0} = \rho v_* \sqrt{\frac{\nu}{\pi t}} \quad , \tag{7.2.8}$$

where μ is the dynamic viscosity and ρ is the density of the fluid; by definition $\nu = \mu/\rho$.

An interesting application of Eq. 7.2.2 in the field of bacteriology is given in the following section.

7.2.2 An Illustration: Bacterial Motility

An experimental investigation featuring our above Problem 1 was carried out by Segel, Chet and Henis (1977) to demonstrate that the motility of bacterial populations can be quantitatively described by the kind of diffusion coefficient we have been considering.

These investigators inserted a tube of extremely small diameter and several centimeters length, sealed at one end and containing clear liquid, into a solution with known bacteria concentration, N_*. At time $t = 0$ the unsealed end of the tube was suddenly opened; bacterial motility produced a diffusion of the organisms into the tube.

After known periods of time, t_n, ranging from 2 to 20 minutes, the tube was removed from the test tank and the number of bacteria, P, in the tube was determined. Observed values of P extended from 1800 to 8000 in the series of experiments. Magnitudes of N_* ranged from 2.5×10^7 to 12.0×10^7 bacteria/ml.

The basis of analysis for the investigation was Eq. 7.2.2. The total number of bacteria diffusing into the tube is given by the expression

$$P = A \int_0^\infty N(x,t)dx \quad , \tag{7.2.9}$$

where A is the cross-section area of the tube. Substitution of Eq. 7.2.2 into this equation and integrating yields $P = 2AN_*\sqrt{Dt_n/\pi}$. From this result, an expression is obtained for the direct computation of the effective diffusion coefficient, or motility, D. For the particular organism being studied (*Pseudomonas fluorescens*), the investigators determined the average value to be $D = 0.25\mathrm{cm}^2/\mathrm{hr}$.

7.2.3 Rectilinear Diffusion with Variable Boundary Condition

Our next problem features a boundary condition which changes sinusoidally with time. However, instead of considering the diffusion of a substance, we examine the analogous problem of conduction of heat. Other exact analogies of this problem are those of the "oscillating plate" analyzed by Batchelor (1967) and the flow of water in coastal aquifers due to tidal motion examined by Bear (1979).

We begin by considering Fourier's law of heat conduction expressed by Eq. 7.0.2. A heat balance on an elemental volume yields the following differential equation

$$\frac{\partial \theta}{\partial t} = \epsilon \frac{\partial^2 \theta}{\partial x^2} \quad , \tag{7.2.10}$$

where θ is the temperature and $\epsilon = K/\rho c$ is the thermal diffusivity; K is the thermal conductivity, ρ is the density, and c is the specific heat of the conducting medium. We note that Eq. 7.2.10 and Eq. 7.2.1 have the same mathematical form.

For the boundary condition we take

$$\theta(0,t) = \theta_* + \theta_m \sin \omega t \quad . \tag{7.2.11}$$

We assume that the initial temperature of the conducting medium is zero.

The solution to Eqs. 7.2.10 and 7.2.11, after an initial transient term has vanished, is

$$\theta = \theta_* + \theta_m e^{-mx} \sin(\omega t - mx) \quad , \tag{7.2.12}$$

in which $m = \sqrt{\omega/2\epsilon}$. This expression represents a damped wave moving in the x-direction with

$$\omega = \frac{2\pi}{T} = \text{wave frequency} \quad ; \qquad m = \frac{2\pi}{L} = \text{wave number} \tag{7.2.13}$$

where T and L are the wave period and wavelength, respectively. The wave velocity is $c = L/T = \omega/m = \sqrt{2\epsilon\omega}$. A damping or penetration distance is defined as $\lambda = 1/m = \sqrt{2\epsilon/\omega}$. This distance is the value of x which reduces the amplitude of oscillation by an amount $1/e = 0.368$ of the amplitude at the boundary.

The flux of heat through unit area of the boundary is

$$f(0,t) = \theta_m \sqrt{\frac{\omega}{\epsilon}} K \sin(\omega t + \pi/4) \quad . \tag{7.2.14}$$

Integrating this equation, gives the total amount of heat stored in the medium during one-half a cycle

$$H = \frac{2\theta_m K}{\sqrt{\epsilon\omega}} \quad . \tag{7.2.15}$$

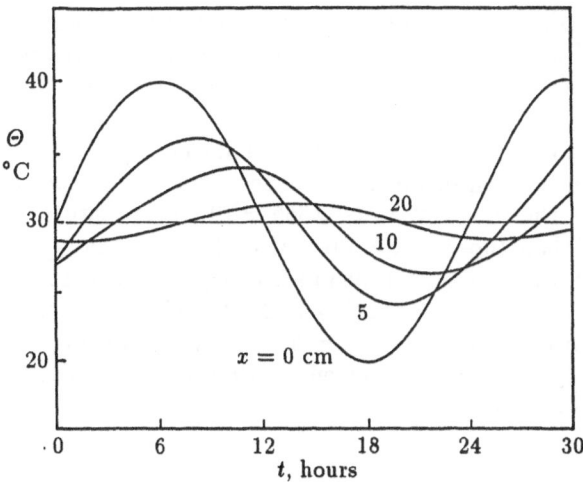

Fig. 7.2.2 Variation of temperature at three depths in a soil due to diurnal temperature oscillation at the surface

7.2.4 An Illustration: Temperature Distribution in the Soil

We now look at a numerical example similar to one presented by Monteith (1973). Our medium is a sandy soil through which there is heat conduction due to a diurnal temperature cycle caused by the sun. The following are the numerical values we need

$$\theta_* = 30°C \qquad \theta_m = 10°C$$
$$T = 24\,h = 86\,400\,s \qquad \omega = 72.72 \times 10^{-6}\,s^{-1}$$
$$\rho = 1.65 \times 10^3\,kg/m^3 \qquad c = 1.01 \times 10^3\,joule/kg\,°C$$
$$K = 0.60\,joule/s\,m\,°C \qquad \epsilon = 0.36 \times 10^{-6}\,m^2/s$$

Substitution of these quantities into Eq 7.2.12 yields the temperature distribution curves shown in Fig. 7.2.2 for values of $x = 0$, 5, 10, and 20 centimeters. From the above relationships, including Eqs. 7.2.14 and 7.2.15, we also obtain the following information

$$\begin{aligned}
\text{Wave number} \qquad & m = 10.0\,m^{-1} \\
\text{Wavelength} \qquad & L = 0.628\,m \\
\text{Penetration distance} \qquad & \lambda = 0.1\,m \\
\text{Wave velocity} \qquad & c = 0.628\,m/day \\
\text{Maximum surface flux} \qquad & f_m = 85.3\,joule/s\,m^2 \\
\text{Heat stored} \qquad & H = 2.35 \times 106\,joule/m^2
\end{aligned}$$

We note that a distance $x = \lambda = 10\,cm$ below the ground surface, the maximum temperature variation from the mean, $\theta_* = 30°$, is about 3.7°C. At a distance of 20 cm, the maximum variation is approximately 1.3°C, and at a distance of 30 cm, only about 0.5°C. We conclude that diurnal temperature changes in the soil due to heating by the sun are confined to a relatively thin layer.

Finally, we take a quick look at the annual, instead of diurnal, temperature cycle. We note that the penetration distance, λ, varies as the square root of the period, T. Accordingly, the values of λ are $\sqrt{365} = 19.1$ times larger for the annual cycle problem than those of the diurnal cycle. Using the same numerical values as before, the penetration distance is now $\lambda = 1.91$ m. At this depth the maximum temperature variation is 3.7°C. The variation is reduced to 0.5°C at a depth of about $3(1.91) = 5.73$ m. We note that the annual temperature cycle has a much greater effect, than does the diurnal cycle, on sub-surface soil temperatures.

7.2.5 Radial Diffusion

By their very nature, many phenomena of spatial diffusion and dispersion are inherently "radial" in complexion. An epidemic tends to spread radially from a geographical center, a city grows radially, water flows radially to a well in an aquifer. We now examine two problems involving radial diffusion.

The basic differential equation for both of these problems is given by Eq. 7.0.6

$$\frac{\partial N}{\partial t} = D\left(\frac{\partial^2 N}{\partial r^2} + \frac{1}{r}\frac{\partial N}{\partial r}\right) \quad . \tag{7.2.16}$$

The initial value is $N(r,0) = 0$, and the solution is bounded as $r \to \infty$.

Problem 1. The boundary condition at r_0 is $N(r_0,t) = N_*$. The solution to this problem is not easy; Carslaw and Jaeger (1959) give the following answer

$$\frac{N}{N_*} = 1 + \frac{2}{\pi}\int_0^\infty e^{-D\xi^2 t}\left(\frac{J_0(\xi r)Y_0(\xi r_0) - Y_0(\xi r)J_0(\xi r_0)}{J_0^2(\xi r_0) + Y_0^2(\xi r_0)}\right)\frac{d\xi}{\xi} \quad , \tag{7.2.17}$$

where $J_0(z)$ and $Y_0(z)$ are, respectively, the Bessel functions of the first and second kinds of zero order. This impossible-looking answer has been numerically integrated; a graphical display of the solution is shown in Fig. 7.2.3. The curves of this figure show the distribution of N with respect to radial distance, r, and time, t. Note that the abscissa is logarithmically scaled.

The flux of N at $r = r_0$ is

$$q = \frac{4N_*D}{\pi^2 r_0}\int_0^\infty e^{-D\xi^2 t}\frac{d\xi}{\xi\left[J_0^2(\xi r_0) + Y_0^2(\xi r_0)\right]} \quad . \tag{7.2.18}$$

The solution to this equation is shown in the log-log plot of Fig. 7.2.4.

Before proceeding to Problem 2, we consider the publication of Jansen and Metz (1979) with the intriguing title "How many victims will a pitfall

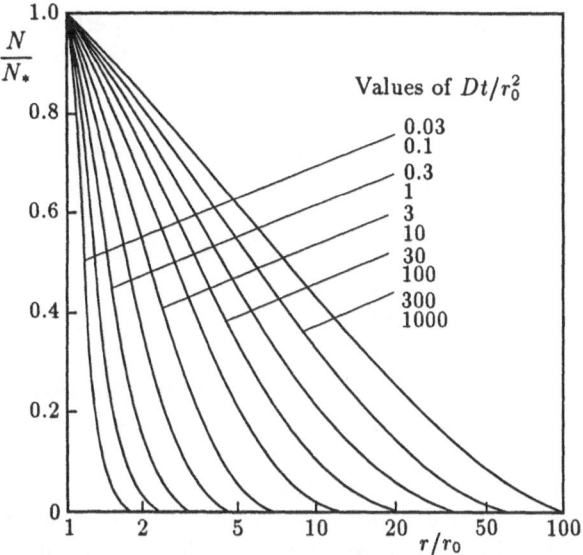

Fig. 7.2.3 Variation of N in an infinite region bounded internally by a cylindrical surface of radius r_0. From Carslaw and Jaeger (1959)

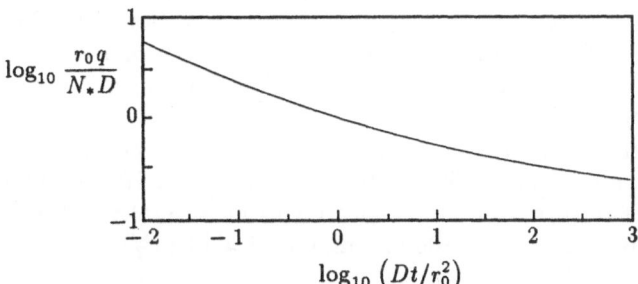

Fig. 7.2.4 Flux into an infinite region through an internal cylindrical surface of radius r_0. From Carslaw and Jaeger (1959)

make?" These investigators devised a mathematical model dealing with the trapping of animals in a circular pit. Their model, based on Eq. 7.2.16, includes analyses of infinite and finite circular regions with several selections of initial and boundary conditions. As would be expected, the analytical solutions given by these researchers contain arrays of Bessel functions similar to those of Eqs. 7.2.17 and 7.2.18.

Laboratory and field experiments were carried out by Jansen and Metz. The laboratory tests involved larvae of fruit flies (*Drosophila melanogaster*) moving on yeast films. The numbers of larvae entering the pit per unit time were measured. The field experiments featured surface-dwelling Collembola (springtails; small wingless jumping insects). Again, the numbers of species entering the pitfall trap per unit time were recorded. From the experimental data the authors were able to determine numerical values of initial density and diffusivity or motility.

Problem 2. For this problem we suppose that a line source along the vertical z-axis emits an amount of substance q per unit time per unit length into the medium. Radial diffusion ensues.

The solution to this problem, again given by Carslaw and Jaeger is

$$N = \frac{q}{4\pi D} \int_{r^2/4Dt}^{\infty} \frac{e^{-\xi}}{\xi} d\xi = \frac{q}{4\pi D} E_1 \left(\frac{r^2}{4Dt} \right) \quad , \tag{7.2.19}$$

in which $E_1(z)$ is the exponential integral; we considered this function in Section 5.3 in connection with linearly decreasing carrying capacities. For small values of z, $E_1(z) = -\gamma - \log_e z + z$, where $\gamma = 0.5772$ is Euler's constant. For large values of z, $E_1(z) = (1 - 1/z)\exp(-z)/z$.

The expression given by Eq. 7.2.19 describes the temperature distribution in an infinite solid due to heating by a thin wire carrying an electric current. In their studies on hemolytic plaque assays in the field of immunology, Perelson and Segel (1978) obtain Eq. 7.2.19 for the determination of the concentration of free antibodies and the rate of plaque growth in thin homogeneous media.

With respect to the Darcy law analogy, mentioned in Section 7.0.2, Eq. 7.2.19 also describes the "drawdown" of the ground water table in an aquifer caused by pumping at a constant rate from a well. This phenomenon is discussed by Bear (1979).

A dimensionless plot of Eq. 7.2.19 is shown as curve (a) in Fig. 7.2.5. A numerical example is presented in Fig. 7.2.6 based on Eq. 7.2.19. The plot shows the variation of N (g/m^3) with distance r (m) for several values of t (h). Values of the constants are: $q = 100$ g/h m and $D = 1.0$ m^2/h.

Next, we integrate Eq. 7.2.19 as follows

$$P = \int_0^r N(r,t) 2\pi r dr = \frac{q}{2D} \int_0^r E_1 \left(\frac{r^2}{4Dt} \right) r dr \quad . \tag{7.2.20}$$

The answer is

$$P = qt \left[1 + \frac{r^2}{4Dt} E_1 \left(\frac{r^2}{4Dt} \right) - \exp\left(-\frac{r^2}{4Dt} \right) \right] \quad . \tag{7.2.21}$$

The quantity P is the total amount of substance contained within the region of radius r at time t. If $r \to \infty$, then $P = P_T = qt$ as we would expect. A plot of Eq. 7.2.21 is shown as curve (b) in Fig. 7.2.5.

Now let us define the radius R of the expanding diffusion circle as that which corresponds to $P/P_T = 0.99$. From Eq. 7.2.21 we obtain $R^2/4Dt = 3.06$. Consequently, the area A of the expanding circle is $A = 38.45 Dt$. Similar expressions are given by Perelson and Segel (1978) for plaque area growth and by Pielou (1977) for the growth of insect dispersal areas.

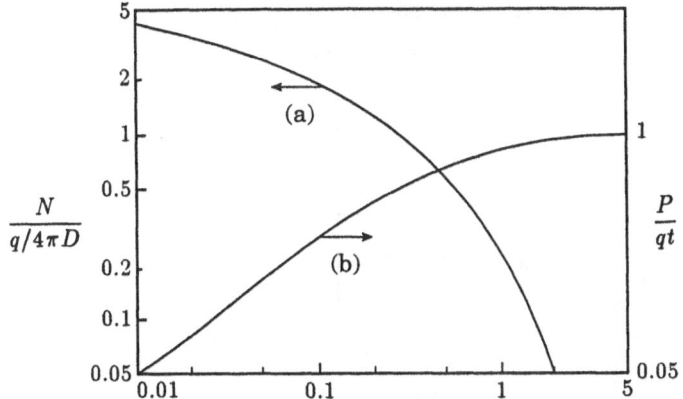

Fig. 7.2.5 Dimensionless plots for radial diffusion from a continuous line source

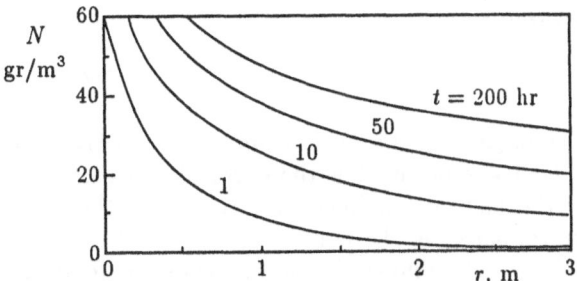

Fig. 7.2.6 Example of radial diffusion from a continuous line source. Parameters are: $D = 1.0\,\mathrm{m}^2/\mathrm{h}$, $q = 10\mathrm{g}/\mathrm{h\,m}$

On this basis, the radius of the circle is $R = 3.50\sqrt{Dt}$ and hence the velocity of the front is $u_* = dR/dt = 1.75\sqrt{D/t}$. In the above numerical example, illustrated in Fig. 7.2.6, when $t = 1.0\mathrm{h}$, the front is moving at a velocity $u_* = 1.75\mathrm{m/h}$; when $t = 200\mathrm{h}$, $u_* = 0.124\mathrm{m/h}$.

7.2.6 An Illustration: Unsteady Fluid Flow in an Aquifer

In order to determine the water resources available in underground aquifers, various types of pumping tests are carried out. As shown in Fig. 7.2.7, one type of test is to drill a well to a depth sufficient to fully penetrate the aquifer being studied. This is the pumping or production well. A second well, for observation, is drilled a distance r from the production well to a depth sufficient to intersect the ground water table by an appreciable distance. A time $t = 0$ pumping is commenced from the production well at a constant rate, Q. At successive times, t, the amount of the "drawdown", H, is measured in the observation well. As we shall see, from the (H,t) data it is possible to determine important characteristics of the aquifer.

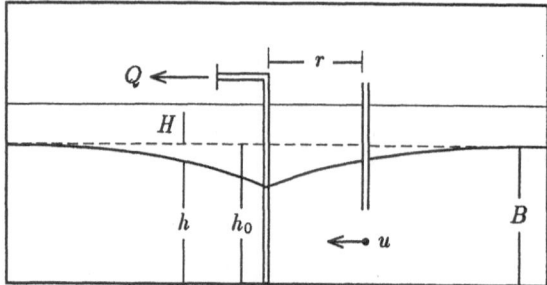

Fig. 7.2.7 Field test configuration to determine aquifer characteristics

The analysis begins with Darcy's law

$$u = -k\frac{\partial h}{\partial r} \quad , \tag{7.2.22}$$

which, we observe, is similar in form to Fick's law and Fourier's law. The velocity of the water is $u(r,t)$. The quantity k is the permeability coefficient (dimensions: velocity); h is the piezometric head or elevation of the water table above a certain datum (dimensions: distance).

A mass balance on an elemental cylindrical volume yields the following differential equation

$$\frac{\partial H}{\partial t} = \frac{T}{S}\left(\frac{\partial^2 H}{\partial r^2} + \frac{1}{r}\frac{\partial H}{\partial r}\right) \quad , \tag{7.2.23}$$

where $H = h_0 - h$. The quantity $T = kB$ is termed the transmissivity; B is the thickness of the aquifer. The dimensionless quantity S is the specific yield; it is essentially the same as the porosity of the aquifer.

We note that Eq. 7.2.23 is identical to Eq. 7.2.16. Furthermore, since $H(r,0) = 0$ and $Q = qB$ is a constant flux at $r = 0$, the initial and boundary conditions are the same as before. Accordingly, the solution to Eq. 7.2.23 is

$$H = \frac{Q}{4\pi T}E_1\left(\frac{Sr^2}{4Tt}\right) \quad . \tag{7.2.24}$$

This expression provides the framework for determination of the aquifer characteristics. The following numerical example illustrates the methodology.

A production well completely penetrates an aquifer of thickness $B = 100$ ft. An observation well is drilled a distance $r = 50$ ft from the production well. At time $t = 0$ pumping is commenced and maintained at a constant volumetric flow rate $Q = 1000$ gal/min $= 2.23 \, \text{ft}^3/\text{s}$. The observed values of drawdown, H, for increasing values of time, t, are given in Table 7.2.1.

Table 7.2.1 Drawdown data from field test on unsteady flow in an aquifer; $B = 100$ ft, $r = 50$ ft, $Q = 1000$ gal/min $= 2.23\,\text{ft}^3/\text{s}$

t [min]	r^2/t [ft^2/s]	H [ft]	t [min]	r^2/t [ft^2/s]	H [ft]
10	4.17	0.14	270	0.15	2.73
30	1.39	0.75	600	0.069	3.92
60	0.69	1.24	1320	0.032	4.75
150	0.28	2.28	2700	0.015	5.57

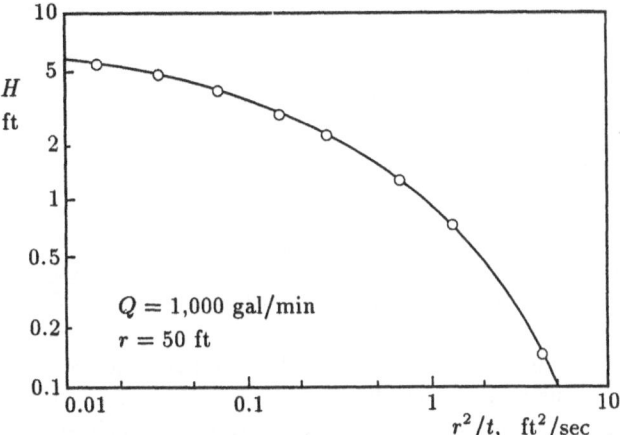

Fig. 7.2.8 Drawdown data from a field test on unsteady flow in an aquifer

A simple and fairly accurate method to determine T and S is to prepare a log-log plot of the quantities r^2/t and H, with the identical scale as the theoretical curve shown in Fig. 7.2.5. Such a plot is presented in Fig. 7.2.8. This plot, superimposed on the theoretical plot, is translated until the data points coincide with the theoretical curve. In our example, a reasonable matching point is (a) $r^2/t = 1.0$, $Sr^2/4Tt = 0.33$ and (b) $H = 1.0$, $E_1(Sr^2/4Tt) = 0.84$. Utilizing Eq. 7.2.24 with $Q = 2.23\,\text{ft}^3/\text{s}$, condition (b) gives $T = 0.150\,\text{ft}^2/\text{s} = 97\,000$ gal/day ft. In turn, utilizing condition (a) yields $S = 0.20$.

To summarize, the transmissivity of the aquifer is $T = 97\,000$ gal/day ft. Since $T = kB$ and $B = 100$ ft, the permeability coefficient of the aquifer is $k = 0.0015$ ft/sec. The specific yield is $S = 0.20$. Neglecting the effects of surface tension, this result indicates that the porosity of the aquifer is $n = 0.20$.

7.2.7 Rectilinear Diffusion with Convection

We conclude this section with the topic of rectilinear convection with diffusion from a continuous source. Examples of this phenomenon are the dispersion of a conservative substance in a waste discharging continuously into a river, salinity intrusion in coastal aquifers, dispersion of gaseous pollutants from continuous plane sources and the like. This problem is readily extended to a line source–cylindrical system and a point source–spherical system.

The basic differential equation is

$$\frac{\partial N}{\partial t} + u_0 \frac{\partial N}{\partial x} = D \frac{\partial^2 N}{\partial x^2} \quad , \tag{7.2.25}$$

where u_0 is the constant convective velocity. Stipulating that $N(x,0) = 0$ and $N(0,t) = N_*$, the bounded solution of Eq. 7.2.25 is

$$\frac{N}{N_*} = \frac{1}{2}\left[\text{erfc}\left(\frac{x - u_0 t}{2\sqrt{Dt}}\right) + \exp\left(\frac{u_0 x}{D}\right)\text{erfc}\left(\frac{x + u_0 t}{2\sqrt{Dt}}\right)\right] \quad . \tag{7.2.26}$$

A graphical display of Eq. 7.2.26, prepared by Ogata (1970), is presented in Fig. 7.2.9. Ogata showed that when $D/u_0 x$ is less than about 0.002, the second term of Eq. 7.2.26 is negligible. We note that when $u_0 = 0$, Eq. 7.2.26 reduces to Eq. 7.2.2, as we would expect.

Numerous investigators, in studies aimed at the determination of dispersion coefficients in porous media flow, have employed Eq. 7.2.26 as the framework for data analysis. A comprehensive bibliography of such studies is provided by Fried and Combarnous (1971). Some typical experimental results, acquired by Ogata, are shown in Fig. 7.2.10. In his experiments the second term of Eq. 7.2.26 was negligibly small. Accordingly, the equation of the solid line in Fig. 7.2.10 is

$$\frac{N}{N_*} = \frac{1}{2}\text{erfc}\left(\frac{x - ut}{2\sqrt{Dt}}\right) \quad , \tag{7.2.27}$$

which is in good agreement with the experimental data. Once again, we observe an S-shaped curve.

Now suppose we have in mind, not dispersion in porous media but rather a question of technology transfer. Consider a certain value of time, t. If the spatial coordinate, x, is smaller than $u_0 t$ then, from Fig 7.2.10 or Eq. 7.2.27, the adoption percentage is large. Alternatively, if x is larger than $u_0 t$, then the percentage is small. These observations are compatible with our presumption that at a particular time, the adoption of a technology is decreased as the distance from the source of the technology becomes larger. We shall consider this topic at some length in Section 7.6.

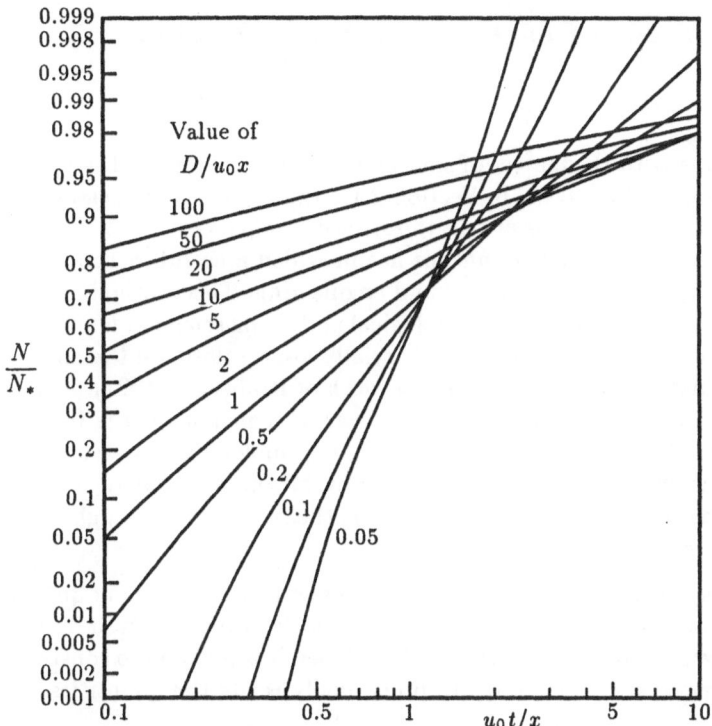

Fig. 7.2.9 Graphical display of the solution to the equation for rectilinear diffusion with convection. From Ogata (1970)

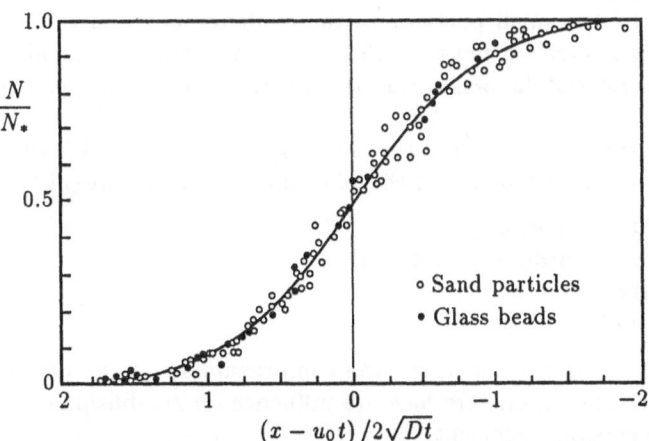

Fig. 7.2.10 Experimental results for dispersion in rectilinear flow in porous media. From Ogata (1970)

7.3 Diffusion with Reaction in a Finite Region

7.3.1 Dimensional Analysis

Before proceeding with the matter of spatial diffusion, we examine briefly the subject of dimensional analysis. In Section 2.1.1 the topic of dimensions and units was raised; we now elaborate on those earlier remarks.

For many years, dimensional analysis has provided a useful methodology in many areas of the physical sciences and engineering. This methodology is finding increasing applications in the biological and social sciences. Numerous references on the subject are available; especially appropriate are the concise book by Barenblatt (1987) and the historic work of Bridgman (1922). Aimed especially at the subject of dimensional analysis in economics is the text by de Jong and Quade (1967); various applications in biology are presented by Vogel (1981) and in mathematical physiology by Mazumdar (1989).

The main feature of dimensional analysis is the so-called Buckingham Pi Theorem which goes as follows

> Consider a system in which there are m independent dimensional quantities which collectively affect the system. There are n fundamental dimensions represented among these m variables. Then it is possible to construct $(m - n)$ dimensionless Π groups to functionally relate the quantities.

The following example illustrates the application of this theorem.

Suppose we have a chemical solute in a liquid entering an ion exchange column or being discharged through an injection well into a porous medium. The mixing created by the small particles of the medium causes distortion of the originally sharp interface between the clear liquid and the ensuing solute liquid. This interfacial distortion can be characterized by a dispersion coefficient, D.

Next comes the hard part; after that the analysis is just a recipe. We suppose that the magnitude of D depends on the following dimensional quantities

- Mean velocity of the liquid, u
- Mean diameter of the medium particles, d
- Mass density of the liquid, ρ
- Dynamic viscosity of the liquid, μ

We hope that the surface tension, σ , the compressibility of the liquid, K, and other conceivable parameters have no influence on establishing the magnitude of the dispersion coefficient.

If that is the case, we can write

$$D = f(u, d, \rho, \mu) \quad . \tag{7.3.1}$$

In this problem we have three fundamental dimensions: mass (M), length (L), and time (T). The dimension of force (F) could be utilized instead of mass (M) since these two quantities are related by Newton's second law,

$F = MLT^{-2}$. If a particular problem were to involve, for example, heat transfer, then the fundamental dimension of temperature would be involved. A problem in which electromagnetic phenomena play a role would introduce the fundamental dimension of electric current or electric charge.

We rewrite Eq. 7.3.1 in implicit form

$$\phi(D, u, d, \rho, \mu) = 0 \quad . \tag{7.3.2}$$

Clearly, we have $m = 5$ independent dimensional quantities and $n = 3$ fundamental dimensions. Consequently, from the Pi theorem, we anticipate $(m - n) = 2$ dimensionless groups relating the quantities. That is

$$\Phi(\Pi_1, \Pi_2) = 0 \quad . \tag{7.3.3}$$

At this point it is helpful to construct a table such as that of Table 7.3.1. We select as many "repeating variables" as there are fundamental dimensions. If the analysis involves fluid motion it is advisable to select a characteristic velocity, length, and fluid property as the repeating variables.

Table 7.3.1 Table of dimensions for dimensional analysis of dispersion in porous media flow

Dimension	Repeating variables			Π group variables	
	u	d	ρ	D	μ
	a	b	c	1	2
Mass M	0	0	1	0	1
Length L	1	1	-3	2	-1
Time T	-1	0	0	-1	-1

We go into detail for the computation of Π_1. First we write

$$\Pi_1 = u^{a_1} d^{b_1} \rho^{c_1} D \quad . \tag{7.3.4}$$

We require Π_1 to be dimensionless. Accordingly, from the table

$$M^0 L^0 T^0 = (LT^{-1})^{a_1}(L)^{b_1}(ML^{-3})^{c_1}(L^2 T^{-1}) \quad .$$

Equating the exponents of M, L, and T

$$M: \quad 0 = c_1$$
$$L: \quad 0 = a_1 + b_1 - 3c_1 + 2$$
$$T: \quad 0 = -a_1 - 1$$

From these three equations we obtain $a_1 = -1$, $b_1 = -1$, and $c_1 = 0$. Therefore

$$\Pi_1 = u^{-1}d^{-1}D = \frac{D}{ud} \quad , \tag{7.3.5}$$

which is, of course, a dimensionless parameter. For historic reasons, the custom has been to utilize the reciprocal of this particular quantity and to give it a special name

Peclet number $\quad Pe = ud/D$. $\tag{7.3.6}$

Determination of Π_2, featuring μ, is carried out in the same fashion. Once again, selecting the reciprocal of the generated dimensionless group we obtain

Reynolds number $\quad Re = \rho ud/\mu$. $\tag{7.3.7}$

Consequently, from Eq 7.3.3 we have

$$\Phi\left(\frac{ud}{D}, \frac{\rho ud}{\mu}\right) = 0 \quad , \tag{7.3.8}$$

or, in explicit form

$$\frac{ud}{D} = f\left(\frac{\rho ud}{\mu}\right) \quad . \tag{7.3.9}$$

This is the final result and it represents all we can obtain from dimensional analysis. It is necessary to resort to straightforward mathematical analysis or experimental studies to obtain more information about the precise functional relationship of Eq. 7.3.9.

On this point, Fried and Combarnous (1971) give the results of considerable experimental work involving dispersion in porous media. Taylor (1953, 1954) presents data concerning the corresponding problem for flow in circular pipes and Sooky (1969) gives the same kind of information dealing with dispersion in rivers and canals.

7.3.2 Exponential Growth in a Finite Region

An interesting problem involving diffusion and reaction is one analyzed years ago by Skellam (1951) and Kierstead and Slobodkin (1953). More recently, this and similar types of problems have been examined at length by Okubo (1980) and Murray (1989).

Specifically, the problem features the exponential growth of an organism, e.g., phytoplankton, with simultaneous rectilinear diffusion of the organism in a channel of length L. In addition, the problem specifies that a hostile environment exists at the open ends of the channel; organisms diffusing through the ends are destroyed.

In mathematical symbols, the problem is expressed by the differential equation

$$\frac{\partial N}{\partial t} = D\frac{\partial^2 N}{\partial x^2} + aN \tag{7.3.10}$$

with the boundary conditions: $N(0,t) = 0$ and $N(L,t) = 0$. For the initial condition we take $N(x,0) = N_0(x)$. As Kierstead and Slobodkin show, the general solution to Eq. 7.3.10 is

$$N(x,t) = \sum_{n=1}^{\infty} A_n e^{(a-n^2\pi^2 D/L^2)t} \sin\left(\frac{n\pi x}{L}\right) \quad, \tag{7.3.11}$$

where

$$A_n = \frac{2}{L}\int_0^L N_0(x)\sin\left(\frac{n\pi x}{L}\right)dx \quad. \tag{7.3.12}$$

The particular solution, of course, depends on the initial condition, $N_0(x)$.

For the moment, we are not interested in the complete answer. Instead, we want to know the minimum length of the channel for which an increase in the magnitude of N is possible. If the length is less than this minimum then the gain in magnitude due to exponential growth, which is proportional to the volume of the channel, is less than the loss due to diffusion through the two ends. In this case, the organism declines to extinction. A critical length, L_c, occurs when the gain and loss are in equilibrium.

This is the kind of problem in which dimensional analysis can provide a certain amount of information. We start by surmising that the critical length, L_c, depends only on the growth coefficient, a, and the diffusion coefficient, D. That is, $L_c = f(a, D)$ or $\phi(L_c, a, D) = 0$. With reference to the previous section on dimensional analysis, we note that $m = 3$ and $n = 2$ (length, L, and time, T). Accordingly, from the Pi theorem, only one dimensionless group is generated , viz., $aL_c^2/D =$ constant. From this result we obtain

$$L_c = k\sqrt{\frac{D}{a}} \quad. \tag{7.3.13}$$

However, to determine the value of k, it is necessary to return to mathematical analysis.

With regard to Eq. 7.3.11 and knowing that $a > 0$, we see that N will increase or decrease with time depending on which term is larger in the exponential quantity, $(a - n^2\pi^2 D/L^2)$. We select $n = 1$ as the crucial value; successive terms in the infinite series, for which $n > 1$, will certainly descend to zero as t becomes infinite. Therefore, to stipulate equilibrium between gain due to growth and loss due to diffusion, we set the exponential quantity equal to zero. This gives the value of the critical length

$$L_c = \pi\sqrt{\frac{D}{a}} \quad. \tag{7.3.14}$$

This answer is compatible with the result obtained from dimensional analysis. Furthermore, it determines that $k = \pi$ in Eq. 7.3.13.

We pursue the mathematical solution somewhat further. Suppose that the initial condition is $N_0(x) = N_0 = \text{constant}$. Substituting this condition into Eq. 7.3.12 gives

$$A_n = \frac{2N_0}{\pi}\frac{1}{n}(1 - \cos n\pi) \quad . \tag{7.3.15}$$

Substituting this result, along with $L = L_c$, into Eq. 7.3.11, and simplifying yields

$$\frac{N}{N_0} = \frac{4}{\pi}\sum_{n=1}^{\infty}\frac{1}{2n-1}e^{-4n(n-1)T}\sin[(2n-1)\xi] \quad , \tag{7.3.16}$$

where $T = at$ and $\xi = x/\sqrt{D/a}$. We note that when $T \to \infty$

$$\frac{N}{N_0} = \frac{4}{\pi}\sin\xi \quad . \tag{7.3.17}$$

A plot of Eq. 7.3.16 is shown in Fig. 7.3.1 for various values of T. At $T = 0$, the total number of organisms per unit cross-section area of the channel is $P_0 = N_0 L_c$. Integration of Eq. 7.3.17 indicates that the total number at $T = \infty$ is $P = 8N_0 L_c/\pi^2$.

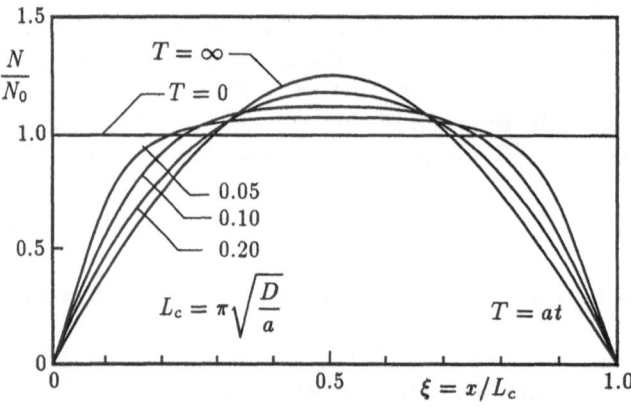

Fig. 7.3.1 Distribution of $N(\xi, t)$ in a rectilinear channel of finite length with diffusion and exponential growth

We could have obtained some of the answers to this problem had we considered only the "steady state" problem, i.e., $\partial N/\partial t = 0$. In this case, Eq. 7.3.10 becomes

$$\frac{d^2 N}{dx^2} + \frac{a}{D}N = 0 \quad . \tag{7.3.18}$$

From this second order ordinary differential equation we obtain

$$N = N_m \sin\left(\sqrt{\frac{a}{D}}x\right) \quad , \tag{7.3.19}$$

where N_m is an undetermined maximum. When $x = 0$ then $N = 0$. Further, when $x = L_c$ then again $N = 0$; this yields $L_c = \pi\sqrt{D/a}$ as before.

Numerous investigators have examined more complicated models of this critical size problem. For example, the effects of convection, density-dependent diffusion, and spatially variable growth coefficient have been studied. Our problem has featured the "absorbing" boundary condition, $N = 0$ at $x = 0$ and L. Other boundary conditions can be applied, including the "reflecting" condition, $\partial N/\partial x = 0$, and the "radiation" condition, $\partial N/\partial x + hN = 0$.

Typical of studies involving these more complex models are those of Gurney and Nisbet (1975) and Shigesada (1980). Okubo (1984) presents a generalized solution to the problem of spatially variable growth coefficients and summarizes the results obtained by other researchers along these lines.

A similar analysis can be carried out for the case of radial diffusion and growth in a circular region instead of a straight channel. We want to determine the minimum radius, r_c, necessary to maintain the organism.

The appropriate differential equation is

$$\frac{\partial N}{\partial t} = D\left(\frac{\partial^2 N}{\partial r^2} + \frac{1}{r}\frac{\partial N}{\partial r}\right) + aN \quad . \tag{7.3.20}$$

The complete solution, $N(r,t)$, is provided by Skellam (1951); we will settle for the solution to the steady state equation

$$\frac{d^2 N}{dr^2} + \frac{1}{r}\frac{dN}{dr} + \frac{a}{D}N = 0 \quad . \tag{7.3.21}$$

This is the Bessel differential equation. A solution which fits the boundary conditions of our problem is

$$N = N_m J_0\left(\sqrt{\frac{a}{D}}r\right) \quad , \tag{7.3.22}$$

where $J_0(z)$ is the Bessel function of the first kind of zero order. If we require $N(r_c) = 0$ then we determine, from a table of Bessel functions, that the first zero of $J_0(z_c) = 2.405$. Accordingly, with $d_c = 2r_c$, we obtain

$$d_c = 4.810\sqrt{\frac{D}{a}} \quad . \tag{7.3.23}$$

This expression gives the minimum diameter of the circular region in which the organism will survive.

A numerical example. In Section 7.1.5, in our illustration involving muskrat dispersal, we determined that $aD = 32.3\,(\text{km/yr})^2$. Estimating that

$a = 2.65\,\mathrm{yr}^{-1}$, then $D = 12.2\,\mathrm{km^2/yr}$. Substituting these numbers into Eq. 7.3.23 yields $d_c = 10.32\,\mathrm{km}$. This result indicates that had the dispersing muskrats been destroyed by a hostile external environment before they spread beyond a 10 kilometer diameter circle centered at the original source, the entire group would have eventually become extinct.

Finally, for completeness, we solve the same problem of diffusion and growth in a sphere. The differential equation for $N(r,t)$ is

$$\frac{\partial N}{\partial t} = D\left(\frac{\partial^2 N}{\partial r^2} + \frac{2}{r}\frac{\partial N}{\partial r}\right) + aN \quad, \tag{7.3.24}$$

which reduces to the following steady state equation

$$\frac{d^2 N}{dr^2} + \frac{2}{r}\frac{dN}{dt} + \frac{a}{D}N = 0 \quad. \tag{7.3.25}$$

The solution to this spherical Bessel equation that satisfies our boundary conditions is

$$N = N_m\sqrt{\frac{D}{a}}\frac{1}{r}\sin\left(\sqrt{\frac{a}{D}}r\right) \quad. \tag{7.3.26}$$

Stipulating that $N(r_c) = 0$, we obtain from this equation the result

$$d_c = 2\pi\sqrt{\frac{D}{a}} \quad, \tag{7.3.27}$$

which is the minimum diameter of the spherical region needed to sustain the organism.

A comparison is made in Fig. 7.3.2 of the rectilinear, cylindrical, and spherical steady state distributions. In the spherical system an inflection point occurs at $r/r_c = 0.662$.

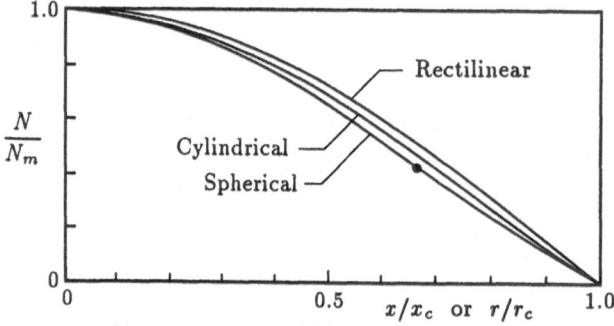

Fig. 7.3.2 Steady state distributions of N in finite rectilinear, cylindrical, and spherical regions with diffusion and exponential growth

7.3.3 Power Law Exponential Growth in a Finite Region

In our consideration of power law exponential growth in Section 2.1.5, we examined the case of hyperbolic or so-called coalition growth. We took another look at this topic in Section 3.4.6. Specifically, we considered the behavior of the equation

$$\frac{dN}{dt} = \frac{a}{N_0} N^2 \quad , \tag{7.3.28}$$

where the initial value, N_0, was introduced for dimensional consistency.

It is of interest to examine some aspects of this growth relationship in the context of our diffusion model. That is

$$\frac{\partial N}{\partial t} = D \frac{\partial^2 N}{\partial x^2} + \frac{a}{N_0} N^2 \quad . \tag{7.3.29}$$

As before, we direct our attention to the steady state equation

$$\frac{d^2 N}{dx^2} + \frac{a}{D N_0} N^2 = 0 \quad . \tag{7.3.30}$$

The boundary conditions are (a) $N = 0$ at $x = 0$ and (b) $N = N_m$ and $dN/dx = 0$ at $x = L_c/2$. This equation is easily integrated with the substitutions: $w = dN/dx$ and $dw/dx = w\, dw/dN$. The solution is

$$\frac{L_c}{\sqrt{D/a}} = \sqrt{\frac{6 N_0}{N_m}} \int_0^1 \frac{d\zeta}{\sqrt{1 - \zeta^3}} \quad . \tag{7.3.31}$$

In our examination of the hyperlogistic equation in Section 3.4.2 we introduced the beta function

$$B(p, q) = \int_0^1 z^{p-1}(1 - z)^{q-1} dz = \frac{\Gamma(p)\Gamma(q)}{\Gamma(p + q)} \quad , \tag{7.3.32}$$

where $\Gamma(z)$ is the gamma function. Letting $z = \zeta^3$, $p = 1/3$, $q = 1/2$ and utilizing various gamma function relationships including $\Gamma(1/2) = \sqrt{\pi}$, Eq. 7.3.31 yields the solution

$$\frac{L_c}{\sqrt{D/a}} = \sqrt{\frac{N_0}{N_m} \frac{[\Gamma(1/3)]^3}{\pi 2^{5/6}}} = 3.4345 \sqrt{\frac{N_0}{N_m}} \quad . \tag{7.3.33}$$

This expression corresponds to the case of coalition growth ($r = 2$). We might as well seek the solution to the general case

$$\frac{d^2 N}{dx^2} + \frac{a}{D N_0^{r-1}} N^r = 0 \quad . \tag{7.3.34}$$

Proceeding in the same fashion as before, we obtain the following answer for power law exponential growth with diffusion in a rectilinear channel with hostile boundaries

$$\frac{L_c}{\sqrt{D/a}} = \left(\frac{N_0}{N_m}\right)^{\frac{r-1}{2}} \sqrt{\frac{2\pi}{r+1}} \frac{\Gamma\left(\frac{1}{r+1}\right)}{\Gamma\left(\frac{1}{r+1} + \frac{1}{2}\right)} \ . \tag{7.3.35}$$

At this point, an example may be helpful, especially with respect to the rather awkward quantity N_0/N_m in the above equation. In the previous section featuring exponential growth ($r = 1$) with initial condition $N(0) = N_0 = $ constant, we determined the steady state solution to be

$$\frac{N}{N_0} = \frac{4}{\pi} \sin\left(\pi\frac{x}{L_c}\right) \ . \tag{7.3.36}$$

Clearly, at $x = L_c/2$, we have the maximum value, $N_m/N_0 = 4/\pi$.

Now we want to determine the values of the critical length, L_c, required to attain this same maximum value for the cases $r \neq 1$. Accordingly we substitute $N_0/N_m = \pi/4$ into Eq. 7.3.35 and compute the critical length of channel for various values of r. The results of computations are presented in Table 7.3.2.

Table **7.3.2** Power law exponential growth with diffusion. Values of critical length parameter $L_c/\sqrt{D/a} = k$

r	k	r	k
0	3.1915	2	3.0437
1/2	3.1734	3	2.9123
2/3	3.1642	4	2.7595
1	π	5	2.5949

We note from the table that the critical length is reduced as the value of r is increased; inherently, this seems logical. However, the critical length is rather insensitive to r; a reduction in length of less than five percent exists between the linear growth case ($r = 0$) and the hyperbolic or coalition growth case ($r = 2$).

7.3.4 Logistic Growth in a Finite Region

Our next problem is that in which diffusion is combined with logistic growth in a finite region. Consequently, we add the crowding term to Eq. 7.3.10 to obtain

$$\frac{\partial N}{\partial t} = D\frac{\partial^2 N}{\partial x^2} + aN - bN^2 \quad . \tag{7.3.37}$$

This expression was devised years ago by Fisher (1937) in connection with his studies on genetic propagation. At about the same time, a more generalized version of the equation was examined by Kolmogorov, Petrovsky and Piskounov (1937).

Though both of these early studies were addressed primarily to biological phenomena, expressions of the form of Eq.7.3.37 arise in present day problems of neurophysiology, chemical reaction theory, and traffic flow. Over the years, this equation has been studied extensively; reviews of these studies are given by Fife (1979), Levin (1981), Mollison (1977) and Murray (1989).

We return to Eq. 7.3.37 later on. For the moment we confine our interest to the steady state form of this non-linear differential equation in order to determine the critical length, as we did in the previous sections. Accordingly, Eq. 7.3.37 becomes

$$\frac{d^2 N}{dx^2} + \frac{a}{D}N - \frac{b}{D}N^2 = 0 \quad . \tag{7.3.38}$$

Again absorbing boundary conditions are imposed at the ends of the channel and, this time, the mid-point of the channel is placed at $x = 0$. Accordingly, $N = 0$ at $x = L_c/2$ and, by symmetry, $dN/dx = 0$ at $x = 0$. We put Eq. 7.3.38 into dimensionless form by letting $N_* = a/b$, $U = N/N_*$, and $\xi = x/\sqrt{D/a}$. These substitutions yield

$$\frac{d^2 U}{d\xi^2} + U(1 - U) = 0 \quad . \tag{7.3.39}$$

We make the additional substitutions

$$V = \frac{U_m - U}{U_m(1 - U_m)} \quad ; \qquad \sigma = 2U_m - 1 \quad , \tag{7.3.40}$$

where $U_m = U(0)$. So the differential equation becomes

$$\frac{d^2 V}{d\xi^2} = 1 + \sigma V - \frac{1}{4}(1 - \sigma^2)V^2 \quad . \tag{7.3.41}$$

Letting $w = dV/d\xi$ gives

$$w\frac{dw}{dV} = 1 + \sigma V - \frac{1}{4}(1 - \sigma^2)V^2 \quad . \tag{7.3.42}$$

The boundary conditions are (a) at $\xi = 0$, $w = 0$, $U = U_m$, $V = 0$ and (b) at $\xi = \xi_c$, $U = 0$, $V = V_0 = 1/(1 - U_m)$. Then Eq. 7.3.42 is integrated to yield

$$2\xi_c = \int_0^{V_0} \frac{dz}{\sqrt{z(1 - \alpha^2 z)(1 + \beta^2 z)}} \quad , \tag{7.3.43}$$

where

$$\alpha^2 = \frac{1}{4}\left(\sqrt{\frac{4-\sigma^2}{3}} - \sigma\right) \quad ; \quad \beta^2 = \frac{1}{4}\left(\sqrt{\frac{4-\sigma^2}{3}} + \sigma\right) \quad . \tag{7.3.44}$$

Finally, we let $\zeta = \arccos(\alpha\sqrt{z})$. This substitution and the definition $\xi_c = L_c/2\sqrt{D/a}$ provide the following final answer

$$\frac{L_c}{\sqrt{D/a}} = \frac{2\sqrt{2}}{\sqrt{\alpha^2 + \beta^2}}\left[F\left(\frac{\pi}{2}, m\right) - F\left(\arccos(\alpha\sqrt{V_0}), m\right)\right] \quad , \tag{7.3.45}$$

where $m = \beta/\sqrt{\alpha^2 + \beta^2}$. The quantity $F(\gamma, m)$ is the elliptic integral of the first kind; it is tabulated in Abramowitz and Stegun (1965). The result expressed by Eq. 7.3.45 is essentially the same as that provided by Skellam (1951).

Tracing back through the various substitutions, we note that Eq. 7.3.45 indicates that

$$\frac{L_c}{\sqrt{D/a}} = f\left(\frac{N_m}{N_*}\right) \quad , \tag{7.3.46}$$

in which N_m is the value of N at $x = 0$; this is the maximum value. If $N_m = 0$, then Eq. 7.3.45 gives $L_c = \pi\sqrt{D/a}$; if $N_m = N_*$, then $L_c = \infty$. The first result indicates that the minimum channel length necessary to maintain even a vanishingly small N_m is the same as that for exponential growth. The second result shows that an infinitely long channel is necessary if the magnitude of N_m is to be equal to the carrying capacity, N_*. A plot of Eq. 7.3.45 is presented in Fig. 7.3.3.

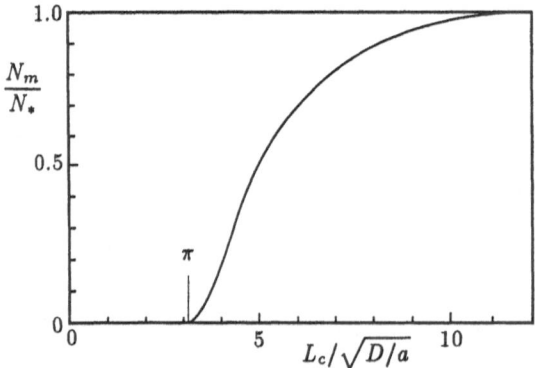

Fig. 7.3.3 Magnitude of the critical length in a rectilinear channel with diffusion and logistic growth

The corresponding problem of steady state radial diffusion and logistic growth in a circular region has been examined by Skellam (1951).

The more complicated problem of unsteady state radial diffusion with logistic growth was considered by McMurtrie (1978). In this instance the differential equation is

$$\frac{\partial N}{\partial t} = D\left(\frac{\partial^2 N}{\partial r^2} + \frac{1}{r}\frac{\partial N}{\partial r}\right) + aN\left(1 - \frac{N}{N_*}\right) \quad .$$ (7.3.47)

This equation was solved by McMurtrie for two cases: (a) $r_0 = 2.5\sqrt{D/a}$ and (b) $r_0 = 2.0\sqrt{D/a}$, where r_0 is the radius of the absorbing boundary. We recall from Eq. 7.3.23 that $r_c = 2.405\sqrt{D/a}$ is the critical radius for the problem of radial diffusion with exponential growth. Thus, the cases considered by McMurtrie were slightly more and somewhat less than the critical value, $k = 2.405$.

The results of McMurtrie's analysis are presented in Fig. 7.3.4. Dimensionless total populations, obtained by spatial integration of $N(r,t)$, are shown in Fig. 7.3.4(a) as functions of dimensionless time, at. When the absorbing radius corresponds to $k = 2.5$, the total population attains a stable equilibrium. On the other hand, when $k = 2.0$ defines the absorbing radius, the total population increases to a maximum value and then declines to extinction. Fig. 7.3.4(b) shows the steady state distribution, $N(r)$, for the $k = 2.5$ case. It is noted that the maximum value, N_m, occurring at $r = 0$, is about nine percent of the carrying capacity, N_*.

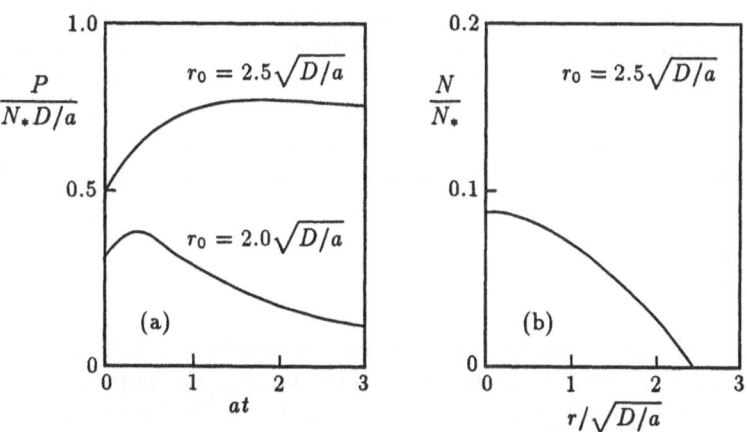

Fig. 7.3.4 Diffusion with logistic growth in a cylindrical system with a hostile boundary at r_0. (a) Total population, $P(r,t)$, and (b) steady state distribution, $N(r)$. From McMurtrie (1978)

7.3.5 An Illustration: Zone of Regulated Fishing

For an illustration of salient features of the preceding analysis, we select the following problem given by Murray (1989).

Suppose fishing is regulated within a zone x_c km from a country's shore (taken to be a straight line) but outside of this zone over-fishing is so excessive that the population is effectively zero. Assume that the fish population grows logistically, disperses by diffusion, and within the zone is harvested with an effort, E.

We utilize the following differential equation to describe the problem

$$\frac{\partial N}{\partial t} = D\frac{\partial^2 N}{\partial x^2} + aN - bN^2 - EN \quad . \tag{7.3.48}$$

The steady state equation and boundary conditions are

$$\frac{d^2 N}{dx^2} + \frac{a - E}{D}N - \frac{b}{D}N^2 = 0 \quad ; \tag{7.3.49}$$

$$\frac{dN}{dx} = 0 \quad \text{at} \quad x = 0 \quad ; \qquad N = 0 \quad \text{at} \quad x = x_c \quad . \tag{7.3.50}$$

It is necessary that $E < a$. The first boundary condition specifies no flux across the shore (reflecting boundary condition); the second specifies a hostile environment (absorbing boundary condition).

We observe that Eq. 7.3.49 is identical to Eq. 7.3.38 with the growth rate, a, replaced by a modified growth rate, $a - E$. The boundary conditions are the same. Consequently, Eq. 7.3.45 provides the complete solution to our problem. As before, a graphical display of the solution is presented in Fig. 7.3.3 with the modifications: $N_* = (a - E)/b$ and $x_c = L_c/2$.

From the above it is immediately established that the minimum width of the zone of regulated fishing is

$$x_c = \frac{\pi}{2}\sqrt{\frac{D}{a - E}} \quad . \tag{7.3.51}$$

As before, x_c is the minimum distance to maintain equilibrium between gain due to growth and loss due to diffusion. In this case N_m, the population at $x = 0$, is zero. As Fig. 7.3.3 indicates, for N_m to be larger than zero, x_c must be larger than $(\pi/2)\sqrt{D/(a - E)}$.

Now some numbers. In a study involving yellowfin tuna in the eastern tropical Atlantic, Conrad (1986) used $a = 1.288\,\mathrm{yr}^{-1}$ and $N_* = a/b = 351.2 \times 10^3$ metric tons. For our example, we select $a = 1.5\,\mathrm{yr}^{-1}$. Arbitrarily we also take $E = a/2 = 0.75$.

There is considerable information available concerning numerical values of the turbulent diffusion coefficient, D. If we consider simply the phenomenon of turbulent diffusion in a horizontal plane at or near the surface of a large body of water, then an analysis by Kolmogorov and Obuknov provides a suitable framework. A summary of this analysis is given by Landau and Lifschitz (1959). The result of this approach, as Okubo (1971, 1980) indicates, leads to the relationship

$$D = k\lambda^{4/3} \quad , \tag{7.3.52}$$

in which D is the horizontal turbulent diffusion coefficient and λ is the "scale" of the phenomenon. The constant has the approximate value $k = 0.0020$ in (cm, sec) units and $k = 30$ in (km, yr) units.

Two problems. The first: what is the proper linear dimension to describe the scale, λ, of our regulated fishing zone. The second: the diffusion coefficient given by Eq. 7.3.52 is that associated with the fluid mechanics of single-phase turbulent incompressible fluid motion. It does not necessarily describe the mechanics of dispersal of a two-phase fluid in which the dispersed phase may range in size from phytoplankton to blue whales.

In his computer studies of the drift of anchovy in the California current, Power (1986) used the value $D = 1.0 \times 10^6 \, \mathrm{cm}^2/\mathrm{sec}$. Possingham and Rough-garden (1990), in their mathematical models of the near-coastline population dynamics of the barnacle *Balanus glandula*, employed $D = 1.0 \times 10^5 \, \mathrm{cm}^2/\mathrm{sec}$. Talbot (1974) carried out field studies involving diffusion of oyster larvae in tidal estuaries on the North Sea coast of England. He obtained values of D ranging from 1.15×10^5 to $9.1 \times 10^5 \, \mathrm{cm}^2/\mathrm{sec}$. For our illustration, we shall use $D = 1.0 \times 10^5 \, \mathrm{cm}^2/\mathrm{sec} = 315 \, \mathrm{km}^2/\mathrm{yr}$.

Accordingly, substituting the values $a = 1.5 \, \mathrm{yr}^{-1}$, $E = 0.75 \, \mathrm{yr}^{-1}$, and $D = 315 \, \mathrm{km}^2/\mathrm{yr}$ into Eq. 7.3.51, we obtain the answer to our problem, $x_c = 32.2 \, \mathrm{km}$. This distance, of course, is based on $E = 0.75$. If $E = 0$, then $x_{c0} = 22.8 \, \mathrm{km}$. Finally, with $D = 315 \, \mathrm{km}^2/\mathrm{yr}$, then from Eq. 7.3.52, the scale is $\lambda = 5.83 \, \mathrm{km}$.

It is apparent that there are numerous implications involving economics in the above illustration and much more complicated versions of it. Topics of bioeconomics and optimal harvesting have been examined at length by Clark (1976, 1981), Conrad (1986), Getz and Haight (1989) and MacCall (1990).

7.4 Diffusion with Convection and Reaction

7.4.1 The Differential Equation and Its Solution

This section begins with brief descriptions of a few examples of the topic we shall now examine.

Suppose we have a river in which the velocity of the water, u_0, is reasonably constant. The turbulence in the water produces a longitudinal dispersion coefficient, D. At some point along the river, $x = 0$, a liquid waste suddenly begins to flow at a constant rate into the river and is quickly mixed with the river water. The waste contains a chemical or biological substance which tends to grow at a rate a or decay at a rate a_*. The concentration of the biochemical substance in the river is zero before the waste begins to enter. After the waste flow commences, the concentration at the well-mixed cross section, $x = 0$, is $C_* =$ constant. The question is: what is the downstream concentration, $C(x,t)$, for $t > 0$?

Or suppose we have a long hot steel rod of constant diameter moving along a conveyor rack at velocity u_0. The rod has a temperature θ_* the instant it leaves the extrusion machine. Heat conduction occurs along the rod with thermal diffusivity κ. In addition, heat is transferred from the rod to the atmosphere at a rate described by a_*. We want to determine the temperature distribution in the rod, $\theta(x,t)$.

Or to be a little extreme, suppose we have a marathon race with an enormous number of participants who, at $x = 0$, $t = 0$, begin to jog, trot or run at an overall average speed, u_0. There are so many participants that their density at the starting line, $x = 0$, is always P_*. Some runners fall by the wayside (a_*); others illegally enter the race (a). Again the question is: what is the participant density $P(x,t)$?

Solutions to these kinds of problems are obtained by utilizing Eq. 7.0.9

$$\frac{\partial N}{\partial t} + u_0 \frac{\partial N}{\partial x} = D \frac{\partial^2 N}{\partial x^2} + F(N) \tag{7.4.1}$$

with the initial and boundary conditions: $N(x,0) = 0$, $N(0,t) = N_*$, and $N(\infty,t) = 0$. For the reaction term we select (a) $F(N) = aN$ for growth (or immigration or stocking) and (b) $F(N) = -a_*N$ for decay (or emigration or harvesting).

We use the reaction term, $F(N) = aN$, to determine the solution to Eq 7.4.1 for growth phenomena. We then replace the parameter a by the parameter $-a_*$ to acquire the answers for decay phenomena. For the growth problem, Eq. 7.4.1 becomes

$$\frac{\partial N}{\partial t} + u_0 \frac{\partial N}{\partial x} = D \frac{\partial^2 N}{\partial x^2} + aN \quad . \tag{7.4.2}$$

The Laplace transformation is a direct and simple method to solve problems of this kind. Taking the Laplace transform of Eq. 7.4.2, yields

$$s\overline{N} - N(x,0) + u_0 \frac{d\overline{N}}{dx} = D \frac{D^2 \overline{N}}{dx^2} + a\overline{N} \quad . \tag{7.4.3}$$

where s is the transform variable. Since the initial condition is $N(x,0) = 0$, this expression becomes

$$\frac{d^2 \overline{N}}{dx^2} - \frac{u_0}{D} \frac{d\overline{N}}{dx} - \left(\frac{s-a}{D} \right) \overline{N} = 0 \quad , \tag{7.4.4}$$

which is a second order linear homogeneous differential equation with constant coefficients. Its solution is not difficult; we considered a similar equation in Section 2.9.5. The two integration constants which arise are easily evaluated from the two boundary conditions. For this relatively easy problem, the inverse transformation is obtained directly from a table to provide the solution $N(x,t)$.

As we did in Section 7.1.2, we could have simplified Eq. 7.4.2 by changing the dependent variable with the substitution

$$N(x,t) = \Gamma(x,t) \exp\left[\frac{u_0 x}{2D} - \left(\frac{u_0^2}{4D} - a\right)t\right] \quad . \tag{7.4.5}$$

This substitution yields the ordinary diffusion or heat conduction equation

$$\frac{\partial \Gamma}{\partial t} = D\frac{\partial^2 \Gamma}{\partial x^2} \quad . \tag{7.4.6}$$

Of course, the price we pay for this simplification is to complicate the boundary condition, $\Gamma(0,t)$.

Alternatively Eq. 7.4.2 can be simplified by changing the independent variable as we also did in Section 7.1.2. In this case, with $z = x - u_0 t$, Eq. 7.4.2 becomes

$$\frac{\partial N}{\partial t} = D\frac{\partial^2 N}{\partial z^2} + aN \quad . \tag{7.4.7}$$

If we like we could simplify this expression even further with the addition substitution, $N(z,t) = C(z,t)\exp(at)$ to obtain

$$\frac{\partial C}{\partial t} = D\frac{\partial^2 C}{\partial z^2} \quad . \tag{7.4.8}$$

Several kinds of substitutions are given by Cannon (1984) to reduce the diffusion equation containing convection and reaction terms to the simple diffusion equation. He also gives transformations needed to simplify equations possessing time dependent and density dependent diffusion coefficients.

A word of caution: as substitutions or transformations are made in variables, we must make certain that corresponding changes are made in the initial and boundary conditions.

As mentioned, the most straightforward way to solve Eq. 7.4.2 is the Laplace transformation. However, there are other ways; one of these is called the method of separation of variables. In this case, with regard to Eq. 7.4.2, we assume that

$$N(x,t) = \phi(x)\psi(t) \quad . \tag{7.4.9}$$

Substituting the various derivatives obtained from this expression into Eq. 7.4.2 and re-arranging gives

$$\frac{1}{\psi}\frac{d\psi}{dt} = \frac{D}{\phi}\frac{d^2\phi}{dx^2} - \frac{u_0}{\phi}\frac{d\phi}{dx} + a = -k \quad . \tag{7.4.10}$$

The left hand member of this equation is a function of t alone and the right hand member is a function of x alone. The only way this can be true is that both members be equal to a constant, say, $-k$. Consequently, we have two ordinary differential equations

$$\frac{d\psi}{dt} + k\psi = 0 \tag{7.4.11}$$

and

$$\frac{d^2\phi}{dx^2} - \frac{u_0}{D}\frac{d\phi}{dx} + \left(\frac{k+a}{D}\right)\phi = 0 \quad . \tag{7.4.12}$$

These equations are easily integrated and the initial and boundary conditions are utilized to determine the value of k and the two integration constants. The desired final solution follows from Eq. 7.4.9.

Another method utilizes Duhamel's theorem. For our purpose, this theorem states that in a semi-infinite region the solution to the problem

$$\frac{\partial U}{\partial t} = D\frac{\partial^2 U}{\partial x^2} \tag{7.4.13}$$

with the conditions: $U(x,0) = 0$ and $U(0,t) = \tau(t)$ is

$$U = \frac{2}{\sqrt{\pi}} \int_{x/2\sqrt{Dt}}^{\infty} \tau\left(t - \frac{x^2}{4Dz^2}\right) e^{-z^2} dz \quad . \tag{7.4.14}$$

For example, if $\tau(t) = 1$, then

$$U = \frac{2}{\sqrt{\pi}} \int_{x/2\sqrt{Dt}}^{\infty} e^{-z^2} dz = \mathrm{erfc}\,\frac{x}{2\sqrt{Dt}} \tag{7.4.15}$$

as we obtained in Eq. 7.2.2. Clearly, Duhamel's method could be utilized to obtain solutions to above Eqs. 7.4.6 and 7.4.8.

These various methods of solution are discussed by Carslaw and Jaeger (1959), Boyce and DiPrima (1986), Farlow (1982), and in other books dealing with partial differential equations.

Regardless of the method we utilize, the solution to Eq. 7.4.2 is

$$\frac{N}{N_*} = \frac{1}{2}\left[e^{(u_0-u_r)x/2D}\,\mathrm{erfc}\left(\frac{x-u_rt}{2\sqrt{Dt}}\right) + e^{(u_0+u_r)x/2D}\,\mathrm{erfc}\left(\frac{x+u_rt}{2\sqrt{Dt}}\right)\right] \quad . \tag{7.4.16}$$

This solution is valid for both the growth equation, $F(N) = aN$, and the decay equation, $F(N) = -a_*N$. The difference in the two appears in the definition of the reference velocity, u_r

$$\text{Growth}: \quad u_r = \sqrt{u_0^2 - 4aD} \quad , \tag{7.4.17}$$

$$\text{Decay}: \quad u_r = \sqrt{u_0^2 + 4a_*D} \quad , \tag{7.4.18}$$

It is noted that if $a = a_* = 0$, then $u_r = u_0$ and Eq. 7.4.16 reduces to Eq. 7.2.26 as would be expected.

It is appropriate to examine the growth and decay cases separately.

7.4.2 Exponential Growth with Convection and Diffusion

In Eq. 7.4.17 we see a complication in the case of growth which does not appear in Eq. 7.4.18 for the case of decay. That is, if $u_0^2 < 4aD$ then u_r becomes imaginary and the complementary error functions in Eq. 7.4.16 possess arguments of the form, $x + iy$, where $i = \sqrt{-1}$. Methods for dealing with error functions having complex arguments are considered by Carslaw and Jaeger (1959). We shall not get into that; we confine our interest to the regime: $u_0^2 > 4aD$.

Now suppose a substance is growing exponentially as it is being convected but there is no diffusion. Then $D = 0$ and we obtain, most easily from Eq. 7.4.2, the result

$$\frac{N}{N_*} = \begin{cases} \exp(ax/u_0); & x < u_0 t \\ 0; & x > u_0 t \end{cases} \tag{7.4.19}$$

This result indicates that in the absence of diffusion, a "piston" front or interface moves in the x-direction with velocity u_0. The relative concentration, N/N_*, is zero ahead of the front; behind the front, the concentration grows exponentially.

For the purpose of graphically displaying the general solution, Eq. 7.4.16, it is helpful to express it in dimensionless form. Making the substitutions

$$U = \frac{N}{N_*} \quad ; \quad \xi = \sqrt{\frac{a}{D}} x \quad ; \quad T = at \quad ; \quad \sigma = \frac{u_0}{2\sqrt{aD}} \quad ;$$

$$\alpha = (\sigma - \sqrt{\sigma^2 - 1}) \quad ; \quad \beta = (\sigma + \sqrt{\sigma^2 - 1}) \quad ; \quad \gamma = 2\sqrt{\sigma^2 - 1} \quad ; \tag{7.4.20}$$

we obtain

$$U = \frac{1}{2}\left[e^{\alpha\xi}\text{erfc}\left(\frac{\xi - \gamma T}{2\sqrt{T}}\right) + e^{\beta\xi}\text{erfc}\left(\frac{\xi + \gamma T}{2\sqrt{T}}\right)\right] \quad . \tag{7.4.21}$$

In Fig. 7.4.1, plots are presented of Eq. 7.4.21 for values of $\sigma = 1$ and $\sigma = 2$ for several values of T. In the limiting case, $T = \infty$, Eq. 7.4.21 becomes $U = \exp(\alpha\xi)$.

It is observed in the figure that maximum values, U_m, occur for particular values of ξ and T. We recall that the initial condition is $U(\xi, 0) = 0$; i.e., the $T = 0$ line coincides with the ξ-axis. Now the substance being transported is growing exponentially; the combined effects of convection and diffusion translate and spread out the step function, $U(0, T) = 1$, imposed at the origin. These effects establish the maxima, delay the inevitable ascent to

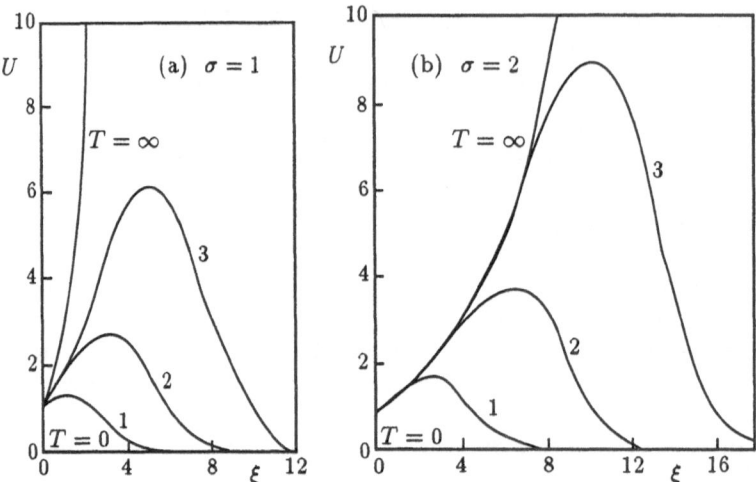

Fig. 7.4.1 Distribution of $U(\xi, T)$ for the case of exponential growth with convection and diffusion; $\sigma = u_0/2\sqrt{aD}$. (a) $\sigma = 1$ and (b) $\sigma = 2$

infinite values of U, and create a kind of diffusion wave or front moving from left to right.

7.4.3 Exponential Decay with Convection and Diffusion

Most of the analysis we carried out and results we obtained for the case of growth are applicable for decay by replacing the quantity a by the quantity $-a_*$.

The differential equation corresponding to exponential decay is

$$\frac{\partial N}{\partial t} + u_0 \frac{\partial N}{\partial x} = D \frac{\partial^2 N}{\partial x^2} - a_* N \quad . \tag{7.4.22}$$

With the same initial and boundary conditions as before, the solution is given by Eqs. 7.4.16 and 7.4.18.

If there is convection but no diffusion, then $D = 0$ and the position of the piston front is

$$\frac{N}{N_*} = \begin{cases} \exp(-a_* x/u_0); & x < u_0 t \\ 0; & x > u_0 t \end{cases} \quad . \tag{7.4.23}$$

In the analysis of a decaying transportable substance, there are no maxima and no imaginary nor complex quantities. As $t \to \infty$, i.e., the steady state case, we obtain the result

$$\frac{N}{N_*} = \exp\left[\left(1 - \sqrt{1 + \frac{4a_* D}{u_0^2}}\right) \frac{u_0 x}{2D}\right] \quad . \tag{7.4.24}$$

If $D = 0$, this expression reduces to the result given by Eq. 7.4.23. In Section 2.1.9, in our illustration involving oxygen distribution in a river, we obtained this same $D = 0$ result for the decay of biochemical oxygen demand (BOD). We note that Eq. 7.4.24 describes the distribution of BOD in a river in which longitudinal dispersion is not neglected.

In dimensionless form, the solution to the decay equation is

$$U = \frac{1}{2}\left[e^{\alpha_* \cdot \xi}\operatorname{erfc}\left(\frac{\xi - \gamma_* T}{2\sqrt{T}}\right) + e^{\beta_* \cdot \xi}\operatorname{erfc}\left(\frac{\xi + \gamma_* T}{2\sqrt{T}}\right)\right] \qquad (7.4.25)$$

in which

$$U = \frac{N}{N_*} \quad ; \quad \xi = \sqrt{\frac{a_*}{D}}x \quad ; \quad T = a_* t \quad ; \quad \sigma = \frac{u_0}{2\sqrt{a_* D}}$$

$$\alpha_* = (\sigma - \sqrt{\sigma^2 + 1}) \quad ; \quad \beta_* = (\sigma + \sqrt{\sigma^2 + 1}) \quad ; \quad \gamma_* = 2\sqrt{\sigma^2 + 1} \ . \tag{7.4.26}$$

In Fig. 7.4.2, plots are shown of the distribution $U(\xi, T)$, for $\sigma = 1$ and $\sigma = 2$ for various values of T.

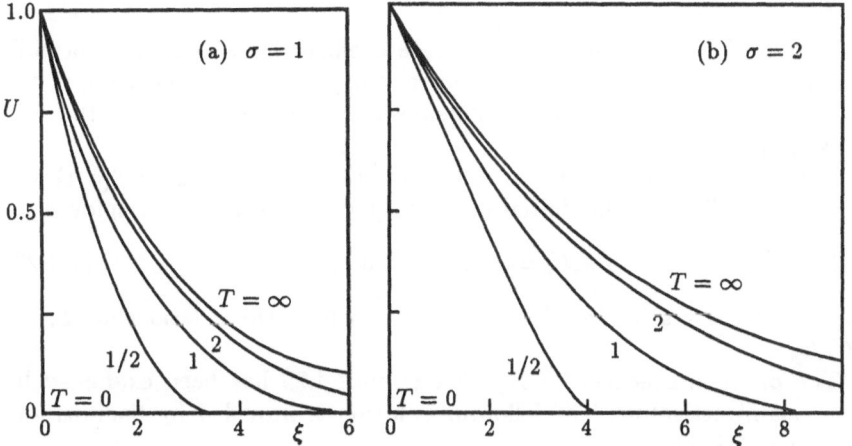

Fig. 7.4.2 Distribution of $U(\xi, T)$ for the case of exponential decay with convection and diffusion; $\sigma = u_0/2\sqrt{a_* D}$. (a) $\sigma = 1$ and (b) $\sigma = 2$

7.4.4 Convection and Diffusion with Interphase Transfer

Without straying too far from our main direction, we generalize the analysis of the preceding section to a certain extent. To illustrate this generalization we consider the following example.

Suppose a fluid containing a chemical substance is conveyed into one end of a long column which is filled with small spherical particles. The particles possess a property which allows them to adsorb the chemical substance we wish to remove from the fluid. That is, there is a transfer of mass from the fluid phase to the solid phase. Such a process occurs in an ion exchange column or an adsorption tower.

Let C be the amount of chemical per unit volume of fluid and S be the amount per unit volume of solid. The porosity of the particle-filled column is n. A mass balance on a horizontal elemental volume of the vertical adsorption column yields the following differential equation

$$\frac{\partial C}{\partial t} + \frac{1-n}{n}\frac{\partial S}{\partial t} + u_0\frac{\partial C}{\partial x} = D\frac{\partial^2 C}{\partial x^2} \ . \tag{7.4.27}$$

The first two terms on the left hand side of this equation represent, respectively, the accumulation of the chemical substance in the fluid phase and in the solid phase. The third term is the convection term and the right hand member is the diffusion term.

Since there are two dependent variables, C and S, we need a second equation to proceed with the problem. For this particular illustration, physical chemistry provides the following relationship for interphase mass transfer

$$\frac{\partial S}{\partial t} = a_*C - b_*S \ , \tag{7.4.28}$$

where a_* and b_* are mass transfer or rate coefficients. This expression indicates that the rate at which the chemical substance is accumulating in the solid phase is proportional to the difference in concentration between the fluid and solid phases.

In principle, we can now answer the question: what are $C = C(x,t)$ and $S = S(x,t)$? The following initial and boundary conditions are employed

$$C(x,0) = 0 \quad ; \quad S(x,0) = 0 \quad ; \quad C(0,t) = C_* \ . \tag{7.4.29}$$

The equilibrium solution of Eq. 7.4.28 provides the relationship $S_* = a_*C_*/b_*$.

The problem specified by Eqs. 7.4.27 to 7.4.29 has been examined by numerous investigators. The following solution is provided by Lapidus and Amundson (1952)

$$\frac{C}{C_*} = \exp\left(\frac{u_0 x}{2D}\right)\left(G(t) + b_*\int_0^t G(t)dt\right) \ , \tag{7.4.30}$$

in which

$$G(t) = \exp\left(-b_* t\right)\int_0^t I_0\left(2\sqrt{\frac{1-n}{n}a_*b_*z(t-z)}\right)$$

$$\times \frac{x}{2\sqrt{\pi D z^3}}\exp\left(-\delta z - \frac{x^2}{4Dz}\right)dz \ , \tag{7.4.31}$$

where $I_0(z)$ is the modified Bessel function of the first kind and

$$\delta = \frac{u_0^2}{4D} + \frac{1-n}{n}a_* - b_* \quad . \tag{7.4.32}$$

Theoretically, knowing $C(x,t)$ we can determine $S(x,t)$ from Eq. 7.4.28.

Even though the above solution contains only well-known functions it is still a horrendous answer. So we examine two simplified special cases.

Case 1. $a_* \neq 0$, $b_* = 0$, $D \neq 0$. By setting $b_* = 0$ in the mass transfer relationship of Eq. 7.4.28 we obtain

$$\frac{\partial S}{\partial t} = a_* C \quad , \tag{7.4.33}$$

which, when substituted into Eq. 7.4.27, gives

$$\frac{\partial C}{\partial t} + u_0 \frac{\partial C}{\partial x} = D \frac{\partial^2 C}{\partial x^2} - \frac{1-n}{n}a_* C \quad . \tag{7.4.34}$$

This expression is essentially the same as Eq. 7.4.22 involving exponential decay with convection and diffusion. So already we have the answer to this special case.

Case 2. $a_* \neq 0$, $b_* \neq 0$, $D = 0$. If we set $D = 0$ in Eq. 7.4.27, our problem is again greatly simplified. In this instance, we obtain the following solutions for $C(x,t)$ and $S(x,t)$

$$U = \begin{cases} J(\phi,\psi); & x < u_0 t \\ 0; & x > u_0 t \end{cases} \tag{7.4.35}$$

$$V = \begin{cases} 1 - J(\psi,\phi); & x < u_0 t \\ 0; & x > u_0 t \end{cases} \tag{7.4.36}$$

where

$$U = \frac{C}{C_*} \quad ; \quad V = \frac{S}{S_*} \quad ;$$

$$\phi = \frac{1-n}{n}\frac{a_*}{u_0}x \quad ; \quad \psi = \frac{b_*}{u_0}(u_0 t - x) \quad , \tag{7.4.37}$$

and

$$J(\phi,\psi) = 1 - e^{-\psi} \int_0^\phi e^{-z} I_0(2\sqrt{\psi z})dz \quad . \tag{7.4.38}$$

These answers were obtained by Goldstein (1953) who carried out an extensive analysis of the problem. He provides numerous properties of the so-called J-function including the following

$$J(\phi,\psi) + J(\psi,\phi) = 1 + e^{-(\phi+\psi)}I_0(2\sqrt{\phi\psi}) \quad ;$$

$$J(\phi,0) = e^{-\phi} \quad ; \quad J(0,\psi) = 1 \quad ; \tag{7.4.39}$$

$$\lim_{\psi \to \infty} J(\phi, \psi) = 1 \quad ; \qquad \lim_{\phi \to \infty} J(\phi, \psi) = 0 \quad .$$

He also indicates that $J(\phi, \psi)$ has the following first derivatives

$$\frac{\partial J}{\partial \phi} = -e^{-(\phi+\psi)} I_0(2\sqrt{\phi\psi}) \quad , \tag{7.4.40}$$

and

$$\frac{\partial J}{\partial \psi} = e^{-(\phi+\psi)} \sqrt{\frac{\phi}{\psi}} I_1(2\sqrt{\phi\psi}) \quad , \tag{7.4.41}$$

where $I_0(z)$ and $I_1(z)$ are the modified Bessel functions of order zero and one respectively. A plot of $J(\phi, \psi)$ is shown in Fig. 7.4.3. We observe that the abscissa and ordinate, respectively, have logarithmic and probability scales.

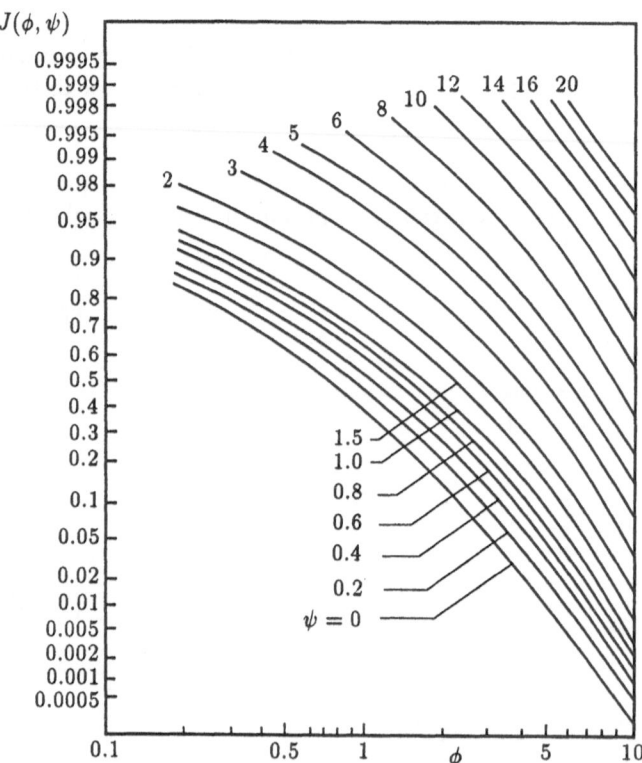

Fig. 7.4.3 A plot of the function $J(\phi, \psi)$ which characterizes a convection-interphase transfer process

Noting that the independent variable ψ contains the time dimension, we examine briefly the relationship between the growing quantity, $U = C/C_*$, and t, as we have done in the preceding sections.

Taking the second derivative of Eq. 7.4.35 with respect to ψ and setting the result equal to zero yields the abscissa, ψ_i, of the inflection point

$$\sqrt{\psi_i} I_1(2\sqrt{\phi\psi_i}) = \sqrt{\phi} I_2(2\sqrt{\phi\psi_i}) \quad . \tag{7.4.42}$$

Iterated solutions of this equation yield the value of ψ_i for a specified value of ϕ. Substitution into Eqs. 7.4.37 and 7.4.35 provide respectively the corresponding values of t_i and U_i.

The relationships for U and V given by Eqs. 7.4.35 and 7.4.36 may be useful in the analysis of other kinds of convection-diffusion-reaction phenomena. For example, these equations might be relevant in quantitative descriptions of temporal-spatial spreading of innovations, technologies, epidemics and other phenomena involving exchange processes.

Another and quite different application of the J-function is one which Goldstein gives concerning the radial diffusion of mass and conduction of heat.

For example, consider a solid circular cylinder of radius r_0 surrounded by a solid infinite region with the same thermal properties. Initially the internal cylinder has a temperature, θ_*; the external region is initially at zero temperature. The problem is to determine the temperature at any value of r and time $t > 0$. Clearly, the internal cylinder cools off as the external region warms up.

The differential equation is

$$\frac{\partial \theta}{\partial t} = \epsilon \left(\frac{\partial^2 \theta}{\partial r^2} + \frac{1}{r} \frac{\partial \theta}{\partial r} \right) \quad , \tag{7.4.43}$$

where ϵ is the thermal diffusivity. The solution given by Goldstein is

$$\frac{\theta}{\theta_*} = 1 - J\left(\frac{r_0^2}{4\epsilon t}, \frac{r^2}{4\epsilon t} \right) \quad . \tag{7.4.44}$$

Utilizing Fourier's law given by Eq. 7.0.2 and the derivative relationship of Eq. 7.4.41, the total flux of heat across the boundary, $r = r_0$, per unit length of cylinder is

$$F = 2\pi K \theta_* \left(\zeta e^{-\zeta} I_1(\zeta) \right) \quad , \tag{7.4.45}$$

in which K is the thermal conductivity and $\zeta = r_0^2/2\epsilon t$. Integration of this equation shows that after an infinite time the total flux of heat across the boundary is $F_T = \pi r_0^2 \rho c \theta_*$ which, of course, is the amount of heat contained initially in the internal cylinder.

The analogous and much simpler rectilinear problem is solved by Crank (1975).

It might be worthwhile to utilize this model as a framework for description of radial dispersion of animal and insect populations.

7.4.5 An Illustration: Chemical Solute Removal by Adsorption

Suppose we have a liquid which contains a chemical substance we wish to remove. One method utilized to a great extent in chemical engineering to attain this objective is to pass the liquid through an adsorption column or ion exchange column.

The column contains a porous material, e.g., activated carbon, which adsorbs the chemical solute from the liquid. Of course, after a certain period of operating time, during which period the solute is removed from the liquid, the adsorbing material becomes saturated and it is necessary to recharge or regenerate the material.

In our analysis of this problem the assumption is made that dispersion can be neglected. Hence, in this $D = 0$ case we utilize the J-function introduced in the previous section. We select the following numerical values: $u_0 = 1.0\,\text{m/hr}$, $n = 0.50$, $a_* = 1.0\,\text{hr}^{-1}$, and $b_* = 0.25\,\text{hr}^{-1}$. Substituting these values into the relationships of Eq.7.4.37 yields: (a) $\phi = x$ and (b) $\psi = 0.25(t - x)$. For selected values of time, t, we utilize Eqs. 7.4.35 and 7.4.36 to calculate $U = C/C_*$ and $V = S/S_*$ as functions of distance, x. Computed results are presented in Fig. 7.4.4.

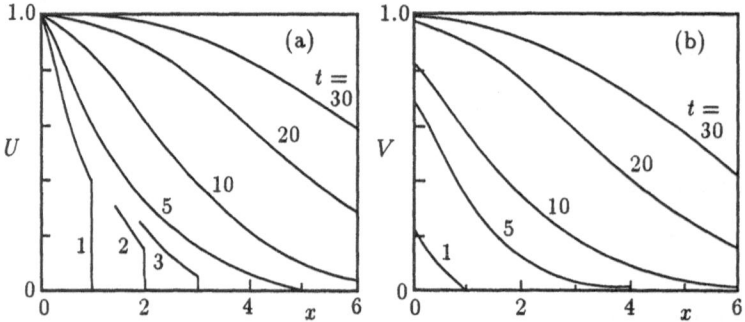

Fig. 7.4.4 Solute concentrations in an adsorption column as functions of distance x for various values of time t, (a) in the liquid phase, U, and (b) in the solid phase, V

The concentration of the solute in the liquid phase, $U = C/C_*$, is shown in Fig. 7.4.4(a).

In the absence of dispersion, there is a sharp interface as the front moves in the x-direction. The concentration at the front is $U(\phi, 0) = \exp(-\phi)$. Now suppose the adsorption column has a length, $L = 5\,\text{m}$. After $t = 5\,\text{h}$, the front reaches $x = L = 5$; the concentration at this breakthrough time is $U = 0.00674$. Subsequently, the effluent concentration increases; when $t = 10\,\text{h}$, $U = 0.11$ and when $t = 20\,\text{h}$, $U = 0.44$.

The concentration of the solute in the solid phase, $V = S/S_*$, is indicated in Fig. 7.4.4(b). At $x = 5$, $t = 5$, we determine that $V = 0$. Thereafter, $V = 0.03$ when $t = 10$ and $V = 0.28$ when $t = 20$.

And so on. It is apparent that there are numerous aspects of the problem which relate to optimal design and operation of adsorption columns. For example, we would want the operating time to be as long as possible and the maximum adsorption capacity of the particles to be utilized before commencing the regeneration cycle. Considerable information dealing with these matters is given by Vermuelen et al. (1973).

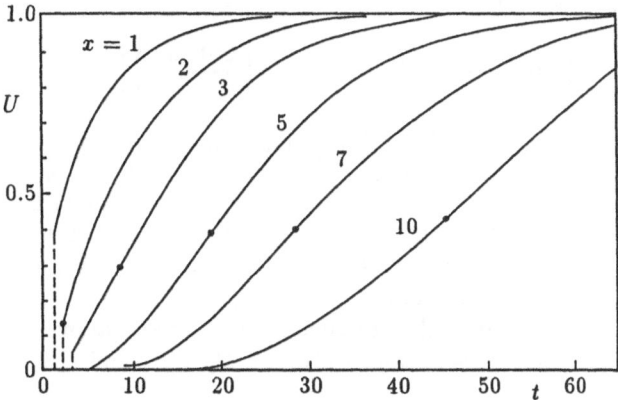

Fig. 7.4.5 Solute concentrations in the liquid phase, U, in an adsorption column as functions of time t for various values of distance x

Finally, we return to our custom of plotting $U = C/C_*$ versus time. Using the above numbers, some computed results are shown in Fig. 7.4.5 for an indefinitely long column. Inflection points for the various $x = $ constant curves are shown by solid dots.

7.5 Diffusion with Confined Exponential Growth

7.5.1 Rectilinear Diffusion

In this final section involving linear differential equations, a rather general case will be examined. We start with the expression

$$\frac{\partial N}{\partial t} + u_0 \frac{\partial N}{\partial x} = D \frac{\partial^2 N}{\partial x^2} + a_*(N_e - N) \quad . \tag{7.5.1}$$

The initial and boundary conditions are: $N(x,0) = N_0$ and $N(0,t) = N_*$; as before, we also require that the solution be bounded for large values of x.

A number of differential equations considered in previous sections are special cases of Eq. 7.5.1. For example, if $u_0 = D = 0$ and $N_e = N_*$, then we obtain the equation for confined exponential growth which we examined in Section 2.3. The above expression provides the framework for an important class of linear phenomena. As we shall see, a similar equation dealing with logistic growth is non-linear; in that instance it is not possible to obtain exact answers except in very special cases.

The solution to Eq. 7.5.1 with the indicated initial and boundary conditions is

$$N = N_0 e^{-a_* t} + N_e(1 - e^{-a_* t})$$

$$+ \tfrac{1}{2}(N_e - N_0)e^{-a_* t}\left[\operatorname{erfc}\left(\frac{x - u_0 t}{2\sqrt{Dt}}\right) + e^{\frac{u_0 x}{D}}\operatorname{erfc}\left(\frac{x + u_0 t}{2\sqrt{Dt}}\right)\right]$$

$$- \tfrac{1}{2}(N_e - N_*)\left[e^{(u_0 - u_r)\frac{x}{2D}}\operatorname{erfc}\left(\frac{x - u_r t}{2\sqrt{Dt}}\right) + e^{(u_0 + u_r)\frac{x}{2D}}\operatorname{erfc}\left(\frac{x + u_r t}{2\sqrt{Dt}}\right)\right]$$

$$(7.5.2)$$

in which $u_r = (u_0^2 + 4a_* D)^{1/2}$. Although this answer looks somewhat formidable it is really quite straightforward; the functions are well- tabulated and the various terms are easily computed.

Several solutions obtained previously are special cases of the above answer. A remarkably simple solution is obtained if we let $N_e = N_*$ and $u_0 = 0$. These substitutions yield

$$N = N_* - (N_* - N_0)e^{-a_* t}\operatorname{erf}\left(\frac{x}{2\sqrt{Dt}}\right) \quad . \qquad (7.5.3)$$

We note that this equation describes the spatial diffusion of a quantity whose temporal growth is given by a confined exponential. This answer reduces further: (a) if $a_* = 0$, $N_0 = 0$, we obtain Eq. 7.2.2

$$N = N_*\operatorname{erfc}\left(\frac{x}{2\sqrt{Dt}}\right) \qquad (7.5.4)$$

and (b) if $D = 0$, we acquire Eq. 2.3.2

$$N = N_* - (N_* - N_0)e^{-a_* t} \quad . \qquad (7.5.5)$$

Plots of Eq. 7.5.3 are shown in Fig. 7.5.1.

7.5.2 An Illustration: Heat Transfer from a River

An environmental problem in numerous rivers and lakes has to do with "thermal pollution". In most cases, such pollution is created at a thermal power plant where cold water from a river is pumped into the plant, utilized for

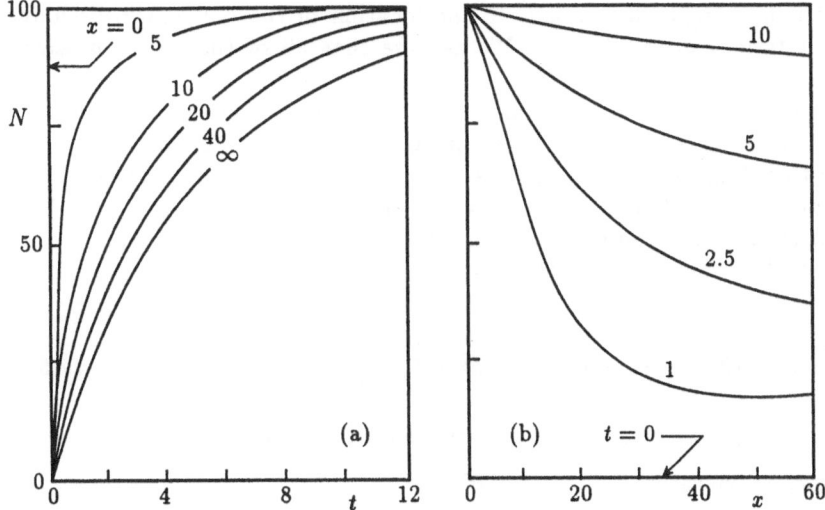

Fig. 7.5.1 Confined exponential growth with diffusion. $a_* = 0.20$, $N_0 = 0$, $N_* = 100$, $D = 100$. (a) N vs t for various x and (b) N vs x for various t

exhaust steam condensation, and returned to the river at an elevated temperature. Not infrequently, this relatively hot water is detrimental to aquatic life in the river.

We use this topic as our next illustration and employ Eqs. 7.5.1 and 7.5.2 as the framework for analysis and computation.

In these two equations, the following changes and definitions are made in the symbols

$N \longrightarrow \theta$ temperature, °C

$D \longrightarrow \epsilon = K/\rho c$

   ϵ thermal diffusivity, m²/s
   ρ density, kg/m³
   c specific heat, joule/kg °C

$a_* \longrightarrow \gamma = h/\rho c H$

   γ surface transfer coefficient, 1/s
   h heat transfer coefficient, joule/s m² °C
   H water depth, m

Subscripts

   θ_0 water temperature, $t = 0$
   θ_* water temperature, $x = 0$
   θ_e air temperature

The origin, $x = 0$, corresponds to the location of the hot water discharge conduit from the power plant. We assume that all of the above-indicated parameters are constant.

If we were interested in obtaining a solution to the entire transient problem, $\theta = \theta(x,t)$, it would be necessary to utilize Eq. 7.5.2. However, much easier to obtain and indeed more relevant is the steady state solution. This solution can be obtained in two ways: (a) let $t \to \infty$ in Eq. 7.5.2 or (b) set $\partial N/\partial t$ in Eq. 7.5.1 and solve the resulting ordinary differential equation for $N(x)$. Either way we obtain the answer

$$\theta = \theta_e + (\theta_* - \theta_e)\exp\left[(u_0 - \sqrt{u_0^2 + 4\gamma\epsilon})\frac{x}{2\epsilon}\right] \quad . \tag{7.5.6}$$

The following numerical values are selected for our computations:

Entrance temperature	$\theta_* = 30°C$
Air temperature	$\theta_e = 20°C$
River velocity	$u_0 = 0.30\,\text{m/s}$
River depth	$H = 3.0\,\text{m}$
Thermal conductivity	$K = 0.59\,\text{joule/s m}°C$
Density	$\rho = 1000\,\text{kg/m}^3$
Specific heat	$c = 4180\,\text{joule/kg}°C$
Heat transfer coefficient	$h = 250\,\text{joule/s m}^2\,°C$
	(The magnitude of h depends primarily on the river velocity. Hindley and Miner (1972) provide information on this topic)
Surface transfer coefficient	$\gamma = 2.0 \times 10^{-5}\,\text{s}^{-1}$
Thermal diffusivity	$\epsilon = 25.0\,\text{m}^2/\text{s}$
	(In this instance, the thermal diffusivity, perhaps better termed the thermal dispersion coefficient, is due to the turbulence and velocity distribution of the fluid flow. Taylor (1954) and Fischer (1968) give information on this subject.)

Substituting these numerical values into Eq. 7.5.6 yields the temperature profile shown in Fig. 7.5.2. We note that the temperature drops from $30°C$ to about $25°C$ after 10 kilometers. After 25 kilometers, the temperature has fallen to around $22°C$ and at 50 kilometers the water temperature is approximately equal to the air temperature.

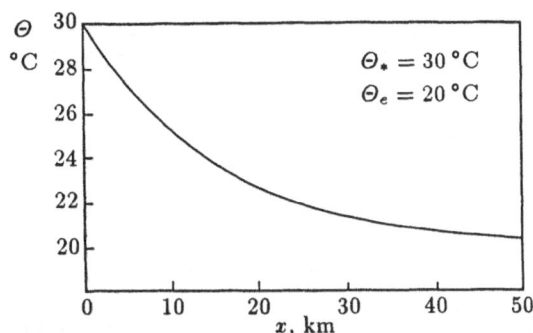

Fig. 7.5.2 Temperature distribution in a river due to heat transfer from the river water to the atmosphere

7.5.3 Radial Diffusion

Next we consider the problem of confined exponential growth in a radial coordinate system. This time we discard convection and re-define the reaction or growth term as indicated in the equation

$$\frac{\partial N}{\partial t} = D\left(\frac{\partial^2 N}{\partial r^2} + \frac{1}{r}\frac{\partial N}{\partial r}\right) + w - a_* N \quad . \tag{7.5.7}$$

In this expression, w is a constant migration term; it may be positive (immigration) or negative (emigration).

For the initial condition we take $N(r,0) = 0$. This time, for a change, we select a constant flux, q, as the boundary condition at $r = 0$. Again, the outer boundary condition stipulates a finite answer.

The solution to Eq. 7.5.7 is

$$N = \frac{w}{a_*}\left(1 - e^{-a_* t}\right) + \frac{q}{4\pi D}\int_{r^2/4Dt}^{\infty} \exp\left(-\zeta - \frac{a_* r^2}{4D\zeta}\right)\frac{d\zeta}{\zeta} \quad . \tag{7.5.8}$$

If $a_* = 0$ in the above expression we obtain

$$N = wt + \frac{q}{4\pi D}\int_{r^2/4Dt}^{\infty} e^{-\zeta}\frac{d\zeta}{\zeta} \quad , \tag{7.5.9}$$

where the second term is the exponential integral defined by Eq. 7.2.19.

If $w = a_* N_*$ in Eq. 7.5.8 we get

$$N = N_*\left(1 - e^{-a_* t}\right) + \frac{q}{4\pi D}\int_{r^2/4Dt}^{\infty} \exp\left(-\zeta - \frac{a_* r^2}{4D\zeta}\right)\frac{d\zeta}{\zeta} \quad . \tag{7.5.10}$$

With this same substitution, $w = a_* N_*$, in the differential equation Eq. 7.5.7, we note that Eq. 7.5.10 provides the solution to the problem of diffusion with confined exponential growth in a radial system with constant flux at the origin.

It would be worthwhile to pursue various topics concerning the complete spatial-temporal solution expressed by Eq. 7.5.8 although the indefinite integral would be troublesome. Instead, we examine the steady state solution of Eq. 7.5.8. So letting $t \to \infty$ we obtain the answer

$$N = \frac{w}{a_*} + \frac{q}{2\pi D} K_0 \left(\sqrt{\frac{a_*}{D}} r \right) \quad , \tag{7.5.11}$$

in which $K_0(z)$ is the modified Bessel function of the second kind and order zero.

Carrying out an area integration of Eq. 7.5.11, we obtain the total population, $P(r)$, contained in a circular region between the origin and any radius r. The answer is

$$P = \frac{\pi w r^2}{a_*} + \frac{q}{a_*} \left[1 - \sqrt{\frac{a_*}{D}} r K_1 \left(\sqrt{\frac{a_*}{D}} r \right) \right] \quad . \tag{7.5.12}$$

At this point, for simplicity, we set $w = 0$. Then, in dimensionless form, Eq. 7.5.11 becomes

$$\frac{N}{q/2\pi D} = K_0(\xi) \quad , \tag{7.5.13}$$

where $\xi = \sqrt{a_*/D}\, r$. We also obtain

$$\frac{P}{q/a_*} = 1 - \xi K_1(\xi) \quad . \tag{7.5.14}$$

A semi-logarithmic plot of Eq. 7.5.13 is shown in Fig. 7.5.3. For large values of ξ, it can be established that

$$\frac{N}{q/2\pi D} = \sqrt{\frac{\pi}{2\xi}} e^{-\xi} \quad . \tag{7.5.15}$$

For comparison, the function $f(\xi) = exp(-\xi)$ is shown in the figure.

For the total population relationship of Eq. 7.5.14, it is convenient to replace the radius with the area, $A = \pi r^2$. Then letting $\phi = A/(D/a_*)$, Eq. 7.5.14 becomes

$$\frac{P}{q/a_*} = 1 - \sqrt{\frac{\phi}{\pi}} K_1 \left(\sqrt{\frac{\phi}{\pi}} \right) \quad . \tag{7.5.16}$$

A log–log plot of this equation is presented in Fig. 7.5.6. For the moment we disregard the data points shown in the figure. When $\phi = 1$ or less, we could show that

$$\frac{P}{q/a_*} = \frac{\phi}{4\pi} \log_e \frac{4\pi}{\phi} \quad . \tag{7.5.17}$$

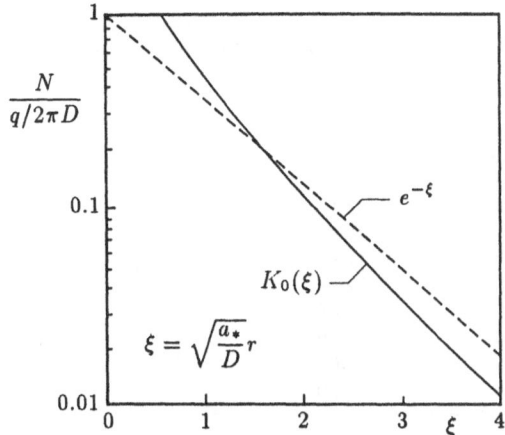

Fig. 7.5.3 Steady state distribution $N(\xi)$ in a radial coordinate system with diffusion, decay, and constant flux, q, at the origin

7.5.4 An Illustration: Population in Cities

Over the years, urban demographers have carried out numerous studies concerning population densities and distributions in cities. In many cases, such studies have sought to determine the population density as a function of distance from the center of a city. Almost invariably and not surprisingly it has been found that density is a maximum at or near the city center and declines with distance from the center.

One of the most comprehensive studies along these lines was that of Clark (1951). To describe the functional relationship between density and distance, Clark utilized the empirical expression

$$N = \alpha \exp(-\beta r) \quad , \tag{7.5.18}$$

where r is the distance from the city center. He determined the values of the constants α and β for many cities of the world ranging from Paris for the year 1817 ($\alpha = 450\,000/\text{mi}^2$; $\beta = 2.35/\text{mi}$) to Los Angeles for 1940 ($\alpha = 30\,000$; $\beta = 0.25$).

Included in Clark's tabulation was a plot of the observed population density distribution for the city of Boston for the year 1940; his data are listed in Table 7.5.1. We use this topic as the first part of our illustration involving population in cities.

Table 7.5.1 Distances from the city center and average population densities, city of Boston, 1940. From Clark (1951)

Distance, r [miles]	Density, N [people/mi^2]	Distance, r [miles]	Density, N [people/mi^2]
0.5	26 300	8.5	3200
1.5	25 100	9.5	2300
2.5	19 900	10.5	1700
3.5	15 500	11.5	1200
4.5	11 500	12.5	900
5.5	9 800	13.5	700
6.5	5 200	14.5	600
7.5	4 600	15.5	500

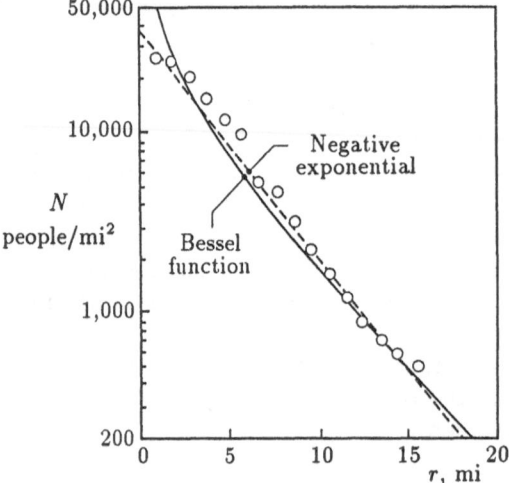

Fig. 7.5.4 Steady state density distribution $N(r)$ in a radial system. Data points refer to city of Boston for the year 1940. From Clark (1951)

The data of Table 7.5.1 are shown in the semi-logarithmic plot of Fig. 7.5.4. It is clear that the negative exponential relationship of Eq. 7.5.18 describes the data very well. A least squares computation of the quantities listed in the table gives $\alpha = 37\,500$, $\beta = 0.286$. The resulting correlation is shown as the dashed line in Fig. 7.5.4.

We now turn to the analytical framework we devised above. Recall that the main result, Eq. 7.5.13, is the steady state solution with constant flux, q, at the origin, features no migration ($w = 0$), and incorporates a decay term ($-a_*N$). Accordingly, our differential equation is

$$D\left(\frac{d^2N}{dr^2} + \frac{1}{r}\frac{dN}{dr}\right) - a_*N = 0 \quad . \tag{7.5.19}$$

This relationship indicates that the net diffusion into and out of an elemental cylindrical volume is equal to the rate of decrease within the volume. Rewriting this equation we obtain

$$\frac{d^2N}{dr^2} + \frac{1}{r}\frac{dN}{dr} = \beta^2 N \quad , \tag{7.5.20}$$

where $\beta^2 = a_*/D$. The empirical relationship employed by Clark, i.e., Eq. 7.5.18, yields

$$\frac{d^2N}{dr^2} = \beta^2 N \quad . \tag{7.5.21}$$

The close similarity between Eqs. 7.5.20 and 7.5.21 is noted; in a rectilinear system the two expressions would be identical.

Comparing our solution, Eq. 7.5.13, with the expression used by Clark, Eq. 7.5.18, suggests the functional correspondences

$$\frac{q}{2\pi D} \to \alpha \quad ; \quad \sqrt{\frac{a_*}{D}} \to \beta \quad ; \quad \frac{q}{a_*} \to \frac{2\pi\alpha}{\beta^2} \quad . \tag{7.5.22}$$

As these expressions indicate, in this model two parameters (α and β, say) are needed to specify the density distribution. The source flux parameter, $q/2\pi D$, provides the mechanism for establishing the maximum density (α). The diffusion length parameter, $\sqrt{a_*/D}$, defines the degree of radial attenuation (β).

Utilizing the asymptotic expression of Eq. 7.5.15, we rewrite the solution, Eq. 7.5.13, in the form

$$\log_e(N\sqrt{r}) = \log_e\left[\frac{\pi}{2}\left(\frac{q}{2\pi D}\right)\left(\sqrt{\frac{D}{a_*}}\right)^{1/2}\right] - \sqrt{\frac{a_*}{D}}r \quad . \tag{7.5.23}$$

A least squares computation of the data of Table 7.5.1 for values of $r = 2.5$ and larger yields: $q/D = 137\,400$, $a_*/D = 0.0493$, and $q/a_* = 2\,792\,000$.

In Eq. 7.5.14, the quantity $\xi K_1(\xi) = 0$ as ξ goes to infinity. Accordingly, the above value of q/a_* specifies the total 1940 population of the entire Boston area. This value is reasonably close to the metropolitan population of $2\,209\,600$ reported in the 1940 census.

Substituting the above indicated numerical values into Eq.7.5.13 yields the solid curve shown in Fig. 7.5.4.

We extend our illustration concerning population in cities to the topic of the relationship between the area, A, of a city and its population, P. Our diffusion-source flux model yielded Eq. 7.5.16 which provides a functional relationship between P and A.

For data relevant to this topic we examine the density–distance relationships reported by Muth (1969). In his study of spatial patterns of population densities in U.S. cities, Muth determined the (α, β) values of 46 cities for the year 1950, ranging in population from Chicago (4 920 800) to Utica, New York (117 400). Information concerning land area and population of the 46 urbanized areas was acquired from the 1950 U.S. Census reports.

The magnitudes of (α, β) were calculated by Muth on the basis of the negative exponential distribution of Eq.7.5.18. The functional relationships connecting α and β with the parameters of our Bessel function distribution (viz., q, a_*, and D) are indicated by Eq. 7.5.22. However, since the two mathematical frameworks are different, it is necessary to alter the numerical values when going from one framework to the other. We confronted the same situation in Section 2.5 in connection with the logistic and normal probability distribution functions.

For our present purpose we convert the (α, β) data in the following way. Utilizing the negative exponential relationship of Eq. 7.5.18 and carrying out an area integration, gives the following expression for the total population contained in a circular region of radius r

$$\frac{P}{2\pi\alpha/\beta^2} = 1 - (1 + \beta r)e^{-\beta r} \quad . \tag{7.5.24}$$

Setting the second derivative of this equation equal to zero provides the inflection point

$$\beta r_{\mathrm{i}} = 1 \quad ; \qquad \frac{P_{\mathrm{i}}}{2\pi\alpha/\beta^2} = 1 - \frac{2}{\mathrm{e}} = 0.264 \quad . \tag{7.5.25}$$

In a similar fashion, we integrate the Bessel function distribution given by Eq. 7.5.13 to obtain the total population relationship

$$\frac{P}{q/a_*} = 1 - \left(\sqrt{\frac{a_*}{D}}r\right) K_1 \left(\sqrt{\frac{a_*}{D}}r\right) \quad . \tag{7.5.26}$$

This time we obtain the following inflection point coordinates

$$\sqrt{\frac{a_*}{D}}r_{\mathrm{i}} = 0.594 \quad ; \qquad \frac{P_{\mathrm{i}}}{q/a_*} = 0.215 \quad . \tag{7.5.27}$$

The two distribution functions, and their inflection points, are shown in Fig. 7.5.5. To obtain a relationship for converting values of (α, β) into values of $(\sqrt{a_*/D}, q/a_*)$, we arbitrarily equate the coordinates of the inflection points. This yields the following results

$$\sqrt{\frac{a_*}{D}} = 0.594\beta \quad ; \qquad \frac{q}{a_*} = 1.229\frac{2\pi\alpha}{\beta^2} \quad . \tag{7.5.28}$$

We can now utilize the data of Muth with the main result of our mathematical model: the plot of Eq. 7.5.16 shown as the solid curve in Fig. 7.5.6. Employing the above-indicated values of the conversion constants, we obtain the following expressions for calculating the data point coordinates of Fig. 7.5.6

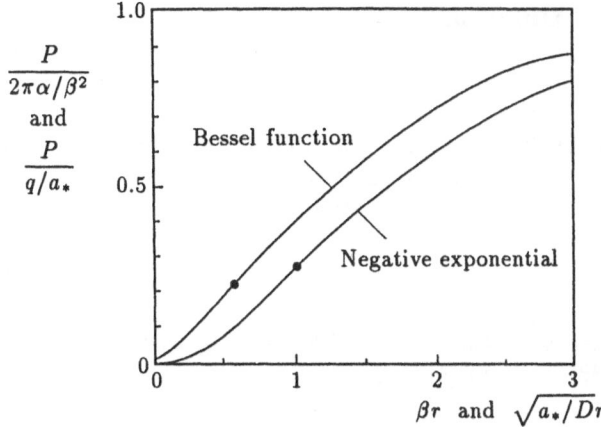

Fig. 7.5.5 Comparison of the cumulative population distributions corresponding to the Bessel function and negative exponential density distributions

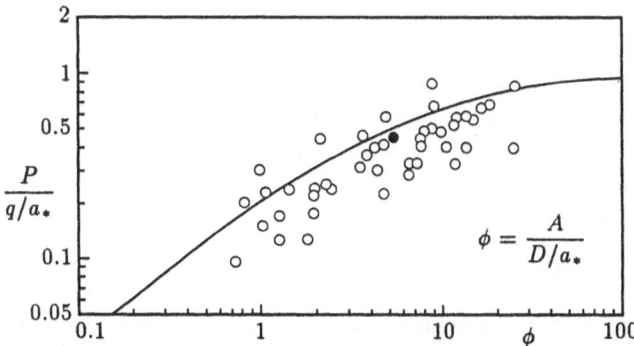

Fig. 7.5.6 Relationship between the population of a city and its area. Comparison of theoretical curve with measured values of 46 cities for year 1950. Solid dot identifies city of Louisville. Data of Muth (1969)

$$\phi = 0.353\beta^2 A \quad ; \qquad \frac{P}{q/a_*} = 0.130\frac{\beta^2}{\alpha}P \quad . \tag{7.5.29}$$

For example, for the city of Louisville, Muth reports $\alpha = 29\,000$ and $\beta = 0.47$. The 1950 Census indicates a population, $P = 472\,700$, and a land area, $A = 66.6$ square miles, for the urbanized area of Louisville. Substituting these numbers into Eqs. 7.5.29 gives $\phi = 5.19$ and $P/(q/a_*) = 0.468$. These coordinates identify the solid dot shown in Fig. 7.5.6.

More precise methods of matching the negative exponential and Bessel function distributions could be utilized; however, they would be somewhat more complicated.

We observe in Fig. 7.5.6 that there is fairly good agreement between the theoretical curve and the measured data.

7.5.5 Temporal–Spatial Diffusion

At this point we examine a problem of temporal–spatial diffusion which combines some of the features considered in the immediately preceding sections. As we shall see in an illustration involving the population of London, this model provides a framework for describing the growth of cities.

We begin with the following diffusion equation

$$\frac{\partial N}{\partial t} = D\frac{\partial^2 N}{\partial x^2} - a_* N \quad . \tag{7.5.30}$$

Now if N does not change with time, i.e., if $\partial N/\partial t = 0$, then Eq. 7.3.30 reduces to the ordinary differential equation

$$\frac{d^2 N}{dx^2} - \frac{a_*}{D}N = 0 \quad , \tag{7.5.31}$$

whose solution is

$$N = \alpha\exp(-\beta x) \quad , \tag{7.5.32}$$

where $\alpha = N(0)$ and $\beta = \sqrt{a_*/D}$. This equation is identical to the empirical expression of Eq. 7.5.18. Hence the selection of Eq.7.5.30 as the vehicle for an unsteady state analysis has a good measure of practical justification.

For the initial condition we take $N(x,0) = 0$. For reasons we discuss later on, we select the following boundary condition

$$N(0,t) = N_*(t) = N_{*0}(gt)e^{-gt} \quad . \tag{7.5.33}$$

This is a gamma distribution of the type we examined in Section 3.5.5. With the following dimensionless variables

$$U = \frac{N}{N_{*0}} \quad ; \qquad T = a_* t \quad ; \qquad \xi = \sqrt{\frac{a_*}{D}}\,x \quad ; \qquad \rho = \frac{g}{a_*} \quad . \tag{7.5.34}$$

The differential equation and initial and boundary conditions become

$$\frac{\partial U}{\partial T} = \frac{\partial^2 U}{\partial \xi^2} - U \quad ; \tag{7.5.35}$$

$$U(\xi,0) = 0 \quad ; \qquad U(0,T) = U_*(T) = \rho T e^{-\rho T} \quad . \tag{7.5.36}$$

The easiest way to solve Eq. 7.5.35 is to make the substitution: $U(\xi,T) = V(\xi,T)\exp(-T)$. This yields

$$\frac{\partial V}{\partial T} = \frac{\partial^2 V}{\partial \xi^2} \quad ; \qquad V(\xi,0) = 0 \quad ; \qquad V(0,T) = \rho T e^{(1-\rho)T} \quad . \tag{7.5.37}$$

Employing the Laplace transformation method, the solution to the transformed equation is

$$\overline{V} = \frac{\rho}{[s-(1-\rho)]^2}e^{-x\sqrt{s}} \quad . \tag{7.5.38}$$

The inverse transform of this expression is well-tabulated. Converting from $V(\xi,T)$ to $U(\xi,T)$, we obtain the final answer

$$U = \frac{1}{2}\frac{\rho T e^{-\rho T}}{m\sqrt{T}} \left[e^{m\xi}(F\,\text{erfc}\,F) - e^{-m\xi}(G\,\text{erfc}\,G)\right] \quad , \tag{7.5.39}$$

in which erfc(z) is the complementary error function and

$$m = \sqrt{1-\rho} \quad ; \qquad \binom{F}{G} = \frac{\xi \pm 2mT}{2\sqrt{T}} \quad . \tag{7.5.40}$$

Plots of Eq. 7.5.39 are shown in Figs. 7.5.7 and 7.5.8 for the case, $\rho = 3/4$. With this value of ρ, we can show that the boundary condition, $U_*(T) = (\rho T)\exp(-\rho T)$, attains its maximum when $T = 4/3$. Thus, in Fig. 7.5.7, the curves are rising with increasing T and in Fig. 7.5.8 they are descending.

Plots of Eq. 7.5.39 are displayed in Fig. 7.5.9 for $\rho = 3/4$ with time, T, as the abscissa.

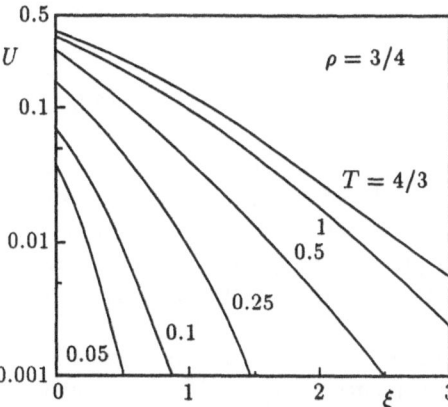

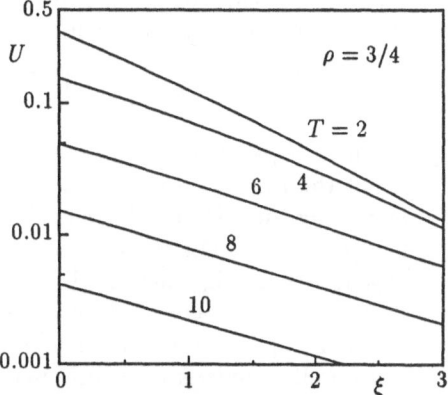

Fig. 7.5.7 Dimensionless density distribution as a function of distance with time as the parameter. Increasing values of boundary density

Fig. 7.5.8 Dimensionless density distribution as a function of distance with time as the parameter. Decreasing values of boundary density

We note in Eq. 7.5.39 that the group $(\rho T)\exp(-\rho T)$ is simply the time-dependent boundary condition, $U_*(T)$. Also, we observe that Eq. 7.5.40 contains the parameter which describes transmission of a wave in the positive and negative ξ-directions with a velocity $c = 2\sqrt{1-\rho}$.

Taken together, these two observations indicate that indeed a wave or front is transmitted in the ξ-direction, moving at constant velocity with an amplitude initially increasing to a maximum and thereafter decreasing. This wave, apparent in Figs. 7.5.7 and 7.5.8, is seen more clearly in Fig. 7.5.9.

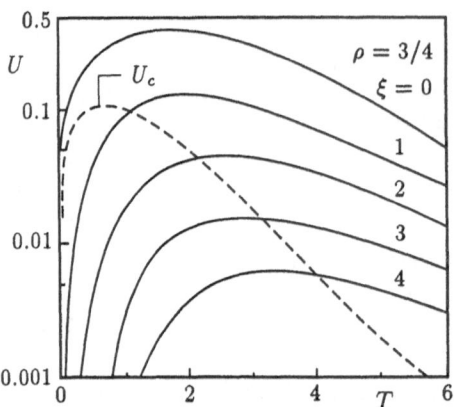

Fig. 7.5.9 Dimensionless density distribution as a function of distance with time as the parameter. Dashed line identifies magnitude of density measured from reference frame moving at the wave velocity

Continuing the topic of wave propagation, suppose that $\xi_c = 2\sqrt{1-\rho}\,T$. This corresponds to the case in which an observer travels with the wave. In this instance Eq. 7.5.39 becomes

$$U_c = \rho T e^{(2-3\rho)T} \operatorname{erfc}\left(2\sqrt{(1-\rho)T}\right) \quad , \tag{7.5.41}$$

in which U_c is the magnitude of U seen by the moving observer. Initially, U_c is not the maximum value, U_m. However, as Kendall (1948) indicates, with increasing time, U_c and U_m coincide and the velocity of the crest of the wave is $c = 2\sqrt{1-\rho}$. For the case $\rho = 3/4$, a plot of Eq. 7.5.41 is shown as the dashed line in Fig. 7.5.9. A display of Eq. 7.5.41 is presented in Fig. 7.5.10 for values of ρ ranging from $\rho = 0.1$ to 1.

In the light of Figs. 7.5.7 and 7.5.8, and looking ahead to our illustration concerning population densities in cities, we return to Eq. 7.5.32, the negative exponential distribution.

The determination of the parameters α and β is a task frequently faced by urban demographers and land economists. In this regard and with reference to Fig. 7.5.7, suppose that at a certain time, e.g., $T = 1$, the (ξ, U) points at $\Delta\xi = 0.5$ intervals represent actual data of population densities, U, measured at various distances, ξ, from the center of a city. Then, as computed from Eq. 7.5.39 with $\rho = 3/4$, we would have the following hypothetical field data

ξ:	0.0	0.5	1.0	1.5	2.0	2.5	3.0
U:	0.3540	0.1870	0.0918	0.0420	0.0176	0.0068	0.0025

Since our formulation is in terms of dimensionless quantities, we write the dimensionless equivalent of Eq. 7.5.32 in the form

$$U = \alpha_* \exp(-\beta_*\xi) \quad . \tag{7.5.42}$$

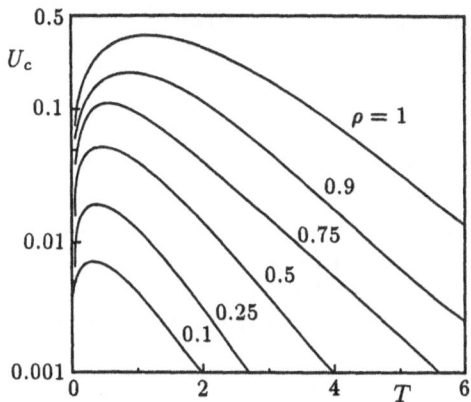

Fig. 7.5.10 Magnitude of density measured from a reference frame moving at the wave velocity for various values of $\rho = g/a_*$

Dropping the subscripts and taking logarithms, we obtain

$$\log_e U = \log_e \alpha - \beta \xi \quad .$$

A least squares computation with the seven data points yields: $\alpha = 0.428$ and $\beta = 1.65$. A similar analysis carried out at time $T = 2$ gives $\alpha = 0.358$ and $\beta = 1.10$.

Values of α and β for values of T ranging from $T = 0.10$ to 8 are shown in Fig. 7.5.11. Not surprisingly, the shape of the $\alpha(T)$ curve resembles a gamma distribution; the $\beta(T)$ curve behaves something like a negative exponential.

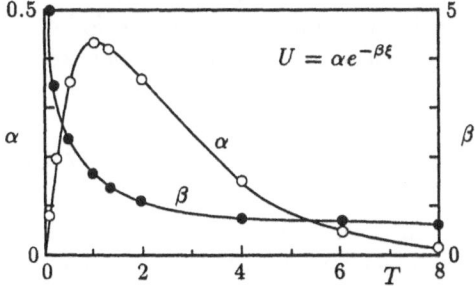

Fig. 7.5.11 Plots of equivalent values of the negative exponential distribution parameters α and β

With respect to the basic solution, Eq. 7.5.39, it is possible to obtain a simpler answer for the case of large values of T. By this we mean that in Eq. 7.5.40, ξ is very small with respect to $2\sqrt{1-\rho} \, T$. In this instance, Eq. 7.5.39 becomes

$$U = \rho T e^{-\rho T} [\cosh m\xi - \operatorname{erf}(m\sqrt{T}) \sinh m\xi] \quad , \tag{7.5.43}$$

where $m = \sqrt{1-\rho}$. If T becomes still larger, so that $\text{erf}(m\sqrt{T}) = 1$, then Eq. 7.5.43 reduces to the very simple expression

$$U = (\rho T e^{-\rho T})e^{-m\xi} \quad . \tag{7.5.44}$$

Clearly, this is another form of the negative exponential distribution.

Throughout this section we have considered $U(\xi, T)$ to be a dimensionless population density distribution. We continue this connotation as we examine the topic of cumulative or total population distribution.

If the geometry of our problem were such that we had strictly rectilinear diffusion then the total population in a strip of width B and length ξ would be

$$\psi = \frac{P}{B\sqrt{D/a_*}\,N_{*0}} = \int_0^\xi U(\xi, T)d\xi \quad . \tag{7.5.45}$$

Substitution of Eq. 7.5.39, or its large T approximations, Eqs. 7.5.43 or 7.5.44, into this expression and carrying out the integration would give an answer of the form $\psi = \psi(\xi, T)$. With the implication that we are considering populations of cities, this model might be appropriate for "rectilinear" cities, such as Bombay or Caracas, but not for the customary "radial" city.

From here on we consider the radial problem. To be precise, we really should have solved the $U = U(\xi, T)$ problem in a polar coordinate system. Had we done so, we would have acquired very complicated integrals of Bessel functions virtually impossible to handle with any ease. More importantly, utilization of the solution to the rectilinear density distribution problem in an ensuing radial cumulative distribution problem is the long-established approach taken by urban geographers, regional planners and others.

Consequently, with a circular area integration we obtain the following expression for the total population contained in a circle of dimensionless radius ξ:

$$\psi = \frac{P}{2\pi(D/a_*)N_{*0}} = \int_0^\xi \xi U(\xi, T)d\xi \quad . \tag{7.5.46}$$

It is implied that r now replaces x in the definition of ξ. A closed-form integration of this equation giving the exact answer for $\psi(\xi, T)$ is not possible. However, numerical integration is straightforward. Fig. 7.5.12 displays dimensionless plots of $\psi(\xi, T)$ based on $\rho = 3/4$.

It is not difficult to obtain answers for the cumulative population for the large T approximation of Eq. 7.5.43 and the very large T approximation of Eq. 7.5.44. For the latter we obtain

$$\psi = \frac{\rho T e^{-\rho T}}{1-\rho}\left[1 - (1 + m\xi)e^{-m\xi}\right] \quad , \tag{7.5.47}$$

in which $m = \sqrt{1-\rho}$.

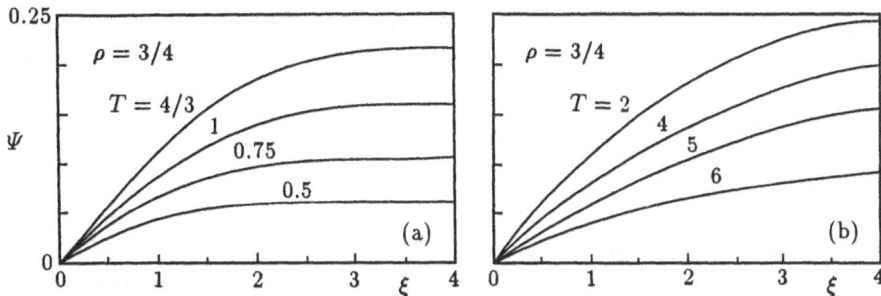

Fig. 7.5.12 Dimensionless cumulative distributions as a function of distance with time as the parameter. (a) Increasing values of boundary density and (b) decareasing values of boundary density

7.5.6 An Illustration: Population of London

Over the years, specialists in numerous disciplines – demography, economics, geography, sociology – have studied the spatial distribution of population in cities, metropolitan areas, and regions.

As mentioned above, many of these studies have been concerned with the determination of the most appropriate density distribution function, $N = N(x)$, to utilize in describing population distribution and total population. Numerous distribution functions have been examined: negative exponential, negative square root exponential, normal probability, polynomial exponential, lognormal, Pareto, gamma, and power law. The contributions of Casetti (1969), Papageorgiou (1971), and Nairn and O'Neill (1988) examine these investigations and summarize their results.

These numerous studies of urban and regional population distributions have two things in common. First, they have been directed primarily toward the problem of spatial distribution. The examination of aspects of temporal distribution has featured the determination of numerical values of parameters (α and β, say) and reporting changes of these values from one decade to the next. Second, the appropriateness of a particular density distribution function has been made largely on the basis of its statistical significance with a particular set of data.

On this latter point, the combined attributes of statistical accuracy and mathematical simplicity have convinced many urban specialists that the negative exponential density distribution function is the most suitable. According to Mogridge (1985) this function was first utilized by Clark (1951) in his study of the population distributions, including determination of temporal changes of α and β, in 20 cities of the world.

The negative exponential distribution function, given by Eq. 7.5.18 or Eq. 7.5.32, has been utilized by numerous investigators. A study by Blumenfeld (1954) of the population distribution of Philadelphia contains data showing an increase and subsequent decline of α and a continual decrease of β over the period 1900 to 1950. The analysis of Winsborough (1962) of the density distribution of Chicago from 1860 to 1950 shows the same rise and fall of α and steady decline of β. The study of Guest (1973) showed similar characteristics of the population distribution of Cleveland as did the work of Edmonston and Davies (1976) involving 37 metropolitan areas in the western region of the United States.

The comprehensive report of Bussière (1976) of the density distributions of Paris and London provides values of α and β at decade intervals commencing in the mid-1850s. The studies of Mogridge, also of the population distributions of London and Paris, are analyzed on the basis of the negative exponential function. He utilizes the results for analysis of topics concerning transport, land use, and energy policy and planning. Finally, Parr (1985) utilizes several distribution functions, including the negative exponential, in analyses of the regional spatial structure of a number of urban regions in the United Kingdom and North America.

Now to our illustration concerning London. Based on data acquired by Bussière (1976), the values of α and β for the London region over the period 1801 to 1971 are reported by Mogridge (1985) and are listed in Table 7.5.2 and displayed in Fig. 7.5.13. These values agree reasonably well with the more limited data of Clark (1951).

Table 7.5.2 Values of the negative exponential distribution parameters for London, 1801 to 1971. From Bussière (1976) and Mogridge (1985)

Year	α [10^3/km^2]	β [1/km]	Year	α [10^3/km^2]	β [1/km]
1801	56.5	0.64	1891	66.9	0.29
1811	62.5	0.60	1901	63.5	0.25
1821	70.3	0.58	1911	55.3	0.22
1831	78.6	0.54	1921	53.8	0.21
1841	94.1	0.56	1931	44.6	0.18
1851	98.9	0.51	1941	35.9	0.15
1861	94.1	0.46	1951	26.7	0.13
1871	82.0	0.39	1961	22.3	0.12
1881	75.7	0.33	1971	15.5	0.10

At this point, we look at the differential equation and the initial and boundary conditions, Eqs. 7.5.35 and 7.5.36. First, we do not know what year corresponds to $T = 0$. Second, even if we did, a very awkward initial condition, $U(\xi, 0)$, for use in the Laplace transform would have to be utilized as the initial density distribution. Third, we know that we need a boundary condition, $U_*(T)$, that initially rises and subsequently declines.

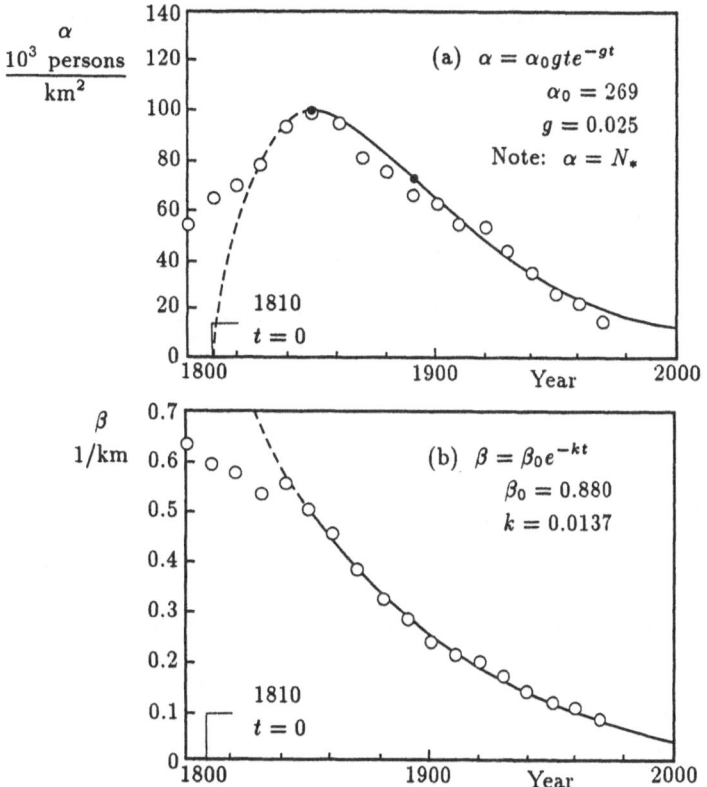

$$\frac{\alpha}{10^3 \frac{\text{persons}}{\text{km}^2}}$$

(a) $\alpha = \alpha_0 g t e^{-gt}$

$\alpha_0 = 269$

$g = 0.025$

Note: $\alpha = N_*$

1810
$t = 0$

$\frac{\beta}{1/\text{km}}$

(b) $\beta = \beta_0 e^{-kt}$

$\beta_0 = 0.880$

$k = 0.0137$

1810
$t = 0$

Fig. 7.5.13 Population of London. Measured values of the negative exponential distribution parameters: (a) magnitude of α and (b) magnitude of β. Data from Mogridge (1985)

An easy way out of these difficulties is to utilize a gamma distribution boundary condition; this we did with Eq.7.5.36. So, with good fortune, we fit the data shown in Fig. 7.5.13(a) to the gamma distribution and, at the same time, establish the time origin. As we shall see, the only drawback to this approach is that we are compelled to assume that the population of London was zero prior to 1810. This means they had only five years to prepare for Waterloo.

Iterated computations involving the data of Table 7.5.2, with a minimum least squares solution criterion, establish the following values for use in Eq. 7.5.33: $N_{*0} = 269$, $g = 0.025$, and $t = 0$ in year 1810. These numerical values determine the magnitude of $\alpha(t)$. With the time datum established, the magnitude of $\beta(t)$ is determined as indicated in Fig. 7.5.13(b). In the remainder of our analysis we are not interested in events prior to year 1850 ($t = 40$).

The final solution, Eq. 7.5.39, is expressed in the following dimensional form

$$N = \frac{N_{*0}gte^{-gt}}{2\sqrt{\lambda t}}\left[e^{\sqrt{\frac{\lambda}{D}}x}(F \operatorname{erfc} F) - e^{-\sqrt{\frac{\lambda}{D}}x}(G \operatorname{erfc} G)\right] \quad , \tag{7.5.48}$$

where

$$\lambda = a_* - g \quad ; \qquad \binom{F}{G} = \frac{x \pm 2\sqrt{\lambda D}\,t}{2\sqrt{Dt}} \quad . \tag{7.5.49}$$

Our original density–distance relationship, Eq. 7.5.32, is written in the form

$$N = \alpha(t)e^{-\beta(t)x} \quad . \tag{7.5.50}$$

Comparing this to Eq. 7.5.48 we note that

$$\alpha(t) = N_{*0}(gt)e^{-gt} \quad . \tag{7.5.51}$$

Accordingly, the quantity $\exp[-\beta(t)x]$ of Eq. 7.5.50 corresponds to the quite complicated remaining part of Eq. 7.5.48.

The 1939 observed population density distribution of London, reported by Clark (1951), is shown in the semi-logarithmic plot of Fig. 7.5.14(a). From these data it is established that $a_* = 0.0333\,1/\text{yr}$ and $D = 0.952\,\text{km}^2/\text{yr}$. From Eq. 7.5.49, we observe that the wave velocity is $c_* = 2\sqrt{(a_* - g)D}$. Substituting numerical values into this relationship gives $c_* = 0.178\,\text{km/yr}$.

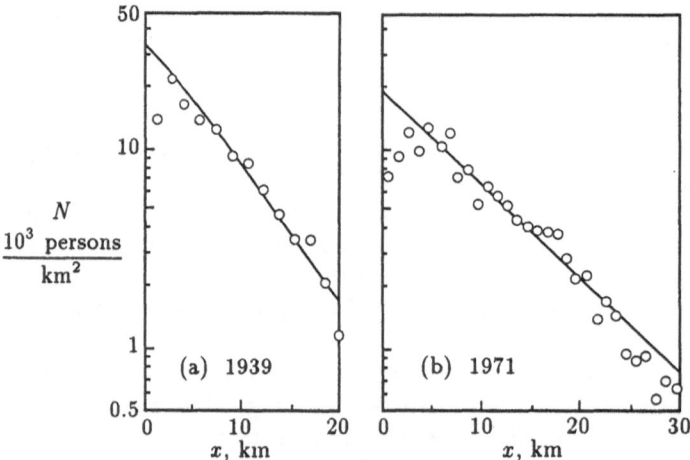

Fig. 7.5.14 Measured density distributions of London. (a) For year 1939; from Clark (1951); (b) for year 1971; from Mogridge (1985)

In his article with the interesting title, "The tidal wave of metropolitan expansion", Blumenfeld (1954) reported: "In the Philadelphia area the crest of the wave of metropolitan expansion moved outward at the rate of one mile per decade during the first half of the 20th century." This result, $c_* = 0.10\,\text{mi/yr} = 0.16\,\text{km/yr}$, is surprisingly close to the above indicated value for London.

The 1971 observed density distribution, indicated by Mogridge (1985), is displayed in Fig. 7.5.14(b). In this case, with $a_* = 0.0333\,1/\text{yr}$, the diffusion coefficient is $D = 1.601\,\text{km}^2/\text{yr}$ and the wave velocity is $c_* = 0.231\,\text{km/yr}$.

Computed curves for the years 1980 and 2000 are shown in Fig. 7.5.15. We note the continual decrease of the effective values of α and β in the plots of London's population density distribution.

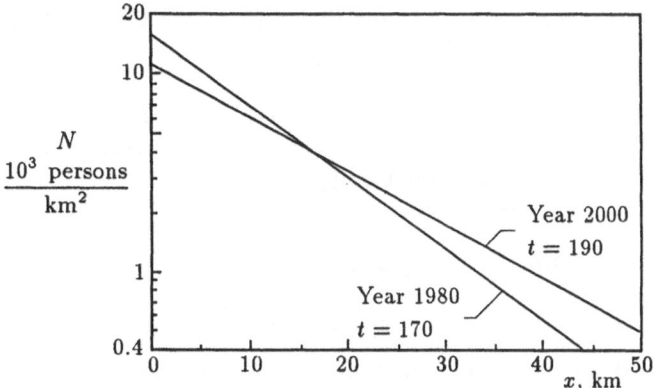

Fig. 7.5.15 Computed density distributions of London for years 1980 and 2000

Finally, we examine the matter of the total population of London. This information is acquired by numerically integrating Eq.7.5.46. Such an integration is accomplished by utilizing incremental annular areas of 2.0 kilometer widths with corresponding average densities provided by the density distribution plots.

For the year 1971, the cumulative population of the metropolitan regional area of London, within a circle of 30 kilometer radius, was $P = 9\,128\,900$. The average population density was 3230 persons/km². The maximum incremental population of 894 000 was in the annular region defined by inner and outer radii of 8 and 10 kilometers.

For the year 2000, the cumulative population of greater London, again defined by a 30 kilometer radius, will be $P = 10\,220\,000$. The average population density will be 3615 persons/km². The maximum incremental population of 829 000 will be in the annular zone of inner and outer radii of 14 and 16 kilometers.

On this basis, the crest of the maximum population wave will have moved 6.0 kilometers during the 29 year period. Accordingly, the average velocity of the wave is approximately 0.20 kilometers per year.

Our illustration concerning the population of London concludes with another look at the negative exponential density distribution

$$N(x,t) = \alpha(t)e^{-\beta(t)x} \tag{7.5.52}$$

in which

$$\alpha(t) = \alpha_0 g t e^{-gt} \tag{7.5.53}$$

and

$$\beta(t) = \beta_0 e^{-kt} \quad . \tag{7.5.54}$$

In other words, we return to the model initially proposed by Clark (1951) and utilized extensively by Bussière (1976), Mogridge (1985) and numerous other investigators.

Plots of Eqs. 7.5.53 and 7.5.54, for London, are shown in Fig. 7.5.13. Recall that $t = 0$ corresponds to the year 1810; it is noted that the correlations are not accurate for values of t less than about 30 (year 1840).

We have the following numerical values

$\alpha_0 = 269 \times 10^3$ persons/km^2 $g = 0.025$ 1/yr

$\beta_0 = 0.880$ 1/km $k = 0.0137$ 1/yr

The total population is obtained from an annular area integration of Eq. 7.5.52. Replacing x by r, we obtain

$$p(r,t) = \frac{2\pi\alpha}{\beta^2} \left[1 - (1 + \beta r)e^{-\beta r}\right] \quad . \tag{7.5.55}$$

When $r = \infty$, this expression becomes

$$p_*(t) = 2\pi\alpha/\beta^2 \quad . \tag{7.5.56}$$

Over a period of a great many years, maps have been devised which show the real areal expanse of London. From a number of such maps and other sources of information, Fig. 7.5.16 has been prepared. Recognizing the impossibility of depicting in detail all features of the city's boundaries, this figure displays the growth of the "built-up" area of London from 1840 to 1981. The Thames River is shown; the solid dot identifies the "City".

Utilizing this map, along with Eqs. 7.5.53 through 7.5.56, our analysis proceeds as follows. The areas between each of the contours of Fig. 7.5.16 are determined and then summed to give the built-up area, A, for the indicated years. The radius of an equivalent circle is calculated from $r_e = \sqrt{A/\pi}$.

Next, for each of the years, the numerical values of α and β are computed from Eqs. 7.5.53 and 7.5.54. We then calculate the magnitude of the "infinite area" population, P_*, from Eq. 7.5.56. Letting $r = r_e$ in Eq. 7.5.55, the

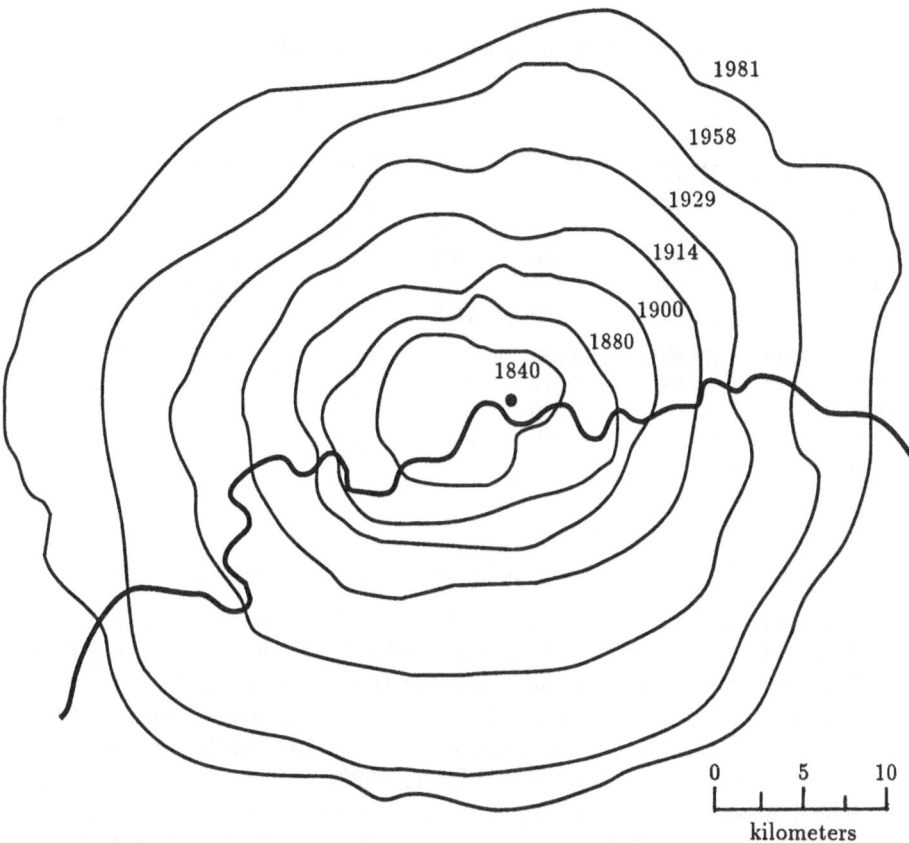

Fig. 7.5.16 Growth of the built-up area of London, 1840 to 1981

"built-up" area population, P, is computed. The average population density is provided by P/A.

It is desirable to make forecasts for the near-term future. To accomplish this, values of the area $A(t)$ were fitted to a logistic curve with the following result

$$A = \frac{A_*}{1 + \left(\dfrac{A_*}{A_0} - 1\right) e^{-at}} \qquad (7.5.57)$$

in which $A_* = 2950\,\mathrm{km}^2$, $A_0 = 20.0\,\mathrm{km}^2$ and $a = 0.0314$ 1/yr.

The results of computation are presented in Table 7.5.3.

Table 7.5.3 Population, area and population density of the built-up regions of London, 1840 to 2010

Year	t [yr]	P_* [million]	A [km^2]	r_e [km]	P [million]	P/A [1/km^2]
1840	30	1.76	63	4.5	1.30	20 600
1880	70	4.53	150	6.9	3.06	20 400
1900	90	6.12	250	8.9	4.06	16 200
1914	104	7.26	420	11.6	5.11	12 100
1929	119	8.69	750	15.5	6.47	8 600
1958	148	11.49	1 230	19.8	7.68	6 200
1981	171	13.91	1 670	23.1	8.12	4 900
1990	180	15.08	1 940	24.8	8.37	4 300
2000	190	16.36	2 140	26.1	8.28	3 900
2010	200	17.60	2 310	27.1	8.04	3 500

The following information is obtained from the table

- The ratio of the built-up area population to the infinite area population, P/P_*, ranged from about 74 percent in 1840 to around 58 percent in 1981. This ratio will decline to approximately 46 percent in 2010.
- The annual rate of population increase attained a maximum during the years 1914 to 1929; its magnitude averaged around 90 000 per year. The peak period seems to have been 1925 to 1929.
- The boundary of the built-up area can be regarded as a wave or front moving radially outward with velocity c_*. Its magnitude, calculated from $c_* = \Delta r_e / \Delta t$, ranged from 0.06 km/yr during the period 1840–1880 to 0.26 km/yr during 1914–1929. The overall average was about 0.15 km/yr.

Finally, in Fig 7.5.17, a comparison is made of the computed populations, P_* and P, with the measured population of the boroughs comprising Greater London over the period 1840 to 1990. Although no precise agreement between computed and measured values is expected, their similarities are noteworthy.

7.6 Diffusion with Logistic Growth

7.6.1 Traveling Wave Solutions

In this section we shall examine some of the features of logistic growth with concurrent diffusion. We omit convection, restrict our attention to one-dimensional rectilinear considerations, and assume that the growth and diffusion coefficients are constants.

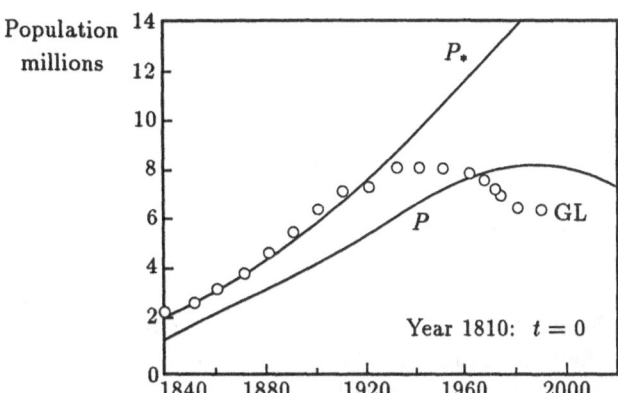

Fig. 7.5.17 Comparison of the measured population of Greater London (GL) with the computed "infinite area" population, P_*, and computed "built-up area" population, P

The differential equation which describes the logistic growth–diffusion process is

$$\frac{\partial U}{\partial t} = D\frac{\partial^2 U}{\partial x^2} + aU(1 - U) \quad , \tag{7.6.1}$$

whose solution, $U = U(x,t)$, depends on specified initial and boundary conditions. As mentioned in Section 7.3.4, this equation has been studied by numerous investigators over many years commencing with the early work of Kolmogorov, Petrovsky and Piskounov (1937) and Fisher (1937). Indeed Eq. 7.6.1 is frequently identified as the Fisher differential equation. In addition to the references indicated in the earlier section, we cite the classic publications by Kendall (1948) and Skellam (1951) and the more recent contributions by Aronson and Weinberger (1975), Hoppensteadt (1975), Kennedy and Aris (1980), Newman (1980), and Rosen (1980).

Our analysis begins with the approach taken by Kendall. Instead of dealing with the non-linear relationship of Eq. 7.6.1 involving a logistic, we consider the following linear expression featuring exponential growth

$$\frac{\partial U}{\partial t} = D\frac{\partial^2 U}{\partial x^2} + aU \quad . \tag{7.6.2}$$

We seek a solution to this equation in the form of a progressive or traveling wave. That is, we want a solution of the type $U = U(x,t) = U(z)$, where $z = x - c_* t$, in which c_* is the speed of a wave whose shape does not change with x nor t. Since the partial derivatives are

$$\frac{\partial U}{\partial t} = -c_*\frac{\partial U}{\partial z} \quad ; \quad \frac{\partial^2 U}{\partial x^2} = \frac{\partial^2 U}{\partial z^2} \quad , \tag{7.6.3}$$

then Eq. 7.6.2 becomes

$$D\frac{d^2U}{dz^2} + c_*\frac{dU}{dz} + aU = 0 \quad . \tag{7.6.4}$$

We observe that we have replaced a partial differential equation, $U = U(x,t)$, by an ordinary differential equation, $U = U(z)$. The roots of Eq. 7.6.4 are given by the characteristic equation

$$m = \frac{c_*}{2D}\left(-1 \pm \sqrt{1 - \frac{4aD}{c_*^2}}\right) \quad . \tag{7.6.5}$$

Some observations. First, if the growth coefficient, a, is replaced by a decay coefficient, $-a_*$, then the square root term in Eq. 7.6.5 is always a real quantity. In this case, we have no difficulty; solutions of Eq. 7.2.4 are monotonic and all wave velocities, c_*, are possible. Second, if $4aD > c_*^2$, then the square root term becomes an imaginary quantity. In this case, U oscillates and hence negative values of U may occur. Accordingly, in order to assure that U will always be a positive quantity, the roots specified by Eq. 7.6.5 must be real not complex. For this to occur, we must have $c_*^2 \geq 4aD$ or $c_* \geq 2\sqrt{aD}$.

This criterion for the speed of a traveling wave was obtained by the other cited investigators including those who extracted the same information from the Fisher equation and even more complicated versions of it. The precise magnitude of the wave speed, above the minimum value, $c_0 = 2\sqrt{aD}$, depends on the initial condition of the problem; this matter is discussed by Kendall (1948), Mollison (1977), Okubo (1980) and others.

Returning to our basic expression of Eq.7.6.1 and making the substitutions, $\xi = (\sqrt{a/D})x$ and $T = at$, we obtain the dimensionless equation

$$\frac{\partial U}{\partial T} = \frac{\partial^2 U}{\partial \xi^2} + U(1 - U) \quad . \tag{7.6.6}$$

The following initial and boundary conditions are imposed

$$U(\xi,0) = \begin{cases} 1 & ; \quad \xi < 0 \\ 0 & ; \quad \xi > 0 \end{cases} \qquad \begin{matrix} U(-\infty,T) = 1 \\ U(\infty,T) = 0 \end{matrix} \tag{7.6.7}$$

This expression indicates that a step function defines the condition at $T = 0$. Now, if there were no growth term (i.e., $a = 0$) in the dimensional expression, Eq.7.6.1, and the initial and boundary conditions were the dimensional equivalent of those of Eq. 7.6.7, then the process of diffusion would increase the magnitude of U in the region $x > 0$ and decrease it in $x < 0$. In this case, as Crank, (1975) shows, the spreading out would be symmetrical about $x = 0$.

However, if in addition to diffusion there is logistic growth, the phenomenon is substantially altered. We recall that the growth rate of the logistic is a maximum when $U = 1/2$ and minimal near $U = 0$ and $U = 1$. This effect tends to increase the magnitude of U in the neighborhood of the origin relative to the magnitudes in the two tails. Consequently, the gradient contained in Fick's law, or its equivalent, increases the diffusion flux in the positive ξ-direction over what it would have been in the absence of concurrent logistic growth. The outcome of this interaction between diffusion and growth is to produce a non-symmetrical distribution about the origin and a wave which moves in the positive ξ-direction.

Again we seek a traveling wave solution. In Eq. 7.6.6, we let $U(\xi, T) = U(\zeta)$, where $\zeta = \xi - cT$ and c is a dimensionless wave velocity. This gives

$$\frac{d^2U}{d\zeta^2} + c\frac{dU}{d\zeta} + U(1 - U) = 0 \quad .$$ (7.6.8)

Letting $dU/d\zeta = V$ yields two first order equations

$$\frac{dU}{d\zeta} = V \quad ; \qquad \frac{dV}{d\zeta} = -cV - U(1 - U) \quad .$$ (7.6.9)

Dividing the second equation by the first gives

$$\frac{dV}{dU} = -\frac{cV + U(1 - U)}{V} \quad .$$ (7.6.10)

This is the mathematical form we would want if we wished to carry out a phase plane (U, V) analysis of the differential equation. We will not go into this subject. Comprehensive treatments of phase plane analyses are given by Conolly (1980), Odell (1980), Jones and Sleeman (1983), Hayashi (1985), Murray (1989), and numerous texts on control theory.

We start with an example given by Jones and Sleeman (1983). Consider the reaction-diffusion equation

$$\frac{\partial U}{\partial T} = \frac{\partial^2 U}{\partial \xi^2} + F(U) \quad .$$ (7.6.11)

Putting this equation into traveling wave form we obtain

$$\frac{d^2U}{d\zeta^2} + c\frac{dU}{d\zeta} + F(U) = 0 \quad .$$ (7.6.12)

For the reaction or growth term, $F(U)$, we take

$$F(U) = \begin{cases} U \quad ; & A : 0 \leq U \leq 1/2 \\ 1 - U \quad ; & B : 1/2 \leq U \leq 1 \end{cases}$$ (7.6.13)

The boundary conditions are: $U(-\infty) = 1$ and $U(\infty) = 0$.

Case I : $c \neq 2$. In region A, Eq. 7.6.12 becomes

$$\frac{d^2U}{d\zeta^2} + c\frac{dU}{d\zeta} + U = 0 \quad .$$ (7.6.14)

The characteristic equation yields the following roots

$$m_1, m_2 = \tfrac{1}{2}(-c \pm \sqrt{c^2 - 4}) \quad .$$ (7.6.15)

We note that the roots are real and distinct as long as $c \geq 2$. Accordingly, the solution to Eq. 7.6.14 is

$$U = A_1 e^{m_1 \zeta} + A_2 e^{m_2 \zeta} \quad .$$ (7.6.16)

In the same way, in region B, the differential equation becomes

$$\frac{d^2U}{d\zeta^2} + c\frac{dU}{d\zeta} + 1 - U = 0 \quad .$$ (7.6.17)

The corresponding characteristic equation gives

$$n_1, n_2 = \tfrac{1}{2}(-c \pm \sqrt{c^2 + 4}) \quad ,$$ (7.6.18)

and so the solution to Eq. 7.6.17 is

$$U = 1 + B_1 e^{n_1 \zeta} + B_2 e^{n_2 \zeta} \quad .$$ (7.6.19)

In order to determine the values of the four constants, A_1, A_2, B_1, and B_2 we utilize the following

a. The value of $U = 1/2$ at $\zeta = 0$.
b. The slopes, i.e., first derivatives of U, corresponding to the two regions must be equal at $\zeta = 0$.
c. The boundary conditions must be satisfied.

Some algebra gives us the following answers.

Region A: $\zeta > 0$

$$U = \frac{1}{4\sqrt{c^2 - 4}}\left(\sqrt{c^2 - 4} + 2c - \sqrt{c^2 + 4}\right) e^{-\frac{1}{2}(c - \sqrt{c^2 - 4})\zeta}$$

$$+ \frac{1}{4\sqrt{c^2 - 4}}\left(\sqrt{c^2 - 4} - 2c + \sqrt{c^2 + 4}\right) e^{-\frac{1}{2}(c + \sqrt{c^2 + 4})\zeta} \quad .$$ (7.6.20)

Region B: $\zeta < 0$

$$U = 1 - \tfrac{1}{2}e^{-\frac{1}{2}(c - \sqrt{c^2 + 4})\zeta} \quad .$$ (7.6.21)

Case II: $c = 2$. This problem is handled the same way though we start by substituting $c = 2$ in Eq. 7.6.14 for Region A and in Eq. 7.6.17 for Region B. For Region A the roots are real and equal. The respective final solutions are

Region A: $\zeta > 0$

$$U = \tfrac{1}{2}e^{-\zeta} + \left(1 - \frac{1}{\sqrt{2}}\right)\zeta e^{-\zeta} \quad . \tag{7.6.22}$$

Region B: $\zeta < 0$

$$U = 1 - \tfrac{1}{2}e^{(\sqrt{2}-1)\zeta} \quad . \tag{7.6.23}$$

Utilizing the solutions presented above, profiles of the traveling wave, $U = U(\zeta)$, corresponding to values of $c = 2$ and $c = 2.5$, are shown in Fig. 7.6.1. We note that the slope decreases as the wave velocity increases. Also, the curves are not symmetrical about $\zeta = 0$.

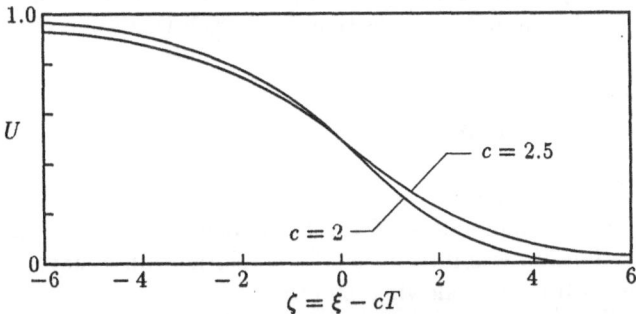

Fig. 7.6.1 A piecewise traveling wave solution of the logistic equation with diffusion

Another relevant analysis of the traveling wave differential equation, Eq. 7.6.8, was carried out by Canosa (1973) and described in detail by Hoppensteadt (1975) and Murray (1989). Again stipulating the boundary conditions, $U(-\infty) = 1$ and $U(\infty) = 0$, perturbation series solutions were obtained for the regions, $\zeta > 0$ and $\zeta < 0$. The outcome of the analysis is the expression

$$U(\zeta) = \frac{1}{1 + e^{\zeta/c}} + \frac{e^{\zeta/c}}{c^2(1 + e^{\zeta/c})^2}\log_e\left(\frac{4e^{\zeta/c}}{(1 + e^{\zeta/c})^2}\right) \tag{7.6.24}$$

plus some higher order terms behaving like $1/c^4$; as we have seen, the minimum value of the dimensionless wave velocity is $c_0 = 2$. A plot of Eq. 7.6.24 is shown in Fig. 7.6.2 for values of $c = 2$ and $c = 2.5$; over most of the range of ζ, the second term of Eq. 7.6.24 is small compared to the first.

We note, in Fig. 7.6.2, the similarity of the wave profiles to those displayed in Fig. 7.6.1. Again, the wave velocity increases as the steepness of the wave

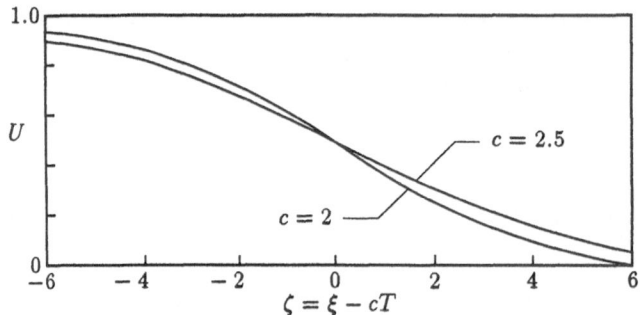

Fig. 7.6.2 A perturbation series traveling wave solution of the logistic equation with diffusion

decreases. It is also observed that if the direction of the ζ-axis is reversed, the wave profile, $U(\zeta)$, resembles a logistic curve. Furthermore, the first term of Eq. 7.6.24 has the form of a logistic and the second term contains two groups which look like the density distribution function, i.e., the first derivative, of a logistic distribution. In the following section we pursue this interesting situation.

7.6.2 A Power Law Traveling Wave Solution

In Section 2.6, in our consideration of the power law logistic equation, and again in Section 3.4, in our examination of the hyperlogistic equation, the Bertalanffy growth relationship was introduced. This relationship is now attached to the diffusion term in the following expression

$$\frac{\partial U}{\partial T} = \frac{\partial^2 U}{\partial \xi^2} + U^r(1 - U^s) \tag{7.6.25}$$

where r and s are positive constants. Setting $r = 1$ gives

$$\frac{\partial U}{\partial T} = \frac{\partial^2 U}{\partial \xi^2} + U(1 - U^s) \quad . \tag{7.6.26}$$

This is the expression which Murray (1989) examines in detail. His solution, outlined below, provides an extremely useful and surprisingly simple framework for the analysis of some very practical problems.

As before, we transform Eq. 7.6.26 into the form

$$\frac{d^2 U}{d\zeta^2} + c\frac{dU}{d\zeta} + U(1 - U^s) = 0 \tag{7.6.27}$$

by letting $U(\xi, T) = U(\zeta)$ where $\zeta = \xi - cT$. Again, the boundary conditions are: $U(-\infty) = 1$ and $U(\infty) = 0$.

In light of the above-described perturbation series analysis, culminating in the result of Eq. 7.6.24, Murray endeavored to obtain a traveling wave solution to Eq. 7.6.27 of the form

$$U(\zeta) = \frac{1}{(1 + \alpha e^{\beta \zeta})^{\gamma}} \quad , \qquad (7.6.28)$$

where α, β and γ are positive constants. We note immediately that Eq. 7.6.28 satisfies the boundary conditions. Substitution of Eq. 7.6.28 and its derivatives into Eq. 7.6.27 yields the following results

$$\gamma = \frac{2}{s} \quad ; \qquad \beta = \frac{s}{\sqrt{2(s+2)}} \quad ; \qquad c = \frac{s+4}{\sqrt{2(s+2)}} \quad ;$$

$$U(0) = U_* \quad ; \qquad \alpha = \frac{1}{U_*^{s/2}} - 1 \quad . \qquad (7.6.29)$$

The second derivative of Eq. 7.6.28 provides the inflection point

$$\zeta_i = \frac{1}{\beta} \log_e \left(\frac{s}{2\alpha} \right) \quad ; \qquad U_i = \frac{1}{(1 + s/2)^{2/s}} \quad . \qquad (7.6.30)$$

It is noted that we have obtained a traveling wave solution not only to the ordinary logistic with diffusion but, remarkably, to the power law logistic with diffusion. For the important case, $s = 1$, Eq. 7.6.26 reduces to the Fisher equation. In this instance, selecting $U_* = 1/2$ to determine the location parameter, α, we have

$$\gamma = 2 \quad ; \qquad \alpha = \sqrt{2} - 1 \quad ; \qquad \beta = \frac{1}{\sqrt{6}} \quad ; \qquad c = \frac{5}{\sqrt{6}} \quad ;$$

$$\zeta_i = \sqrt{6} \log_e \left(\frac{\sqrt{2}+1}{2} \right) \quad ; \qquad U_i = \frac{4}{9} \quad . \qquad (7.6.31)$$

In passing, we note that if $s = 2$, then $\gamma = 1$. In this interesting case, the traveling wave solution, Eq. 7.6.28, itself reduces to the ordinary logistic equation

$$U(\zeta) = \frac{1}{1 + \alpha e^{\beta \zeta}} \qquad (7.6.32)$$

and $\beta = 1/\sqrt{2}$, $c = 3/\sqrt{2}$, $\alpha = 1$, $\zeta_i = 0$, and $U_i = 1/2$.

With reference to the displays of Figs. 7.6.1 and 7.6.2 as well as to the logistic-like form of Eq. 7.6.28, we now reverse the direction of plotting U versus ζ. Negative values of ζ are plotted to the right and positive to the left. We utilize Eq. 7.6.28, and the S-shaped curves it generates, as the framework for examination of two illustrations in the field of economic geography. The definition sketches of Fig. 7.6.3 may be helpful.

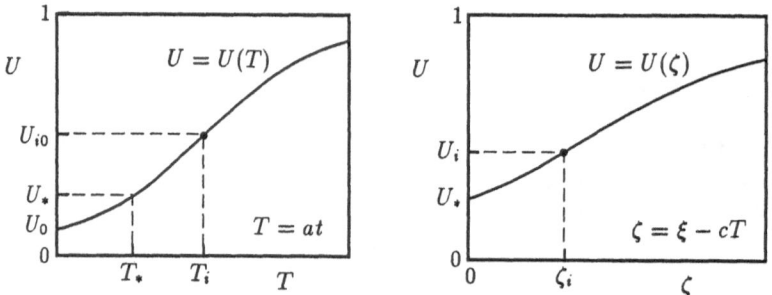

Fig. 7.6.3 Definition sketches for application of the traveling wave solution

7.6.3 An Illustration: Diffusion of Tractor Utilization

As the first illustration of an application of the traveling wave solution obtained above, we utilize the data obtained by Casetti and Semple (1969) involving the adoption of tractors in midwest America.

This comprehensive study yielded information concerning the percentage of farms utilizing tractors in 25 states during the period from 1920 to 1964. The values of percent utilization, U, were reported for the years: 1920, 1925, 1930, 1940, 1945, 1950, 1954, 1959, and 1964. Virtually all of the North Central, South Central, and Mountain states were included in the study. North Dakota was selected as the "diffusion pole" (i.e., $x = 0$). Distances from the center of North Dakota to the center of each state provided the values of the spatial variable, x. The year 1920 was taken to be the time origin (i.e., $t = 0$).

The model utilized by Casetti and Semple to solve the spatial-time problem, $U = U(x, t)$, was to assume a solution of the form

$$U = \frac{1}{1 + \exp[f(x,t)]} \quad , \tag{7.6.33}$$

where

$$f(x, t) = a_0 + a_1 x + a_2 x^2 + b_0 t + b_1 xt + b_2 x^2 t \quad . \tag{7.6.34}$$

Carrying out regression analyses on the field data, they obtained the results: $a_0 = 1.85$, $a_2 = 3.4 \times 10^{-6}$, and $b_0 = -0.114$; the remaining constants of Eq. 7.6.34 were found to be statistically insignificant.

If we write the ordinary logistic equation in its usual form

$$U = \frac{1}{1 + \left(\frac{1}{U_0} - 1\right) e^{b_0 t}} \quad , \tag{7.6.35}$$

and retain only the indicated significant terms of Eq. 7.6.34, we obtain

$$U_0(x) = \frac{1}{1 + e^{a_0 + a_2 x^2}} \quad , \tag{7.6.36}$$

which indicates that this particular distance–time formulation is tantamount to imposing, on the ordinary logistic equation, an x-distribution of the initial value, U_0. This is indeed a straightforward method for solving a practical problem.

This "expansion method", featuring the incorporation of polynomial variables in functional relationships, such as the logistic equation, is discussed by Casetti (1986) in a generalized way. A number of investigators have examined the tractor diffusion data of Casetti and Semple; these include the studies of Cliff and Ord (1975), Mahajan and Peterson (1979, 1985), Sonis (1983), Morrill (1985) and Morrill et al. (1988).

The objective of our illustration is to analyze and display these $U(x,t)$ tractor adoption data in the framework of the traveling wave model. The first step is to examine the data of each state separately. Assuming ordinary logistic growth in each state, we determine the value of the growth coefficient, a, and the initial ($t = 0$) adoption fraction, U_0. The results obtained for all 25 states are listed in Table 7.6.1. Plots of observed field data for the states of South Dakota, Oklahoma and Louisiana are shown in Fig. 7.6.4; the corresponding ordinary logistic distributions are shown as dashed curves. For the moment we disregard the solid curves in the figure.

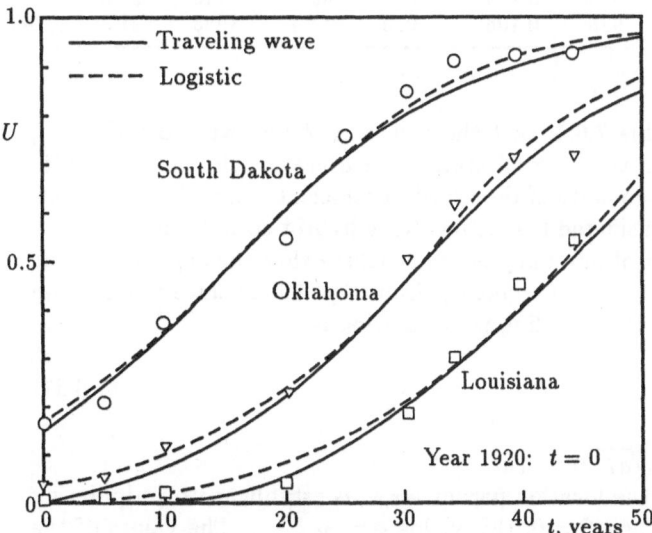

Fig. 7.6.4 Diffusion of tractor utilization in the states South Dakota, Oklahoma, and Louisiana. Data of Casetti and Semple (1969)

Table 7.6.1 Numerical values of parameters of diffusion of tractor utilization. Data of Casetti and Semple (1969)

No.	State	x	a	U_0	t_*	D	c_*	x_*
1	South Dakota	140	0.105	0.160	15.7	180	8.9	41.3
2	Minnesota	200	0.124	0.084	19.4	207	10.3	40.9
3	Montana	300	0.108	0.112	19.2	545	15.7	71.0
4	Nebraska	300	0.118	0.098	18.9	517	15.9	66.3
5	Wyoming	332	0.114	0.065	23.3	426	14.2	61.1
6	Iowa	365	0.114	0.109	18.5	819	19.7	84.9
7	Wisconsin	380	0.131	0.065	20.4	638	18.6	69.9
8	Kansas	450	0.103	0.124	19.1	1301	23.6	112.7
9	Colorado	482	0.108	0.073	23.5	931	20.5	92.9
10	Idaho	525	0.135	0.030	26.3	708	19.9	72.5
11	Missouri	540	0.116	0.026	31.2	618	17.3	72.9
12	Illinois	545	0.105	0.114	19.6	1772	27.8	130.1
13	Michigan	560	0.135	0.042	23.3	1034	24.1	87.7
14	Utah	595	0.133	0.016	31.0	663	19.2	70.6
15	Oklahoma	630	0.105	0.035	31.4	918	20.1	93.3
16	Indiana	641	0.112	0.062	24.3	1493	26.3	115.7
17	New Mexico	715	0.112	0.020	35.1	893	20.4	89.5
18	Nevada	733	0.117	0.035	28.6	1356	25.7	107.9
19	Ohio	740	0.112	0.060	24.5	1950	30.2	132.0
20	Arkansas	750	0.132	0.005	39.9	643	18.8	69.8
21	Texas	800	0.119	0.021	32.2	1238	24.8	102.0
22	Arizona	815	0.078	0.067	33.7	1797	24.2	151.8
23	Mississippi	900	0.135	0.003	43.9	748	20.5	74.5
24	Louisiana	925	0.106	0.010	43.4	1033	21.3	98.9

We observe, in Table 7.6.1, that the values of the growth coefficient, a, are approximately the same for all states; the average value is $a = 0.116$. Incidentally, this near-equality of the growth coefficient is not the case in our second illustration, considered below, involving hybrid corn diffusion.

From this first part of our analysis, we conclude that the exponent of the power law logistic is $s = 1$. Consequently, from Eqs. 7.6.29 since $\gamma = 2/s$, we have $\gamma = 2$. Therefore the traveling wave solution is

$$U = \frac{1}{(1 + \alpha e^{\beta \zeta})^2} , \tag{7.6.37}$$

where $\zeta = \xi - cT = (\sqrt{a/D})x - cat$.

The magnitude of the location parameter, α, is established by stipulating that $U = U_* = 1/2$ when $\zeta = 0$; this yields $\alpha = \sqrt{2} - 1$. The values of the shape parameter, β, and dimensionless wave velocity, c, are provided by Eqs. 7.6.29 with $s = 1$; we obtain $\beta = 1/\sqrt{6}$ and $c = 5/\sqrt{6}$.

The arrival time, t_*, of the traveling wave at the center of each state is defined, in this first illustration, as the time corresponding to $U_* = 1/2$. Since the growth patterns for all the states are ordinary logistics, the arrival time is the same as the inflection point time, i.e., $t_* = t_i$. Further, when $U = 1/2$, we have $\zeta = 0$. Accordingly, since $\zeta = \xi - cT = (\sqrt{a/D})x - cat$, we obtain

$$D = \frac{x^2}{c^2 a t_*^2} \quad ; \qquad \zeta = ca(t_* - t) \quad . \tag{7.6.38}$$

From the first of these relationships we calculate the diffusion coefficient, D, corresponding to each state. In addition, we compute the diffusion velocity, $c_* = c\sqrt{aD}$, and a diffusion length, $x_* = \sqrt{D/a}$. The numerical values of these parameters are listed in Table 7.6.1.

From the second relationship of Eq. 7.6.38, we calculate the values of the wave parameter, ζ, for all values of time, t, and distance, x.

Our mathematical framework predicts that all data points should fall on the traveling wave profile described by Eq. 7.6.37. The results of computations are displayed in Fig. 7.6.5. Since there were so many data points, only about a third of them are shown. The coordinates of the inflection point, given by Eq. 7.6.31, are: $\zeta_i = 0.461$, $U_i = 0.444$. This point is identified by the solid dot in the figure.

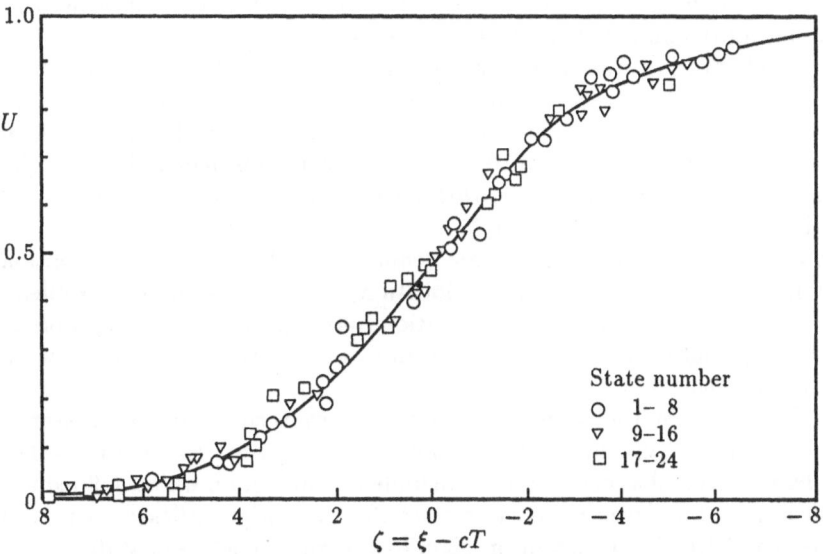

Fig. 7.6.5 Diffusion of tractor utilization in 25 states. Comparison of observed values with traveling wave solution. Data of Casetti and Semple (1969)

Finally, knowing all the numerical values of all the parameters, we compute the $U(t)$ relationships for the states of South Dakota, Oklahoma and Louisiana from the traveling wave solution, Eq. 7.6.37. The results are shown as the solid curves in Fig. 7.6.4. We note fairly good overall agreement between the traveling wave solutions and the ordinary logistic solutions.

We conclude the illustration with a few observations. First, we recall that the mechanism for the provision of spatial diffusion in our model is the spatial derivative, $D(\partial^2 U/\partial x^2)$. We recall further that this quantity is based on Fick's law of diffusion. Clearly, there is certainly no reason to expect that the diffusion of tractor utilization across states obeys the same law as the diffusion of chemicals through liquids. Nevertheless, in this case, the spatial diffusion term does appear to provide an acceptable framework.

Second, as Murray (1989) emphasizes, the traveling wave is only one solution to the Fisher equation and it may not even be the most meaningful one. Finally, the traveling wave formulation, discussed for a long time by Hagerstrand (1967), Morrill (1968), Hanham and Brown (1976) and other investigators, seems to provide a rational framework for combining spatial and time variables for diffusion of a technology when there is a clearly identified diffusion pole.

7.6.4 An Illustration: Adoption of Hybrid Corn Revisited

In Section 2.6, where we examined the power law logistic distribution, we considered an illustration involving the adoption of hybrid corn on America's farms during the period from about 1935 to 1960.

That illustration was based on an extensive study by Griliches (1957, 1960); our example of Section 2.6.6 featured the adoption of hybrid corn in the state of Alabama. The main result, we recall, was that the exponent in the power law logistic, Eq. 2.6.10, for Alabama is approximately $s = 2/3$. We also obtained values for the growth coefficient, $a = 0.522$, and the initial adoption fraction, $U_0 = 1.50 \times 10^{-4}$.

In that earlier illustration we were dealing with the variation of adoption fraction with time, t, at a particular location, x, i.e., the temporal problem, $U = U(t)$. We now enlarge our consideration to the spatial–temporal problem, $U = U(x,t)$, and again utilize the traveling wave analysis as the framework for our analysis.

In his study of the diffusion of hybrid corn technology in 31 states, Griliches generally divided each state into nine crop reporting districts. For each district, as well as each state, he included an index, m, which defined the $U = 0.10$ adoption level; these indices set the time origins, based on $m = 0$ for the year 1940, for the adoption diagrams of all districts and states.

The values of m for the districts of eastern Iowa and northwestern Illinois were $m = -5.15$ and $m = -4.81$, respectively. These were the largest values reported in the study. The average value, $m = -4.98$, is now rounded to $m = -5.0$ and we select, as the diffusion pole, the city of Clinton, Iowa on the Mississippi River. The value $m = -5.0$ indicates that in the year 1935.0, hybrid corn adoption in the Clinton area was at the 10 percent level. This sets the time origin (year 1935 : $t = 0$) of our problem; the location of Clinton establishes the spatial origin, $x = 0$.

A plot of the adoption history for five states is shown in Fig. 7.6.6; we examine three of these: Iowa, Kentucky, and Alabama.

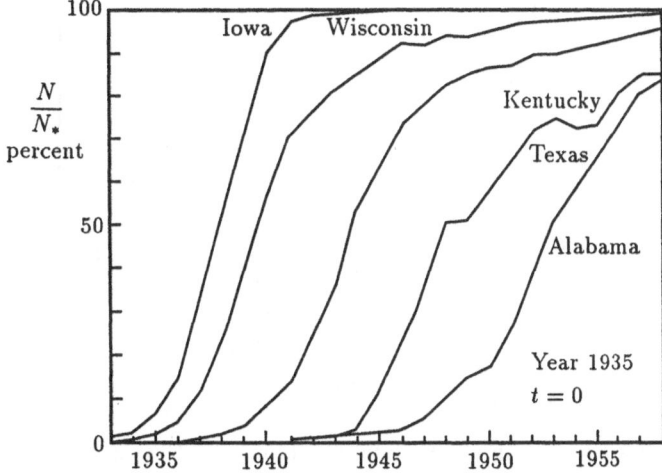

Fig. 7.6.6 Adoption of hybrid corn in five midwestern and southern states. From Griliches (1960)

It may be helpful to examine our problem in two parts: A. Local temporal analysis and B. Regional temporal–spatial analysis.

A. Local temporal analysis. As we did in Section 2.6.6 for Alabama, a power law logistic is fitted to one year time interval points obtained from Fig. 7.6.6. Rounding results to simple fractions gives $s = 1$ for Iowa, $s = 3/4$ for Kentucky, and $s = 2/3$ for Alabama. With values of the exponent s established, we express the power law logistic, Eq. 2.6.10, in the form

$$\log_e \left(\frac{1}{U^s} - 1 \right) = \log_e \left(\frac{1}{U_0^s} - 1 \right) - ast \quad . \tag{7.6.39}$$

Least squares computations involving measured values of U at times, t, yield the values of the growth coefficient, a, and the initial adoption fraction, U_0.

So for each of the three states we have equations of the form $U = U(t)$. Computed curves are compared to the observed data in Fig. 7.6.7. We note that the calculated curve for Kentucky does not agree very well with the data for $U > 0.80$; the Iowa and Alabama curves are in good agreement with observed values. The solid dots on the curves show the inflection points.

B. Regional temporal–spatial analysis. From Eq. 7.6.39 we calculate the arrival times, t_*, corresponding to $U_* = 0.10$. Next, from a map we measure, as accurately as we can, the distance x from the diffusion pole – Clinton, Iowa – to the center of each of the three states. Dividing this

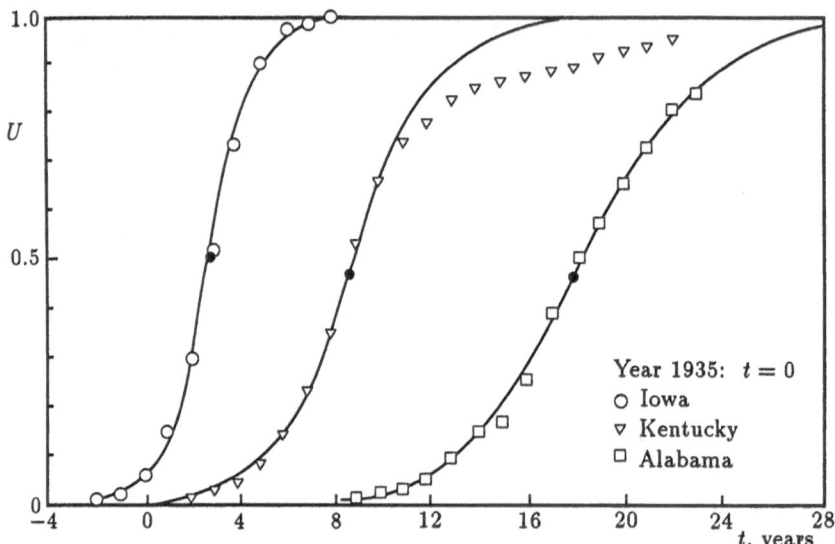

Fig. 7.6.7 Power law logistic growth of adoption of hybrid corn. Data of Griliches (1960)

distance, x, by the arrival time, t_*, gives the velocity of the diffusion wave, c_*. For example, for Alabama, $s = 2/3$, $a = 0.522$ and $U_0 = 1.50 \times 10^{-4}$. These quantities yield $t_* = 13.16$ (year: 1948.16). We determine that the distance from Clinton to the center of Alabama is $x = 670$ miles. This gives $c_* = 670/13.16 = 50.91$ miles/year.

As the framework for the next aspect of our analysis we utilize Eq. 7.6.28. The values of γ, β, and c are provided by Eqs. 7.6.29; α is determined by requiring that $U_* = 0.10$ when $\zeta = 0$. The diffusion coefficient, D, is determined from the condition $\zeta = 0$;

$$\xi = cT_* \quad ; \qquad \sqrt{\frac{a}{D}}x = cat_* \quad . \tag{7.6.40}$$

For example, for Alabama, $s = 2/3$, $\gamma = 3$, $\beta = 1/2\sqrt{3} = 0.289$, and $c = 7/2\sqrt{3} = 2.021$; in addition, $\alpha = 1.154$. Substituting known values of c, a, and t_*, into Eq. 7.6.40 gives $D = 1220\,\mathrm{mi}^2/\mathrm{yr}$. Incidentally, we recall that $c_* = c\sqrt{aD}$; so again, $c_* = 50.91\,\mathrm{mi}/\mathrm{yr}$.

With values of all the parameters now ascertained, we compute the curve, $U = U(\lambda)$, where $\lambda = \beta\zeta$, and, in Fig. 7.6.8, compare the results with observed data for the three states. Points for values of $U > 0.85$ for Kentucky are not shown. Inflection points are identified by the solid dots. As we would expect, there is a slight difference between the curves for $s = 2/3$ and $s = 1$; the one for $s = 3/4$ is in between. Essentially, these curves are the profiles of the traveling waves. We easily establish that an increase in the value of s

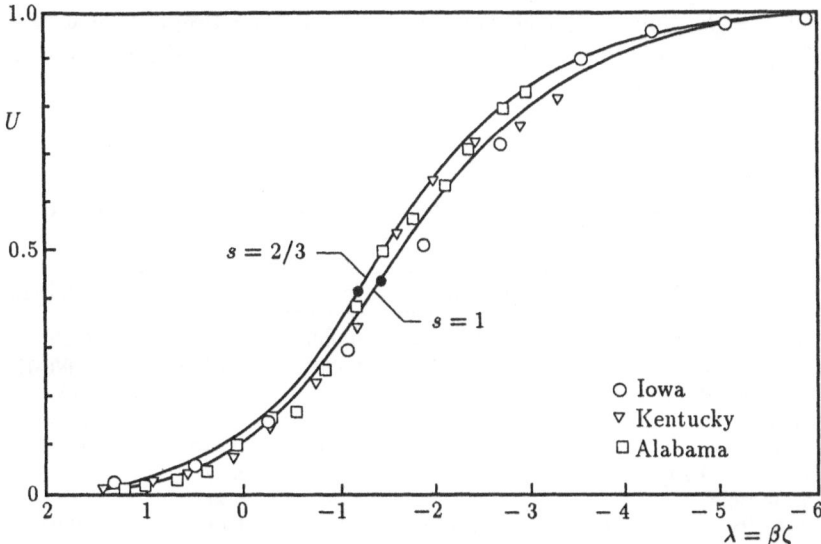

Fig. 7.6.8 Traveling wave display of hybrid corn adoption. Data of Griliches (1960)

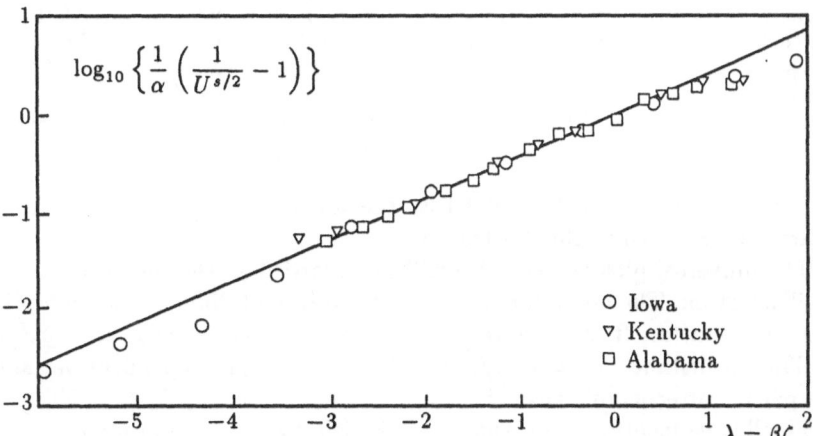

Fig. 7.6.9 Universal plot of hybrid corn adoption. Data of Griliches (1960)

increases the dimensionless wave velocity, c, and decreases the steepness of the wave profile at the inflection point.

All of the data can be shown in another fashion by writing Eq. 7.6.28 in the form

$$\frac{1}{\alpha}\left(\frac{1}{U^{s/2}} - 1\right) = e^{\lambda} \quad , \tag{7.6.41}$$

where $\lambda = \beta\zeta$. Such a plot is shown in the semi-logarithmic display of Fig. 7.6.9. The predicted linear relationship is evident except in the region of large λ (i.e., large x, small t).

Table **7.6.2** Numerical values of parameters of adoption of hybrid corn. Data of Griliches (1957, 1960)

Parameter	Symbol	Units	Iowa	Kentucky	Alabama
Growth coefficient	a	1/yr	0.959	0.658	0.522
Initial value $(t = 0)$	U_0	–	0.0602	0.00414	0.000150
Power law exponent	s	–	1	3/4	2/3
Inflection point	t_i	yr	2.87	8.89	18.5
Inflection point	U_{io}	–	0.500	0.474	0.465
Arrival time $(U = 0.10)$	t_*	yr	0.60	5.21	13.16
Distance from pole	x	mi	140	430	670
Diffusion velocity	c_*	mi/yr	234.5	82.6	50.91
Initial value $(\zeta = 0)$	α	–	2.162	1.371	1.154
Wave profile exponent	γ	–	2	8/3	3
Wave shape coefficient	β	–	0.408	0.320	0.289
Wave velocity	c	–	2.041	2.025	2.021
Diffusion coefficient	D	mi^2/yr	13 770	2530	1220
Inflection point	λ_i	–	−1.46	−1.30	−1.24
Inflection point	U_i	–	0.444	0.428	0.420

For completeness, Table 7.6.2 lists the numerical values of the various parameters involved in the illustration.

The universal plot of Fig. 7.6.9 will be regarded as the final objective of our illustration. The economics relating to the rate of diffusion and adoption is another matter. It is conceivable that the diffusion velocity, $c_* = c\sqrt{aD}$, or diffusion distance, $x_* = \sqrt{D/a}$, could be meaningful dependent variables in a particular economic analysis.

Finally, we mention again that the surprisingly simple solution expressed by Eq. 7.6.28 is only one solution to the Fisher reaction-diffusion equation and indeed may not even be the most relevant one. Murray (1989) cautions that a careful phase plane analysis of the differential equation should be carried out for a precise analysis of a problem.

Nevertheless, for our illustrations involving spatial diffusion of tractor utilization and hybrid corn adoption, the above-outlined model appears to provide a simple and rational framework for handling a very complex problem.

8. Conclusion

We devote this final section to short elaborations of several subjects we have examined to varying degrees and to brief introductions to other subjects we have not considered at all. We close with a discussion of some topics that various investigators are currently studying or that may warrant investigation in the future.

In Chapters 4 and 5 considerable attention was given to the subject of time dependent growth coefficients and carrying capacities. Our basic differential equation

$$\frac{dN}{dt} = a(t)N \left(1 - \frac{N}{N_*(t)}\right) \tag{8.1}$$

yielded the solution

$$N = \frac{N_0 \exp\left(\int_0^t a(\xi)d\xi\right)}{1 + N_0 \int_0^t \frac{a(\xi)}{N_*(\xi)} \left(\exp \int_0^\xi a(\eta)d\eta\right) d\xi} . \tag{8.2}$$

We examined several forms of $a_*(t)$ with $N_*(t) =$ constant and several kinds of $N_*(t)$ with $a(t) =$ constant. Clearly, a large number of simultaneous combinations of $a(t)$ and $N_*(t)$ are available for analysis; most of these would probably not be very interesting nor useful.

Noting that so many phenomena are cyclical in nature, it is not surprising that considerable attention has been given to the problem of sinusoidal variations of the growth coefficient and carrying capacity. An extensive analysis of this case was carried out by Nisbet and Gurney (1976). Their study was somewhat generalized with the inclusion of a phase angle, ϕ, in the growth coefficient term and a discrete time delay in the resource term. This problem of periodic forms $a(t)$ and $N_*(t)$ has also been examined at length by Coleman (1979), Arrigoni and Steiner (1985), Cushing (1986), Vance (1990) and numerous others.

As we would anticipate, such cyclic type solutions to the logistic equation find application in biology and ecology because of annual and daily periodic changes of climatological variables. It is reasonable to expect such applications in demography and economics and perhaps even to seasonal things like travel and tourism, marketing and sales, and manufacturing and construction.

In this category of time dependent logistic parameters, an example is provided in a study carried out by Robinson and Deano (1985). These investigators examined the degree to which acid rain alters fish survival in wooded regions. Specifically they determined the effects of pH, aluminum and decomposed leaves (pine, oak, cypress, willow) on the survival of golden shriners (*Notemiqonus crystoleucas*). They established that small values of pH (4.5 or less) and large values of aluminum concentration are detrimental to fish survival but decomposed leafwater is beneficial. Because of annual cyclic changes of climate and hydrology, an appropriate framework for analysis of this phenomenon might be the utilization of sinusoidal forms for both $a(t)$ and $N_*(t)$.

In Chapter 6 we considered some problems and illustrations involving discrete and distributed time delays. We saw that even in the very simple case of exponential growth, the introduction of a discrete time delay produces a rather complicated problem. Our method of solution features the use of Laplace transforms and the consequent appearance of an infinite series of exponential terms involving complex quantities.

An alternative way to solve linear delay differential equations is the method of steps. The methodology is described by Driver (1977), El'sgol'ts and Norkin (1973) and Zwillinger (1989). The procedure features successive integrations over time intervals imposed by the initial interval condition. A drawback to the method is that analytical expressions become very complicated after a few interval integrations. Driver provides an illustration involving mixing of chemical solutions in a tank, a process of great importance in chemical engineering and biotechnology.

Some interesting points are raised by MacDonald (1989) in connection with discrete time delays. Starting with Newton's second law of motion, we derive the following linear differential equation for a damped oscillating system

$$m\frac{d^2y}{dt^2} + c\frac{dy}{dt} + ky = f(t) \quad , \qquad (8.3)$$

in which y is the displacement of a mass, m, from its equilibrium position at time t, and c and k are respectively, a damping coefficient and a spring constant; an external driving force is $f(t)$. Incidentally, an exact electric circuit analogy is

$$L\frac{d^2I}{dt^2} + R\frac{dI}{dt} + \frac{1}{C}I = f(t) \quad , \qquad (8.4)$$

where I is the current and L, R, and C are the inductance, resistance and capacitance.

We note that Eq. 8.3 is identical to Eq. 2.9.17 if we take $f(t)$ and $\beta_* = 0$. So we already have the solution to the homogeneous form of Eq. 8.3. The roots of the characteristic equation are

$$n = \frac{1}{2}\left(-\frac{c}{m} \pm \frac{1}{m}\sqrt{c^2 - 4km}\right) \quad . \tag{8.5}$$

As in Section 2.9.5, we obtain the following solution regimes of Eq. 8.3

1. $c^2 = 4\,km$ critical damping

2. $c^2 > 4\,km$ overdamping   (8.6)

3. $c^2 < 4\,km$ underdamping

Case 3 gives rise to attenuated (amplified) oscillations when c is positive (negative).

We recall Taylor series from Eq. 6.1.2

$$y(t - \tau) = y(t) - \tau\frac{dy(t)}{dt} + \frac{1}{2}\tau^2\frac{d^2y(t)}{dt^2} - \cdots \quad . \tag{8.7}$$

To illustrate the effect of discrete time delay on the second order linear relationship of Eq. 8.3, MacDonald supposes that the restoring force due to the spring, i.e., the third term in Eq. 8.3, is delayed by time, τ. Truncating the series after the linear term gives the following form of Eq. 8.3

$$m\frac{d^2y}{dt^2} + (c - m\tau)\frac{dy}{dt} + ky = f(t) \quad . \tag{8.8}$$

We note that a delay in the restoring force reduces the damping force. If c were a small positive number at the outset, it is seen that the discrete time delay could change attenuated oscillations into amplified oscillations.

MacDonald carries the analysis a step further by truncating the Taylor series after the quadratic term. This yields

$$\left(m + \frac{k\tau^2}{2}\right)\frac{d^2y}{dt^2} + (c - k\tau)\frac{dy}{dt} + ky = f(t) \quad . \tag{8.9}$$

In this instance we observe that the mass of the system is effectively increased by a small amount. Consequently the solution regimes indicated in Eq. 8.6 might well be altered because of the time delay.

A similar kind of illustrative analysis is given by Driver (1977) involving a ship rolling due to wave forces. We could interpret the changes in the behavior of the electric circuit described by Eq. 8.4 in the same way. We also observe, in our Section 2.9.6 illustration of a national economy model, that the relationships involving supply, demand and investment could be substantially altered by discrete time delays.

In the field of structural dynamics, time delays caused by vortex shedding due to wind action may produce undesirable vibration and oscillation. Tall slender chimneys and highly flexible suspension bridges are especially vulnerable.

A long list of applications of discrete delay differential equations is given by Driver in the subjects of hydrodynamics and electrodynamics, nuclear reactors and plasma jets, rocket engines and transistor circuits, photo emulsions and transmission lines, epidemiology and neurology, business cycles and economic growth, and so on.

Even though it is not directly related to discrete time delays, it is appropriate to raise another subject involving second order ordinary differential equations. Over the years several investigators have indicated that first order equations, such as the ordinary logistic and its Lotka-Volterra extensions for multi-component phenomena, are inadequate.

Invariably second order equations are proposed as the more suitable alternative. In some instances, Newton's second law is utilized to provide, as in Eq. 8.3, a second derivative with time; damping and restoring force terms are then devised. In spite of its long-recognized shortcomings, it appears that the logistic equation is far too entrenched to be replaced. In any event, these second order endeavors are examined by Clark (1971), Innis (1972), Fletcher (1974), Yee (1980) and Bojadziev (1983).

The preceding remarks refer primarily to discrete time delays. With respect to distributed time delays, we have considered two illustrations, one involving growth and decline of populations of central cities and metropolitan areas and one involving growth and decline of bacterial populations.

Beyond these and other applications in demography and biology there are many phenomena in other disciplines in which distributed time delays make appearances. Rabotnov (1980) gives extensive treatment to the subject of "hereditary mechanics". Lal et al. (1983) examine the problem of innovation diffusion in a social system involving both externally and internally influenced information transfer mechanisms and with distributed time delays. For a long time, hydrologists have employed the methodology of such time delays in analyses relating rainfall to runoff.

It would be interesting to carry out (a) an economic analysis, involving distributed time delays, of the rise and fall of, for example, the Dow-Jones index during the period from, say, 1920 to 1940, (b) a geologic analysis of the relationship between forest fires and rainfall and eventually ensuing mud slides and debris flows, and (c) an historic analysis involving rises and falls of earlier civilizations.

Although we have devoted nearly all of our attention to single-species matters, it should be mentioned that Volterra (1926) gave much attention to the two species Lotka-Volterra problem with distributed time delays. (For reference, the relevant equations without the time delay terms are given by Eqs. 5.5.20 and 5.5.21.) This problem is examined at length by Davis (1961) and by numerous investigators listed in Section 6.3.

In connection with two-species Lotka-Volterra phenomena, we mention the study by Bhargava (1989) concerning technology substitution; the old technology is the prey, the new is the predator. In a similar vein, in their work on urban economics and demography, Dendrinos and Mullally (1985) consider population and income to be the predator and prey. Borsellino and Torre (1974) essentially regard population and pollution to be the competitors. An interesting study would feature the competition between an environment as a prey and development as a predator.

In Chapter 7 our entire effort was devoted to topics of spatial–temporal diffusion. We hardly scratched the surface. Along with a large number of future investigations involving time dependent parameters and time delay phenomena in growth processes, there are a great many types of diffusion problems yet to be analyzed. In principle, we would want to superimpose these three considerations.

With regard to diffusion, we start with the following equation

$$\frac{\partial U}{\partial t} = D\frac{\partial}{\partial x}\left(U^n\frac{\partial U}{\partial x}\right) + aU^r\left(1 - U^s\right) \quad . \tag{8.10}$$

In this relationship we have omitted the convection term, $u_0\partial U/\partial x$, and have selected the hyperlogistic as the reaction term. Also, for the first time in our overall analysis, we indicate that diffusion is density dependent.

A solution for the case of no reaction (i.e., $a = 0$) is provided by Pattle (1959). For the case in which $s = 1$, Newman (1980) gives a number of traveling wave solutions, $U(z)$, for various values of r and n; as before, $z = x - c_*t$. Newman shows that when $n > 0$, the density $U(z) = 0$ at a prescribed positive value, z_c. He also shows that when $n = 0$ and $r = 2$ the density distribution is an ordinary logistic. Newman points out applications not only in growth phenomena but also in population genetics and combustion processes.

An exact traveling wave solution of Eq. 8.10 is given by Kaliappan (1984) for the case in which $n = 0$, $r = 1$. His answer is the same as that obtained by Murray (1989) given by Eq. 7.6.28. A number of solutions are presented by Montroll and West (1973) for various problems involving logistic growth and diffusion.

As before we write Eq. 8.10 in dimensionless form by letting $T = at$ and $\xi = (\sqrt{a/D})x$. Accordingly

$$\frac{\partial U}{\partial T} = \frac{\partial}{\partial \xi}\left(U^n\frac{\partial U}{\partial \xi}\right) + U^r\left(1 - U^s\right) \quad . \tag{8.11}$$

In traveling wave form this relationship becomes, with $\zeta = \xi - cT$,

$$\frac{d}{d\zeta}\left(U^n\frac{dU}{d\zeta}\right) + c\frac{dU}{d\zeta} + U^r\left(1 - U^s\right) = 0 \quad . \tag{8.12}$$

The boundary conditions are: $U(-\infty) = 1$ and $U(\infty) = 0$. Three very useful solutions to this equation are provided by Murray.

Case 1. $n = 0$, $r = 1$, $s = 1$. This is the case we examined in Section 7.6.2.

Case 2. $n = 0$, $r = s + 1$. In this instance we have

$$\frac{\partial U}{\partial T} = \frac{\partial^2 U}{\partial \xi^2} + U^{s+1}(1 - U^s) \quad , \tag{8.13}$$

whose solution is

$$U(\zeta) = \frac{1}{(1 + \alpha e^{\beta \zeta})^\gamma} \quad , \tag{8.14}$$

where

$$\gamma = \frac{1}{s} \quad ; \quad \beta = \frac{s}{\sqrt{s+1}} \quad ; \quad c = \frac{1}{\sqrt{s+1}} \quad ; \tag{8.15}$$

and, as before, α is a location constant. If we take $s = 1$ then we have

$$\frac{\partial U}{\partial T} = \frac{\partial^2 U}{\partial \xi^2} + U^2(1 - U) \quad , \tag{8.16}$$

and $\gamma = 1$, $\beta = 1/\sqrt{2}$, and $c = 1/\sqrt{2}$. The solution is simply

$$U = \frac{1}{1 + \alpha e^{\zeta/\sqrt{2}}} \quad . \tag{8.17}$$

As indicated above, this answer is also given by Newman. As he points out, it describes the important case in population genetics in which a heterozygote (aA) is as viable as one of the homozygotes (aa, AA). We note that this solution also represents the case in which the reaction term is modified coalition growth instead of logistic growth.

Case 3. $n = 1$, $r = 1$, $s = 1$. This case describes density dependent diffusion with logistic reaction. That is

$$\frac{\partial U}{\partial T} = \frac{\partial}{\partial \xi}\left(U\frac{\partial U}{\partial \xi}\right) + U(1 - U) \quad . \tag{8.18}$$

The traveling wave solution, with $c = 1/\sqrt{2}$, is

$$U(\zeta) = \begin{cases} 1 - \exp[(\zeta - \zeta_c)/\sqrt{2}] & ; \quad \zeta < \zeta_c \\ 0 & ; \quad \zeta > \zeta_c \end{cases} \tag{8.19}$$

in which ζ_c is the front of the wave; it can be handled as a location constant.

The subject of density dependent diffusion has been examined by numerous investigators including Aronson (1980), Crank (1975), Fife (1979), Ghez (1988), Gurney and Nisbet (1975), Gurtin and MacCamy (1977) and Shigesada (1980).

Now to some of the subjects of spatial diffusion we have not considered at all. Many or most in this category are simply too complicated or too different to handle with the straightforward Fickian type frameworks we utilized above. In this category one finds the kinds of problems facing epidemiologists, geographers, regional planners, and sociologists. For example, (a) the rate of adoption of a new strain of rice in Indonesia, developed in the Philippines, (b) the creation of new towns and the disappearance of old, located near or far from large cities, and (c) the spread of HIV/AIDS in diverse geographical settings, are examples of spatial-temporal phenomena confronting investigators in the various disciplines.

In recent years, much attention has been given to these kinds of problems and this activity continues. Representative analyses and comprehensive surveys are given by Bartholomew (1982), Brown (1981), Chorley and Haggett (1967), Cliff and Ord (1975), and Morrill, Gaile and Thrall (1988).

These references cite the many advances made by numerous investigators on spatial diffusion. We limit our attention to two very useful and relevant studies: (a) Rushton and Mautner (1955) who present a deterministic model of the spread of a simple epidemic in a cluster of communities and (b) Hägerstand (1965, 1967) who devised a spatial– temporal model for diffusion of technological innovations and applied a Monte Carlo computation methodology to obtain solutions.

The important contribution made by Rushton and Mautner (1955) is in connection with logistic-type growth in a community and consequent concurrent growth in related communities. They examined the problem of a simple epidemic (or two-component innovation transfer) in which the numbers of infectives (or adopters) is x and susceptibles (or potential adopters) is y; the total number in a community is $n = x + y$.

They start with the set of equations

$$\frac{dy_i}{dt} = -y_i \left[\alpha_i(n_i - y_i) + \sum_j \beta_{ij}(n_j - y_j) \right] , \tag{8.20}$$

where $i = 1, \ldots, m$ and $j \neq i$, refer to the i-th and j-th communities; α and β are, respectively, the intra- and inter-community infection (or transfer) rates. Substantial simplification is made if we let $n_i = n$, $\alpha_i = \alpha$, and $\beta_{ij} = \alpha\gamma$. The solutions to Eqs. 8.20 are determined from the initial conditions, $y_i(0)$. Even with these simplifications, computations become very complicated.

A numerical example is presented by Rushton and Mautner for the special case in which $m = 2$, $\gamma = 0.2$, $x_1(0) = 0.05$ and $x_2(0) = 0$. Solution of a greatly simplified Eq. 8.20 provides the answers $x_1(t)$ and $x_2(t)$. Plots of $x_1(t)$ and $x_2(t)$ resemble ordinary logistic curves. They are similar in shape to the curves shown in Fig. 5.2.2 in which $x_1(t)$ corresponds to the logistically increasing carrying capacity, $N_*(t)$, and $x_2(t)$ to the growth curve, $N(t)$. For this special case, $m = 2$ and $\gamma = 0.20$, Rushton and Mautner compute $x_1(t)$ and $x_2(t)$ for $x_1(0) = 0.01, 0.02, 0.05, 0.10, 0.20, 0.30$, and 0.50 all with $x_2(0) = 0$.

Even for this relatively simple $m = 2$ case, analysis and computation is not easy. Further, the scheme does not explicitly introduce a specified spatial parameter. Nevertheless, the method does provide a solution to the deterministic simple epidemic (or innovation transfer) problem involving two or more related communities.

Another classic contribution in the matter of spatial diffusion is that of Hägerstrand (1965, 1967). His use of Monte Carlo simulation to provide a methodology for solving problems in spatial diffusion is noteworthy.

From about 1925 to 1950, Hägerstrand carried out studies concerning a region of southern Sweden involving three agricultural innovations: (a) improved pasture subsidies, (b) bovine tuberculosis control, and (c) soil mapping. We shall look at the diffusion of an innovation featuring the first of these: government subsidies to small farmers to improve their farms and woodlands.

Up to a certain point, the methodology is something like the analysis of a simple epidemic or adopter logistic growth. Hägerstrand introduces the term "neighborhood effect" to characterize the higher probability of contact and consequent information transfer between a teller and receiver that are in close proximity. This gives rise to a relationship between migrant density and distance from an information source which, in turn, generates a so-called mean information field.

Next, a grid system composed of $n \times n$ squares is superimposed on a map of the particular geographical area. The mean information field assigns to each cell of the grid a probability of expected migrants. For the initial condition, Hägerstrand utilized the observed number of farms who had already adopted the farm subsidy by 1929.

At this point, Monte Carlo simulation begins. This computational methodology is a procedure involving statistical sampling techniques to obtain a probabilistic approximation to the solution of a mathematical or physical problem. The ideas behind the method are far from new; the well-known Buffon problem – tossing a needle on a sheet of lined paper to determine the value of π – is an early ancestor. However, development of present-day methodology began with the work of von Neumann and Ulam in the late 1940s. In the ensuing years, Monte Carlo simulation was greatly advanced, especially in the fields of quantum and statistical mechanics. Elementary treatments of the subject are given by Beckmann (1971) and Sobol (1975); more advanced treatments are provided by Hammersley and Handscomb (1964) and Kalos and Whitlock (1986).

Back to farm subsidies. Hägerstrand carried out three simulation series to obtain the spatial-temporal distribution of this particular problem of innovation diffusion. The results of each series were essentially the same and were in close agreement with observed patterns of adoption for the years 1930, 1931 and 1932. Plots of $U = U(r,t)$ qualitatively resemble those shown in Fig. 7.2.6. Some of the properties of Hägerstrand's model are examined by Gale (1972). Extensions and generalizations of the methodology are described by Morrill et al (1988).

In addition to the references cited earlier, phenomena of spatial diffusion in the field of economic geography have been analyzed by Casetti and Semple (1969), Webber and Joseph (1978, 1979) and Haining (1983).

A few final remarks about a very complicated phenomenon: spatial diffusion of epidemics. The 1957 book of Bailey, revised in Bailey (1975), provides benchmark analyses of the subject. Further advances are presented in Bailey (1980) along with an extensive list of references on spatial models.

A comprehensive and lucid treatment of the geographic spread of epidemics is given by Murray (1989). He begins with the inclusion of spatial diffusion terms in the epidemic relationships of Eqs. 2.8.2 and 2.8.3

$$\frac{\partial S}{\partial t} = D\frac{\partial^2 S}{\partial x^2} - \beta SI \quad ; \tag{8.21}$$

$$\frac{\partial I}{\partial t} = D\frac{\partial^2 I}{\partial x^2} + \beta SI - \gamma I \quad . \tag{8.22}$$

As in the case of the logistic equation with diffusion, Eq. 7.6.1, Murray obtains traveling wave solutions to Eqs. 8.21 and 8.22. One interesting outcome is that when $\lambda = \gamma/\beta S_0$ is less than unity then an epidemic wave will be propagated; when λ is larger than unity there will be no wave. We note that this is the same threshold condition we had in Section 2.8.2 in the absence of diffusion. The velocity of the epidemic wave is

$$c_* = 2\left[\beta S_0 D\left(1 - \frac{\gamma}{\beta S_0}\right)\right]^{1/2} \quad . \tag{8.23}$$

This analysis is extended by Murray to the two-dimensional case and illustrated with the problem of spatial spread of rabies among foxes. A relatively non-technical presentation of this problem is given by Murray (1987). The results of other investigations of the phenomenon of geographical spread of rabies are given by Anderson et al. (1981) and by Källén et al. (1985).

Probably the most famous of all epidemics is the Black Death, the plague that killed a quarter of the people of Europe during the period 1347 to 1350. A descriptive account of this catastrophic event is given by Langer (1964); more quantitative treatments are provided by Noble (1974) and Murray (1989).

In a category somewhere between dispersing muskrats and geographically spreading rabid foxes is the recent behavior of so-called Africanized honey bees (*Apis mellifera scutellata*). This bee is the hybrid offspring of European honey bees and African queen bees brought to Brazil in 1956. Evidently the African queens escaped into the Brazilian countryside and began interbreeding with native honey bees.

As in the case of the muskrats and gypsy moths, which we considered in Section 7.1.5, the Africanized bees began to disperse geographically. By

1971 they had spread from their original source near Rio de Janeiro to north-west Brazil and to northern Argentina. By 1980 the bees had reached west-ern Venezuela and in 1982 they were observed in Panama. They continued their northward dispersion: to southern Mexico by 1985 and to the lower Rio Grande valley of Texas in late 1990.

From the observed dispersion distance, Panama to Texas during the pe-riod 1982 to 1990, we estimate the front velocity to be $c_* = 2\sqrt{aD} = 400\,\text{km/yr}$. Accordingly, the value of the growth-dispersion parameter is $aD = 4.0 \times 10^4\,(\text{km/yr})^2$. From Section 7.1.5, the average value of c_* for the muskrat, gypsy moth and Japanese beetle dispersals is about 10 km/yr. On this basis, the Africanized bees disperse 40 times faster and the growth-dispersion parameter is 1600 times larger. This phenomenon of Africanized bee dispersion is examined at length by Taylor (1985).

Finally, we look briefly at the extremely important phenomenon of spatial-temporal spread of the human immunodeficiency virus (HIV) and the as-sociated acquired immune deficiency syndrome (AIDS). Many regard this epidemic as the most devastating epidemic since the Black Death plague of the 14th century. Our consideration of this problem is limited to a listing of some of the important studies on the population dynamics of the HIV/AIDS epidemic.

Though not directly concerned with AIDS, comprehensive coverage of the population biology of infectious diseases is given by Anderson and May (1979) and May and Anderson (1979). The same authors, May and Anderson (1987), examine the transmission dynamics of HIV infection. Further treat-ment of transmission dynamics is given by Dietz (1988) and mathematical models for the AIDS epidemic are developed by Hyman and Stanley (1988). The demographic consequences of HIV/AIDS epidemics are presented by May, Anderson and McLean (1988). Their considerations are continued in the conference proceedings edited by Castillo-Chavez et al. (1989) devoted in large measure to problems in epidemiology. In closing, a lucid presentation is given by Murray (1989) of the problem of mathematical modeling of the mechanisms for transmitting HIV/AIDS.

References

Abramowitz, M., Stegun, I.A. (Eds.) (1965): Handbook of Mathematical Functions, Dover, New York

Allen, R.G.D. (1965): Mathematical Economics, 2nd edn., Macmillan, New York

Anderson, R.M.; May, R.M. (1979): Population biology of infectious diseases: Part I, Nature **280**, 361–367

Anderson, R.M., Jackson, H.C., May, R.M., Smith, A.M. (1981): Population dynamics of fox rabies in Europe, Nature **289**, 765–771

Andrewartha, H.G., Birch, L.C. (1964): The Distribution and Abundance of Animals, The University of Chicago Press, Chicago

Aronson, D.G. (1980): Density-dependent interaction-diffusion systems. Dynamics and Modelling of Reactive Systems (Eds. W.E. Stewart, W.H. Ray, C.C. Conley), 161–176, Academic Press, New York

Aronson, D.G., Weinberger, H.F. (1975) Nonlinear diffusion in population genetics, combustion, and nerve propagation. Partial Differential Equations and Related Topics, Lecture Notes in Mathematics, Vol. 446, 5–49, (Ed. J.A. Goldstein), Springer, Berlin, Heidelberg

Arrigoni, M., Steiner, A. (1985): Logistisches Wachstum in fluktuierender Umwelt, Journal of Mathematical Biology **21**, 237–241

Ashton, W.D. (1972): The Logit Transformation , Charles Griffin, London

Atkins, P.W. (1982): Physical Chemistry, 2nd edn., W.H. Freeman, San Francisco

Austin, A., Brewer, J.W. (1971): World population growth and related technical problems, Technological Forecasting and Social Change **3**, 23-49

Ayres, R.U. (1969): Technological Forecasting and Long-Range Planning, McGraw-Hill, New York

Bailey, N.T.J. (1975): The Mathematical Theory of Infectious Diseases and its Applications, 2nd edn., Griffin, London

Bailey, N.T.J. (1980): Spatial models of the epidemiology of infectious diseases, Biological Growth and Spread (Eds. W. Jager, H. Rost, P. Tantu), Lecture Notes in Biomathematics, Vol. 38, 233–261, Springer, Berlin, Heidelberg

Barclay, G.W. (1958): Techniques of Population Analysis, Wiley, New York

Barenblatt, G.I. (1987): Dimensional Analysis, Gordon and Breach, New York

Bartholomew, D.J. (1981): Mathematical Models in Social Science, Wiley, Chichester

Bartholomew, D.J. (1982): Stochastic Models for Social Processes, 3rd edn., Wiley, Chichester

Baserga, R. (1981): The cell cycle, New England Journal of Medicine **304**, 453–459

Batchelor, G.K. (1967): An Introduction to Fluid Dynamics, Cambridge University Press, Cambridge

Batschelet, E. (1979): Introduction to Mathematics for Life Scientists, 3rd edn., Springer, Berlin, Heidelberg

Bear, J. (1979): Hydraulics of Groundwater, McGraw-Hill, New York

Beck, K. (1982): A model of the population genetics of cystic fibrosis in the United States, Mathematical Biosciences **58**, 243–257

Becking, L.G.M. (1948): On the analysis of sigmoid curves, Acta Biotheoretica **8**, 42–59

Beckmann, M.J. (1970): The analysis of spatial diffusion processes, Papers of the Regional Science Association **25**, 109–117

Beckmann, P. (1971): A History of Pi, St. Martin's Press, New York

Bellman, R. and Cooke, K.L. (1963): Differential-difference Equations, Academic Press, New York

Berkson, J. (1944): Application of the logistic function to bio-assay, Journal of the American Statistical Association **39**, 357–365

Berkson, J. (1951): Why I prefer logits to probits, Biometrics **7**, 327–339

Berkson, J. (1953): A statistically precise and relatively simple method of estimating the bioassay with quantal response, based on the logistic function, American Statistical Association Journal **48**, 565–599

Bertalanffy, L. von (1957): Quantitative laws in metabolism and growth, The Quarterly Review of Biology **32**, 217–231

Bertalanffy, L. von (1968): General System Theory, George Braziller, New York

Bhargava, S.C. (1989): Generalized Lotka-Volterra equations and the mechanism of technological substitution, Technological Forecasting and Social Change **35**, 319–326

Bird, R.B., Stewart, H.E., Lightfoot, E.M. (1960): Heat, Mass and Momentum Transfer, Prentice-Hall, Englewood Cliffs, New Jersey

Blackman, A.W. (1972): A mathematical model for trend forecasts, Technological Forecasting and Social Change **3**, 441–152

Bliss, C.I. (1935): The calculation of the dosage-mortality curve, Annals of Applied Biology **22**, 134–167

Blumenfeld, H. (1954): The tidal wave of metropolitan expansion, Journal of the American Institute of Planners **20**, 3–14

Bogin, B. (1988): Patterns of Human Growth, Cambridge University Press, Cambridge

Bojadziev, G. (1983): Population modelling by higher order differential equations, Population Biology, (Eds. H.T. Freeman and C. Strobeck), Lecture Notes in Biomathematics, Vol. 52, 141–145, Springer, Berlin, Heidelberg

Borsellino, A., Torre, V. (1974): Limits to growth from Volterra theory of population, Kybernetik **16**, 113–118

Boyce, W.E., DiPrima, R.C. (1986): Elementary Differential Equations and Boundary Value Problems, 4th edn., Wiley, New York

Brandt, S. (1976): Statistical and Computational Methods in Data Analysis, North Holland, Amsterdam

Brauer, F. (1976): Constant rate harvesting of populations governed by Volterra integral equations, Journal of Mathematical Analysis and Applications **56**, 18–27.

Brauer, F., Sanchez, D.A. (1975): Constant rate population harvesting: equilibrium and stability, Theoretical Population Biology **8**, 12–30

Braun, M. (1982): The spread of technological innovations, Chapter 6, Differential Equation Models, (Eds. M. Braun, C.S. Coleman, D.A. Drew), Springer, New York, 91–97

Bridgeman, P.W. (1922): Dimensional Analysis, Yale University Press, New Haven, Connecticut

Brody, S. (1945): Bioenergetics and Growth, Reinhold, New York

Brown, L.A. (1981): Innovation Diffusion, Methuen, London

Burghes, D.N., Wood, A.D. (1984): Mathematical Models in the Social, Management and Life Sciences, Wiley, New York

Bush, R.R., Mosteller, R. (1955): Stochastic Models for Learning, Wiley, New York

Bussière, R. (1976): Interactions urbaines: le modele residentiel du CRU, Annales. Centre de Recherche d'Urbanisme, Paris

Byerlee, D., Hesse de Polanco E. (1982) The rate and sequence of adoption of improved cereal technologies: a case of rainfed barley in the Mexican altiplano, Working Paper 82/4, Centro Internacional de Majoramiento de Maiz y Trigo (CIMMYT), Mexico, D.F., Mexico

Camp, T.R. (1962): Water and Its Impurities, Reinhold, New York

Cannon, J.R. (1984): The One-Dimensional Heat Equation, Encyclopedia of Mathematics and Its Applications, Vol. 23, (Ed. G.C. Rota), Addison- Wesley, Menlo Park, California

Canosa, J. (1973): On a nonlinear diffusion equation describing population growth, IBM Journal of Research and Development 17, 307-313

Carslaw, H.S., Jaeger, J.C. (1959): Conduction of Heat in Solids, 2nd edn., Clarendon Press, Oxford

Casetti, E. (1969): Alternate urban population density models: an analytical comparison of their validity range, Studies in Regional Science, (Ed. A.J. Scott), 105-116, Pion Limited, London

Casetti, E. (1986): The dual expansion method: an application for evaluating the effects of population growth on development, IEEE Transactions on Systems, Man and Cybernetics SMC-16, 29-39

Casetti, E., Semple R.K. (1969): Concerning the testing of spatial diffusion hypotheses, Geographical Analysis 1, 254-259

Castillo-Chavez, C., Levin, S.A., Shoemaker, C.A. (Eds.) (1989): Mathematical Approaches to Problems in Resource Management and Epidemiology, Lecture Notes in Biomathematics, Vol. 81, Springer, Berlin, Heidelberg

Caughley, G. (1970): Eruption of ungulate populations, with emphasis on Himalayan thar in New Zealand, Ecology 51, 53-71

Caughley, G. (1976): Wildlife management and the dynamics of ungulate populations, Applied Biology, (Ed. T.H. Coaker) 1, 183-246

Causton, D.R., Venus, J.C. (1981): The Biometry of Plant Growth, Edward Arnold Publishers, London

Central Statistical Office (1976): Annual Abstract of Statistics, 1975 edn., Her Majesty's Stationery Office, London

Chakravarti, I.M., Laha, R.G., Roy, J. (1967): Handbook of Methods of Applied Statistics, Vol. I, Wiley, New York

Chorley, R.J., Haggett, P. (1967): Models in Geography, Methuen, London

Clark, C. (1951): Urban population densities, Journal of the Royal Statistical Society 64, 490-496

Clark, C.W. (1976): Mathematical Bioeconomics: The Optimal Management of Renewable Resources, Wiley-Interscience, New York

Clark, C.W. (1981): Bioeconomics, Chapter 16, Theoretical Ecology, (Ed. R.M. May), 2nd edn., 387-418, Blackwell, Oxford

Clark, J.P. (1971): The second derivative and population modeling, Ecology 52, 606-613

Clements, D.L. (1984): An Introduction to Mathematical Models in Economic Dynamics, North Oxford Academic Publishing, Oxford

Cliff, A.D., Ord, J.K. (1975): Space time modelling with an application to regional forecasting, Transactions of the Institute of British Geographers 64, 119-128

Coale, A.J. (1972): The Growth and Structure of Human Populations, Princeton University Press, Princeton

Cohen, D.S., Coutsias, E., Neu, J.C. (1979): Stable oscillations in single species growth models with heredity effects, Mathematical Biosciences **44**, 255–268

Cole, L.C. (1957): Sketches of general and comparative demography, Cold Water Harbor Symposium on Quantitative Biology **22**, 1–15

Coleman, B.D. (1979): Nonautonomous logistic equations as models of the adjustment of populations to environmental change, Mathematical Biosciences - **45**, 159–173

Conolly, B. (1980): Techniques in Operational Research: Volume 2: Models, Search and Randomization, Wiley, New York

Conrad, J.M. (1986): Bioeconomics and the management of renewable resources, Mathematical Ecology, (Eds. T.G. Hallam and S.A. Levin), 381–403, Springer, Berlin, Heidelberg

Conrad, J.M., Clark, C.W. (1987): Natural Resource Economics: Notes and Problems, Cambridge University Press, Cambridge

Cooper, R.A., Weekes, A.J. (1983): Data, Models and Statistical Analysis, Philip Allen Publishers, Oxford

Cox, D.R. (1967): Renewal Theory, Methuen, London

Cox, D.R. (1970): The Analysis of Binary Data, Methuen, London

Cox, P.R. (1976): Demography, 5th edn., Cambridge University Press, Cambridge

Crank, J. (1975): The Mathematics of Diffusion, 2nd edn., Clarendon Press, Oxford

Cunningham, W.J. (1954): A nonlinear differential-difference equation of growth, Proceedings of the National Academy of Science **40**, 708–713

Cunningham, W.J. (1958): Introduction to Nonlinear Analysis, McGraw-Hill, New York

Cushing, J.M. (1977): Integrodifferential equations and delay models in population dynamics, Lecture Notes in Biomathematics, Vol. 20, Springer, Berlin, Heidelberg

Cushing, J.M. (1979): Volterra integrodifferential equations in population dynamics, Mathematics of Biology, 81–148, Centro Internazionale Matematico Estivo, Naples

Cushing, J.M. (1986): Oscillatory population growth in periodic environments, Theoretical Population Biology **30**, 289–308

Davis, H.T. (1961): Introduction to Nonlinear Differential and Integral Equations, Dover, New York

de Jong, F.J., Quade, W. (1967): Dimensional Analysis for Economists, North Holland, Amsterdam

Dendrinos, D.S., Mullally, H. (1985): Urban Evaluation Studies in the Mathematical Ecology of Cities, Oxford University Press, Oxford

Dietz, K. (1967): Epidemics and rumors: a survey, Journal of the Royal Statistical Society **130 A**, 505–528

Dietz, K. (1988): On the transmission dynamics of HIV, Mathematical Biosciences **90**, 397–414

Dixon, R. (1980): Hybrid corn revisited, Econometrica **48**, 1451–1461

Dobson, A.J. (1983): Introduction to Statistical Modelling, Chapman and Hall, London

Draper, N.R., Smith,. H. (1981): Applied Regression Analysis, Wiley, New York

Driver, R.D. (1977): Ordinary and Delay Differential Equations, Springer, Berlin, Heidelberg

Eason, G., Coles, C.W., Gettinby, G. (1980): Mathematics and Statistics for the Bio-Sciences, Wiley, New York

Edmonston, B., Davies, O. (1976): Population suburbanization in the western region of the United States: 1900–1970, Land Economics **52**, 393–403

Elderton, W.P., Johnson, N.L. (1969): Systems of Frequency Curves, Cambridge University Press, Cambridge

El'sgol'ts, L.E., Norkin, S.B. (1973): Introduction to the Theory and Application of Differential Equations with Deviating Arguments, Academic Press, New York

Elton, C.S. (1958): The Ecology of Invasions by Animals and Plants, Methuen, London

Estes, W.K (1950): Toward a statistical theory of learning, Psychological Review **57**, 94–107

Falkner, F., Tanner, J.M. (Eds.) (1978): Human Growth: A Comprehensive Treatise, Plenum Press, New York

Farlow, S.J. (1982): Partial Differential Equations for Scientists and Engineers, Wiley, New York

Feller, W. (1940): On the logistic law of growth and its empirical verifications in biology, Chapter 4, Applicable Mathematics of Non-Physical Phenomena, (Eds. F. Olivera-Pinto and B.W. Conolly), 123–138, Ellis Horwood, Chichester, 1982

Feller, W. (1941): On the integral equation of renewal theory, Annals of Mathematical Statistics **12**, 243–267

Fife, P.C. (1979): Mathematical Aspects of Reacting and Diffusing Systems, Lecture Notes in Biomathematics, Vol. 28, Springer, Berlin, Heidelberg

Finney, D.J. (1971): Probit Analysis, 3rd edn., Cambridge University Press, Cambridge

Fischer, H.B. (1968): Dispersion predictions in natural streams, Journal of the Sanitary Engineering Division, Proceedings of the American Society of Civil Engineers **94**, 927–942

Fischer, R.A. (1983): Wheat, Proceedings of Symposium on Potential Productivity of Field Crops under Different Environments, International Rice Research Institute, (IRRI), Los Banos, Philippines

Fisher, J.C., Pry, R.H. (1971): A simple substitution model of technological change, Technological Forecasting and Social Change, **3**, 75–88

Fisher, R.A. (1937): The wave of advance of advantageous genes, Annals of Eugenics **7**, 355–369

Fletcher, R.I. (1974): The quadratic law of damped exponential growth, Biometrics **30**, 111–124

France, J.,Thornley, J.H.M. (1989) Mathematical Models in Agriculture, Butterworths, London

Frauenthal, J.C. (1980): Introduction to Population Modeling, Birkhauser, Boston

Fried, J.J., Combarnous, M.A. (1971): Dispersion in porous media, Advances in Hydroscience, (Ed. V.T. Chow), Vol. 7, 170–282, Academic Press, New York

Frisch, R., Holme, H. (1935): The characteristic solutions of a mixed difference and differential equation occurring in economic dynamics, Econometrica **3**, 225–239

Gale, S. (1972): Some formal properties of Hagerstrand's model of spatial interactions, Journal of Regional Science **12**, 199–217

Gandolfo, G. (1971): Mathematical Methods and Models in Economic Dynamics, North Holland, Amsterdam

Garcia, O. (1983): A stochastic differential equation model for the height growth of forest stands, Biometrics **39**, 1059–1072

Gause, G.F. (1934): The Struggle for Existence, Hafner, New York

Getz, W.M., Haight, R.G. (1989): Population Harvesting: Demographic Models of Fish, Forest, and Animal Resources, Princeton University Press, Princeton, New Jersey

Ghez, R. (1988): A Primer of Diffusion Problems, Wiley, New York

Gilpin, M.E., Ayala, F.J. (1973): Global models of growth and competition, Proceedings National Academy of Science, USA **70**, 3590–3593

Goel, N.S., Maitra, S.C., Montroll, E.M. (1971): On the Volterra and other nonlinear models of interacting populations, Reviews of Modern Physics **43**, 231–276

Goldstein, S. (1953): On the mathematics of exchange processes in fixed columns. I. Mathematical solutions and asymptotic expansions, Proceedings of the Royal Society of London **A219**, 151–171

Goldsworthy, P.R., Colegrove, M. (1974): Growth and yield of highland maize in Mexico, Journal of Agricultural Science **83**, 213–221

Gompertz, B. (1825): On the nature of the function expressive of the law of mortality, Philosophical Transactions **27**, 513–585

Gopalsamy, K. (1985): Nonoscillaton in a delay-logistic equation, Quarterly of Applied Mathematics **43**, 189–197

Gopalsamy, K. (1986): Oscillations in a delay-logistic equation, Quarterly of Applied Mathematics **44**, 447–461

Greenkorn, R.A. (1983): Flow Phenomena in Porous Media, Marcel Dekker, New York

Griliches, Z. (1957): Hybrid corn: an exploration in the economics of technological change, Econometrica **25**, 501–522

Griliches, Z. (1960): Hybrid corn and the economics of innovation, Science **132**, 275–280

Guest, A.M. (1973): Urban growth and population densities, Demography **10**, 53–69

Gumbel, E.J. (1958): Statistics of Extremes, Columbia University Press, New York

Gumowski, I. (1981): Qualitative properties of some dynamic systems with pure delay, Modèles Mathématiques en Biologie, (Eds. C. Chevalet and A. Micali), Lecture Notes in Biomathematics, Vol. 41, Springer, Berlin, Heidelberg

Gurney, W.S.C., Nisbet, R.M. (1975): The regulation of inhomogeneous populations, Journal of Theoretical Biology **52**, 441–457

Gurney, W.S.C., Blythe, S.P., Nisbet, R.M. (1980) Nicholson's blowflies revisited, Nature **287**, 17–21

Gurtin, M.E., MacCamy, R.C. (1977): On the diffusion of biological populations, Mathematical Biosciences **33**, 35–49

Hägerstrand, T. (1965): A Monte Carlo approach to diffusion, European Journal of Sociology **6**, 43–67

Hägerstrand, T. (1967): Innovation Diffusion as a Spatial Process, University of Chicago Press, Chicago

Haining, R. (1983): Spatial and spatial-temporal interaction models and the analysis of patterns of diffusion, Transactions of the Institute of British Geographers **8**, 158–186

Hallam, T.G. (1986): Population dynamics in a homogeneous environment, Mathematical Ecology, (Eds. T.G. Hallam and S.A. Levin), 61–94, Springer, Berlin, Heidelberg

Hammersley, J.M., Handscomb, D.C. (1964): Monte Carlo Methods, Meuthen, London

Hanham, R.Q., Brown, L.A. (1976): Diffusion waves within the context of regional economic development, Journal of Regional Science **16**, 65–70

Hastings, N.A.J., Peacock, J.B. (1974): Statistical Distributions, Wiley, New York

Hauspie, R.C., Wachholder, A., Baron, G. (1980): A comparative study of the fit of four different functions to longitudinal data of growth in height of Belgian girls, Annals of Human Biology **7**, 347–358

Hayashi, C. (1985): Nonlinear Oscillations in Physical Systems, Princeton University Press, Princeton, New Jersey

Haynes, K.E., Mahajan, V., White, G.M. (1977): Innovation diffusion: a deterministic model of space-time integration with physical analog, Social-Economic Planning Sciences **11**, 25–29

Hoppensteadt, F. (1975): Mathematical Theories of Populations: Demographics, Genetics and Epidemics, Society for Industrial and Applied Mathematics, Philadelphia

Hull, D.E., Kent, J.W. (1952): Radioactive tracers to mark interfaces and measure intermixing in pipelines, Industrial and Engineering Chemistry 44, 2745–2750

Hutchinson, C.E. (1948): Circular causal systems in ecology, Annals New York Academy of Science 50, 221–246

Hutchinson, C.E. (1979): An Introduction to Population Ecology, Yale University Press, New Haven, Connecticut

Hyman, J.M., Stanley, E.A. (1988): Using mathematical models to understand the AIDS epidemic, Mathematical Biosciences 90, 415–473

Impagliazzo, J. (1985): Deterministic Aspects of Mathematical Demography, Springer, Berlin, Heidelberg

Innis, G. (1972): The second derivative and population modeling: another view, Ecology 53, 720–723

Jansen, M.J.W., Metz, J.A.J. (1979): How many victims will a pitfall make? Acta Biotheoretica 28, 98–122

Jarvis, L.S. (1981): Predicting the diffusion of improved pastures in Uruguay, American Journal of Agricultural Economics 63, 495–502

Jeffreys, H. (1967): Theory of Probability, 3rd edn., Oxford University Press, Oxford

Jones, G.S. (1961): Asymptotic behavior and periodic solutions of a nonlinear differential-difference equation, Proceedings of the National Academy of Science 47, 879–882

Jones, G.S. (1962): On the nonlinear differential-difference equation $f'(x) = -\alpha f(x - 1)[1 + f(x)]$, Journal of Mathematical Analysis and Applications 4, 440–469

Jones, D.S., Sleeman, B.D. (1983): Differential Equations and Mathematical Biology, Allen & Unwin, London

Jordan, D.W., Smith, P. (1977): Nonlinear Ordinary Differential Equations, Oxford University Press, Oxford

Jost, W. (1952): Diffusion in Solids, Liquids, Gases, Academic Press, New York

Kakutani, S., Marcus, L. (1958): On the non-linear difference-differential equation $y'(t) = [A - By(t - \tau)]y(t)$, Contributions to the Theory of Nonlinear Oscillatons, Vol. IV, (Ed. S. Lefschetz), Princeton University Press, Princeton, New Jersey

Kalecki, M. (1935): A macrodynamic theory of business cycles, Econometrica 3, 327–344

Kaliappan, P. (1984): An exact solution for travelling waves of $u_t = Du_{xx} + u - u^k$, Physica D 11, 368–374

Källén, A., Arcuri, P., Murray, J.D. (1985) A simple model for the spatial spread and control of rabies, Journal of Theoretical Biology 116, 377–393

Kalos, M.H., Whitlock, P.A. (1986): Monte Carlo Methods, Wiley, New York

Kendall, D.G. (1948): A form of wave propagation associated with the equation of heat conduction, Proceedings of the Cambridge Philosophical Society 44, 591–594

Kendall, D.G. (1952): On the role of variable generation time in the development of a stochastic birth process, Biometrica 35, 316–330

Kennedy, C.R., Aris, R. (1980): Traveling waves in a simple population model involving growth and death, Bulletin of Mathematical Biology 42, 397–429

Kermack, W.O., McKendrick, A.G. (1927): A contribution to the mathematical theory of epidemics, Proceedings of the Royal Society of London 115 A, 700–721; also 138, 55–83 (1932); 141, 94–122 (1933)

Keyfitz, N. (1968): Introduction to the Mathematics of Population, Addison-Wesley, Reading, Massachusetts

Kierstead, H., Slobodkin, L.B. (1953): The size of water masses containing plankton blooms, Journal of Marine Research **12**, 141–147

Kindlmann, P. (1985): A model of aphid population with age structure, Mathematics in Biology and Medicine, (Eds. V. Capasso, E. Grosso and S.L. Paveri-Fontana), Lecture Notes in Biomathematics, Vol. 57, 72–77, Springer, Berlin, Heidelberg

Kingsland, S.E. (1985): Modeling Nature: Episodes in the History of Population Ecology, University of Chicago Press, Chicago

Klein, D.R. (1968): The introduction, increase, and crash of reindeer on St. Matthew Island, Journal of Wildlife Management **32**, 350–367

Knolle, H. (1988): Cell Kinetic Modelling and the Chemotherapy of Cancer, Lecture Notes in Biomathematics, Vol. 75, Springer, Berlin, Heidelberg

Kolmogorov, A., Petrovsky, I., Piskounov, N. (1937): Study of the diffusion equation with growth of the quantity of matter and its application to a biological problem, Chapter 7, Applicable Mathematics of Non-Physical Phenomena, (Eds. F. Olivera-Pinto and B.W. Conolly), 169–184, Ellis Horwood, Chichester, 1982

Kostitzin, V.A. (1937): Biologie Mathématique, Librarie Armand Colin, Paris

Kostitzin, V.A. (1939): The integro-differential equations for the toxic contamination of a medium, The Golden Age of Theoretical Ecology: 1923–1940, (Eds. F.M. Scudo and J.R. Ziegler), Lecture Notes in Biomathematics, Vol. 22, 50–53, Springer, Berlin

Krebs, C.J. (1978): Ecology: The Experimental Analysis of Distribution and Abundance, 2nd edn., Harper & Row, New York

Kreith, F. (1958): Principles of Heat Transfer, International Textbook Co., Scranton, Pennsylvania

Kreyszig, E. (1983): Advanced Engineering Mathematics, 5th edn., Wiley, New York

Laird, A.K. (1965): Dynamics of tumor growth: Comparison of growth rates and extrapolation of growth curve to one cell, British Journal of Cancer **19**, 278–291

Lal, V.B., Karmeshu, Kaicker, S. (1988) Modeling innovation diffusion with distributed time lag, Technological Forecasting and Social Change **34**, 103–113

Lancaster, K. (1968): Mathematical Economics, Dover, New York

Landau, L.D., Lifshitz, E.M. (1959) Fluid Mechanics, Pergamon Press, New York

Lanford, H.W. (1972): Technological Forecasting Methodologies, American Management Association, New York

Langer, W.L. (1964): The Black Death, Scientific American **210**, 114–121

Lapidus, L., Amundsen, N.R. (1952): Mathematics of adsorption in beds. VI. The effect of longitudinal diffusion in ion exchange and chromotographic columns, Journal of Physical Chemistry **56**, 984–988

Leach, D. (1981): Re-evaluation of the logistic curve for human populations, Journal of the Royal Statistical Society **144**, 94–103

Lebowitz, J.L., Rubinow, S.I. (1974): A theory for the age and generation time distribution of a microbial population, Journal of Mathematical Biology **1**, 17–36

Lekvall, P., Wahlbin, C. (1973): A study of some assumptions underlying innovation diffusion functions, Swedish Journal of Economics **75**, 362–377

Leslie, P.H. (1945): On the uses of matrices in certain population mathematics, Biometrika **33**, 183–212

Levin, S. (1981): Models of population dispersal, Differential Equations and Applications in Ecology, Epidemics, and Population Problems, (Eds. S.N. Busenberg and K.L. Cooke), 1–18, Academic Press, New York

Lewis, D. (1966): Quantitative Methods in Psychology, University of Iowa Press, Iowa City, Iowa

Linstone, H.A., Sahal, D. (Eds.) (1976): Technological Substitution. Forecasting Techniques and Applications, Elsevier, New York

Linz, P. (1985): Analytical and Numerical Methods for Volterra Equations, Society for Industrial and Applied Mathematics, Philadelphia

Lotka, A.J. (1924): Elements of Physical Biology, Williams and Wilkins Co., Baltimore, Published as Elements of Mathematical Biology, Dover, New York, 1956

Luce, R.D. (1959): Individual Choice Behavior: A Theoretical Analysis, Wiley, New York

MacCall, A.D. (1990): Dynamic Geography of Marine Fish Populations, University of Washington Press, Seattle

MacDonald, N. (1978): Time Lags in Biological Models, Lecture Notes in Biomathematics, Vol. 27, Springer, Berlin, Heidelberg

MacDonald, N. (1989): Biological Delay Systems: Linear Stability Theory, Cambridge, University Press, Cambridge

Mahajan, V., Peterson, R.A. (1979): Integrating time and space in technological substitution models, Technological Forecasting and Social Change 14, 231–241

Mahajan, V., Peterson, R.A. (1985): Models for Innovation Diffusion, Sage Publications, Beverly Hills, California

Malthus, T. (1798): An Essay on the Principle of Population, Reprinted 1970, Penguin Books, Harmondsworth, England

Mansfield, E. (1961): Technical change and the rate of imitation, Econometrica 29, 741–766

Mansfield, E., Hensley, C. (1960): The logistic process: tables of the stochastic epidemic curve and applications, Journal of the Royal Statistical Society 22B, 332–337

Marr, A.G., Painter, P.R.,, Nilson, E.H. (1969): Microbial Growth, Cambridge University Press, Cambridge

Matthews, G.A. (1984): Pest Management, Longman, London

Mattingly, P.F. (1987): Patterns of horse devolution and tractor diffusion in Illinois, 1920–1982, Professional Geographer 39, 298–309

May R.M., (1974): Stability and Complexity in Model Ecosystems, Princeton University Press, Princeton, New Jersey

May R.M., (Ed.) (1981): Theoretical Ecology. Principles and Applications. 2nd edn., Blackwell Scientific Publications, Oxford

May, R.M., Conway, G.R., Hassell, M.P., Southwood, T.R.E. (1974): Time delays, density dependence, and single species oscillations, Journal of Animal Ecology 43, 747–770

May, R.M., Anderson, R.M. (1979): Population biology of infectious diseases: Part II, Nature 280, 455–461

May, R.M., Anderson, R.M. (1987): Transmission dynamics of HIV infection, Nature 326, 137–142

May, R.M., Anderson, R.M., McLean, A.R. (1988): Possible demographic consequences of HIV/AIDS epidemics: I Assuming HIV infection always leads to AIDS, Mathematical Biosciences 90, 475–505

Maynard Smith, J. (1968): Mathematical Ideas in Biology, Cambridge University Press, Cambridge

Maynard Smith, J. (1974): Models in Ecology, Cambridge University Press, Cambridge

Mazanov, A., Tognetti, K.P. (1974): Taylor series expansion of delay differential equations – a warning! Journal of Theoretical Biology 46, 271–282

Mazumdar, J. (1989): An Introduction to Mathematical Physiology and Biology, Cambridge University Press, Cambridge

McCullagh, P., Nelder, J.A. (1983): Generalized Linear Models, Chapman and Hall, London

McKendrick, A.G. (1926): Applications of mathematics to medical problems, Proceedings of the Edinburgh Mathematical Society **44**, 98–130

McMurtrie, R. (1978): Persistence and stability of single-species and prey–predator systems in spatially heterogeneous environments, Mathematical Biosciences **39**, 11–51

Miller, R.K. (1971): Nonlinear Volterra Integral Equations, W.A. Benjamin, Inc., Menlo Park, California

Miller, R.S., Botkin, D.B. (1974): Endangered species; models and prediction, American Scientist **62**, 172–181

Moelwyn-Hughes, E.A. (1961): Physical Chemistry, 2nd edn., Pergamon Press, London

Mogridge, M.J.H. (1985): Transport, land use and energy interaction, Urban Studies **22**, 481–492

Mollison, D. (1977): Spatial contact models for ecological and epidemic spread, Journal of the Royal Statistical Society **39 B**, 283–326

Monteith, J.L. (1973): Principles of Environmental Physics, Edward Arnold, London

Montroll, E.W., West, B.J. (1973) Models of population growth, diffusion, competition and rearrangement, Synergetics (Ed. H. Haken), 143–156, B.G. Teubner, Stuttgart

Morrill, R. (1968): Waves of spatial diffusion, Journal of Regional Science **8**, 1–18

Morrill, R. (1985): The diffusion of the use of tractors again, Geographical Analysis **17**, 88–94

Morrill, R., Gaile, G.L., Thrall, G.I. (1988): Spatial Diffusion, Sage Publications, Newbury Park, California

Moss, R., Watson, A., Ollason, J. (1982): Animal Population Dynamics, Chapman and Hall, London

Murphy, G.M. (1960): Ordinary Differential Equations and their Solutions, D. Van Nostrand, Princeton, New Jersey

Murray, J.D. (1987): Modeling the spread of rabies, American Scientist **75**, 280–284

Murray, J.D. (1989): Mathematical Biology, Springer, Berlin, Heidelberg

Muth, R.F. (1969): Cities and Housing, University of Chicago Press, Chicago

Nair, K.R. (1954): The fitting of growth curves, Statistics and Mathematics in Biology, Iowa State College Press, Ames, Iowa, 119–132

Nairn, A.G.M., O'Neill, G.J. (1988): Population density functions: a differential equation approach, Journal of Regional Science **28**, 89–102

Nelder, J.A. (1961): The fitting of a generalization of the logistic curve, Biometrics **17**, 89–110

Newman, W.I. (1980): Some exact solutions to a non-linear diffusion problem in population genetics and combustion, Journal of Theoretical Biology **85**, 325–334

Nicholson, A.J. (1954): An outline of the dynamics of animal populations, Australian Journal of Zoology **2**, 9–65

Nicholson, A.J. (1957): The self-adjustment of populations to change, Cold Spring Harbor Symposium on Quantitative Biology **22**, 153–173

Nisbet, R.M., Gurney, W.S.C. (1976): Population dynamics in a periodically varying environment, Journal of Theoretical Biology **56**, 459–475

Noble, J.V. (1974): Geographic and temporal development of plagues, Nature **250**, 726–729

Odell, G.M. (1980): Qualitative theory of systems of ordinary differential equations, including phase plane analysis and the use of the Hopf bifurcation theorem, Mathematical Models in Molecular and Cellular Biology, (Ed. L.A. Segel), 649–727, Cambridge University Press, Cambridge

Ogata, A. (1970): Theory of dispersion in a granular medium, Geological Survey Professional Paper 411-I, U.S. Government Printing Office, Washington

Okubo, A. (1971): Oceanic diffusion diagrams, Deep-Sea Research, 18, 789–802

Okubo, A. (1980): Diffusion and Ecological Problems: Mathematical Models, Springer, Berlin, Heidelberg

Okubo, A. (1984): Critical patch size for plankton and patchiness, Mathematical Ecology, Lecture Notes in Biomathematics, Vol. 54, (Eds. S.A. Levin and T.G. Hallam) 456–477, Springer, Berlin, Heidelberg

Oliveira-Pinto, F., Conolly, B.W. (Eds.) (1982) Applicable Mathematics of Non-Physical Phenomena, Ellis Horwood, Chichester

Oliver, F.R. (1964): Methods of estimating the logistic growth function, Applied Statistics 13, 57–66

Papageorgiou, G.J. (1971): A theoretical evaluation of the existing population density gradient functions, Economic Geography 47, 21–26

Parr, J.B. (1985): A population-density approach to regional spatial structure, Urban Studies 22, 289–303

Pattle, R.E. (1959): Diffusion from an instantaneous point source with a concentration-dependent coefficient, Quarterly Journal of Mechanics and Applied Mathematics 12, 407–409

Pearl, R. (1927): The growth of populations, Quarterly Review of Biology 2, 532–548

Pearl, R., Reed, L.J. (1920): On the rate of growth of the population of the United States and its mathematical representation, Proceedings of the National Academy of Sciences 6, 275–288

Pearson, K. (1913): Systematic fitting of curves to observations, Biometrika 1, 265–288

Pearson, K. (1934): Tables of the Incomplete Gamma Function, Cambridge University Press, Cambridge

Pearson, K. (1968): Tables of the Incomplete Beta Function, 2nd edn., Cambridge University Press, Cambridge

Perelson, A.S., Segel, L.A. (1978) A singular perturbation approach to diffusion reaction equations containing a point source, with application to the hemolytic plaque assay, Journal of Mathematical Biology 6, 75–85

Peschel, M., Mende, W. (1986): The Predator–Prey Model, Springer, Vienna

Pielou, E.C. (1977): Mathematical Ecology, Wiley, New York

Pinney, E. (1958): Ordinary Difference-Differential Equations, University of California Press, Berkeley

Pollard, J.H. (1973): Mathematical Models for the Growth of Human Populations, Cambridge University Press, Cambridge

Possingham, H.P., Roughgarden, J. (1990): Spatial population dynamics of a marine organism with a complex life cycle, Ecology 71, 973–985

Powell, E.O. (1956): Growth rate and generation time of bacteria, with special reference to a continuous culture, Journal of General Microbiology 15, 492–511

Powell, E.O., Errington, F.P. (1963): Generation times of individual bacteria: some corroborative measurements, Journal of General Microbiology 31, 315–327

Power, J.H. (1986): A model of the drift of northern anchovy, *Engraulis mordax*, larvae in the California current, Fishery Bulletin (U.S.) 84, 585–603

Preece, M.A., Baines, M.J. (1978): A new family of mathematical models describing the human growth curve, Annals of Human Biology 5, 1–24

Prentice, R.L. (1976): Generalization of the probit and logit methods for dose response curves, Biometrics **32**, 761–768

Prescott, D.M. (1959): Variations in the individual generation times of *Tetrahymena geleii* HS, Experimental Cell Research **16**, 279–284

Pruitt, K.M., Turner, M.E., Boackle, R.J. (1974): A kinetic model for the quantitative analysis of complement, Journal of Theoretical Biology **44**, 207–217

Rabotnov, Y.N. (1980): Elements of Hereditary Solid Mechanics, Mir, Moscow

Ralston, B. (1980): The dynamics of communication. Evolving Geographical Structures: Mathematical Models and Theories for Space-Time Processes. (Eds. D.A. Griffin and A.C. Lea), Martinus Nijhoff, The Hague

Ratkowsky, D.A. (1983): Nonlinear Regression Modeling, Marcel Dekker, New York

Régnier, J., Lambin, S. (1938): Etude sur le croit microbien en fonction de la quantite de substance nutritive de milieux de culture, Comptes Rendus d'Academie de Sciences **207**, 1263–1266

Reid, A.T. (1952): Note on the growth of bacterial populations, Bulletin of Mathematical Biophysics **14**, 313–316

Robertson, T.B. (1923): The Chemical Basis of Growth and Senescence, Lippincott, Philadelphia

Robinson, J.W., Deano, P.M. (1986): Acid rain: the effects of pH, aluminum, and leaf decomposition products on fish survival, American Laboratory, 1-7

Rogers, E.M. (1983): Diffusion of Innovations, 3rd edn., Free Press, New York

Rohsenow, M., Hartnett, J.P. (1973): Handbook of Heat Transfer, McGraw-Hill, New York

Rosen, G. (1980): On the Fisher and the cubic-polynomial equations for the propagation of species properties, Bulletin of Mathematical Biology **42**, 95–106

Ross, R. (1911): The Prevention of Malaria, 2nd ed., Murray, London

Rubinow, S.I (1968): A maturity-time representation for cell populations, Biophysical Journal **8**, 1055–1073

Rubinow, S.I. (1973): Mathematical Problems in the Biological Sciences, Society for Industrial and Applied Mathematics, Philadelphia

Rubinow, S.I. (1980): Cell Kinetics, Mathematical Models in Molecular and Cellular Biology, (Ed. L.E. Segel), Cambridge University Press, Cambridge

Rushton, S., Mautner, A.J. (1955): The deterministic model of a simple epidemic for more than one community, Biometrika **42**, 126–132

Saaty, T.L. (1981): Modern Nonlinear Equations, Dover, New York

Samuelson, P.A. (1983): Foundations of Economic Analysis, Harvard University Press, Cambridge, Massachusetts

Scudo, F.M. (1971): Vito Volterra and theoretical ecology, Theoretical Population Biology **2**, 1–23

Scudo, F.M., Ziegler, J.R. (1976): Vladimir Aleksandrovich Kostitzin and theoretical ecology, Theoretical Population Biology **10**, 395–412

Scudo, F.M., Ziegler, J.R. (Eds.) (1978): The Golden Age of Theoretical Ecology: 1923-1940, Lecture Notes in Biomathematics, Vol. 22, Springer, Berlin, Heidelberg

Segel, L.A., Chet, I., Henis, Y. (1977): A simple quantitative assay for bacterial motility, Journal of General Microbiology **98**, 329–337

Segel, L.A. (Ed.) (1980): Mathematical Models in Molecular and Cellular Biology, Cambridge University Press, Cambridge

Sharif, M.N. (1983): Management of Technology Transfer and Development, UN/-ESCAP Regional Centre for Technology Transfer, Bangalore, India

Sharif, M.N., Islam, M.N. (1980): The Weibull distribution as a general model for forecasting technological change, Technological Forecasting and Social Change **18**, 247–256

Sharif, M.N., Ramanathan, K. (1981): Binomial innovation diffusion models with dynamic potential adoptor population, Technological Forecasting and Social Change **20**, 63–87

Sharif, M.N., Ramanathan, K. (1982): Polynomial innovation diffusion models, Technological Forecasting and Social Change **21**, 301–323

Sharif, M.N., Ramanathan, K. (1984): Temporal models of innovation diffusion, IEEE Transactions on Engineering Management **EM-31**, 76–86

Sharpe, F.R., Lotka, A.J. (1911): A problem in age-distribution, Philosophical Magazine **21**, 435–438

Shigesada, N. (1980): Spatial distribution of dispersing animals, Journal of Mathematical Biology **9**, 85–96

Skellam, J.G. (1951): Random dispersal in theoretical populations, Biometrika **38**, 196–218

Slobodkin, L.B. (1961): Growth and Regulation of Animal Populations, Holt, Rinehart and Winston, New York

Small, R.D. (1987): Population growth in a closed system, Mathematical Modelling: Classroom Notes in Applied Mathematics, (Ed., M.S. Klamkin), 317–320, Society for Industrial and Applied Mathematics, Philadelphia

Smith, D., Keyfitz, N. (Eds.) (1977): Mathematical Demography, Springer, Berlin, Heidelberg

Smith, F.E. (1963): Population dynamics in *Daphnia magna* and a new model for population growth, Ecology **44**, 651–663

Sobol, I.M. (1975): The Monte Carlo Method, Mir, Moscow

Song, J., Yu, J. (1988): Population System Control, Springer, Berlin, Heidelberg

Sonis, M. (1983): Spatio-temporal spread of competitive innovations: an ecological approach, Papers of the Regional Science Association **52**, 159–174

Sooky, A.A. (1969): Longitudinal dispersion in open channels, Journal of the Hydraulics Division, Proceedings of the American Society of Civil Engineers **95**, 1327–1346

Spiegel, M.R (1965): Theory and Problems of Laplace Transforms, McGraw- Hill, New York

Spiegel, M.R. (1981): Applied Differential Equations, 3rd edn., Prentice- Hall, Englewood Cliffs, New Jersey

Spiegelman, M. (1968): Introduction to Demography, 2nd edn., Harvard University Press, Cambridge, Massachusetts

Stapleton, E. (1976): A normal distribution as a model of technological substitution, Technological Forecasting and Social Change **8**, 325–334

Stoudt, H.W., Damon, A., McFarland, R.A. (1960) Heights and weights of white Americans, Human Biology **32**, 331–341

Sutton, O.G. (1953): Micrometeorology, McGraw-Hill, New York

Sweeny, B.M. (1969): Rhythmic Phenomena in Plants, Academic Press, New York

Takahashi, M. (1966): Theoretical basis for cell cycle analysis I, Journal of Theoretical Biology **13**, 202–211

Takahashi, M. (1968): Theoretical basis for cell cycle analysis II, Journal of Theoretical Biology **18**, 195–209

Talbot, J.W. (1974): Diffusion studies in fisheries biology, Sea Fisheries Research, (Ed., F.R. Harden), Paul Elek Scientific Books, London

Taylor, G.I. (1953): Dispersion of soluble matter in solvent flowing slowly through a tube, Proceedings of the Royal Society of London **A 219**, 186–203

Taylor, G.I. (1954): The dispersion of matter in turbulent flow through a pipe, Proceedings of the Royal Society of London **A 223**, 446–468

Taylor, Jr., O.R. (1985): African bees: potential impact in the United States, Bulletin of Entomology Society of America, Winter 1985, 15–24

Thomas, H.A., Jr. (1950): Graphical determination of BOD curve constants, Water and Sewage Works **97**, 123–129

Thompson, D.W. (1963): On Growth and Form, 2nd edn., Cambridge University Press, Cambridge

Tinbergen, J. (1931): A shipbuilding cycle, Jan Tinbergen Selected Papers, (Eds., L.H. Klaasen, L.M. Kyck, and H.J. Witteveen), North Holland, Amsterdam, 1959

Turner, M.E., Blumenstein, B.A., Sebaugh, J.L. (1969): A generalization of the logistic law of growth, Biometrics **25**, 577–580

Turner, M.E., Bradley, E.L., Kirk, K.A. (1976): A theory of growth, Mathematical Biosciences **29**, 367–373

Usher, J.R. (1980): Mathematical derivation of optimal uniform treatment schedules for the fractional irradiation of human tumors, Mathematical Biosciences **49**, 157–184

Valentine, H.T. (1985): Tree-growth models: derivations employing the pipe-model theory, Journal of Theoretical Biology **117**, 579–585

Vance, R.R. (1990): Population growth in a time-varying environment, Theoretical Population Biology **37**, 438–454

van der Ploeg, F. (1984): Mathematical Methods in Economics, Wiley, New York

van Iwaarden, J.L. (1985): Ordinary Differential Equations with Numerical Techniques, Harcourt Brace Jovanovitch, Orlando, Florida

Verhulst, P.F. (1838): Notice sur la loi que la population suit dans son accroissement, Correspondance Mathématique et Physique Publiee par A. Quetelet, Brussels **10**, 113–121

Vermeulen, T., Klein, G., Heister, N.K. (1973): Adsorption and Ion exchange, Section 16, Chemical Engineers Handbook, (Ed., J.H. Perry), McGraw-Hill, New York

Vogel, S. (1981): Life in Moving Fluids, Willard Grant Press, Boston

Volterra, V. (1926): Variations and fluctuations in the numbers of coexisting animal species, The Golden Age of Theoretical Ecology: 1923–1940, (Eds. F.M. Scudo and J.R. Ziegler), Lecture Notes in Biomathematics, Vol. 22, 65–236, Springer, Berlin, Heidelberg

Volterra, V. (1934): Comments on the note by Mr. Régnier and Miss Lambin, Study of a case of microbial competition (*Bacillus coli–Staphylococcus aureus*), The Golden Age of Theoretical Ecology: 1923–1940, (Eds. F.M. Scudo and J.R. Ziegler), Lecture Notes in Biomathematics, Vol. 22, 47–49, Springer, Berlin, Heidelberg

Volterra, V., Kostitzin, V.A. (1939): Comments on the toxic action of the medium relative to the note by Mr. Régnier and Miss Lambin, The Golden Age of Theoretical Ecology: 1923-1940, (Eds. F.M. Scudo and J.R. Ziegler), Lecture Notes in Biomathematics, Vol. 22, 54–56, Springer, Berlin, Heidelberg

von Foerster, H. (1959): Some remarks on changing populations, The Kinetics of Cellular Proliferation, (Ed., F. Stohlman, Jr.), Grune and Stratton, New York

von Foerster, H., Mora, P.M., Amiot, L.W., (1960): Doomsday: Friday, 13 November, A.D. 2026, Science **132**, 1291–1295

Waltman, P. (1974): Deterministic Threshold Models in the Theory of Epidemics, Lecture Notes in Biomathematics, Vol. 1, Springer, Berlin, Heidelberg

Webber, M.J., Joseph, A.E. (1978): Spatial diffusion processes 1: A model and an approximate method, Environment and Planning A **10**, 651–665

Webber, M.J., Joseph, A.E. (1979): Spatial diffusion methods 2: numerical analysis, Environment and Planning A **11**, 335–347

Westoff, C.F. (1986): Fertility in the United States, Science **234**, 554–559

Wilson, A.G., Kirkby, M.J. (1980): Mathematics for Geographers and Planners, 2nd edn., Oxford University Press, Oxford

Wilson, L.L. (1964): Catalog of Cycles, Part I – Economics, Foundation for the Study of Cycles, Irvine, California

Winker, C.A., Hinshelwood, C.N. (1935): The thermal decomposition of acetone vapour, Proceedings of the Royal Society of London A 149, 340–359

Winsborough, H.H. (1962): An ecological approach to the theory of suburbanization, American Journal of Sociology 68, 565–570

Winsor, C.P. (1932a): The Gompertz curve as a growth curve, Proceedings of the National Academy of Sciences 18, 1–8

Winsor, C.P. (1932b): A comparison of certain symmetrical growth functions, Journal of the Washington Academy of Sciences 22, 73–83

Wright, E.M. (1946): The non-linear difference-differential equation, Quarterly Journal of Mathematics 17, 245–252

Wright, E.M. (1955): A non-linear difference-differential equation, J. Reine Angew. Math. 194, 66–67

Yee, J. (1980): A nonlinear, second-order population model, Theoretical Population Biology 18, 175–191

Yule, G.U. (1925): The growth of population and the factors which control it, Journal of the Royal Statistical Society 88, 1–62

Zeger, S.L., Harlow, S.D. (1987): Mathematical models from laws of growth to tools for biological analysis: fifty years of Growth, Growth 51, 1–21

Zwillinger, D. (1989): Handbook of Differential Equations, Academic Press, San Diego

Author Index

Abramowitz, M.   95, 101, 158, 176, 177, 234, 256, 364
Allen, R.G.D.   144, 146
Amiot, L.W.   19, 186
Amundsen, N.R.   374
Anderson, R.M.   427, 428
Andrewartha, H.G.   224
Arcuri, P.   427
Aris, R.   403
Aronson, D.G.   403, 424
Arrigoni, M.   209, 419
Ashton, W.D.   48, 101
Atkins, P.W.   80
Austin, A.   19, 186, 190
Ayala, F.J.   106
Ayres, R.U.   35

Bailey, N.T.J.   125, 427
Baines, M.J.   73
Barclay, G.W.   285
Barenblatt, G.I.   354
Baron, G.   73
Bartholomew, D.J.   52, 78, 125, 425
Baserga, R.   231
Batchelor, G.K.   343, 344
Batschelet, E.   52, 144, 222, 223
Bear, J.   344, 348
Beck, K.   249
Becking, L.G.M.   96
Beckmann, M.J.   342, 426
Bellman, R.   290, 291
Berkson, J.   51, 96, 100, 236
Bertalanffy, L. von   105, 175, 227, 229
Bhargava, S.C.   35, 422
Birch, L.C.   224
Bird, R.B.   52
Blackman, A.W.   35
Bliss, C.I.   101, 172, 173
Blumenfeld, H.   395, 398
Blumenstein, B.A.   175, 249, 283
Blythe, S.P.   303

Boackle, R.J.   176
Bogin, B.   69, 73
Bojadziev, G.   422
Borsellino, A.   307, 317, 423
Botkin, D.B.   123, 125
Boyce, W.E.   144, 290, 370
Bradley, E.L.   106, 175
Brandt, S.   93, 101
Brauer, F.   117, 125, 298, 304
Braun, M.   114
Brewer, J.W.   19, 186, 190
Bridgman, P.W.   354
Brody, S.   73, 88
Brown, L.A.   52, 414, 425
Burghes, D.N.   144
Bush, R.R.   81
Bussiere, R.   396, 400
Byerlee, D.   103, 104

Camp, T.R.   25
Cannon, J.R.   329, 369
Canosa, J.   407
Carslaw, H.S.   95, 329, 341, 346, 347, 370, 371
Casetti, E.   395, 410, 411, 412, 413, 427
Castillo-Chavez, C.   428
Caughley, G.   298, 325
Causton, D.R.   110, 151
Central Statistical Office   206
Chakravarti, I.M.   48
Chet, I.   343
Chorley, R.J.   425
Clark, C.   385, 386, 395, 396, 398, 400
Clark, C.W.   117, 123, 367
Clark, J.P.   422
Clements, D.L.   144
Cliff, A.D.   411, 425
Coale, A.J.   234
Cohen, D.S.   317
Cole, L.C.   2
Colegrove, M.   52

Coleman, B.D.   209, 419
Coles, C.W.   144
Combarnous, M.A.   333, 352, 356
Conolly, B.W.   2, 405
Conrad, J.M.   123, 366, 367
Conway. G.R.   298
Cooke, K.L.   290, 291
Cooper, R.A.   201
Coutsias, E.   317
Cox, D.R.   51, 156, 171
Cox, P.R.   285
Crank, J.   95, 329, 341, 377, 404, 424
Cunningham, W.J.   141, 290, 298, 299
Cushing, J.M.   209, 28, 304, 419

Damon, A.   69, 70, 71
Davies, O.   396
Davis, H.T.   141, 274, 317, 422
Deano, P.M.   420
de Jong, F.J.   354
Dendrinos, D.S.   423
Dietz, K.   125, 428
DiPrima, R.C.   144, 290, 370
Dixon, R.   114
Dobson, A.J.   51, 172
Draper, N.R.   48, 51
Driver, R.D.   290, 420, 421

Eason, G.   144
Edmonston, B.   396
Elderton, W.P.   196, 201
El'sgol'ts, L.E.   420
Elton, C.S.   339, 340
Errington, F.P.   202, 236
Estes, W.K.   80

Falkner, F.   73
Farlow, S.J.   341, 370
Feller, W.   2, 96, 195, 234
Fife, P.C.   363, 424
Finney, D.J.   101
Fischer, H.B.   382
Fischer, R.A.   263, 264, 332
Fisher, J.C.   35, 36, 37
Fisher, R.A.   2, 363, 403
Fletcher, R.I.   422
France, J.   155
Frauenthal, J.C.   267, 304
Fried, J.J.   333, 352, 356
Frisch, R.   297

Gaile, G.L.   411, 425, 426
Gale, S.   426

Gandolfo, G.   144
Garcia, O.   112, 113
Gause, G.F.   2, 202
Gettinby, G.   144
Getz, W.M.   367
Ghez, R.   342, 424
Gilpin, M.E.   106
Goel, N.S.   290
Goldstein, S.   375
Goldsworthy, P.R.   52
Gompertz, B.   2
Gopalsamy, K.   298
Greenkorn, R.A.   333
Griliches, Z.   114, 115, 414, 415, 416, 417, 418
Guest, A.M.   396
Gumbel, E.J.   171
Gumowski, I.   293
Gurney, W.S.C.   209, 298, 303, 359, 419, 424
Gurtin, M.E.   336, 424

Hagerstrand, T.   414, 425, 426
Haggett, P.   425
Haight, R.G.   367
Haining, R.   427
Haining, R.   42
Hallam, T.G.   117
Hammersley, J.M.   426
Handscomb, D.C.   426
Hanham, R.Q.   414
Harlow, S.D.   69
Hartnett, J.P.   64
Hassell, M.P.   298
Hastings, N.A.J.   201
Hauspie, R.C.   73
Hayashi, C.   405
Haynes, K.E.   78
Heister, N.K.   379
Henis, Y.   343
Hensley, C.   126
Hesse de Polanco, E.   103, 104
Hinshelwood, C.N.   19
Holme, H.   297
Hoppensteadt, F.   125, 234, 403, 407
Hull, D.E.   333, 335
Hutchinson, C.E.   297, 300, 301
Hyman, J.M.   428

Impagliazzo, J.   151, 222, 234
Innis, G.   422
Islam, M.N.   159, 160, 161

Jackson, H.C.   427
Jaeger, J.C.   95, 329, 341, 346, 347, 370, 371
Jansen, M.J.W.   346
Jarvis, L.S.   38, 39
Jeffreys, H.   196
Johnson, N.L.   196, 201
Jones, D.S.   144, 405
Jones, G.S.   297, 298, 299
Jordan, D.W.   141
Joseph, A.E.   427
Jost, W.   329

Kaicker, S.   35, 304, 422
Kakutani, S.   297
Kalecki, M.   297
Kaliappan, P.   423
Kallen, A.   427
Kalos, M.H.   426
Karmeshu   35, 304, 422
Kendall, D.G.   202, 336, 392, 403, 404
Kennedy, C.R.   403
Kent, J.W.   333, 335
Kermack, W.O.   125, 135
Keyfitz, N.   2, 16, 151, 224, 234, 271
Kierstead, H.   356
Kindlmann, P.   324
Kingsland, S.E.   3, 80
Kirk, K.A.   106, 175
Kirkby, M.J.   78, 181
Klein, D.R.   324
Klein, G.   379
Knolle, H.   231, 234, 236
Kolmogorov, A.   2, 363, 403
Kostitzin, V.A.   2, 316, 317, 319, 322
Krebs, C.J.   234
Kreith, F.   52
Kreyszig, E.   144

Laha, R.G.   48
Laird, A.K.   155
Lal, V.B.   35, 304, 422
Lambin, S.   322, 323, 325
Lancaster, K.   144
Landau, L.D.   367
Lanford, H.W.   35, 214
Langer, W.L.   427
Lapidus, L.   374
Leach, D.   206
Lebowitz, J.L.   234
Lekvall, P.   78
Leslie, P.H.   234
Levin, S.A.   363, 428

Lewis, D.   80
Lifshitz, E.M.   367
Lightfoot, E.M.   52
Linstone, H.A.   35
Linz, P.   305
Lotka, A.J.   2, 48, 234, 271, 282
Luce, R.D.   81

MacCall, A.D.   367
MacCamy, R.C.   336, 424
MacDonald, N.   233, 289, 298, 304, 309, 317, 420
Mahajan, V.   35, 78, 411
Maitra, S.C.   290
Malthus, T.R.   2, 5
Mansfield, E.   35, 126
Marcus, L.   297
Marr, A.G.   202, 235
Matthews, G.A.   174
Mattingly, P.F.   214
Mautner, A.J.   425
May, R.M.   139, 267, 270, 288, 298, 300, 301, 302, 304, 342, 427, 428
Maynard Smith, J.   271, 293
Mazanov, A.   298
Mazumdar, J.   354
McCullagh, P.   51, 156
McFarland, R.A.   69, 70, 71
McKendrick, A.G.   125, 135, 234
McLean, A.R.   428
McMurtrie, R.   365
Mende, W.   174, 196
Metz, J.A.J.   346
Miller, R.K.   305
Miller, R.S.   123, 125
Moelwyn-Hughes, E.A.   19
Mogridge, M.J.H.   395, 396, 397, 398, 399, 400
Mollison, D.   125, 363, 404
Monteith, J.L.   345
Montroll, E.M.   290, 423
Mora, P.M.   19, 186
Morrill, R.   411, 414, 425, 426
Moss, R.   139
Mosteller, R.   81
Mullally, H.   423
Murphy, G.M.   144
Murray, J.D.   125, 289, 356, 363, 366, 405, 407, 408, 414, 418, 423, 427, 428
Muth, R.F.   387, 389

Nair, K.R.   48
Nairn, A.G.M.   395

Nelder, J.A.   48, 51, 156, 175
Neu, J.C.   317
Newman, W.I.   403
Nicholson, A.J.   302
Nilson, E.H.   202, 235
Nisbet, R.M.   209, 298, 303, 359, 419, 424
Noble, J.V.   427
Norkin, S.B.   420

Odell, G.M.   405
Ogata, A.   333, 352, 353
Okubo, A.   336, 338, 356, 359, 367, 404
Oliveira-Pinto, F.   2
Oliver, F.R.   48
Ollason, J.   139
O'Neill, G.J.   395
Ord, J.K.   411, 425

Painter, P.R.   202, 235
Papageorgiou, G.J.   395
Parr, J.B.   396
Pattle, R.E.   423
Peacock, J.B.   201
Pearl, R.   2, 48, 281
Pearson, K.   176, 196, 199, 234
Perelson, A.S.   348
Peschel, M.   174, 196
Peterson, R.A.   35, 411
Petrovsky, I.   2, 363, 403
Pielou, E.C.   139, 224, 234, 348
Pinney, E.   290, 293, 294, 296
Piskounov, N.   2, 363, 403
Pollard, J.H.   234, 304
Possingham, H.P.   367
Powell, E.O.   202
Power, J.H.   367
Preece, M.A.   73
Preece, M.A.   7
Prentice, R.L.   96, 162, 172
Prescott, D.M.   202, 203, 233, 234
Pruitt, K.M.   176
Pry, R.H.   35, 36, 37

Quade, W.   354

Rabotnov, Y.N.   305, 307, 422
Ralston, B.   52, 78
Ramanathan, K.   20, 62, 77, 78, 126, 249
Ratkowsky, D.A.   51
Reed, L.J.   48, 281
Regnier, J.   322, 323, 325

Reid, A.T.   309
Robertson, T.B.   80
Rogers, E.M.   42
Rohsenow, M.   64
Rosen, G.   403
Ross, R.   125
Roughgarden, J. 367
Roy, J.   48
Rubinow, S.I.   202, 203, 233, 234
Rushton, S.   425

Saaty, T.L.   290, 298
Sahal, D.   35
Samuelson, P.A.   144
Sanchez, D.A.   117, 125, 298
Scudo, F.M.   2, 3
Sebaugh, J.L.   175, 249, 283
Segel, L.A.   144, 342, 343, 348
Semple, R.K.   410, 411, 412, 413, 427
Sharif, M.N.   20, 35, 52, 62, 77, 78, 126, 156, 159, 160, 161, 249
Sharpe, F.R.   234
Shigesada, N.   336, 359, 424
Shoemaker, C.A.   428
Skellam, J.G.   336, 337, 356, 359, 364, 403
Sleeman, B.D.   144, 405
Slobodkin, L.B.   303, 356
Small, R.D.   309, 317
Smith, A.M.   427
Smith, D.   2, 151
Smith, F.E.   138
Smith, H.   48, 51
Smith, P.   11
Sobol, I.M.   426
Song, J.   234
Sonis, M.   411
Sooky, A.A.   332, 356
Southwood, T.R.E.   298
Spiegel, M.R.   144, 290
Spiegelman, M.   285
Stanley, E.A.   428
Stapleton, E.   37
Stegun, I.A.   95, 101, 158, 176, 177, 234, 256, 364
Steiner, A.   209, 419
Stewart, H.E.   52
Stoudt, H.W.   69, 70, 71
Sutton, O.G.   57
Sweeny, B.M.   224

Takahashi, M.   202
Talbot, J.W.   367

Tanner, J.M.   73
Taylor, G.I.   332, 356, 382
Taylor, O.R., Jr.   428
Thomas, H.A., Jr.   67
Thompson, D.W.   2, 69
Thornley, J.H.M.   155
Thrall, G.I.   411, 425, 426
Tinbergen, J.   295, 297
Tognetti, K.P.   298
Torre, V.   307, 317, 423
Turner, M.E.   106, 175, 176, 249, 283

Usher, J.R.   110

Valentine, H.T.   52, 139
Vance, R.R.   419
van der Ploeg, F.   224, 271
van Iwaarden, J.L.   290
Venus, J.C.   110, 151
Verhulst, P.F.   2, 5, 48
Vermeulen, T.   379
Vogel, S.   354
Volterra, V.   2, 271, 304, 309, 317, 319, 322, 422
von Foerster, H.   19, 186

Wachholder, A.   73
Wahlbin, C.   78
Waltman, P.   125
Watson, A.   139
Webber, M.J.   427
Weekes, A.J.   201
Weinberger, H.F.   403
West, B.J.   423
Westoff, C.F.   285
White, G.M.   78
Whitlock, P.A.   426
Wilson, A.G.   78, 181
Wilson, L.L.   304
Winkler, C.A.   19
Winsborough, H.H.   396
Winsor, C.P.   96, 151, 196
Wood, A.D.   144
Wright, E.M.   297, 298

Yee, J.   422
Yu, J.   234
Yule, G.U.   48

Zeger, S.L.   69
Ziegler, J.R.   2, 3
Zwillinger, D.   144, 420

Subject Index

accelerator   145
adopter population   43
adoption of herbicides   103–105
adoption of hybrid corn   114–116, 414–418
Africanized honey bees (*Apis mellifera scutellata*)   427–428
age dependent growth   234
age structure   233
AIDS (acquired immune defficiency syndrome)   428
animal growth   105, 175, 224
arctangent distribution   96, 197, 198
arctangent–exponential distribution   195, 198, 205–207, 236
autocatalytic reaction   80

bacterial motility (*Pseudomonas fluorescens*)   343–345
Bernoulli differential equation   106
Bertalanffy–Richards model   112
Bessel functions   346, 359, 376, 384, 388–389, 394
beta function   162, 176
beta response-strength model   81
biochemical oxygen demand (BOD)   24, 67–69
bioeconomics   123
biological dispersion   336
– *Ondatra zibethica* (muskrats)   336–338
– *Lymantria dispar* (gypsy moth)   338–340
– *Popilia japonica* (Japanese beetle)   340–341
Black Death   427
Bombay plague of 1905–1906   135–136

carrying capacity   5, 13, 40, 53, 150, 209

Cauchy distribution   197
cell
– cycle   231
– generation times   203
– kinetics   202, 234
Chanter growth function   155
chemical kinetics   19–23, 79
chemical solute removal by adsorption   378–379
coefficient of
– kurtosis   205
– skewness   204
– variation   158, 204
coalition growth   19, 176, 185–188
Coleman model   77, 81, 134
Collembola springtails   347
combined exponential–confined exponential distribution   88
combined logistic–confined exponential distribution   76, 183
community innovation   63
complement mediated lysis   176
confined exponential distribution   6, 7, 52–55, 71, 76, 88, 114, 140, 175, 177, 183, 201
confined exponential distribution variable equilibrium value   60
– exponentially variable   62
– linearly variable   60
– rectangular pulse   60
– sinusoidally variable   58
consumer innovation   63
correlation coefficient   41, 51, 103
crowding coefficient   27, 150, 209
cumulative distribution   24, 30, 41, 92, 130, 162

Dalton equation   57
Daphnia magna (water flea)   138
decay coefficient   150

density distribution   31, 41, 43, 92, 130, 162
deoxygenation coefficient   25, 69
diffusion equation   327–329
diffusion and convection with interphase transfer   373–377
diffusion from instantaneous sources   329–331
diffusion of
- innovations   37, 42, 52, 62, 78, 126, 426
- news and rumors   78
- tractor utilization   410–414
diffusion (radial)   346–349
diffusion (radial) with exponential growth   335–336
diffusion (rectilinear) with convection   331–332, 352–353
diffusion with confined exponential growth   379–380
diffusion with convection and reaction   367–372
- exponential growth   371–372
- exponential decay   372–373
diffusion with logistic growth   402–408
diffusion with reaction
- in a finite region   354
- exponential growth   356–360
- logistic growth   362–365
- power law exponential growth   361–362
dimensional analysis   354–356
dimensions and units   8
dispersion in pipelines   232–235
Dodd model   77, 81, 134
dose response analysis of beetle mortality   172–174
doubling times   10, 53, 201
Drosophila melanogaster (fruit flies)   347
dynamics of tumor growth   155–156

effective life span   161
enrollments in universities in the U.S.   254–255
epidemics   20, 78
- simple epidemics (SI)   125, 129
- general epidemics (SIR)   125, 129
epidemics and technology transfer   125
equilibrium value   28, 60

error function and complementary error function   94, 96 342, 370, 373, 380, 398
Eulota fruticum (land snail)   227
evaporation from a water surface   57
explosion time   18
exponential function   5, 7, 9, 88
- growth   11, 12, 14, 47, 149, 175, 202
- with migration   13
- logistic distribution   177
exponential integral   266, 348, 350, 383
exponential logistic equation   177
extinction time   14
extreme maximum value distribution   105, 149, 167–169, 173, 218
extreme minimum value distribution   105, 169–172, 174, 220

farm population of the U.S.   245–248
Fick (diffusion phenomena)   1
finite difference method   46
fish cultivation and harvesting   122–123
Fisher differential equation   403
fluoridated water in the U.S.   63
frequency response   64

gamma distribution   24, 199, 201, 202, 231, 397
- function   24, 73, 162, 176, 232
generalized distribution   162
- exponential equation   210
- logistic equation   210
- symmetrical function   163–167
generation time of cells   201–205, 232
Gompertz distribution   2, 73, 105, 114, 149–151, 169, 225
growth and self-contanimation of bacteria   322–325
growth coefficient   5, 7, 13, 27, 40, 53
growth rate   5, 7, 82, 97, 137, 168, 171
- of wheat plant components   263–266
growth of
- cell populations   231–236
- humans   69–73
- land snails (Eulota fruticum)   227–230
- pine trees   112–116
- plant leaves   151–155

– public debt   191–194
– water fleas and trees   138–140
growth and decline of
– Australian blowflies (*Lucilia cuprina*)
  302–304
– populations of Northeast and East
  North Central American cities
  310–316
– U.S. sailing vessels   212–214

half–life time   53
harvesting (emigration)   13
harvesting rate   14
heat conduction   95
– transfer   64
heat transfer from a river   380–383
HIV (human immuno deficiency virus)
  428
horses and mules on U.S. farms
  258–260
hyperbolic growth   16–19
hyperlogistic distribution   45, 149,
  174–182, 191, 237

improved pasture technology   38–44
incomplete beta function   177
– gamma function   199, 235
infection rate   127
information diffusion   57
– transfer coefficient   58
innovation diffusion   37, 42, 52, 62, 78,
  126

kinematics of growth phenomena   42
kurtosis   205

laboratory and field tests with fruit flies
  (*Drosophila melagonaster*) 347–348
land snail (*Eulota fruticum*)   227
Laplace transformation   291, 368, 390
least squares method   11, 36, 40, 49,
  153, 174, 203
linear–operator model in learning   81
logistic distribution   5, 7, 24, 27–34,
  47, 77, 96–102, 114, 173, 175, 178, 183,
  198, 205, 219, 235
variable growth coefficient   209
– exponentially decreasing (I)   225
– exponentially decreasing (II)   230

– exponentially variable   218
– hyperbolically variable   214
– linearly variable   210
– sinusoidally variable   223
logistic growth with
– constant harvesting   119
– constant stocking   117
– migration   117, 134
– variable harvesting   121
logits   96, 100, 104
Lotk-aVolterra equations   271

Malthus (growth phenomena)   1, 2, 5
Malthusian (exponential) growth
  equation   8, 149, 202
maximum correlation coefficient
  method   51
McKendrick-von Foerster equation
  233
mean value (expectation)   32, 93, 157,
  203
mechanism for transfer
– internally influenced interaction
  model   52, 79, 81, 83, 134, 183
– externally influenced source model
  52, 73, 78, 81, 83, 134, 183
mitosis phase (cells)   232, 236
Mitscherlich equation   52, 140
modal value   204
modified coalition growth   19,
  185–188, 192
– logistic distribution   137, 140
molecular diffusion   95
monomolecular reaction   79
Monte Carlo simulation   426
multiplier–accelerator model of a
  national economy 144–147

neighborhood effect   426
Newton's law of cooling   64
normal probability (Gaussian)
  distribution   34, 37, 92–95, 96–102,
  165, 173, 198, 212

oscillatory phenomena in
– animal and plant life   224
– demography   224
– ecology   224
– economics and business   224
oxygen balance   65
– sag equations   25
– sag curve in a river   24–27

oxygen balance
- transfer across a water surface 65–67

Paramecium caudatum   202
patents issued for inventions   237–240
Pearson Type III (gamma) distribution 199, 202
- Type VII distribution   96, 196, 198
perfect memory model   234
period of oscillation 143
phase plane display   126, 128, 270–275
point matching method   47
population growth of
- California   89–91
- Great Britain   205–207
- Great Plains States   216–218
- London   395–402
- United States   11–13, 281–284
- World   16–19, 45, 188–190
population in cities   385–389
population wave   398–399, 402
power law exponential function   15, 175, 215, 237
- logistic distribution   45, 105–109, 114, 140, 175
prey–predator phenomena   37
prime mover horsepower   44–52
probits   96, 101, 104
public interest in a news event   73–76
pulse function   167

quadratic least squares method   49
quantal response   172

radial diffusion   383–385
railroads   216
railway mileage in the U.S.   275–279
reaeration coefficient   19, 22
relative removal rate   127
removal rate   127
Richards function   110
Riccati differential equation   29, 83, 209

sale of development property   110–112, 140
sandhill crane   123–125

simple harmonic growth coefficient 224
SIR (epidemic) model 20, 126
skewness   109, 204
spatial diffusion 327
spatial diffusion of epidemics   427
specific growth rate   2, 5, 7, 82, 97, 137, 151, 168, 171
speed of response   145
standard deviation   33, 43, 93, 158, 203
steam locomotives on U.S. railroads 260–262
step function   166
stocking (immigration)   13
Streeter–Phelps equations   25
substitution of diesel locomotives for steam locomotives 159–162
survival function   222
- of rats   222–223
synthetic fibers   35

takeover time   36
Taylor series   6, 136, 141, 290
technology diffusion   87
- substitution   34, 149, 159, 214
- transfer   20, 37, 52, 77, 125
temperature distribution in the soil 345–349
temporal–spatial diffusion   390–395
Tetrahymena geleii   202, 233
time delay (discrete)   37, 136, 141, 209, 233, 288
- in the exponential equation   289–294
- in the logistic equation   297–301
time delay (distributed)   288
- delay integral in the crowding term 304–310
- delay integral in a pollution term 317–321
Tinbergen's shipbuilding cycle   295–296
tornado warning device   86–87, 182–185
transfer coefficient   37, 39, 53, 103, 209
transient response   60, 64
tree growth model   139
tumor growth   149, 155–156

unsteady fluid flow in an aquifer 349–351

variable carrying capacity   241
- exponentially variable   242
- hyperbolically variable   262
- linearly variable   256
- logistically variable   249
- power law logistic with a power law logistically variable carrying capacity   279
- sinusoidally variable   267
- sinusoidally variable, exponentially changing   275

variance   32, 158, 203
Verhulst, (growth phenomena)   1, 2

water flea (*Daphnia magna*)   134
water quality model   65
Weibull distribution   105, 149, 156–159, 201

zone of regulated fishing   366–367